MONOGRAPHS IN STATISTICAL PHYSICS AND THERMODYNAMICS

Volume 1

Non-Equilibrium Statistical Mechanics

by I. PRIGOGINE

Volume 2

Thermodynamics

With quantum statistical illustrations

by P. T. LANDSBERG

Volume 3

Ionic Solution Theory

Based on cluster expansion methods

by HAROLD L. FRIEDMAN

Volume 4

Statistical Mechanics of Charged Particles

by R. BALESCU

MONOGRAPHS IN STATISTICAL PHYSICS AND THERMODYNAMICS

Editor: I. PRIGOGINE
Professor of Physical Chemistry and of Theoretical Physics
Université Libre, Brussels, Belgium

VOLUME 4

STATISTICAL MECHANICS OF CHARGED PARTICLES

R. BALESCU

Faculté des Sciences

Université Libre, Brussels, Belgium

1963

INTERSCIENCE PUBLISHERS

a division of John Wiley & Sons, Ltd.

London - New York - Sydney

Library of Congress Catalog Card Number 63-17454

PRINTED IN THE NETHERLANDS

BY N.V. DIJKSTRA'S DRUKKERIJ V.H. BOEKDRUKKERIJ GEBROEDERS HOITSEMA

GRONINGEN

Preface

There exists at the present time a constantly growing literature on the theoretical physics of assemblies of charged particles. A first thing is striking when one looks at the existing monographs: their diversity. One group includes what is now called "plasma physics" in a restricted sense: it studies gases of charged particles at very high temperatures either from a macroscopic point of view (magnetohydrodynamics) or from an elementary kinetic point of view. Other groups include more specialized topics, such as electrolyte solutions, discharges in gases, wave propagation in plasmas, ionospheric physics, astrophysics, etc. Most of these books are devoted to the study of ionized gases in conditions such that classical mechanics is a good approximation. Quantum-mechanical gases of charged particles are studied either in general textbooks of quantum statistics, or in monographs on solid-state physics. It therefore seemed desirable to have a textbook where assemblies of charged particles are studied from a unified point of view, regardless of their origin.

A second striking feature of the existing books is the lack of rigor in the justification of the fundamental equations of plasma physics. The latter are "derived" in a partly intuitive or semi-phenomenological way. Indeed the main object of these books is the calculation of physical quantities (transport coefficients, hydrodynamical behavior, thermodynamic functions, . . .) from the fundamental equations, which are accepted more or less *a priori*. Those readers who were interested in the justification and the limitations of these equations had to be referred to the periodical literature. The study of the latter is, however, difficult and time consuming, as is always the study of a field which is still growing.

For the two reasons stated above, it seemed to us that a book dealing with the rigorous statistical-mechanical foundation of plasma physics from a unified point of view would fill a gap in the existing literature. Both rigor and unification are made possible by the recent achievement of a general statistical mechanics of non-equilibrium processes, developed mainly at the Université Libre de Bruxelles by I. Prigogine and his coworkers. Within this framework one can treat with the same tools the short-time behavior or the long-time dissipative evolution of ionized gases, from the highest temperature regions, where the gas is purely classical, down to absolute zero, where the plasma is completely degenerate. The well known difficulties due to the long range of the Coulomb forces can be overcome rather simply, as well in equilibrium as in non-equilibrium situations. The mathematical method used here is modern perturbation theory carried up to all orders in the expansion parameter and supplemented by a diagram technique which helps in classifying and selecting the terms in the perturbation series.

We tried to make this book as much as possible self-contained. In particular, no previous knowledge of general non-equilibrium statistical mechanics is necessary: all the concepts needed for the reading of the book are studied in some detail, mainly in Chapters 1 and 14. However, readers interested in more details about the general theory are referred to the recent book of Prigogine which appeared in the same series.* It is desirable that the reader have some familiarity with plasma physics in order to appreciate the theory.** The quantum-mechanical part requires a previous elementary knowledge of the second quantization formalism. From the mathematical point of view, an extensive use is made of some elementary theorems of complex analysis, of the Fourier and the Laplace transformations and of the theory of Cauchy integrals. The latter is reviewed in detail in Appendix 2.

A book like this is necessarily incomplete. We had to select — and thus restrict — the material carefully in order to keep the

* I. Prigogine, *Non-Equilibrium Statistical Mechanics*, Interscience, New York, 1962.

** These elementary concepts could be acquired from a first reading of the excellent books of L. Spitzer, Jr., (*Physics of Fully Ionized Gases*, Interscience, New York, 1956) or of J.-L. Delcroix (*Introduction to the Theory of Ionized Gases*, Interscience, New York, 1960).

book within a reasonable size. As a general rule, we preferred to take the simplest possible situations or models and study them in great detail, rather than to treat superficially a large variety of systems. The main emphasis is laid on the derivation of the fundamental equations and the exploration of their basic properties rather than on solutions and calculations which can be found elsewhere.

The reader might be surprised by the large number of appendices appearing in this book. Actually these appendices are of three types. Appendices 1 and 2 are intended to review some general mathematical concepts which are extensively used in the text. Appendices 3 to 7 contain particular calculations which are too lengthy to be included in the main text. Finally appendices 8 to 11, which have been added in the proofs, are results obtained after the manuscript has been sent to the printer. We thought that it would be interesting to include these recent results in the book, although their logical place would have been in the main text.

We would like to express our most sincere gratitude to Professor I. Prigogine, whose guidance and help, as well as through direct critical discussion concerning the work presented here, or through his enthusiasm for all scientific activities, had a deep influence upon our scientific and human evolution. Many thanks are also due to Professors R. Brout, S. Prager, H. Taylor, Drs. Fr. Jeener–Henin, P. Résibois, G. Severne, Ph. de Gottal, B. Abraham, A. Kuszell, and S. Fujita, with whom we had many discussions and whose criticisms were invaluable to us. We would also like to acknowledge the partial financial support of the European Office of the Air Research and Development Command, U.S.A.F.

Contents

PART I. CLASSICAL SYSTEMS

1. The General Diagram Method for Classical Gases . . . 1

1. Distribution Functions and the Liouville Equation 1
2. The Fourier Transforms of the Distribution Functions and the Macroscopic Specification of Physical Systems 8
3. Size and Shape of Fourier Components at Time Zero 19
4. The Green's Function and the Formal Solution of the Liouville Equation. 26
5. Perturbation Theory 33
6. The Diagram Technique 41
7. Orders of Magnitude of Diagrams 45
8. Reduced Distribution Functions at Time t 47

2. The Short-Time Behavior of Classical Plasmas — The Vlassov Equation. 55

9. General Discussion of the Properties of Classical Plasmas 55
10. Choice of Diagrams for Short Times 60
11. Derivation of the Vlassov Equation 65
12. Some General Properties of the Vlassov Equation . 69

3. The Theory of Plasma Oscillations 73

13. The Solution of the Initial-Value Problem for the Linearized Vlassov Equation 73
14. Individual Particle Behavior and Collective Bevior 76
15. Individual Particle Behavior 80
16. The "Collective" Behavior 83
17. The Overall Behavior of the Plasma 86

4. The Dispersion Equation 89

18. The Dielectric Constant of the Plasma 89
19. The Stability Criterion. 95
20. Limiting Forms of the Dispersion Equation . . . 101
21. The Dielectric Constant for a Maxwellian Distribution 108
22. An Example of an Unstable Distribution. 112

5. The van Kampen–Case Treatment of the Vlassov Equation 117

23. Eigenfunctions of the Vlassov Operator 117
24. Eigenfunctions of the Adjoint Vlassov Operator. . 120

25. Completeness of the Set of Vlassov Eigenfunctions 122
26. Application of the Eigenvalue Expansion to the Solution of the Initial-Value Problem 126

6. The Long-Time Behavior of Classical Plasmas . . . 129
27. General Structure of the Perturbation Series for Long Times 129
28. The Cycle Diagram in the Case of a Plasma . . . 133
29. Choice of Contributions for a Plasma 140

7. The Landau ("Fokker–Planck") Approximation . . 146
30. Evolution of the Velocity Distribution Function . 146
31. The Concept of Pseudo-Diagonal Fragments . . . 151
32. The Evolution of $\rho_{\mathbf{k}}(\alpha; t)$ and of $f(\alpha; t)$ 156
33. The Boltzmann Equation and its Relation to the Landau Equation 162

8. Properties of the Landau Equation 170
34. Irreversibility and the H-Theorem 170
35. Explicit Form of the Landau Equation 174
36. The Hydrodynamic Equations 178
37. The Fokker–Planck Equation 181
38. Connection with the Theory of Brownian Motion . 186

9. The Ring Approximation 192
39. The Ring Diagrams 192
40. Summation of the Rings 198
41. Solution of the Integral Equation for Plasmas. (Imaginary Part of F) 201
42. Inhomogeneous Systems 207

10. Binary Correlations in the Ring Approximation . . 210
43. Preliminary Discussion of the Diagrams 210
44. Summation of the Creation Diagrams 216
45. Solution of the Integral Equation for Plasmas. (Real Part of F) 220

11. The General Description of a Plasma in the Ring Approximation 223
46. General Properties of the Kinetic Equation. . . . 223
47. Connection between Correlation Function and Kinetic Equation 226
48. Connection between the Ring Equation and the Landau Equation 229
49. Brownian Motion in the Ring Approximation. . . 233
50. Limitations of the Ring Approximation 238

12. The Equilibrium State 242
51. Correlations in Equilibrium. 242
52. Summation of Creation Diagrams in Equilibrium . 245
53. Thermodynamics of a Plasma in Equilibrium . . . 247
54. Mayer's Theory 250

13. Non-Equilibrium Stationary States and the Theory of Transport Coefficients 257
55. Free Relaxation and Forced Relaxation 257
56. Traditional Theories of Transport Coefficients . . 259
57. A General Formula for the Electric Current . . . 268
58. Existence and Stability of a Stationary State . . . 274
59. Stationary State vs. Kinetics of Approach 277
60. The Perfect Lorentz Gas 280

PART II. QUANTUM-STATISTICAL SYSTEMS

14. The General Diagram Method for Quantum Gases 287
61. The Density Matrix and the Wigner Distribution Function 287
62. The Fourier Transform of the Wigner Function . . 294
63. The Equation of Evolution of the Fourier Components of the Wigner Function 303
64. The Limit of Boltzmann Statistics and the Classical Limit 309
65. The Quantum-Statistical Diagram Technique . . . 312

15. Short-Time Behavior of Quantum Plasmas and the Quantum Vlassov Equation 316
66. The Choice of Diagrams 316
67. Derivation and Solution of the Linear Quantum Vlassov Equation 321
68. General Features of the Quantum Vlassov Description 323
69. The Stability Condition for Quantum-Statistical Systems 325
70. The Quantum Dielectric Constant for an Electron Gas at Zero Temperature 328

16. Long-Time Behavior of Quantum Plasmas — The Cycle Approximation 335
71. Choice of Diagrams 335
72. The Quantum-Mechanical Cycle Diagram 336
73. The Quantum "Cycle" Equation 340
74. General Properties of the Cycle Equation 342
75. Motion of a Test Particle in an Electron Gas at Zero Temperature 345

17. The Quantum-Statistical Ring Approximation . . . 353
76. The Quantum-Statistical Ring Diagrams 353
77. Derivation of the Kinetic Equation 356
78. Collective Effects in the Motion of a Test Particle through an Electron Gas at Zero Temperature . . 359

18. Binary Correlations and the Equilibrium State for an Electron Gas 367
79. The Density Correlation Function 367
80. Eigenvalues and Eigenfunctions of the Kernel $G_k(|\tau|)$ 372
81. The Ground-State Energy of an Electron Gas . . . 375

Appendices

1. Properties of the Laplace Transformation 385
2. The Cauchy Integral 390
3. Orthogonality Properties of Vlassov Eigenfunctions 406
4. Derivation of a General Type of Kinetic Equation. 408
5. Miscellaneous Theorems Concerning Normal Products of Creation and Destruction Operators 411
6. Derivation of the Quantum-Statistical Liouville Equation 415
7. Proof of Equation (81.1) 417
8. Non-markoffian Kinetic Equation for a Stable Classical Plasma 419
9. Kinetic Equation for Unstable Classical Plasmas . 431
10. An Alternative Summation Method for the Ring Diagrams 445
11. The Kinetic Equation for a Plasma in a Strong External Field 453

Symbol Index 463

Author Index 469

Subject Index 473

Part I

Classical Systems

CHAPTER 1

The General Diagram Method for Classical Gases

§ 1. Distribution Functions and the Liouville Equation

A plasma is, dynamically speaking, an assembly of N particles, a fraction of which at least is electrically charged. The most general type of plasma consists of positive ions, negative ions, free electrons, and neutral molecules in various states of excitation. Such a system is, of course, very complicated, because of the variety of processes which can occur. In this book we shall restrict ourselves to the study of only a part of these processes which are the most characteristic of a plasma, i.e. the electrostatic interactions. For this purpose, we consider a simplified model consisting of a mixture of charged particles alone. Assume there are s species of particles, characterized by an index μ or ν; there are N_μ particles of species μ, and the total number of particles is $\sum_{\mu=1}^{s} N_\mu = N$. The particles of species μ are characterized electrically by a charge $z_\mu e$, and dynamically by a mass m_μ. Denoting by $\mathbf{x}_{j\mu}$ the position of particle j, of species μ, and its momentum by $\mathbf{p}_{j\mu}$, the gas is described by the following hamiltonian:

$$H_{pl} = \sum_{\mu=1}^{s} \sum_{j=1}^{N_\mu} \frac{p_{j\mu}^2}{2m_\mu} + e^2 \sum_{\nu,\mu=1}^{s} \sum_{j<n} V_{jn}^{(\mu\nu)} \tag{1.1}$$

with

$$V_{jn}^{(\mu\nu)} = z_\mu z_\nu \frac{1}{|\mathbf{x}_{j\mu} - \mathbf{x}_{n\nu}|} \tag{1.2}$$

The hamiltonian (1.1) describes a fully ionized plasma, which can be considered as a limiting state in much the same sense that a perfect gas is a limit of an ordinary gas. However, for computational purposes, even this hamiltonian is rather complicated. The complication, however, is mainly a notational one: the presence of a mixture necessarily introduces very cumbersome expressions in which each letter is affected by many subscripts and superscripts. Such complications are usually unessential and will be avoided by adopting an even more simplified model which describes the plasma as a one-component gas of charged particles. However, in order to be physically realistic, one must assume that the charged particles move in a background which is oppositely charged so as to neutralize exactly the overall charge of the gas.* The hamiltonian of this system will be written as:

$$H = \sum_{j=1}^{N} \frac{p_j^2}{2m} + e^2 \sum_{j<n}\sum V_{jn}(|\mathbf{x}_j - \mathbf{x}_n|) \tag{1.3}$$

This hamiltonian describes a general assembly of particles, interacting through central two-body forces, deriving from a potential e^2V_{jn}. e^2 is the square of the charge (the valence of the particles is assumed to be unity for simplicity). The potential energy function is then:

$$V_{jn} = \frac{1}{|\mathbf{x}_j - \mathbf{x}_n|} \quad \text{(plasma)} \tag{1.4}$$

However, the considerations of Chapter 1 will not be restricted to the case of a plasma, but will apply to any hamiltonian of the form (1.3). In the general case, the parameter e^2 has of course no longer the meaning of an electrical charge: it is then simply some measure of the strength of the interactions. It will, however, prove convenient to keep the notation e^2 also in the general case.

It should be noted that the hamiltonians (1.1) and (1.3) describe a gas (or a plasma) in the absence of any external field. In the present part of this book we shall be concerned only with such simple systems. The formal extension of the theory to cover the case of an external field is usually simple. We shall study the plasma in external fields in Chapter 13.

The hamiltonian (1.1) or (1.3) contains implicitly a complete dynamical description of a plasma. One can derive from it the exact dynamical equations of motion:

* Although most of the calculations in this book are made for this one-component model, there is no essential difficulty in their extension to multicomponent systems. For the sake of completeness the extended formulae will be indicated in the more important places.

$$\begin{cases} \dfrac{\partial H(x, p)}{\partial \mathbf{x}_j} = -\dot{\mathbf{p}}_j \\ \dfrac{\partial H(x, p)}{\partial \mathbf{p}_j} = \dot{\mathbf{x}}_j \end{cases} \tag{1.5}$$

However, we are stopped at the very beginning if we try to solve these equations. We are in the presence of a system of $6N$ non-linear differential equations, N being of the order of 6×10^{23}. It must be realized that the difficulty is not only a practical one. Even if we could imagine a perfect computing machine which could solve eqs. (1.5), the solution would be absolutely useless. Indeed, to give a meaning to eqs. (1.5) they must be supplemented by a set of $6N$ initial conditions. It is absolutely inconceivable for a human observer to measure simultaneously the positions and momenta of 10^{23} particles at a given time, or to prepare the system in such a way that these variables have prescribed values. Therefore, the exact formal solution of (1.5) would be of no use in predicting any physical process.

We need, therefore, a concept which is closer to accessible macroscopic facts. This concept is provided by the idea of an *ensemble*, introduced by Gibbs. Instead of considering a unique system, we study a set of systems which are dynamically identical (i.e. same hamiltonian) but differ in their initial conditions. The natural framework for the description of such a system is a $6N$-dimensional space, called the *phase space*, the coordinates of which are the positions and momenta of the $6N$ particles. A particular system in a given state of motion will be described by a point in the phase space. The dynamical evolution of the complete system is described by the motion of this point along a trajectory in phase space. The ensemble representing the real system therefore corresponds to a cloud of points, which is usually assumed to be continuous. Its mathematical description will be given by a function representing the density of the cloud at each point in phase space:

$$f_N(\mathbf{x}_1, \ldots, \mathbf{x}_N, \mathbf{p}_1, \ldots, \mathbf{p}_N; t)$$

This function will be called the *N-particle distribution function.*

It should perhaps be stressed that the description of the motion is different here from the hamiltonian description implied by (1.5). The positions $\mathbf{x}_i$ and the momenta $\mathbf{p}_i$ are *not* to be regarded as functions of time; the evolution of the system is described by the change in time of the density at a *given* point $(\mathbf{x}_1, \ldots, \mathbf{x}_N, \mathbf{p}_1, \ldots, \mathbf{p}_N)$ in phase space.

We shall often need shorter notations for this function, or, generally, for any function F of the set of N momenta and N positions. The following shorthands will be currently used:

$$F(x, p; t) \equiv F(\{\mathbf{x}_j\}, \{\mathbf{p}_j\}; t) \equiv F(\mathbf{x}_1, \ldots, \mathbf{x}_N, \mathbf{p}_1, \ldots, \mathbf{p}_N; t)$$

Whenever any confusion is impossible the italic p will be used to denote the set of *all* momenta $\mathbf{p}_1$ to $\mathbf{p}_N$.

As a consequence of the well known Liouville theorem of classical mechanics,* the cloud representing the ensemble moves like an incompressible fluid, and therefore obeys a continuity equation in phase space:

$$\frac{\partial f_N}{\partial t} + \sum_{j=1}^{N} \left\{ \frac{\partial f_N}{\partial \mathbf{x}_j} \cdot \dot{\mathbf{x}}_j + \frac{\partial f_N}{\partial \mathbf{p}_j} \cdot \dot{\mathbf{p}}_j \right\} = 0$$

or, using the Hamilton equations (1.5),

$$\frac{\partial f_N}{\partial t} + [f_N, H] = 0 \tag{1.6}$$

where the second term is the Poisson bracket defined as:

$$[h, g] = \sum_j \left(\frac{\partial h}{\partial \mathbf{x}_j} \cdot \frac{\partial g}{\partial \mathbf{p}_j} - \frac{\partial h}{\partial \mathbf{p}_j} \cdot \frac{\partial g}{\partial \mathbf{x}_j} \right) \tag{1.7}$$

Introducing the hamiltonian of a one-component gas (1.3), the following explicit *Liouville equation* is obtained:

$$\frac{\partial f_N}{\partial t} + \sum_j \mathbf{v}_j \cdot \frac{\partial f_N}{\partial \mathbf{x}_j} = e^2 \sum_{j<n}\sum \frac{\partial V_{jn}}{\partial \mathbf{x}_j} \cdot \left(\frac{\partial}{\partial \mathbf{p}_j} - \frac{\partial}{\partial \mathbf{p}_n} \right) f_N \tag{1.8}$$

where $\mathbf{v}_j = \mathbf{p}_j/m$. We now note that, whereas the *momenta* $\mathbf{p}_j$ are natural variables within the framework of the hamiltonian

* See, e.g., H. Goldstein, *Classical Mechanics*, Addison-Wesley Publ. Co., Cambridge, Mass., 1953.

formalism, the *velocities* $\mathbf{v}_j$ are more convenient variables in formulae which connect microscopic and macroscopic quantities. We shall, therefore, use from here on (with the exception of Chapter 14) the variables $\mathbf{v}_j$ instead of $\mathbf{p}_j$. The change is trivial as long as magnetic fields are absent and relativistic effects are neglected. The N-particle distribution function will be regarded as a function of positions, velocities and time:

$$f_N(\mathbf{x}_1, \ldots, \mathbf{x}_N, \mathbf{v}_1, \ldots, \mathbf{v}_N; t)$$

The corresponding change of variables is immediately performed in eq. (1.8). From here on we shall always use the following short notations:

$$\begin{aligned} \partial_t &\equiv \partial/\partial t \\ \boldsymbol{\nabla}_j &\equiv \partial/\partial \mathbf{x}_j \\ \partial_j &\equiv \partial/\partial \mathbf{v}_j \\ \partial_{jn} &\equiv \partial/\partial \mathbf{v}_j - \partial/\partial \mathbf{v}_n \end{aligned} \tag{1.9}$$

With these symbols, the Liouville equation is written as follows:

$$\partial_t f_N + \sum_j \mathbf{v}_j \cdot \boldsymbol{\nabla}_j f_N = (e^2/m) \sum_{j<n} \sum (\boldsymbol{\nabla}_j \cdot V_{jn}) \cdot \partial_{jn} f_N \tag{1.10}$$

Equation (1.10) implies that the integral of f_N over all positions and velocities is constant in time. We may therefore normalize f_N as follows

$$\int (d\mathbf{x})^N (d\mathbf{v})^N f_N = 1 \tag{1.11}$$

After this formal introduction of the Liouville equation, we must state the physical interpretation of the concept of an ensemble. There is some controversy among authors concerning this point, the controversy being essentially centered around the famous ergodic theorem.* We do not wish to enter here into these rather sterile discussions, especially since some recent work has cast some doubt on the relevance of this theorem to real physical systems.** We shall adopt the following inter-

* See, e.g., A. I. Khinchine, *Mathematical Foundations of Statistical Mechanics,* Dover Publ. Co., New York, **1949**.

** P. Résibois and I. Prigogine, *Acad. Roy. Belg. Bull. Classe Sci.,* **46**, 53 (1960).

pretation: it is impossible for a macroscopic observer to ascertain the issue of a single measurement performed on a system whose initial state is "macroscopically" specified (we shall come back to the concept of "macroscopic specification"). The only possible prediction is the average issue of a large number of such measurements performed on the same macroscopic system. We assume that this averaging process is weighted by the distribution function f_N. This function must be constructed at time zero in such a way as to be compatible with our macroscopic information about the system. However, because of the large number of particles in the system, the issue of any measurement will be very close (presumably within an error of order N^{-1}) to the average value predicted by the ensemble. The latter statement has never been proven formally but is a very natural hypothesis.

The conclusion of this discussion is the following. The observable value of any macroscopic dynamical quantity is the average value, weighted with f_N, of the corresponding microscopic quantity

$$A(t)_{\text{macro}} = \int (d\mathbf{x})^N (d\mathbf{v})^N f_N(x, v; t) A(x, v) \tag{1.12}$$

We now note that the information contained in f_N is actually redundant. For all quantities of macroscopic interest, such as the density, the hydrodynamic velocity, etc., the quantity $A(x, v)$ is a function of the positions and velocities of a very small number of particles (say, one, two . . .) as compared to the total number of particles in the system. Therefore the weighting function of A in (1.12) is actually the integral of f_N over all particles except those on which A depends. These integrals are called *reduced distribution functions of s particles*. They will be defined as follows:

$$f_s(\mathbf{x}_1, \ldots, \mathbf{x}_s, \mathbf{v}_1, \ldots, \mathbf{v}_s; t) = \frac{N!}{(N-s)!} \int (d\mathbf{x})^{N-s} (d\mathbf{v})^{N-s} f_N \tag{1.13}$$

The factor $N!/(N-s)!$ is convenient for the following reason. If we interpret f_N as a probability, then f_s defined without the factorials would represent the probability of finding the *specified* particle 1 at $(\mathbf{x}_1, \mathbf{v}_1)$, particle 2 at $(\mathbf{x}_2, \mathbf{v}_2)$, etc. However, in a large physical system of identical particles, all the particles play the same role; a given macroscopic property is determined by a set of particles, regardless of their identity. It is therefore convenient to multiply the integral of f_N by the number of ways one can choose s particles out of the total number N.

In terms of these functions the most important macroscopic quantities are defined as follows:*

Density at point $\mathbf{x}$

$$n(\mathbf{x}; t) = \int d\mathbf{x}_1 d\mathbf{v}_1 \, \delta(\mathbf{x} - \mathbf{x}_1) f_1(\mathbf{x}_1, \mathbf{v}_1; t) \tag{1.14}$$

Local (hydrodynamic) velocity at point $\mathbf{x}$

$$\mathbf{u}(\mathbf{x}; t) = [n(\mathbf{x}; t)]^{-1} \int d\mathbf{x}_1 d\mathbf{v}_1 \, \mathbf{v}_1 \delta(\mathbf{x}-\mathbf{x}_1) f_1(\mathbf{x}_1, \mathbf{v}_1; t) \tag{1.15}$$

Local energy density at point $\mathbf{x}$

$$\begin{aligned} E(\mathbf{x}; t) = [n(\mathbf{x}; t)]^{-1} \Big\{ \int d\mathbf{x}_1 d\mathbf{v}_1 \tfrac{1}{2} m v_1^2 \delta(\mathbf{x}-\mathbf{x}_1) f_1(\mathbf{x}_1, \mathbf{v}_1; t) \\ + \tfrac{1}{2} e^2 \int d\mathbf{x}_1 d\mathbf{x}_2 d\mathbf{v}_1 d\mathbf{v}_2 \, V_{12}(\mathbf{x}_1-\mathbf{x}_2) \delta(\mathbf{x}-\mathbf{x}_1) f_2(\mathbf{x}_1, \mathbf{x}_2, \mathbf{v}_1, \mathbf{v}_2; t) \Big\} \end{aligned} \tag{1.16}$$

Density correlation between points $\mathbf{x}$ and $\mathbf{x}'$

$$\begin{aligned} g(\mathbf{x}, \mathbf{x}'; t) = \int d\mathbf{x}_1 d\mathbf{x}_2 d\mathbf{v}_1 d\mathbf{v}_2 \, \delta(\mathbf{x}_1-\mathbf{x}) \delta(\mathbf{x}_2-\mathbf{x}') \\ \times [f_2(\mathbf{x}_1, \mathbf{x}_2, \mathbf{v}_1, \mathbf{v}_2; t) - f_1(\mathbf{x}_1, \mathbf{v}_1; t) f_1(\mathbf{x}_2, \mathbf{v}_2; t)] \end{aligned} \tag{1.17}$$

Other average quantities will be encountered later.

In multicomponent systems, the reduced distribution functions must be further specified. In a system of s components there are s types of one-particle distribution:

$$f_1^{(\sigma)}(\mathbf{x}_1, \mathbf{v}_1; t), \qquad \sigma = 1, 2, \ldots, s$$

This notation obviously refers to the distribution of particle 1 of type σ. Similarly there are $\frac{1}{2}s(s+1)$ types of two-particle distribution:

$$f_2^{(\sigma,\sigma')}(\mathbf{x}_1, \mathbf{x}_2, \mathbf{v}_1, \mathbf{v}_2; t)$$

This symbol represents the distribution of particle 1 of type σ and of particle 2 of type σ'. The generalizations of the definitions (1.14–16) are:

* See, e.g., J. H. Irving and J. G. Kirkwood, *J. Chem. Phys.*, **18**, 817 (1950); D. Massignon, *Mécanique Statistique des Fluides*, Dunod, Paris, 1957.

Density at point $\mathbf{x}$:

$$n(\mathbf{x}; t) = \sum_{\sigma=1}^{s} \int d\mathbf{x}_1 d\mathbf{v}_1 \delta(\mathbf{x}-\mathbf{x}_1) f_1^{(\sigma)}(\mathbf{x}_1, \mathbf{v}_1; t) = \sum_{\sigma=1}^{s} n^{(\sigma)}(\mathbf{x}; t) \qquad (1.14')$$

Local velocity at point $\mathbf{x}$:

$$\mathbf{u}(\mathbf{x}; t) = [n(\mathbf{x}; t)]^{-1} \sum_{\sigma=1}^{s} \int d\mathbf{x}_1 d\mathbf{v}_1 \mathbf{v}_1 \delta(\mathbf{x}-\mathbf{x}_1) f_1^{(\sigma)}(\mathbf{x}_1, \mathbf{v}_1; t) \qquad (1.15')$$

Energy density at point $\mathbf{x}$:

$$E(\mathbf{x}; t) = [n(\mathbf{x}; t)]^{-1} \Big\{ \sum_{\sigma=1}^{s} \int d\mathbf{x}_1 d\mathbf{v}_1 (\tfrac{1}{2} m v_1^2) \delta(\mathbf{x}-\mathbf{x}_1) f_1^{(\sigma)}(\mathbf{x}_1, \mathbf{v}_1; t)$$
$$+ \tfrac{1}{2} e^2 \sum_{\sigma, \sigma'} \int d\mathbf{x}_1 d\mathbf{x}_2 d\mathbf{v}_1 d\mathbf{v}_2 \, V^{(\sigma\sigma')}(\mathbf{x}_1 - \mathbf{x}_2) \delta(\mathbf{x}-\mathbf{x}_1) f_2^{(\sigma, \sigma')}(\mathbf{x}_1, \mathbf{x}_2, \mathbf{v}_1, \mathbf{v}_2; t) \qquad (1.16')$$

Three other types of reduced distribution function will be introduced: the reduced distribution of s velocities, φ_s; the reduced distribution of s positions, n_s; and the reduced distribution of s velocities and r positions ($s \neq r$). These functions are defined as follows:

$$\varphi_s(\mathbf{v}_1, \ldots, \mathbf{v}_s; t) = \int (d\mathbf{x})^N (d\mathbf{v})^{N-s} f_N \qquad (1.18)$$

$$n_s(\mathbf{x}_1, \ldots, \mathbf{x}_s; t) = \frac{N!}{(N-s)!} \int (d\mathbf{x})^{N-s} (d\mathbf{v})^N f_N \qquad (1.19)$$

$$f_{s,r}(\mathbf{x}_1, \ldots, \mathbf{x}_s, \mathbf{v}_1, \ldots, \mathbf{v}_r; t) = \frac{N!}{(N-s)!} \int (d\mathbf{x})^{N-s} (d\mathbf{v})^{N-r} f_N \qquad (1.20)$$

(Note that, for later convenience, φ_s is normalized to unity.)

§ 2. The Fourier Transforms of the Distribution Functions and the Macroscopic Specification of Physical Systems

It can be seen from the discussion of § 1 that all macroscopic quantities are functionals (or, more precisely, moments) of reduced distribution functions of a small set of particles. It is therefore clear that the specification of a macroscopic system amounts to stating some properties of the reduced distribution functions.

A very general type of macroscopic classification of physical systems at the initial time is one into *homogeneous* (or uniform) and *inhomogeneous* (or non-uniform) systems. A homogeneous gas is one in which physical properties are the same at all points in space. The mathematical property which expresses this fact is the invariance of the N-particle distribution function (and therefore of all reduced distribution functions) with respect to a bulk translation of all positions:

$$f_N(\mathbf{x}_1+\mathbf{a},\ldots,\mathbf{x}_N+\mathbf{a},\mathbf{v}_1,\ldots,\mathbf{v}_N;t)=f_N(\mathbf{x}_1,\ldots,\mathbf{x}_N,\mathbf{v}_1,\ldots,\mathbf{v}_N;t) \tag{2.1}$$

where $\mathbf{a}$ is an arbitrary vector. Formula (2.1) implies in particular that the one-particle reduced distribution function is independent of $\mathbf{x}$, and that the two-particle reduced distribution function only depends on the relative distance of the two particles:

$$\begin{cases} f_1(\mathbf{x}_1,\mathbf{v}_1;t)=c\varphi_1(\mathbf{v}_1;t) & \text{(homogeneous systems)} \\ f_2(\mathbf{x}_1,\mathbf{x}_2,\mathbf{v}_1,\mathbf{v}_2;t)=f_2(\mathbf{x}_1-\mathbf{x}_2,\mathbf{v}_1,\mathbf{v}_2;t) & \end{cases} \tag{2.2}$$

It may be noted that eq. (2.1) can be written for all times, because an initially homogeneous system in the absence of an external field remains so in the course of its dynamical evolution. (As will be seen later, this is a consequence of the central character of the interaction forces.)

The most important requirement on any physical theory will be discussed now. A gas is a system consisting of a *very large* number of particles ($N\sim 6\times 10^{23}$) enclosed in a *very large* volume Ω [the ratio of Ω (say 1 cm^3) to molecular volumes (say $(10^{-8})^3$ cm^3) is of the order of 10^{24}]. From the mathematical point of view this turns out to be an advantage, because we may take the limits $N\to\infty$, $\Omega\to\infty$, and in this limit many simplifications will occur. One must, however, be careful in taking these limits, because they are not independent. The ratio N/Ω, which is the average number density of the gas, is a physical parameter which must remain constant and finite in this limiting process. The exact specification of the limit is therefore

$$\begin{cases} N\to\infty \\ \Omega\to\infty \\ N/\Omega=\text{finite constant}=c \end{cases} \tag{2.3}$$

The preceding discussion introduces quite naturally our general requirement. In a real gas, all *local* properties (such as hydrodynamic variables, intensive thermodynamic properties, correlations, etc.) must have finite values which are independent of the size of the system.

The main assumption in the theory of Prigogine and his collaborators,* first introduced by Prigogine and Balescu**, is the following: *The distribution function f_N which is taken as an initial condition in any physical problem must be compatible with the requirement that all reduced distribution functions f_s of a finite number of particles must be finite in the limit* (2.3). A second point which must come out of the theory, if it is consistent, is that initially finite reduced distributions must remain so in the course of time. This will be proven in § 8.

This fundamental requirement is reflected in the mathematical properties of the Fourier components of the distribution functions. In order to perform a Fourier transformation we must specify the boundary conditions for our system. As the volume will eventually go to infinity, it is not necessary to refine these conditions too much; surface effects, which depend crucially on boundary conditions, will eventually give a negligible relative contribution (of order $1/N^{\frac{1}{3}}$) to any local quantity. We may therefore adopt very simple boundary conditions. Specifically, we assume that the gas is enclosed in a cubic box of volume Ω and imagine that the system is periodically repeated in space along the three directions $(\overline{\mathbf{1}}_x, \overline{\mathbf{1}}_y, \overline{\mathbf{1}}_z)$ parallel to the edges of the box:

$$\begin{aligned} f_N(\{\mathbf{x}_j + \Omega^{\frac{1}{3}}\overline{\mathbf{1}}_x\}, v) &= f_N(\{\mathbf{x}_j + \Omega^{\frac{1}{3}}\overline{\mathbf{1}}_y\}, v) \\ = f_N(\{\mathbf{x}_j + \Omega^{\frac{1}{3}}\overline{\mathbf{1}}_z\}, v) &= f_N(\{\mathbf{x}_j\}, v) \end{aligned} \tag{2.4}$$

The initial distribution function will now be expanded in a Fourier series:

$$f_N(x, v) = \sum_{\mathbf{k}_1} \cdots \sum_{\mathbf{k}_N} \tilde{\rho}_{\mathbf{k}_1, \ldots, \mathbf{k}_N}(\mathbf{v}_1, \ldots, \mathbf{v}_N) \exp\left(i \sum_{j=1}^{N} \mathbf{k}_j \cdot \mathbf{x}_j\right) \tag{2.5}$$

* I. Prigogine, *Non-Equilibrium Statistical Mechanics*, Interscience, New York, **1963**.

** I. Prigogine and R. Balescu, *Physica*, **25**, 281, 302 (1959).

The boundary condition (2.4) requires

$$\mathbf{k}_j = (2\pi/\Omega^{\frac{1}{3}})\mathbf{n}_j \tag{2.6}$$

where $\mathbf{n}_j$ is a vector whose components are integers. The series (2.5) will be called the "*compact Fourier expansion*"; it is useful because it leads to short formulae. However, it does not yet display the full information which is required.

The N-fold summation in (2.5) can be split into a certain number of terms. One isolates $\tilde{\rho}_{0\ldots0}$, then the sum of the terms containing all components with a single non-vanishing wave-vector, corresponding to particle j: $\rho_{0\ldots0,\mathbf{k}_j,0\ldots0}$, etc.:

$$\begin{aligned} f_N = \tilde{\rho}_{0\ldots0} &+ \sum_j \sum_{\mathbf{k}_j}{}' \tilde{\rho}_{0\ldots\mathbf{k}_j\ldots0} \exp(i\mathbf{k}_j \cdot \mathbf{x}_j) \\ &+ \sum_j \sum_n \sum_{\mathbf{k}_j}{}' \sum_{\mathbf{k}_n}{}' \tilde{\rho}_{0\ldots\mathbf{k}_j\ldots\mathbf{k}_n\ldots0} \exp(i\mathbf{k}_j \cdot \mathbf{x}_j + i\mathbf{k}_n \cdot \mathbf{x}_n) + \ldots \end{aligned} \tag{2.7}$$

The prime on the summation symbols means that the value $\mathbf{k}_j = 0$ is excluded (because it is already taken into account elsewhere!).

Among Fourier components with a given number of wave-vectors, those for which the sum of the wave-vectors vanishes play a special role. In a homogeneous system, all Fourier components are of that type. This can be verified by inserting (2.5) into (2.1): for the latter condition to be satisfied it is seen that $\sum \mathbf{k}_j$ must be zero for all Fourier components. As the equilibrium distribution function (in the absence of external fields) is always homogeneous, it is quite natural to study these terms separately. More generally, a set of non-vanishing wave-vectors can be split into subsets such that the sum of the wave-vectors in one or more subsets is zero.

We are now ready to introduce our fundamental requirement of finiteness of the (initial) reduced distribution functions. In order to satisfy this condition the various types of Fourier components described above must have a well-defined dependence on the volume. The result will be quoted, and the finiteness condition will be verified on an example which shows its general validity. The expansion (2.7) will be rewritten in the form:

$$
\begin{aligned}
f_N(x, v; 0) = \Omega^{-N}\{\rho_0(v; 0) &+ (8\pi^3/\Omega) \sum_j \sum_{\mathbf{k}}{}' \rho_{\mathbf{k}}(\mathbf{v}_j \,|\, \ldots; 0) \exp(i\mathbf{k}\cdot\mathbf{x}_j) \\
&+ (8\pi^3/\Omega)^2 \sum_j \sum_n \sum_{\mathbf{k}}{}' \sum_{\substack{\mathbf{k}' \\ (\mathbf{k}+\mathbf{k}'\neq 0)}}{}' \rho_{\mathbf{k}\mathbf{k}'}(\mathbf{v}_j, \mathbf{v}_n \,|\, \ldots; 0) \exp(i\mathbf{k}\cdot\mathbf{x}_j + i\mathbf{k}'\cdot\mathbf{x}_n) \\
&+ (8\pi^3/\Omega) \sum_j \sum_n \sum_{\mathbf{k}}{}' \rho_{\mathbf{k},-\mathbf{k}}(\mathbf{v}_j, \mathbf{v}_n \,|\, \ldots; 0) \exp[i\mathbf{k}\cdot(\mathbf{x}_j - \mathbf{x}_n)] + \ldots \\
&+ (8\pi^3/\Omega)^{r-s} \sum_j \ldots \underbrace{\sum_n \sum_{\mathbf{k}_j}{}' \ldots \sum_{\mathbf{k}_n}{}'}_{\substack{r \text{ non-vanishing} \\ \text{wave-vectors}}} \underbrace{\delta_{\mathbf{k}_a+\ldots+\mathbf{k}_b} \ldots \delta_{\mathbf{k}_c+\ldots+\mathbf{k}_d}}_{s \text{ conditions}} \\
&\rho_{\mathbf{k}_j\ldots\mathbf{k}_n}(\mathbf{v}_j, \ldots, \mathbf{v}_n \,|\, \ldots; 0) \exp(i\mathbf{k}_j\cdot\mathbf{x}_j + \ldots + i\mathbf{k}_n\cdot\mathbf{x}_n) + \ldots\}
\end{aligned}
\tag{2.8}
$$

The symbol $\delta_{\mathbf{k}}$ denotes a product of three Kronecker δ-symbols:

$$\delta_{\mathbf{k}} = \delta_{k_x,0}\,\delta_{k_y,0}\,\delta_{k_z,0}$$

with

$$\delta_{i,j} = \begin{cases} 0 & \text{if } i \neq j \\ 1 & \text{if } i = j \end{cases}$$

The notation chosen for the Fourier components needs some comment. The series (2.8) is identical with the series (2.7), but is written with different symbols. The translation from one series to another is given in Table 2.1. The velocity variables on which a given Fourier component depends are split into two groups, separated by a bar. To the left of the bar we write the velocities of the particles for which the wave-vector is non-zero; these velocities are written in the same order as the corresponding wave-vector indices. To the right of the bar we write all the other velocities. In most cases the latter are irrelevant and are replaced by a series of dots. After the two groups of velocities we write (if necessary) a semi-colon followed by the value of the time. (Note that in the case of ρ_0 there is, of course, no particle in the first group.)

In many cases a simpler notation is sufficient. Instead of writing the velocities $\mathbf{v}_j, \ldots \mathbf{v}_n, \ldots$ of the first group, we simply write the indices of the corresponding particles: $j, n, \ldots$. This notation is given in the second line of the second column in Table 2.1.

We now introduce the BASIC REGULARITY CONDITION: *It is required that the Fourier components ρ appearing in eq.* (2.8) *do not depend on N or Ω otherwise than through the finite ratio N/Ω.*

Table 2.1. Relation between Fourier components of the distribution function in the compact notation (2.5) and in the explicit notation (2.8); initial orders of magnitude in the charge (§ 3); value of the exponent ν, eq. (5.26).

Compact notation		Explicit notation	Order in e^2	ν
$\tilde{\rho}_{\mathbf{0}\ldots\mathbf{0}}(v;0)$		$\Omega^{-N}\rho_{\mathbf{0}}(\|v;0)$ $\Omega^{-N}\rho_{\mathbf{0}}(\|\ldots;0)$	e^0	0
$\tilde{\rho}_{\mathbf{0}\ldots\mathbf{k}_n\ldots\mathbf{0}}(v;0)$		$\Omega^{-N}(8\pi^3/\Omega)\,\rho_{\mathbf{k}}(\mathbf{v}_n\|\ldots;0)$ $\Omega^{-N}(8\pi^3/\Omega)\,\rho_{\mathbf{k}}(n\|\ldots;0)$	—	1
$\tilde{\rho}_{\mathbf{0}\ldots\mathbf{k}_j\ldots\mathbf{k}_n\ldots\mathbf{0}}(v;0)$	$\mathbf{k}_j+\mathbf{k}_n\neq 0$	$\Omega^{-N}(8\pi^3/\Omega)^2\rho_{\mathbf{k}\mathbf{k}'}(\mathbf{v}_j,\mathbf{v}_n\|\ldots;0)$ $\Omega^{-N}(8\pi^3/\Omega)^2\rho_{\mathbf{k}\mathbf{k}'}(j,n\|\ldots;0)$	—	2
$\tilde{\rho}_{\mathbf{0}\ldots\mathbf{k}_j\ldots\mathbf{k}_n\ldots\mathbf{0}}(v;0)$	$\mathbf{k}_j+\mathbf{k}_n=0$	$\Omega^{-N}(8\pi^3/\Omega)\,\rho_{\mathbf{k},-\mathbf{k}}(\mathbf{v}_j,\mathbf{v}_n\|\ldots;0)$ $\Omega^{-N}(8\pi^3/\Omega)\,\rho_{\mathbf{k},-\mathbf{k}}(j,n\|\ldots;0)$	e^2	1
$\tilde{\rho}_{\mathbf{0}\ldots\mathbf{k}_j\ldots\mathbf{k}_n\ldots\mathbf{k}_m\ldots\mathbf{0}}(v;0)$	$\mathbf{k}_j+\mathbf{k}_n+\mathbf{k}_m\neq 0$	$\Omega^{-N}(8\pi^3/\Omega)^3\rho_{\underline{\mathbf{k}\mathbf{k}'\mathbf{k}''}}(\mathbf{v}_j,\mathbf{v}_n,\mathbf{v}_m\|\ldots;0)$	—	3
$\tilde{\rho}_{\mathbf{0}\ldots\mathbf{k}_j\ldots\mathbf{k}_n\ldots\mathbf{k}_m\ldots\mathbf{0}}(v;0)$	$\mathbf{k}_j+\mathbf{k}_n=0$	$\Omega^{-N}(8\pi^3/\Omega)^2\rho_{\underline{\mathbf{k},-\mathbf{k},\mathbf{k}'}}(\mathbf{v}_j,\mathbf{v}_n,\mathbf{v}_m\|\ldots;0)$	—	2
$\tilde{\rho}_{\mathbf{0}\ldots\mathbf{k}_j\ldots\mathbf{k}_n\ldots\mathbf{k}_m\ldots\mathbf{0}}(v;0)$	$\mathbf{k}_j+\mathbf{k}_n+\mathbf{k}_m=0$	$\Omega^{-N}(8\pi^3/\Omega)^2\rho_{\underline{\mathbf{k},\mathbf{k}',-\mathbf{k}-\mathbf{k}'}}(\mathbf{v}_j,\mathbf{v}_n,\mathbf{v}_m\|\ldots;0)$	e^4	2

Moreover, eq. (1.11) induces the following *normalization condition*:

$$\int (d\mathbf{v})^N \rho_0(|v; 0) = 1 \tag{2.9}$$

The regularity condition of the reduced distribution functions will be verified now in the case of $f_1(\mathbf{x}_\alpha, \mathbf{v}_\alpha; 0)$. Using the definition (1.13), formula (2.8) must be integrated over all variables except $\mathbf{x}_\alpha$, $\mathbf{v}_\alpha$, and the result must be multiplied by N. Keeping in mind that the (triple) integral of exp $(i\mathbf{k}\cdot\mathbf{x})$ over the cubic container of the gas equals $\Omega\delta_{\mathbf{k}}$, and that the value $\mathbf{k} = 0$ is excluded from all summations over wave-vectors, it is easily seen that:

All but the first two terms on the r.h.s. of (2.8) vanish after integration over the $N-1$ positions.

In the second term of the r.h.s. of (2.8), only the component $\rho_{\mathbf{k}}(\alpha|\ldots)$ gives a non-vanishing contribution.

We therefore obtain the simple formula:

$$f_1(\mathbf{x}_\alpha, \mathbf{v}_\alpha) = N\Omega^{-1}\{\rho_0(|\mathbf{v}_\alpha) + (8\pi^3/\Omega) \sum_{\mathbf{k}}' \rho_{\mathbf{k}}(\mathbf{v}_\alpha|)e^{i\mathbf{k}\cdot\mathbf{x}_\alpha}\} \tag{2.10}$$

where

$$\rho_0(|\mathbf{v}_\alpha) = \int (d\mathbf{v})^{N-1}\rho_0(|\{\mathbf{v}\})$$

with a corresponding definition holding for $\rho_{\mathbf{k}}(\mathbf{v}_\alpha|)$.

We are now able to perform the limiting process (2.3) explicitly. In the limit $\Omega \to \infty$, the wave-vector $\mathbf{k}$, defined by (2.6), becomes a *continuous* variable, and the following asymptotic formula holds: *

$$(8\pi^3/\Omega) \sum_{\mathbf{k}} \to \int d\mathbf{k} \qquad (\Omega \to \infty) \tag{2.11}$$

ρ_0 and $\rho_{\mathbf{k}}$ have been assumed to be independent of N and Ω, so that (2.10) tends to the *finite* limit:

$$f_1(\mathbf{x}_\alpha, \mathbf{v}_\alpha) = c\{\rho_0(|\mathbf{v}_\alpha) + \int d\mathbf{k}\, \rho_{\mathbf{k}}(\mathbf{v}_\alpha|)e^{i\mathbf{k}\cdot\mathbf{x}_\alpha}\} \tag{2.12}$$

* With the value $\mathbf{k} = 0$ excluded from the sum, the integral is actually a Cauchy principal part; however, this fact usually makes no difference here with the usual interpretation and will generally be disregarded.

It is also seen, on integrating (2.12) over $\mathbf{x}_\alpha$ and dividing by N, that:

$$\varphi_1(\mathbf{v}_\alpha) = \rho_0(|\mathbf{v}_\alpha) \tag{2.13}$$

Moreover, the normalization condition (2.9) induces the following normalization condition for the one-particle momentum distribution:

$$\int d\mathbf{v}_\alpha \varphi_1(\mathbf{v}_\alpha) = 1 \tag{2.14}$$

The coefficient $\rho_\mathbf{k}(\alpha|)$ has also a simple physical meaning: it is the Fourier transform of the function

$$H(\mathbf{x}_\alpha) = c^{-1}[f_1(\mathbf{x}_\alpha, \mathbf{v}_\alpha) - c\varphi_1(\mathbf{v}_\alpha)] \tag{2.15}$$

which will be called the *inhomogeneity factor*. When integrated over $\mathbf{v}_\alpha$, it measures the local deviation of the density from the average density c, and is therefore a measure of the degree of non-uniformity of the system. The latter function will be called the (local) *density excess* and will be denoted by $h(\mathbf{x}_\alpha)$:

$$h(\mathbf{x}_\alpha) = c^{-1}[n(\mathbf{x}_\alpha) - c] \tag{2.16}$$

Its Fourier transform is:

$$h_\mathbf{k} = \int d\mathbf{v}_\alpha \rho_\mathbf{k}(\mathbf{v}_\alpha|) \tag{2.17}$$

The next topic which will be discussed is the nature of *correlations* among particles. Consider for simplicity a homogeneous system.* A calculation entirely similar to the previous one gives the following expression for the two-particle reduced distribution:

$$f_2(\mathbf{x}_\alpha - \mathbf{x}_\beta, \mathbf{v}_\alpha, \mathbf{v}_\beta) = c^2\{\varphi_2(|\mathbf{v}_\alpha, \mathbf{v}_\beta) + \int d\mathbf{k}\, \rho_{\mathbf{k},-\mathbf{k}}(\mathbf{v}_\alpha, \mathbf{v}_\beta|) e^{i\mathbf{k}\cdot(\mathbf{x}_\alpha - \mathbf{x}_\beta)}\} \tag{2.18}$$

We may now introduce the so-called *two-particle correlation function*

$$c^2 G(\mathbf{x}_\alpha, \mathbf{x}_\beta, \mathbf{v}_\alpha, \mathbf{v}_\beta) = \{f_2(\mathbf{x}_\alpha, \mathbf{x}_\beta, \mathbf{v}_\alpha, \mathbf{v}_\beta) - f_1(\mathbf{x}_\alpha, \mathbf{v}_\alpha) f_1(\mathbf{x}_\beta, \mathbf{v}_\beta)\} \tag{2.19}$$

* Correlations in inhomogeneous systems are treated in § 3.

This function is needed in the calculation of the correlation of any local macroscopic quantities at two points [see, for instance, eq. (1.17)] and is a measure of the reciprocal influence of two particles located at $\mathbf{x}_\alpha$ and $\mathbf{x}_\beta$. It is physically obvious that *the correlations must vanish when the particles are widely separated*

$$G(\mathbf{x}_\alpha-\mathbf{x}_\beta, \mathbf{v}_\alpha, \mathbf{v}_\beta) \to 0 \quad \text{for} \quad \mathbf{x}_\alpha-\mathbf{x}_\beta \to \infty \tag{2.20}$$

It should be noted that this requirement can only be realized in a very large system. In an isolated two-body system the two particles are permanently correlated, because their trajectories are exactly fixed by the laws of dynamics. However, in a large system the two specified particles, after a collision, will undergo a large number of irregularly distributed interactions with the particles of the medium; hence when they arrive far apart their mutual correlation is completely destroyed. It should be noted that for the proof of the theorems of this section, the *qualitative* condition (2.20) is sufficient (in particular, we do not need assumptions about the mathematical law of decay of correlations or about the length scale of the correlations). A large part of the theory of homogeneous systems is based only on this qualitative statement. But in other problems, such as the study of inhomogeneous systems, stronger assumptions are necessary; these mainly concern the orders of magnitude of the various length scales appearing in the problem. Such types of assumption are discussed in § 3.

A substitution of (2.18) and (2.2) into (2.19) gives the following expression for G:

$$\begin{aligned} G(\mathbf{x}_\alpha-\mathbf{x}_\beta, \mathbf{v}_\alpha, \mathbf{v}_\beta) &= \{\varphi_2(\mathbf{v}_\alpha, \mathbf{v}_\beta)-\varphi_1(\mathbf{v}_\alpha)\varphi_1(\mathbf{v}_\beta)\} \\ &+ \int d\mathbf{k}\, \rho_{\mathbf{k},-\mathbf{k}}(\mathbf{v}_\alpha, \mathbf{v}_\beta|)e^{i\mathbf{k}\cdot(\mathbf{x}_\alpha-\mathbf{x}_\beta)} \end{aligned} \tag{2.21}$$

Take the limit of this expression for $\mathbf{x}_\alpha-\mathbf{x}_\beta \to \infty$. According to a well known theorem on Fourier transforms (the Riemann–Lebesgue theorem),* if $\rho_{\mathbf{k},-\mathbf{k}}$ is a sufficiently regular function of $\mathbf{k}$ (for any given values of $\mathbf{v}_\alpha$, $\mathbf{v}_\beta$), the integral of the second term on the r.h.s. of (2.21) tends to zero as $\mathbf{x}_\alpha-\mathbf{x}_\beta \to \infty$. As the first bracket on the r.h.s. of this formula is distance-independent, it is seen that condition (2.20) requires:

$$\varphi_2(\mathbf{v}_\alpha, \mathbf{v}_\beta) = \varphi_1(\mathbf{v}_\alpha)\varphi_1(\mathbf{v}_\beta) \tag{2.22}$$

* See, e.g., E. C. Titchmarsh, *Introduction to the Theory of Fourier Integrals*, 2nd. edition, Clarendon Press, Oxford, 1948.

The foregoing argument can easily be generalized to the discussion of more complicated correlation patterns. The physical requirement is that *the reduced distribution function of a set of s particles must be factorized into a product of two reduced distribution functions of lower order whenever the set can be split into two subsets, such that the mutual distance of any particle of the first subset to any particle of the second one is very large.* The consequence of this requirement, which generalizes (2.20), can be expressed by two formulae:

$$\varphi_s(\mathbf{v}_1, \ldots, \mathbf{v}_s) = \prod_{i=1}^{s} \varphi_1(\mathbf{v}_i) \tag{2.23}$$

$$\rho_{\mathbf{k}_1\ldots\mathbf{k}_r}(\mathbf{v}_1, \ldots, \mathbf{v}_r\,|\mathbf{v}_{r+1}, \ldots, \mathbf{v}_s) = \rho_{\mathbf{k}_1\ldots\mathbf{k}_r}(\mathbf{v}_1, \ldots, \mathbf{v}_r|) \prod_{i=r+1}^{s} \varphi_1(\mathbf{v}_i), \quad s > r \tag{2.24}$$

and therefore

$$f_{r,s}(\mathbf{x}_1, \ldots, \mathbf{x}_r, \mathbf{v}_1, \ldots, \mathbf{v}_s) = f_r(\mathbf{x}_1, \ldots, \mathbf{x}_r, \mathbf{v}_1, \ldots, \mathbf{v}_r) \prod_{i=r+1}^{s} \varphi_1(\mathbf{v}_i), \quad s > r \tag{2.25}$$

We may take advantage of the factorization theorems (2.23–25) in order to simplify the notations.

a) Any reduced s-particle velocity distribution is expressed in terms of the one-particle distribution alone. We shall henceforth never more use the notation φ_s for $s \neq 1$, but rather write automatically a product of functions φ_1. Therefore the subscript 1 becomes superfluous and we drop it:

$$\rho_0(|\mathbf{v}_\alpha; t) \equiv \varphi_1(\mathbf{v}_\alpha; t) \equiv \varphi(\mathbf{v}_\alpha; t) \equiv \varphi(\alpha; t) \tag{2.26}$$

b) Any reduced Fourier component other than ρ_0 can be expressed through eq. (2.24) in terms of the Fourier component integrated over all particles except those for which the wave-vectors are non-zero. In the old notation this means that there are no velocities to the right of the bar. The latter becomes superfluous and can be omitted:

$$\rho_{\mathbf{k}_1,\ldots,\mathbf{k}_r}(\mathbf{v}_1, \ldots, \mathbf{v}_r|; t) \equiv \rho_{\mathbf{k}_1,\ldots,\mathbf{k}_r}(\mathbf{v}_1, \ldots, \mathbf{v}_r; t) \equiv \rho_{\mathbf{k}_1,\ldots,\mathbf{k}_r}(1, \ldots, r; t) \tag{2.27}$$

The terminology and notations used in this book for the most

Table 2.2. Nomenclature for the principal functions derived from reduced distribution functions. The Fourier transforms and Fourier–Laplace transforms are called for instance (F-) density excess, (F-L-) density correlation, etc. When no confusion is possible, the letters (F-) or (F-L-) will be ommitted.

Name	Phase-space function	Fourier transform	Fourier–Laplace transform
Inhomogeneity factor	$c^{-1}[f_1(\mathbf{x}_\alpha, \mathbf{v}_\alpha; t) - c\varphi(\mathbf{v}_\alpha; t)]$ $= H(\mathbf{x}_\alpha, \mathbf{v}_\alpha; t)$ $= H(\alpha; t) = H(\alpha)$	$\rho_{\mathbf{k}}(\mathbf{v}_\alpha; t)$ $= \rho_{\mathbf{k}}(\alpha; t)$ $= \rho_{\mathbf{k}}(\alpha)$	$\rho_{\mathbf{k}}(\mathbf{v}_\alpha; z)$ $= \rho_{\mathbf{k}}(\alpha; z)$
Density excess	$c^{-1}[n(\mathbf{x}_\alpha; t) - c]$ $= h(\mathbf{x}_\alpha; t)$	$\int d\mathbf{v}_\alpha \rho_{\mathbf{k}}(\mathbf{v}_\alpha; t)$ $= h_{\mathbf{k}}(t)$	$\int d\mathbf{v}_\alpha \rho_{\mathbf{k}}(\mathbf{v}_\alpha; z)$ $= h_{\mathbf{k}}(z)$
Two-body correlation function (Homogeneous systems)	$c^{-2}[f_2(\alpha, \beta; t)$ $-c^2\varphi(\alpha; t)\varphi(\beta; t)]$ $= G(\lvert\mathbf{x}_\alpha - \mathbf{x}_\beta\rvert, \mathbf{v}_\alpha, \mathbf{v}_\beta; t)$ $= G(\alpha, \beta; t)$	$\rho_{\mathbf{k}, -\mathbf{k}}(\mathbf{v}_\alpha, \mathbf{v}_\beta; t)$ $= \rho_k(\mathbf{v}_\alpha, \mathbf{v}_\beta; t)$ $= \rho_k(\alpha, \beta; t)$	$\rho_k(\mathbf{v}_\alpha, \mathbf{v}_\beta; z)$ $= \rho_k(\alpha, \beta; z)$
Density correlation function (Homogeneous systems)	$\int d\mathbf{v}_\alpha d\mathbf{v}_\beta G(\alpha, \beta; t)$ $= g(\lvert\mathbf{x}_\alpha - \mathbf{x}_\beta\rvert; t)$ $= g(r; t)$	$\int d\mathbf{v}_\alpha d\mathbf{v}_\beta \rho_k(\alpha, \beta; t)$ $= g_k(t)$	$\int d\mathbf{v}_\alpha d\mathbf{v}_\beta \rho_k(\mathbf{v}_\alpha, \mathbf{v}_\beta; z)$ $= g_k(z)$

important concepts related to reduced distribution functions are given in Table 2.2.

We now summarize the main results thus far obtained. Two physical requirements have been introduced:

A) *Finiteness of the reduced distribution functions of any finite set of particles in the limit* $N \to \infty$, $\Omega \to \infty$, $N/\Omega = c$.

B) *Factorization of the reduced distribution functions of a set of two widely separated clusters.*

These conditions led to the two results:

a) *The Fourier expansion* (2.8) *of* f_N, *with components* ρ *independent of* N *and* Ω.

b) *The factorization properties* (2.23)–(2.25).

§ 3. Size and Shape of Fourier Components at Time Zero

The two general assumptions A and B of § 2 are sufficient for the proof of many important properties concerning the evolution of large systems over long periods of time. * However, one is often interested in more specific problems, such as the systematic derivation of kinetic equations for systems characterized by a given order of approximation in some parameters like e^2 or c. In order to perform consistently this type of analysis we need to make some physically reasonable assumptions about the size of the Fourier components at the initial instant of time.

A distinction must first be made between *one-particle* Fourier components [i.e. $\varphi(\mathbf{v})$ and $\rho_{\mathbf{k}}(\mathbf{v})$] and *many-body* Fourier components (i.e. those with more than one non-vanishing wave-vector). This distinction arises directly from the physical origin of these functions. *We have a possibility of external action on the one-body distribution.* We may, for instance, prepare a system with a given velocity distribution by putting together several streams of particles with various velocities. We can construct various types of inhomogeneities by the action of an external field or by putting side by side systems of different concentrations or different tempe-

* I. Prigogine, *Non-Equilibrium Statistical Mechanics*, Interscience, New York, 1963.

ratures, etc. On the other hand *we have absolutely no control over the correlations.* These are produced by the molecular interactions* and adapt themselves to the instantaneous microscopic state of the system according to the laws of dynamics. Actually it will be shown in the next sections that in the Fourier description the exact mechanical evolution of the system appears as a dynamics of correlations: these are continually being created and destroyed according to well defined patterns represented by our diagrams.

In establishing a reasonably general theory, one does not impose a specific form and magnitude on the initial condition adopted, but rather tries to characterize a very large class of initial conditions to which the theory would apply. A very natural choice of this class would be one for which *the initial instant does not play a privileged role*: in this way a shift of the origin of time would not modify the form of the theory. In our present problem we must assume that the correlations existing at time zero have appeared by the same mechanism and thus have the same order of magnitude as those which will appear at later times. This is a simple consequence of our lack of macroscopic control of the correlations. Even if, by an extremely improbable fluctuation, a correlation of a radically different order of magnitude appears at a given time, it will decay very rapidly and be replaced by another of the "normal" order of magnitude.

We first discuss *homogeneous systems*. In order to determine the "normal" order of magnitude, we note that it is the same as *the size of the correlations in thermal equilibrium*. This statement should seem natural in view of the previous discussion. Indeed, as the order of magnitude of the correlations is preserved during the evolution, and as the end-point of the evolution is thermal equilibrium, it is clear that the various correlations will be at all times of the order of magnitude of the equilibrium correlations. The proof of this statement will be very easily made *a posteriori* by the reader once he is familiar with the diagram technique explained in §§ 6–8. We should warn the reader that this assumption does not mean that the theory applies only to systems close

* We exclude here the study of turbulent motion, in which long-range correlations are produced by external macroscopic action.

to equilibrium. The functional form of the one-body functions and of the correlations can be very far from the one prevailing at equilibrium. The assumption expresses only the fact that correlations are produced at all times by molecular interactions and by no other, external, mechanism.

It will now be shown how the normal orders of magnitude can be determined from the equilibrium distribution function. The latter, representing a system in thermal equilibrium, is the well known Maxwell–Boltzmann distribution:

$$f_N^0 = Z_N^{-1} \exp\,(-\beta \sum_j \tfrac{1}{2} m v_j^2 - e^2 \beta \sum_{i<j} V_{ij}) \tag{3.1}$$

where Z_N is the partition function and $\beta = 1/kT$, k being the Boltzmann constant and T the absolute temperature. We now expand the exponential and the partition function in powers of e^2. Calling Z_N^K the kinetic part of the partition function we obtain:

$$\begin{aligned} f_N^0 = (Z_N^K)^{-1} \exp\,(-\tfrac{1}{2}\beta \sum_j m v_j^2)\{1 + e^2 \sum_{i<j} V_{ij} + e^4 [\sum_{i<j} V_{ij} V_{ij} \\ + \sum_{i<j<m} V_{ij} V_{jm} + \sum_{i<j<m<n} V_{ij} V_{mn}] + \ldots\} \end{aligned} \tag{3.2}$$

(We did not write down the terms arising from the expansion of the partition function explicitly.) We may, moreover, expand $V_{ij}(|\mathbf{x}_i - \mathbf{x}_j|)$ in a Fourier series:

$$V_{ij}(|\mathbf{x}_i - \mathbf{x}_j|) = \frac{8\pi^3}{\Omega} \sum_{\mathbf{k}} V_{\mathbf{k},\,-\mathbf{k}}\, e^{i\mathbf{k}\cdot(\mathbf{x}_i - \mathbf{x}_j)}$$

Substituting into (3.2) we obtain

$$\begin{aligned} f_N^0 = (Z_N^K)^{-1} \exp(-\tfrac{1}{2}\beta \sum_j m v_j^2) \Big\{ [1 + e^2 \ldots] \\ + \frac{8\pi^3}{\Omega} \sum_{i<j} \sum_{\mathbf{k}} e^{i\mathbf{k}\cdot(\mathbf{x}_i - \mathbf{x}_j)} [e^2 V_{\mathbf{k},-\mathbf{k}} + e^4 \sum_{\mathbf{k}'} V_{\mathbf{k}',-\mathbf{k}'} V_{\mathbf{k}-\mathbf{k}',\mathbf{k}'-\mathbf{k}} + \ldots] \\ + \left(\frac{8\pi^3}{\Omega}\right) \sum_{i<j<m} \sum_{\mathbf{k}} \sum_{\mathbf{k}'} e^{i\mathbf{k}\cdot\mathbf{x}_i + i(\mathbf{k}'-\mathbf{k})\cdot\mathbf{x}_j - i\mathbf{k}'\cdot\mathbf{x}_m} [e^4 V_{\mathbf{k},-\mathbf{k}} V_{\mathbf{k}',-\mathbf{k}'} + e^6 \ldots] + \ldots \Big\} \end{aligned} \tag{3.3}$$

Comparing this formula with the general expansion (2.8) we can identify corresponding Fourier components. The orders of

magnitude in e^2 are given by the least power of e^2 which appears in the expansion of the Fourier component considered. Thus ρ_0 will be assumed of order e^0, $\rho_{\mathbf{k},-\mathbf{k}}$ of order e^2, $\rho_{\mathbf{k},\mathbf{k}'-\mathbf{k},-\mathbf{k}'}$ of order e^4, etc. These normal orders of magnitude in e^2 are given in Table 2.1 for the first Fourier components.

In order to determine the dependence on c of the equilibrium Fourier components at time 0, one makes a similar expansion of f_N^0 (or rather of the various f_s^0) in powers of c. This type of expansion, analogous to the well known cluster expansion, has been performed explicitly by various authors.* The result is that all Fourier components have a contribution which is independent of c, plus terms proportional to higher powers of c. All initial Fourier components are therefore assumed to be of the same dominant order of magnitude in the concentration, i.e. c^0.

The previous discussion is only useful for determining the initial order of magnitude of the homogeneous Fourier components. Nothing can be inferred from the equilibrium state concerning *inhomogeneous components*, because the equilibrium is a homogeneous state. Another type of consideration is needed here. Initially the inhomogeneity is set up by some external macroscopic device, for instance by increasing the density in some region, or by imposing a temperature gradient, etc. Because of the macroscopic origin of this initial preparation, the length scale over which the value of the local quantities [and therefore of $f_1(\alpha)$] varies is usually very long compared with any characteristic microscopic length. In performing a Fourier transform of f_1, it is therefore evident that this function will have only Fourier components of very long wave-lengths, compared with microscopic lengths (or equivalently, very small wave-vectors compared with inverse microscopic lengths). Let us call L_h the "*hydrodynamic length scale*", which may be defined as the lower bound of $f_1/\nabla f_1$, and L_m the largest characteristic *molecular length scale* (such as the mean free path, or the range of the interactions). We shall summarize the preceding discussion by assuming that

$$L_h \gg L_m \tag{3.4}$$

* See, e.g., J. de Boer, *Rept. Progr. Phys.*, **12**, 305 (1948).

and consequently:

$$\rho_{\mathbf{k}}(\alpha; 0) \begin{cases} \neq 0 & \text{for} \quad \mathbf{k} = O(L_h^{-1}) \\ \simeq 0 & \text{for} \quad \mathbf{k} \gg L_h^{-1} \end{cases} \tag{3.5}$$

In other words, $\rho_{\mathbf{k}}(\alpha; 0)$ as a function of $\mathbf{k}$ is sharply peaked around $\mathbf{k} = 0$, with a width of order L_h^{-1}. It is clear, from its physical origin, that the initial inhomogeneity factor cannot be related to the interactions or to the density, and therefore will be assumed to be of the order e^0c^0 in these parameters.

Let us now discuss inhomogeneous Fourier components with more than one non-vanishing wave-vector, and more particularly $\rho_{\mathbf{kk}'}(\alpha, \beta; 0)$. It is easily seen from the general Fourier expansion (2.8) that

$$\rho_{\mathbf{kk}'}(\alpha, \beta) = (8\pi^3)^{-2} \int d\mathbf{x}_\alpha d\mathbf{x}_\beta \exp(-i\mathbf{k} \cdot \mathbf{x}_\alpha - i\mathbf{k}' \cdot \mathbf{x}_\beta) G'(\alpha, \beta) \tag{3.6}$$

where

$$G'(\alpha, \beta) = c^{-2}\{f_2(\alpha, \beta) - cf_1(\alpha)\varphi(\beta) - cf_1(\beta)\varphi(\alpha) + c^2\varphi(\alpha)\varphi(\beta)\} \tag{3.7}$$

Adding and subtracting the term $f_1(\alpha)f_1(\beta)$ within the brackets in (3.7) this equation can also be written in the form

$$G'(\alpha, \beta) = G(\alpha, \beta) + H(\alpha)H(\beta) \tag{3.8}$$

where $G(\alpha, \beta)$ is the two-particle correlation function defined by (2.19) and $H(\alpha)$ is the inhomogeneity factor defined by (2.15). We stress the fact that the Fourier component $\rho_{\mathbf{k},\mathbf{k}'}(\alpha, \beta)$ with $\mathbf{k}+\mathbf{k}' \neq 0$, unlike $\rho_{\mathbf{k},-\mathbf{k}}(\alpha, \beta)$, contains two distinct terms. One is related to the two-body correlation function and is thus of the same physical nature and origin as $\rho_{\mathbf{k},-\mathbf{k}}(\alpha, \beta)$. It is, however, modified by the inhomogeneity because $G(\alpha, \beta)$ is no longer the same in all places within the system. The second term is a *product* of two inhomogeneity factors and hence is characteristic of inhomogeneous systems.

Following the decomposition (3.8) we may therefore write a corresponding equation in Fourier space

$$\rho_{\mathbf{k},\mathbf{k}'}(\alpha, \beta) = \rho_{[\mathbf{k},\mathbf{k}']}(\alpha, \beta) + \rho_{\mathbf{k}}(\alpha)\rho_{\mathbf{k}'}(\beta) \tag{3.9}$$

The properties of the inhomogeneity factor have been discussed above. It follows from (3.5) that the second term in (3.9) is significantly different from zero only if simultaneously $k = O(L_h^{-1})$ and $k' = O(L_h^{-1})$. If one or both wave-vectors $\mathbf{k}$ and $\mathbf{k}'$ lie within the molecular domain (i.e. have large magnitudes, of order L_m^{-1}) the Fourier component $\rho_{\mathbf{k},\mathbf{k}'}(\alpha, \beta)$ reduces to $\rho_{[\mathbf{k},\mathbf{k}']}(\alpha, \beta)$ and describes pure correlations.

We now study the true correlation coefficient $\rho_{[\mathbf{k},\mathbf{k}']}(\alpha, \beta)$. It is convenient to make the change of variables

$$\mathbf{R} = \tfrac{1}{2}(\mathbf{x}_\alpha + \mathbf{x}_\beta), \qquad \mathbf{r} = \tfrac{1}{2}(\mathbf{x}_\alpha - \mathbf{x}_\beta) \tag{3.10}$$

In terms of these variables

$$\rho_{[\mathbf{k}\mathbf{k}']}(\alpha, \beta) = \tfrac{1}{4}(8\pi^3)^{-2}\int d\mathbf{R}d\mathbf{r}\exp[-i(\mathbf{k}+\mathbf{k}')\cdot\mathbf{R}-i(\mathbf{k}-\mathbf{k}')\cdot\mathbf{r}]G(\mathbf{R},\mathbf{r}) \tag{3.11}$$

and conversely

$$G(\mathbf{R},\mathbf{r}) = \int d\mathbf{k}d\mathbf{k}'\exp[i(\mathbf{k}+\mathbf{k}')\cdot\mathbf{R}+i(\mathbf{k}-\mathbf{k}')\cdot\mathbf{r}]\rho_{[\mathbf{k},\mathbf{k}']}(\alpha, \beta) \tag{3.12}$$

In a non-uniform system we may describe the correlations as follows. Let us first fix the center of mass of a couple of particles α, β; we then have a certain correlation pattern between the two particles, which is described by $G(\mathbf{R}_0, \mathbf{r})$ for given $\mathbf{R} = \mathbf{R}_0$. This correlation pattern changes with the position of the center of mass as a result of the inhomogeneity. Two facts will be important in the discussion.

a) The variation of G with the position $\mathbf{R}$ of the center of mass of the couple is *slow* on the molecular scale L_m. Therefore, if the fundamental condition (3.4) is satisfied, one can conclude from (3.12), as in the case of $\rho_{\mathbf{k}}(\alpha)$, that *$\rho_{[\mathbf{k}\mathbf{k}']}(\alpha, \beta)$ as a function of* $\mathbf{k}+\mathbf{k}'$ *is sharply peaked around* $\mathbf{k} + \mathbf{k}' = 0$ *with a width of order* L_h^{-1}.

b) The correlations must obey the fundamental physical requirement of vanishing whenever the distance $\mathbf{r}$ between the particles exceeds some critical value. As we are interested in systems in which correlations are only due to molecular interactions, this critical length is of the order of L_m or smaller. When $r > L_m$, f_2 becomes a product of f_1's and (2.19) shows that $G(\alpha, \beta) \to 0$.

Consider now formula (3.11) for values of $|\mathbf{k} - \mathbf{k}'|$ of the order of L_h^{-1}. From our previous discussion, **k** and **k'** must also satisfy the condition $|\mathbf{k} + \mathbf{k}'| \approx L_h^{-1}$ [otherwise $\rho_{[\mathbf{k},\mathbf{k}']}(\alpha, \beta) \approx 0$]. We thus consider the case in which *both* **k** *and* **k'** *are of order* L_h^{-1}. In this case the factor $\exp[-i(\mathbf{k} - \mathbf{k}') \cdot \mathbf{r}]$ is a very slowly varying function of **r** so that the effective contributions to the integral come from a very large domain of integration, say a sphere of radius L_h. Let us consider another concentric sphere in the **r**-space whose radius is of the order of L_m. Let us call I the inside of the latter sphere and II the region comprised between the two spheres (see Fig. 3.1). According to our previous discussion the function G vanishes in region II. Because of the very large difference in size between the regions I and II we may replace $G(\alpha, \beta)$ by zero everywhere and hence the r.h.s. of eq. (3.9) reduces to its second term. The error thus committed will be of the order of the ratio between the volumes of regions I and II, i.e. $L_m^3/L_h^3 \ll 1$ [by virtue of (3.4)].

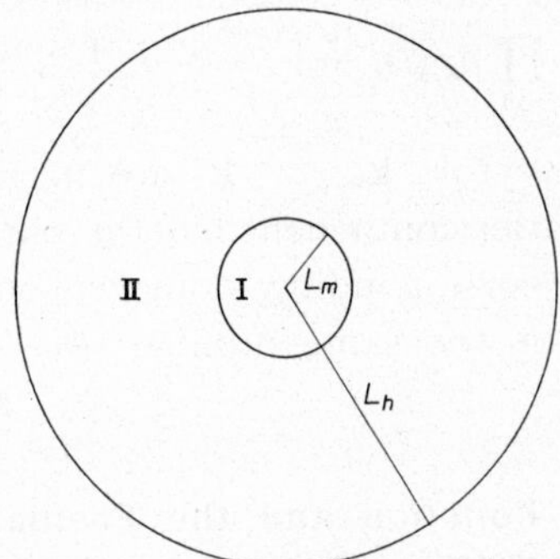

Fig. 3.1. Domains of integration over **r** in eq. (3.11).

We therefore reach the important conclusion:

$$\rho_{\mathbf{kk}'}(\alpha, \beta) = \rho_{\mathbf{k}}(\alpha)\rho_{\mathbf{k}'}(\beta) + O(L_m^3/L_h^3), \quad \text{for} \quad \begin{cases} k \sim L_h^{-1} \\ k' \sim L_h^{-1} \end{cases} \tag{3.13}$$

In the "hydrodynamic" range of values of both wave-vectors **k**, **k'**, the Fourier component $\rho_{\mathbf{kk}'}$ is therefore determined only by the inhomogeneity factors, and is moreover a product of the Fourier transforms of these factors. It follows also from the previous discussion that it is independent of e^2 and of c at the initial time.

In the range $|\mathbf{k} - \mathbf{k}'| \sim L_m^{-1}$, the Fourier component $\rho_{\mathbf{kk}'}(\alpha, \beta)$ represents true correlations and is of course no longer factorized. It has the same physical nature in this range as the component $\rho_{\mathbf{k},-\mathbf{k}}(\alpha, \beta)$ describing correlations in a homogeneous system. It is therefore natural to assume that it has the same dependence on e^2 as the corresponding homogeneous component.

These properties can be generalized to all higher inhomogeneous components, and various factorization theorems can be derived. We shall retain the following three properties of inhomogeneous Fourier components.

If condition (3.4) is satisfied, then:

a) $\rho_{\mathbf{k}_1,\ldots,\mathbf{k}_s}(1, \ldots, s)$, with $\sum_1^s \mathbf{k}_j \neq 0$, as a function of $\sum_1^s \mathbf{k}_j = \mathbf{K}$ is sharply peaked around $\mathbf{K} = 0$, with a width of order L_h^{-1}.

b) If *all* wave-vectors $\mathbf{k}_1, \ldots, \mathbf{k}_s$ are in the range L_h^{-1}, then

$$\rho_{\mathbf{k}_1,\ldots,\mathbf{k}_s}(1, \ldots, s) = \prod_{j=1}^{s} \rho_{\mathbf{k}_j}(j), \quad \text{for } k_j \sim L_h^{-1}, \quad j = 1, \ldots, s \tag{3.14}$$

c) If the wave-vectors $\mathbf{k}_1, \ldots, \mathbf{k}_s$ are in the range L_m^{-1}, the inhomogeneous Fourier component has the same (initial) dependence on e^2 as the corresponding homogeneous component (i.e. the component with the same number of non-vanishing wave-vectors).

§ 4. The Green's Function and the Formal Solution of the Liouville Equation

The whole discussion of §§ 2 and 3 concerned the distribution functions at a fixed instant of time, which may be taken arbitrarily as $t = 0$. The next problem is the study of the evolution of these functions in time.

It has been stressed repeatedly that the distribution functions which are really relevant to the study of physical properties are the reduced distribution functions of a finite small number of particles. It would therefore seem natural to derive equations for f_s from the Liouville equation (1.10) and solve these equations.

This is the approach used in many theories of non-equilibrium statistical mechanics, such as, for instance, Bogoliubov's.* The difficulty, however, lies in the fact that the equations for f_s derived from (1.10) by an integration over a certain number of variables are not closed; they form an infinite system (or hierarchy) in which each f_s is related to f_{s+1} (Yvon–Born–Green–Bogoliubov–Kirkwood, or Y–B–G–B–K, hierarchy):

$$\begin{aligned} &\partial_t f_s(1, \ldots, s) + \sum_{j=1}^{s} \mathbf{v}_j \cdot \boldsymbol{\nabla}_j f_s \\ &\qquad = (e^2/m) \sum_{j=1}^{s} \int d\mathbf{x}_{s+1} d\mathbf{v}_{s+1} (\boldsymbol{\nabla}_j V_{j,s+1}) \cdot \partial_j f_{s+1}(1, \ldots, s+1) \end{aligned} \tag{4.1}$$

In order to solve this hierarchy, an assumption is needed which expresses in some way the higher order reduced distribution functions in terms of lower ones. The justification of such assumptions is always very difficult and usually not very convincing (see also §§ 47 and 50).

The approach used by Prigogine and his school ** is opposite to the previous one: The Liouville equation is first solved formally for $f_N(t)$, and from this solution expressions for the reduced distribution functions are derived by integration. The main advantage lies in the fact that one starts with a *unique linear equation* for f_N which can be solved by using the standard techniques of linear mathematics. Several equivalent methods can be used as a starting point for this investigation, such as the direct iteration solution of the Liouville equation in the "interaction representation" (which was used for the first time in this problem by Brout and Prigogine †) or the resolvent formalism first applied to this problem by Résibois. Both methods are explained in Prigogine's monograph referred to above. We shall

* N. N. Bogoliubov, *Problems of a Dynamical Theory in Statistical Physics*, Moskow, 1946; translated by E. K. Gora, in *Studies in Statistical Mechanics*, vol. 1, edited by J. de Boer and G. E. Uhlenbeck, Interscience, New York, 1962.

** I. Prigogine, *Non-Equilibrium Statistical Mechanics*, Interscience, New York, 1963.

† R. Brout and I. Prigogine, *Physica*, **22**, 621 (1956); see also I. Prigogine and R. Balescu, *Physica*, **25**, 281, 302 (1959).

use in this book essentially the latter method, but shall introduce it by using the *Green's function* of the Liouville equation. This method has the advantage of clarifying some minor points in the usual presentation. Green's functions of the Liouville equation have been studied by Andrews * from a rather different point of view.

Before proceeding with the calculations, we shall outline our program.

A) We first introduce the concepts of the *Green's function* and of the *resolvent* associated with the Liouville equation. These objects are looked at as operators which, acting on the distribution function at time zero, transform it into the distribution function at time t. A knowledge of these operators therefore automatically provides the solution of the initial-value problem of the Liouville equation.

B) We derive an integral equation for these operators and solve it by an *iteration (or perturbation) procedure* carried out to all orders in the coupling parameter e^2. At this point it will appear that the Fourier representation introduced in § 2 provides the most convenient working basis for such calculations.

C) The perturbation series obtained in this way contains an infinity of terms, each of which is generally a very complicated expression. In each physical problem we are led to choose from among these terms a subseries. The latter contains all the contributions which are dominant according to certain criteria which have been fixed in advance and which define the nature of the problem. In order to make this selection process practically feasible, we devise a *diagram technique*. The latter consists in associating with each term of the series a well defined graph. The correspondence is such that the order of magnitude of each term, be it even very complicated, can be read off from a simple inspection of certain simple geometrical features of the corresponding diagram.

D) We have obtained in this way a power-series solution of the Liouville equation, i.e. a formula for $f_N(t)$. From this solution we

* F. C. Andrews, *Acad. Roy. Belg. Bull. Classe Sci.*, **46**, 475 (1960).

obtain very simply expressions for the *reduced distribution functions* by an integration process. It turns out that the diagram technique is substantially simplified by this reduction process.

E) Once the diagram technique is established (this is the purpose of the remaining sections of Chapter 1) the practical working procedure for solving a given problem is as follows:

Decide a criterion of choice from physical considerations.

Select all the diagrams compatible with this criterion and disregard all other terms of the general perturbation series. This choice is made by purely geometrical (or topological) arguments.

Write the mathematical expression corresponding to the selected diagrams and try to sum the subseries obtained in this way.

This procedure will be repeatedly illustrated throughout this book.

We now proceed to realize this program explicitly. Consider first the general inhomogeneous Liouville equation, written in the form:

$$\mathscr{L} f_N(x, v; t) = s(x, v; t) \tag{4.2}$$

where

$$\begin{aligned} \mathscr{L} &= \mathscr{L}^0 + e^2 \mathscr{L}' \\ \mathscr{L}^0 &= \partial_t + \sum_{j=1}^{N} \mathbf{v}_j \cdot \boldsymbol{\nabla}_j \\ \mathscr{L}' &= \sum_{j<n}\sum \mathscr{L}'_{jn} = -m^{-1} \sum_{j<n}\sum (\boldsymbol{\nabla}_j V_{jn}) \cdot \partial_{jn} \end{aligned} \tag{4.3}$$

and $s(x, v; t)$ is an arbitrary function of $\mathbf{x}_1, \ldots, \mathbf{x}_N, \mathbf{v}_1, \ldots, \mathbf{v}_N$ and the time. In many interesting cases, $s \equiv 0$, see (1.10), but we shall not make this restriction here. Consider now two arbitrary functions $\psi_1(x, v; t)$ and $\psi_2(x, v; t)$ which satisfy the following boundary conditions:

a) *Periodic boundary conditions in configuration space* (see 2.4)

$$\begin{aligned} \psi_i(\mathbf{x}_1 + \Omega^{\frac{1}{3}} \overline{1}_x, \ldots, \mathbf{x}_N + \Omega^{\frac{1}{3}} \overline{1}_x, \{v\}; t) \\ = \psi_i(\mathbf{x}_1 + \Omega^{\frac{1}{3}} \overline{1}_y, \ldots, \mathbf{x}_N + \Omega^{\frac{1}{3}} \overline{1}_y, \{v\}; t) \\ = \psi_i(\mathbf{x}_1 + \Omega^{\frac{1}{3}} \overline{1}_z, \ldots, \mathbf{x}_N + \Omega^{\frac{1}{3}} \overline{1}_z, \{v\}; t) \\ = \psi_i(\mathbf{x}_1, \ldots, \mathbf{x}_N, \{v\}; t) \end{aligned} \tag{4.4}$$

b) *Homogeneous boundary conditions at infinity in velocity space*

$$\psi_i \to 0 \quad \text{for} \quad v_j \to \infty, \quad \text{any } j \tag{4.5}$$

For later convenience, we shall even assume that the vanishing is of exponential type, i.e. that all the derivatives of ψ_i vanish as $v_j \to \infty$.

A simple integration by parts then proves the *Generalized Green's formula* for the operator $\mathscr{L}$:

$$\int (d\mathbf{x})^N \int (d\mathbf{v})^N \int_{t_0}^{t_1} dt \, \{\psi_1 \mathscr{L} \psi_2 + \psi_2 \mathscr{L} \psi_1\} = \int (d\mathbf{x})^N \int (d\mathbf{v})^N \{\psi_1(t_1)\psi_2(t_1) - \psi_1(t_0)\psi_2(t_0)\} \tag{4.6}$$

In this formula, the $\mathbf{x}$-integration is over the volume of the cubic container of the system, and the $\mathbf{v}$-integration is over the whole velocity space. Comparing (4.6) with the usual Green's formula,* we see that the *adjoint* $\tilde{\mathscr{L}}$ of the Liouville operator is:

$$\tilde{\mathscr{L}} = -\mathscr{L} \tag{4.7}$$

We now define the *Green's function* (or, more precisely, the *retarded Green's function*) as the solution of the equation:

$$-\tilde{\mathscr{L}} \mathscr{G}(xvt|x'v't') \equiv \mathscr{L} \mathscr{G}(xvt|x'v't') = \delta(x-x')\delta(v-v')\delta(t-t') \tag{4.8}$$

satisfying the "*causality condition*" **

$$\mathscr{G}(xvt|x'v't') = 0 \quad \text{for} \quad t < t' \tag{4.9}$$

It should be noted that x', v', t' are considered as given parameters in this equation; the operator $\tilde{\mathscr{L}}$ (or $-\mathscr{L}$) acts on the set x, v, t.

We may also define an *adjoint Green's function* $\tilde{\mathscr{G}}(xpt|x'p't')$ as the solution of the equation

$$-\mathscr{L} \tilde{\mathscr{G}}(xvt|x'v't') = \delta(x-x')\delta(v-v')\delta(t-t') \tag{4.10}$$

* See, e.g., F. Schlögl, *Handbuch der Physik*, vol. 1, p. 218, Springer Verlag, Berlin, 1956.

** The Green's function, as well as all functions appearing in this and the next section, are assumed to satisfy the boundary conditions (4.4) and (4.5) (unless explicitly stated otherwise).

satisfying the "*anticausal condition*"

$$\tilde{\mathscr{G}}(xvt|x'v't') = 0 \quad \text{for} \quad t < t' \tag{4.11}$$

Applying Green's formula (4.6) to $\psi_1 = \mathscr{G}(xvt|x'v't')$, $\psi_2 = \tilde{\mathscr{G}}(xvt|x''v''t'')$ (with $t' > 0$, $t'' > 0$) and integrating from $t_0 = 0$ to $t_1 = \infty$, the fol- fowing reciprocity relation is obtained

$$\mathscr{G}(xvt|x'v't') = \tilde{\mathscr{G}}(x'v't'|xvt) \tag{4.12}$$

Using this formula, as well as (4.10), it is seen that

$$\mathscr{L}\mathscr{G}(x'v't'|xvt) = \mathscr{L}\tilde{\mathscr{G}}(xvt|x'v't') = -\delta(x-x')\delta(v-v')\delta(t-t') \tag{4.13}$$

We are now ready for the solution of the initial-value problem, which we formulate as follows:

To find the solution of the equation

$$\mathscr{L}\, f_N(x, v; t) = s(x, v; t) \tag{4.14}$$

for $t \geqq 0$, *which reduces at time* 0 *to a prescribed function* $q_N(x, v)$:

$$f_N(x, v; 0) = q_N(x, v) \tag{4.15}$$

Applying formula (4.6) to $\psi_1 = f_N(x, v; t)$, $\psi_2 = \mathscr{G}(x''v''t''|xvt)$, taking $t_0 = 0$, $t_1 = \infty$, and using (4.9), (4.13)–(4.15), one easily obtains

$$\begin{aligned} f_N(x, v; t) &= \int dx'dv'\, \mathscr{G}(xvt|x'v'0)q_N(x', v') \\ &\quad + \int_0^t dt' \int dx'dv'\, \mathscr{G}(xvt|x'v't')s(x', v'; t') \end{aligned} \tag{4.16}$$

This formula shows very clearly how the effect of the initial condition at time $t = 0$ and the effect of the source at all times $t' < t$ are propagated toward time t. It is seen that the causal condition on $\mathscr{G}$ expresses the fact that only the behavior of the source s at times t' *earlier* than t can influence the value of f_N at time t. This property expresses therefore the *causal relationship* between the source and the initial condition, and the effect $f_N(t)$. It should be noted that the causality condition is not an *a priori* assumption, but is imposed [through Green's formula (4.6)] by the fact that we are solving an initial-value problem.

It is also easily seen that, as the coefficients in the Liouville equation are not functions of time, $\mathscr{G}(t|t')$ can only depend on its time variables through the difference $t - t'$: this is an expression of the *homogeneity* in time.

The two properties of causality and homogeneity in time require that $\mathscr{G}$ has the following functional form:

$$\mathscr{G}(x, v, t|x', v', t') = \theta(t - t')\, G(x, v|x', v';\ t - t') \tag{4.17}$$

where the *Heaviside function* $\theta(x)$ is defined by

$$\theta(x) = \begin{cases} 1, & x > 0 \\ 0, & x < 0 \end{cases} \tag{4.18}$$

We now introduce a quantity which will play the central role in this book. It is defined as the Fourier transform of the Green's function with respect to $t - t'$ and is called the *resolvent* $R(xv|\, x'v';\, z)$, or more briefly $R(z)$.

$$R(xv|x'v';\, z) = \int_{-\infty}^{\infty} d(t-t')\mathrm{e}^{iz(t-t')}\mathscr{G}(xvt|x'v't') \tag{4.19}$$

Equation (4.17) implies that (4.19) reduces to:

$$R(xv|x'v';\, z) = \int_0^{\infty} d\tau\, \mathrm{e}^{iz\tau} G(xv|x'v';\, \tau) \tag{4.20}$$

which means that $R(z)$ is actually a *one-sided Laplace transformation*. Its properties can therefore be studied by the general methods of the theory of Laplace transforms (see Appendix 1).

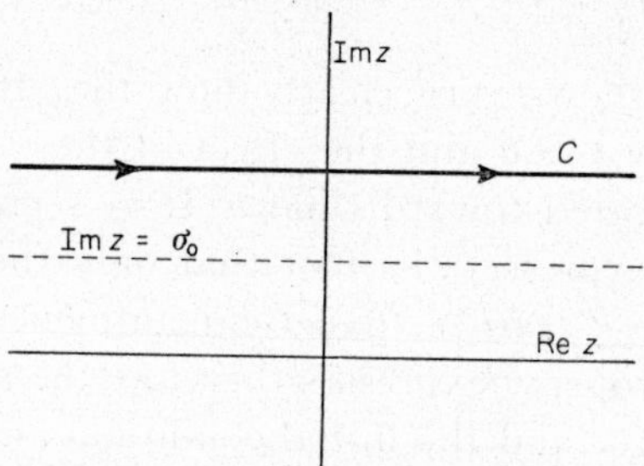

Fig. 4.1. Contour of integration C in (4.21). σ_0 is the ordinate of convergence of the Laplace integral (see Appendix 1).

Formula (4.20) defines $R(z)$ as a *regular function of the complex variable* z for all values of z for which the integral of the r.h.s.

converges. Conversely, $G(\tau)$ can be expressed in terms of $R(z)$ by means of the Laplace inversion formula:

$$G(\tau) = (2\pi)^{-1} \int_C dz \, \mathrm{e}^{-iz\tau} R(z) \tag{4.21}$$

where C is the usual contour of the Laplace inversion formula, i.e. a parallel to the real axis lying above all singularities of $R(z)$ in the complex plane of z (see Fig. 4.1).

Substituting expression (4.21) into eq. (4.16) and taking $s = 0$, we find the following form for the general solution of the initial-value problem of the Liouville equation (1.10):

$$f_N(x, v; t) = (2\pi)^{-1} \int_C dz \, \mathrm{e}^{-izt} \int dx' dv' R(x, v|x', v'; z) f_N(x', v'; 0) \tag{4.22}$$

This is the general form of the solution which will be used as a starting point for the theory.

§ 5. Perturbation Theory

It is, of course, impossible to solve exactly in finite form the Liouville equation for a system of N interacting particles. However, if it is possible to separate the equation into two parts such that the first part can be solved exactly, the solution of the complete equation can be expressed in terms of an infinite series, each term of which can be calculated exactly. This series is a solution of the equation, provided it converges. The convergence of the series will be assumed without proof, on physical arguments.*

In the present case the separation of the Liouville operator is obvious. The simple part is $\mathscr{L}^0$, which represents free particles, whereas the complicated part is $e^2\mathscr{L}'$. In order to set up a perturbation expansion, we first derive an integral equation for the Green's function. Equation (4.8) for the Green's function is written as:

$$\mathscr{L}^0_{y''}\mathscr{G}(y''|y') + e^2\mathscr{L}'_{y''}\mathscr{G}(y''|y') = \delta(y''-y') \tag{5.1}$$

* This convergence assumption is made at the present time in most problems of theoretical physics which are solved by perturbation procedures; we do not try to justify it further.

In order to save space we have denoted the set of variables $\{x, v; t\}$ by the single symbol y. The subscript y'' means that the Liouville operator acts on the double primed variables (for instance: $\mathscr{L}^0_{y''} \equiv \partial_{t''} + \sum \mathbf{v}''_j \cdot \partial/\partial \mathbf{x}''_j$).

Let $\mathscr{G}^{(0)}(y''|y')$ be the causal Green's function of the unperturbed Liouville operator, satisfying the equation:

$$\mathscr{L}^0_{y''}\mathscr{G}^{(0)}(y''|y') = \delta(y''-y') \tag{5.2}$$

Multiply the two sides of eq. (5.1) by $\mathscr{G}^{(0)}(y|y'')$ and integrate over y''. The result is:

$$\begin{aligned}\int dy''\,\mathscr{G}^{(0)}(y|y'')\mathscr{L}^0_{y''}\mathscr{G}(y''|y') + e^2\int dy''\,\mathscr{G}^{(0)}(y|y'')\mathscr{L}'_{y''}\mathscr{G}(y''|y')\\ = \mathscr{G}^{(0)}(y|y')\end{aligned} \tag{5.3}$$

Integrating by parts and using eq. (4.13) the first term is transformed as follows:

$$\begin{aligned}\int dy''\,\mathscr{G}^{(0)}(y|y'')\mathscr{L}^0_{y''}\mathscr{G}(y''|y') &= -\int dy''\,\mathscr{G}(y''|y')\mathscr{L}^0_{y''}\mathscr{G}^{(0)}(y|y'')\\ &= +\int dy''\,\mathscr{G}(y''|y')\delta(y-y'') = \mathscr{G}(y|y')\end{aligned}$$

Hence eq. (5.3) becomes

$$\mathscr{G}(y|y') = \mathscr{G}^{(0)}(y|y') - e^2\int dy''\,\mathscr{G}^{(0)}(y|y'')\mathscr{L}'_{y''}\mathscr{G}(y''|y')$$

or, more explicitly,

$$\begin{aligned}\mathscr{G}(xvt|x'v't') &= \mathscr{G}^{(0)}(xvt|x'v't')\\ &\quad - e^2\int_0^t dt''\int dx''dv''\,\mathscr{G}^{(0)}(xvt|x''v''t'')\mathscr{L}'_{y''}\mathscr{G}(x''v''t''|x'v't')\end{aligned} \tag{5.4}$$

This formula expresses the Green's function as the solution of an integral equation. We now use the fact that the Green's function depends on t and t' only through the difference $t - t'$. Equation (5.4) is therefore a *convolution* equation (see Appendix 1). If we therefore take a Laplace transform, going from the Green's function to the resolvent [see eq. (4.20)], eq. (5.4) becomes:

$$\begin{aligned}R(xv|x'v';z) &= R^0(xv|x'v';z)\\ &\quad - e^2\int dx''dv''\,R^0(xv|x''v'';z)\mathscr{L}'_{y''}R(x''v''|x'v';z)\end{aligned} \tag{5.5}$$

This equation is very easily solved by successive iteration:

$$\begin{aligned} R(xv|x'v';z) = {} & R^0(xv|x'v';z) \\ & -e^2\int dx_1\,dv_1\,R^0(xv|x_1v_1;z)\mathscr{L}'R^0(x_1v_1|x'v';z) \\ & +e^4\int dx_1\,dv_1\,dx_2\,dv_2\,R^0(xv|x_1v_1;z) \\ & \qquad \mathscr{L}'R^0(x_1v_1|x_2v_2;z)\mathscr{L}'R^0(x_2v_2|x'v';z)+\dots \end{aligned} \tag{5.6}$$

This formula provides the perturbation expansion we were looking for.

We now note that the resolvent depends on the *pairs* of variables (x, x') and (v, v'). $R(xv|x'v';z)$ can therefore be looked at as the matrix element $\langle xv|\mathscr{R}(z)|x'v'\rangle$ of an *operator* $\mathscr{R}(z)$ which is a function of the complex variable z. Equation (5.5) can therefore be written in operator form as

$$\mathscr{R}(z) = \mathscr{R}^0(z) - e^2\mathscr{R}^0(z)\mathscr{L}'\mathscr{R}(z) \tag{5.7}$$

The action of the operator $\mathscr{R}(z)$ on a function $F(x, v)$ is defined as usual by a matrix product:

$$\mathscr{R}(z)F \equiv \int dx'\,dv'\langle xv|\mathscr{R}(z)|x'v'\rangle F(x'v') \tag{5.8}$$

In particular, the solution (4.22) can be written in operator form as:

$$\begin{aligned} f_N(t) &= (2\pi)^{-1}\int_C dz\,\mathrm{e}^{-izt}\mathscr{R}(z)f_N(0) \\ &= (2\pi)^{-1}\int_C dz\,\mathrm{e}^{-izt}\sum_{n=0}^{\infty}(-e^2)^n\mathscr{R}^0(z)[\mathscr{L}'\mathscr{R}^0(z)]^n f_N(0) \end{aligned} \tag{5.9}$$

This operator point of view gives to formulae (5.7) and (5.9) a flexibility which we shall presently utilize. In particular, we shall not be bound to the "(x, v)-representation", but may use any convenient change of representation (as in quantum mechanics) which facilitates computations.*

In order to perform calculations explicitly, the first problem

* A specific example showing the passage from the matrix point of view to the operator point of view is given below.

will be to find the *unperturbed Green's function* of the Liouville equation. We rewrite eq. (5.2) explicitly

$$\left\{\frac{\partial}{\partial t}+v\frac{\partial}{\partial x}\right\}\mathscr{G}^{(0)}(xvt|x'v't') = \delta(x-x')\delta(v-v')\delta(t-t') \tag{5.10}$$

We first note that, since v appears in the operator of the l.h.s. only as a multiplicative factor, $\mathscr{G}^{(0)}$ is of the form

$$\mathscr{G}^{(0)}(xvt|x'v't') = \delta(v-v')\Gamma^{(0)}(xt|x't') \tag{5.11}$$

The equation then becomes:

$$\left\{\frac{\partial}{\partial t}+v'\frac{\partial}{\partial x}\right\}\Gamma^{(0)}(xt|x't') = \delta(x-x')\delta(t-t') \tag{5.12}$$

We now use a method of solution suggested by Sokolov and Ivanenko,* which has already been used in the present case by Andrews.** It consists in exploiting the formal operational equation, equivalent to (5.12),

$$\Gamma^{(0)}(xt|x't') = (\mathscr{L}^0)^{-1}\delta(x-x')\delta(t-t') \tag{5.13}$$

This expression is given a meaning in terms of Fourier transforms with respect to time and to space. We use the following representation of the δ-functions:

$$\delta(t-t')\delta(x-x') = \Omega^{-\frac{1}{3}}\sum_k (2\pi)^{-1}\int_{-\infty}^{\infty} d\omega\, e^{i\omega(t-t')}\, e^{ik(x-x')}$$

where k has been defined in (2.6). It is easily seen that, in the limit (2.3) where the set of permitted values of the wave-vector k approaches a dense set, $\Gamma^{(0)}$ is of the following form:

$$\Gamma^{(0)}(xt|x't') = (2\pi)^{-2}\int dk\int d\omega\, e^{i\omega(t-t')+ik(x-x')} \cdot\left\{\frac{\mathscr{P}}{i(\omega+kv')}+\alpha\delta(\omega+kv')\right\} \tag{5.14}$$

Here $\mathscr{P}$ means the Cauchy principal value of the integral:

$$\mathscr{P}\int_{-\infty}^{\infty} dx\, f(x)\,\frac{1}{x} = \lim_{\varepsilon\to 0}\left\{\int_{-\infty}^{-\varepsilon}+\int_{+\varepsilon}^{+\infty}\right\} dx\, f(x)\,\frac{1}{x} \tag{5.15}$$

The occurrence of the δ-function in (5.14) is characteristic of a continuous spectrum. Its contribution is actually the general solution of the homogeneous equation corresponding to (5.12). The arbitrary constant α remaining in (5.14) must still be determined from the initial condition.

* A. Sokolov and D. Ivanenko, *Quantum Theory of Fields* (in Russian), Gos. Izdat. Techn. Teoret. Liter., Moscow, 1952.

** F. Andrews, *Acad. Roy. Belg. Bull. Classe Sci.*, **46**, 475 (1960).

(It should be noted that the boundary conditions in momentum and configuration space are correctly satisfied by $\Gamma^{(0)}$.) We therefore evaluate the ω and k integrals, which are simple*

$$\Gamma^{(0)}(xt|x't') = \tfrac{1}{2}[\varepsilon(t-t')+(\alpha/\pi)]\delta[x-x'-v'(t-t')] \tag{5.16}$$

In order to satisfy the initial condition ($\Gamma^{(0)} = 0,\ t < t'$) the bracketed factor must reduce to the Heaviside function $\theta(t-t')$ [see (4.18)]. This requires that $\alpha = \pi$.

We may now collect the results, going back to the explicit notation

$$\begin{aligned}&\mathscr{G}^{(0)}(\{\mathbf{x}\}, \{\mathbf{v}\}; t|\{\mathbf{x}'\}, \{\mathbf{v}'\}; t')\\ &\qquad = \theta(t-t')\prod_{j=1}^{N}\delta[\mathbf{x}_j-\mathbf{x}_j'-\mathbf{v}_j'(t-t')]\delta(\mathbf{v}_j-\mathbf{v}_j')\end{aligned} \tag{5.17}$$

This result is physically very simple. It expresses the motion of the free particles from x' at t' to x at t, along straight trajectories, with constant velocity, $v = v'$. From the mathematical point of view, however, (5.17) is not yet quite convenient. Regarded as the matrix element of an operator it is seen that $\mathscr{G}^{(0)}$ is diagonal in v, but not in x. It is always convenient in perturbation theory to start with a diagonal unperturbed operator. The representation in which $\mathscr{G}^{(0)}$ is diagonal is easily found. It is sufficient to take as a basis a set of plane waves defined in the same way as in § 2:

$$\psi_{\{\mathbf{k}\}}(\mathbf{x}) = \Omega^{-N/2}\,\mathrm{e}^{i\Sigma\mathbf{k}_j\cdot\mathbf{x}_j} \tag{5.18}$$

The change of representation is performed as is usual in quantum mechanics:

$$\begin{aligned}&\langle k, v|\mathscr{G}^{(0)}(t-t')|k', v'\rangle\\ &\qquad = \Omega^{-N}\int(d\mathbf{x})^N(d\mathbf{x}')^N\,\mathrm{e}^{-i\Sigma\mathbf{k}_j\cdot\mathbf{x}_j}\langle xv|\mathscr{G}^{(0)}(t-t')|x'v'\rangle\\ &\qquad\times\mathrm{e}^{i\Sigma\mathbf{k}_j'\cdot\mathbf{x}_j'} = \theta(t-t')\mathrm{e}^{-i\Sigma\mathbf{k}_j\cdot\mathbf{v}_j'(t-t')}\prod_{s=1}^{N}\delta_{\mathbf{k}_s-\mathbf{k}'_s}\delta(\mathbf{v}_s-\mathbf{v}_s')\end{aligned} \tag{5.19}$$

* Use is made of the following representation of the principal part (Heitler, *Quantum Theory of Radiation*, Clarendon Press, Oxford, 1955)

$$\mathscr{P}\frac{1}{x} = \lim_{T\to\infty}\frac{1-\cos Tx}{x} = \frac{1}{2i}\int_{-\infty}^{\infty}dp\,\varepsilon(p)\mathrm{e}^{ipx}$$

where

$$\varepsilon(p) = \begin{cases}+1, & p > 0\\ -1, & p < 0\end{cases}$$

[The notation $\delta_{\mathbf{k}}$ is defined after formula (2.8).] The Green's operator is diagonal in this representation, as was desired. It is now very easy to calculate the unperturbed resolvent operator in the Fourier representation by using formula (4.20). The result is:

$$\langle kv|\mathscr{R}^0(z)|k'v'\rangle = \frac{1}{i(\sum_j \mathbf{k}_j \cdot \mathbf{v}_j - z)} \prod_{s=1}^{N} \delta_{\mathbf{k}_s - \mathbf{k}'_s} \delta(\mathbf{v}_s - \mathbf{v}'_s) \tag{5.20}$$

In the present representation, the operator $\mathscr{R}^0$ is represented by a matrix whose indices are $\mathbf{k}_1, \ldots, \mathbf{k}_N$; $\mathbf{k}'_1, \ldots, \mathbf{k}'_N$ and $\mathbf{v}_1, \ldots, \mathbf{v}_N$; $\mathbf{v}'_1, \ldots, \mathbf{v}'_N$. It acts on vectors in a functional space which is spanned by the basis

$$\Omega^{-N/2} e^{i\Sigma \mathbf{k}_j \cdot \mathbf{x}_j} \prod_{r=1}^{N} \delta(\mathbf{v}_r - \mathbf{v}'_r)$$

The fact that the basis in velocity space is formed by a system of quite singular functions (or rather, distributions) is not very convenient in calculations. It will prove simpler in most cases not to take matrix elements in velocity space. We then introduce the following (one could say "mixed") representation:

$$\langle k|\mathscr{R}^0(z)|k'\rangle = \frac{1}{i(\sum \mathbf{k}_j \cdot \mathbf{v}_j - z)} \prod_{s=1}^{N} \delta_{\mathbf{k}_s - \mathbf{k}'_s} \tag{5.21}$$

The difference about (5.21) is that the matrix element $\langle k|\mathscr{R}^0(z)|k'\rangle$ is no longer a number (like $\langle kv|\mathscr{R}^0(z)|k'v'\rangle$) but an *operator in velocity space*. In the case of the unperturbed Liouville operator the difference is purely verbal, because $\mathscr{R}^0$ is simply a multiplicative operator in v. In the perturbed case, however, $\langle k|\mathscr{R}(z)|k'\rangle$ will be a differential operator or even an infinite series of differential operators of higher and higher orders.

In order to clarify these ideas we shall consider a very simple mathematical example which is of the type occurring in subsequent problems. We assume that the operator in the "mixed" representation is

$$\langle k|\mathscr{R}|k'\rangle = a\, \partial/\partial v$$

where a is a function of k, k' and z, but not of v. The corresponding matrix element in the $\{k, v\}$ representation is obtained as is usual in quantum mechanics:

$$\langle kv|\mathscr{R}|k'v'\rangle = a\int dv''\,\delta(v-v'')\frac{\partial}{\partial v''}\delta(v''-v')$$
$$= a[\partial\delta(v-v')/\partial v']$$

It is very easily seen that these two representations are equivalent. Indeed, from the first form we obtain (F being an arbitrary function of v):

$$\langle k|\mathscr{R}|k'\rangle F(v) = a\,dF/dv$$

and from the second:

$$\int dv'\langle kv|\mathscr{R}|k'v'\rangle F(v') = a\int dv'[d\delta(v-v')/dv]F(v')$$
$$= -a\int dv'[d\delta(v-v')/dv']F(v') = a\int dv'\delta(v-v')dF(v')/dv' = a dF(v)/dv$$

In order to utilize formula (5.9), we still need to evaluate the matrix elements of the perturbation $\mathscr{L}'$. In order to do so a Fourier expansion of the interaction energy is necessary:

$$V_{jn}(|\mathbf{x}_j-\mathbf{x}_n|) = (8\pi^3/\Omega)\sum_{\mathbf{l}} V_{\mathbf{l}}\,e^{i\mathbf{l}\cdot(\mathbf{x}_j-\mathbf{x}_n)} \tag{5.22}$$

If the interaction forces are central, as we always assume here, the Fourier components $V_{\mathbf{l}}$ must satisfy the condition

$$V_{-\mathbf{l}} = V_{\mathbf{l}} \tag{5.23}$$

In other words, (5.23) implies that $V_{\mathbf{l}}$ only depends on the absolute value l of the vector $\mathbf{l}$. Using (4.3) together with (5.18) and (5.22) we obtain

$$\langle \mathbf{k}_1\ldots\mathbf{k}_N|\mathscr{L}'_{jn}|\mathbf{k}'_1\ldots\mathbf{k}'_N\rangle = \Omega^{-N}(8\pi^3/\Omega)\sum_{\mathbf{l}}\int d\mathbf{x}_1\ldots d\mathbf{x}_N$$
$$e^{-i\Sigma\mathbf{k}_r\cdot\mathbf{x}_r}m^{-1}V_l(-i\mathbf{l}\cdot\partial_{jn})e^{i\mathbf{l}\cdot(\mathbf{x}_j-\mathbf{x}_n)}e^{i\Sigma\mathbf{k}'_r\cdot\mathbf{x}_r}$$
$$= (8\pi^3/\Omega)m^{-1}\sum_{\mathbf{l}}V_l(-i\mathbf{l}\cdot\partial_{jn})\delta_{\mathbf{k}'_j+\mathbf{l}-\mathbf{k}_j}\delta_{\mathbf{k}'_n-\mathbf{l}-\mathbf{k}_n}\prod_{r(\neq j,n)}\delta_{\mathbf{k}'_r-\mathbf{k}_r} \tag{5.24}$$
$$= (8\pi^3/\Omega)m^{-1}V_{|\mathbf{k}'_j-\mathbf{k}_j|}\,i(\mathbf{k}'_j-\mathbf{k}_j)\cdot\partial_{jn}\delta_{\mathbf{k}'_j+\mathbf{k}'_n-\mathbf{k}_j-\mathbf{k}_n}\prod_{r(\neq j,n)}\delta_{\mathbf{k}'_r-\mathbf{k}_r}$$

The main result of formula (5.24) can be expressed as a *fundamental selection rule* for the interactions:

The only non-vanishing matrix elements of the operator $\mathscr{L}'_{jn}$ *are those for which*:

a) *all but two primed wave-vectors are equal to the corresponding unprimed wave-vectors;*

b) *the two primed wave-vectors which are changed in the interaction must be such that their sum equals the sum of the corresponding unprimed ones.*

This selection rule is a direct consequence of the fact that the interactions are binary and central.

It must be stressed again that the matrix elements $\langle k|\mathscr{L}'_{jn}|k'\rangle$ are *operators* in velocity space, acting on functions of the velocities through the factors $\partial_{jn} \equiv \partial/\partial \mathbf{v}_j - \partial/\partial \mathbf{v}_n$. This essential fact should not be forgotten when handling formal expressions.

We may now Fourier transform the Liouville equation (1.10), using eqs. (2.8), (4.3) and (5.24), with the result:

$$\begin{aligned} \partial_t \rho_{\{k\}}(t) + i \sum_j \mathbf{k}_j \cdot \mathbf{v}_j \rho_{\{k\}}(t) \\ = e^2 \sum_{\{k'\}} (8\pi^3/\Omega)^{\nu'-\nu} \langle k|\mathscr{L}'|k'\rangle \rho_{\{k'\}}(t) \end{aligned} \tag{5.25}$$

The exponents ν and ν' are defined as follows:

$$(8\pi^3/\Omega)^{\nu} = \tilde{\rho}_{\{k\}} \Omega^N / \rho_{\{k\}}$$
$$(8\pi^3/\Omega)^{\nu'} = \tilde{\rho}_{\{k'\}} \Omega^N / \rho_{\{k'\}}$$

In the simplest cases these exponents can be read directly from Table 2.1.

We now insert formulae (5.21) and (5.24) into (5.9) in order to obtain the formal solution of the Liouville equation (5.25) as a power series in e^2. Of course, in the representation we have chosen, f_N must be represented by formula (2.8). The result is

$$\begin{aligned} \rho_{\{k\}}(v;t) = (2\pi)^{-1} \sum_{n=0}^{\infty} \int_C dz\, \mathrm{e}^{-izt} (-e^2)^n \sum_{\{k'\}} (8\pi^3/\Omega)^{\nu'-\nu} \\ \langle \{k\}|\mathscr{R}^0(z)[\mathscr{L}'\mathscr{R}^0(z)]^n|\{k'\}\rangle \rho_{\{k'\}}(v;0) \end{aligned} \tag{5.26}$$

Using our previous calculations, a typical term of this series is (to third order in e^2)

$$\begin{aligned} (2\pi)^{-1} \int_C dz\, \mathrm{e}^{-izt} (-e^2)^3 \frac{1}{i(\Sigma\, \mathbf{k}_r \cdot \mathbf{v}_r - z)} \sum_{j<n} \sum_{\{\mathbf{k}''\}} \langle \{\mathbf{k}\}|\mathscr{L}'_{jn}|\{\mathbf{k}''\}\rangle \\ \cdot \frac{1}{i(\Sigma\, \mathbf{k}''_r \cdot \mathbf{v}_r - z)} \sum_{m<p} \sum_{\{\mathbf{k}'''\}} \langle \{\mathbf{k}''\}|\mathscr{L}'_{pm}|\{\mathbf{k}'''\}\rangle \frac{1}{i(\Sigma\, \mathbf{k}'''_r \cdot \mathbf{v}_r - z)} \\ \cdot \sum_{t<n} \sum_{\{\mathbf{k}'\}} \langle \{\mathbf{k}'''\}|\mathscr{L}'_{tn}|\{\mathbf{k}'\}\rangle \frac{1}{i(\Sigma\, \mathbf{k}'_r \cdot \mathbf{v}_r - z)} (8\pi^3/\Omega)^{\nu'-\nu} \rho_{\{\mathbf{k}'\}}(0) \end{aligned} \tag{5.27}$$

The structure of these formulae is very simple. Reading from right to left, one starts with a Fourier component at time 0, then follows a succession of matrix elements of, alternatively, $\mathscr{R}^0(z)$ and $\mathscr{L}'$. The first and last operator of this succession is always a matrix element of $\mathscr{R}^0(z)$. The whole expression is multiplied by $(2\pi)^{-1} \exp(-izt)$ and integrated over z. This formula can be looked on as a new picture of classical dynamics. Using the language of quantum mechanics, one can say that one starts with a given correlation pattern at time 0, this pattern is propagated with $\mathscr{R}^0(z)$ (which is diagonal), then an interaction $\mathscr{L}'$ changes the pattern to a $\rho_{\{\mathbf{k}'''\}}$, which is again propagated, and so on until the final correlation pattern $\rho_{\{\mathbf{k}\}}$ is achieved at time t.

§ 6. The Diagram Technique

The general formula (5.26) is in a suitable form for the application of modern perturbation techniques. We shall now establish precise rules which associate a diagram with each expression of the type (5.27) and conversely with each well constructed diagram, a unique mathematical formula. In this way, when treating a complicated problem, it turns out to be easier to first draw and select the appropriate diagrams, and to begin the computations only after the elimination of all spurious difficulties.

A diagram is a graphical representation of a contribution to $\rho_{\{\mathbf{k}\}}(t)$ due to $\rho_{\{\mathbf{k}'\}}(0)$, according to formula (5.26). This representation is so faithful that the terms "diagram" and "contribution" will very often be used with the same meaning. The two main elements which define the structure of a given term in formula (5.26) are the matrix elements of $\mathscr{R}^0(z)$ and of $\mathscr{L}'_{jn}$, the number and nature of which determine each term.

A) With each matrix element of $\mathscr{R}^0(z)$, $\langle\{\mathbf{k}\}|\mathscr{R}^0(z)|\{\mathbf{k}\}\rangle = 1/i(\sum_j \mathbf{k}_j \cdot \mathbf{v}_j - z)$, we associate a set of superposed *lines* running from the right to the left; the number of lines is equal to the number of *non-vanishing* wave-vectors in the set $\{\mathbf{k}\}$.

B) Each line is labeled with an *index* representing the particle associated with the corresponding wave-vector.

C) With each matrix element of $\mathscr{L}'$, $\langle\{\mathbf{k}'\}|\mathscr{L}'_{jn}|\{\mathbf{k}\}\rangle$, we associate a *vertex* which is the concourse of the lines labeled j and n in the set $\{\mathbf{k}\}$ (if any) and of the lines labeled j and n in the set $\{\mathbf{k}'\}$ (if any).

Before making further comments on the properties of these diagrams, let us illustrate the general rules by an example. Consider one term contained in formula (5.26), which we specify as follows:

$$\begin{aligned}
\rho_{\mathbf{k}_\alpha}(\alpha| \ldots; t) &= (2\pi)^{-1}\int_C dz e^{-izt}(-e^2)^3 \frac{1}{i(\mathbf{k}_\alpha \cdot \mathbf{v}_\alpha - z)} \\
&\times \sum_j \sum_{\mathbf{k}_j''} \sum_{\mathbf{k}_\alpha''} \langle \mathbf{k}_\alpha | \mathscr{L}'_{\alpha j} | \mathbf{k}_\alpha'' \mathbf{k}_j'' \rangle \frac{1}{i(\mathbf{k}_\alpha'' \cdot \mathbf{v}_\alpha + \mathbf{k}_j'' \cdot \mathbf{v}_j - z)} \qquad (6.1) \\
&\times \sum_n \sum_{\mathbf{k}_n'''} \sum_{\mathbf{k}_\alpha'''} \langle \mathbf{k}_\alpha'' \mathbf{k}_j'' | \mathscr{L}'_{jn} | \mathbf{k}_\alpha''' \mathbf{k}_n''' \rangle \frac{1}{i(\mathbf{k}_\alpha''' \cdot \mathbf{v}_\alpha + \mathbf{k}_n''' \cdot \mathbf{v}_n - z)} \sum_{s<r} \sum_{\mathbf{k}_\alpha'} \sum_{\mathbf{k}_n'} \\
&\cdot \sum_{\mathbf{k}_r'} \sum_{\mathbf{k}_s'} \langle \mathbf{k}_\alpha''' \mathbf{k}_n''' | \mathscr{L}'_{rs} | \mathbf{k}_\alpha' \mathbf{k}_n' \mathbf{k}_r' \mathbf{k}_s' \rangle \frac{1}{i(\mathbf{k}_\alpha' \cdot \mathbf{v}_\alpha + \mathbf{k}_n' \cdot \mathbf{v}_n + \mathbf{k}_r' \cdot \mathbf{v}_r + \mathbf{k}_s' \cdot \mathbf{v}_s - z)} \\
&\cdot (8\pi^3/\Omega)^{4-1} \rho_{\mathbf{k}_\alpha' \mathbf{k}_n' \mathbf{k}_r' \mathbf{k}_s'}(\alpha, n, r, s | \ldots; 0)
\end{aligned}$$

We begin here at time zero with a state in which four particles (α, n, r, s) are correlated. An interaction between particles r and s leads to a state where only particles α and n are correlated; we say that the correlation between r and s is *destroyed* by the interaction. By the fundamental conservation law of wave-vectors, this process is only possible if $\mathbf{k}_\alpha''' = \mathbf{k}_\alpha'$, $\mathbf{k}_n''' = \mathbf{k}_n'$, $\mathbf{k}_r' + \mathbf{k}_s' = 0$ [this can be verified by evaluating the first matrix element on the right by using formula (5.24)]. The binary correlation between particles α and n is then transformed by an interaction between m and n into a correlation between α and m. We shall call this an *interchange process*: in this process the conservation law requires $\mathbf{k}_\alpha'' = \mathbf{k}_\alpha'''$, $\mathbf{k}_m'' = \mathbf{k}_n'''$. The last process is again a destruction of correlation, which ends in an inhomogeneity factor. Let us stress that, as $\rho_{\mathbf{k}_\alpha}(\alpha| \ldots; t)$ is the final state in which we are interested, there is, of course, no summation over the wave-vector $\mathbf{k}_\alpha$, and no summation over the particle index α: this is the reason for our using a greek letter to denote this *fixed* particle.

According to our rules, this contribution to $\rho_{\mathbf{k}}(\alpha| \ldots; t)$ will be associated with the diagram in Fig. 6.1. For reasons of clarity we draw a vertex with one incoming line on each side (interchange vertex) as a *loop*. This type of vertex will prove especially important in plasma physics. Note that the diagram of Fig. 6.1 is *disconnected*, which means that starting from any given vertex one cannot join all other vertices by following the lines. We call "*external lines*" all lines which start from a vertex but do not end at another vertex. The *external lines at left* (i.e. those whose free ends point toward the left) describe the correlation pattern at time t [the diagram of Fig. 6.1 has one external line at left, corresponding to $\rho_{\mathbf{k}}(\alpha| \ldots ; t)$]. The *external lines at right* describe the correlation pattern at time zero [in our example there are four external lines at right, corresponding to $\rho_{\mathbf{k}'_\alpha \mathbf{k}'_n \mathbf{k}'_r \mathbf{k}'_s}(\alpha, n, r, s| \ldots ; 0)$.

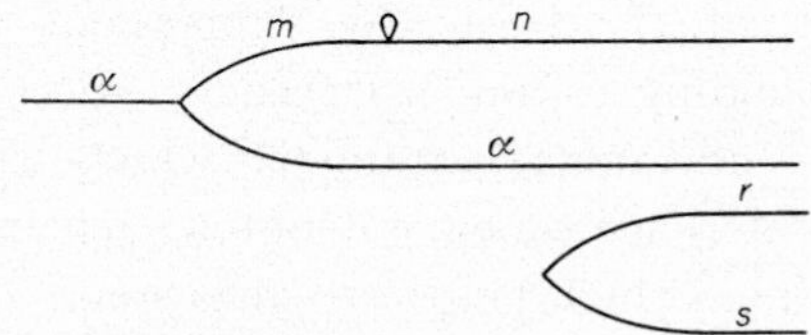

Fig. 6.1. A typical diagram [third-order contribution to $\rho_{\mathbf{k}}(\alpha; t)$].

The simplest diagrams are of course the one-vertex diagrams. It is easily seen that the conservation rules limit the number of possible types of vertices. The definition C of a vertex can be analyzed according to the number of lines joining at each side of the vertex, as follows:

	Number of lines to the left	Number of lines to the right
a	0	0
b		1
c		2
d	1	0
e		1
f		2
g	2	0
h		1
i		2

There cannot be more than two lines entering or leaving a vertex. Vertices of type (b) and (d) are excluded by the conservation rule (there is no transition from a state with zero wave-vectors to a state with a non-zero wave-vector). A direct calculation using (5.24) shows that vertex (a) gives a vanishing contribution. In the case (e) we must distinguish the case in which the lines on both sides bear the same particle index or different indices; the former case also has a vanishing matrix element. We are therefore left with *six basic vertices*, to which correspond the six possible matrix elements of $\mathscr{L}'$. All possible one-vertex diagrams are obtained from the six basic vertices by an adjunction of lines which do not pass through the vertex; in the corresponding contribution these extra lines modify the matrix elements of $\mathscr{R}^0$ (according to rule A), but not those of $\mathscr{L}'$. For the purpose of convenience in actual calculations, we have collected in Table 6.1 at the end of the book the expressions for the matrix elements corresponding to the six basic vertices, together with the values of an additional parameter which will be discussed in § 7. For later reference we have denoted each vertex by a letter and also by a name, which helps visualization.

Table 6.1 also defines our short notations for the matrix elements $\langle\{\mathbf{k}\}|\mathscr{L}'_{jn}|\{\mathbf{k}'\}\rangle$. As stated above, the value of these elements depends only on $\mathbf{k}_j$, $\mathbf{k}_n$, $\mathbf{k}'_j$ and $\mathbf{k}'_n$ (provided of course that all other $\mathbf{k}_r$ equal the corresponding $\mathbf{k}'_r$). We therefore write explicitly only the values of these four wave-vectors. Moreover, if some of these are zero, they are not written down (see B, D, F). If both $\mathbf{k}_j$ and $\mathbf{k}_n$ (or both $\mathbf{k}'_j$ and $\mathbf{k}'_n$) are zero, the corresponding state is simply denoted by $|0\rangle$ (see A, C).

From the six basic vertices one can build up any complex diagram. In practice, the procedure is the following. One is interested in the possible contributions to the value of a given Fourier component at time t: this means that the number of external lines at left as well as their particle labels are given. From the state at time t one can go backwards in time and draw successive vertices chosen among the six basic ones. The successive vertices must of course be chosen at each step in such a way as to be compatible with the given lines. For instance, if one starts with one line at left, the first vertex added cannot be of type A, B or E (Table 6.1), because these have two outgoing lines to the left. One must therefore connect to the line a vertex of type D or F or draw a disconnected vertex of type C (see Fig. 6.2). At the next step the choice of possibilities is much larger. Some of these possibilities are shown in Fig. 6.2. In this process of successive addition of vertices one must be careful not to violate the conservation of wave-vectors (see Fig. 6.2).

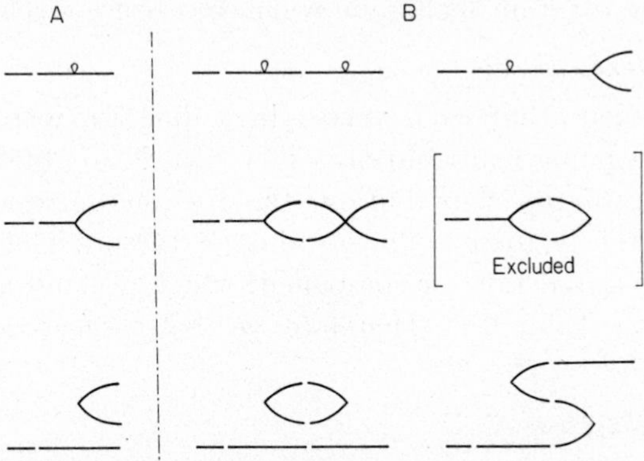

Fig. 6.2. Examples of the construction of complex diagrams by successive additions of vertices to a given external line. Section A shows all possible one-vertex diagrams with one external line at left. Section B shows some of the two-vertex diagrams with one external line at left.

After the drawing of the lines and of the vertices, the lines must be correctly labeled. The practical process of labeling proceeds from left to right. If there are external lines at left, the indices of these lines are given. It has already been noted that these indices are not summed over in the final contribution; we make the convention of always using *greek indices* for such particles. The labeling then proceeds toward the right according to the models drawn in Table 6.1. If a new particle index appears after a given vertex (i.e. if we introduce an interaction with a particle whose label has not yet appeared to the left of that vertex), the new particle is necessarily labeled with a *roman* index, because this index must be summed over in the final expression.

§ 7. Orders of Magnitude of Diagrams

In §§ 2 and 3 the class of conditions describing a reasonable type of initial system has been discussed in terms of the dependence of the Fourier components of $f_N(0)$ on the parameters e^2, N and Ω. §§ 4 to 6 have been devoted to a discussion of the operations by which $f_N(0)$ is transformed into $f_N(t)$. We are now prepared to discuss the dependence on the parameters e^2, N, Ω of the Fourier components at any time $t > 0$. This discussion is made very easy by the diagram technique which has been developed in § 6.

A. Dependence on e^2

It is obvious that each vertex in a diagram introduces into the corresponding contribution a factor of e^2; to obtain the total dependence of the contribution, these e^2 factors must be multiplied by the factor e^{2i}, which indicates the order of magnitude in e^2 of the initial Fourier component which is acted upon by the diagram (see Table 2.1). Therefore, *the e^2-dependence of a diagram with n vertices is* $e^{2(n+i)}$.

B. Dependence on N

This feature is determined by the labeling of the diagram. Each roman index corresponds to a summation over the particles, and therefore to an order of magnitude N. A diagram involving *r roman indices* is therefore of order N^r (for large N).

C. Dependence on Ω

The volume dependence of a diagram is determined by the nature of its vertices. One may classify the 6 vertices of Table 6.1 into two classes according to their volume dependence. We illustrate this fact by two typical examples; we calculate the contributions corresponding to two one-vertex diagrams. Consider first the contribution to $\rho_{\mathbf{k}}(\alpha|\ldots;t)$ due to $\rho_{\mathbf{k}}(j|\ldots;0)$ through a one-vertex transition of type F. Using Table 6.1, formula (5.26) and the "dictionary", Table 2.1, we obtain

$$\rho_{\mathbf{k}}(\alpha|\ldots;t) = (2\pi)^{-1}\int_C dz\,\mathrm{e}^{-izt}\frac{1}{i(\mathbf{k}\cdot\mathbf{v}_\alpha - z)}\sum_j\sum_{\mathbf{k}'}\frac{8\pi^3}{\Omega}\left(-\frac{e^2V_k}{m}\right)(-i\mathbf{k}\cdot\partial_{\alpha j})$$
$$\delta_{\mathbf{k}'-\mathbf{k}}\frac{1}{i(\mathbf{k}'\cdot\mathbf{v}_j - z)}\left(\frac{8\pi^3}{\Omega}\right)^{1-1}\rho_{\mathbf{k}'}(j|\ldots;0) \qquad (7.1)$$
$$= \frac{8\pi^3}{\Omega}\sum_j(2\pi i)^{-1}\int_C dz\,\mathrm{e}^{-izt}\frac{1}{\mathbf{k}\cdot\mathbf{v}_\alpha - z}e^2m^{-1}V_k\mathbf{k}\cdot\partial_{\alpha j}\frac{1}{\mathbf{k}\cdot\mathbf{v}_j - z}\rho_{\mathbf{k}}(j|\ldots;0)$$

This diagram is therefore proportional to Ω^{-1}. Consider now a one-vertex contribution to $\rho_{\mathbf{k},-\mathbf{k}}(\alpha, \beta | \ldots; t)$ from $\rho_0(|\ldots; 0)$ through vertex A.

$$\rho_{\mathbf{k},-\mathbf{k}}(\alpha, \beta | \ldots; t) = \frac{1}{2\pi} \int dz \, e^{-izt} \frac{\Omega}{8\pi^3} \frac{1}{i(\mathbf{k} \cdot \mathbf{g}_{\alpha j} - z)} \frac{8\pi^3}{\Omega} \left(-\frac{e^2 V_k}{m} \right)$$
$$(-i\mathbf{k} \cdot \partial_{\alpha\beta}) \delta_{\mathbf{k}-\mathbf{k}} \frac{1}{-iz} \rho_0(| \ldots; 0) \tag{7.2}$$

This diagram is independent of the volume. Similar calculations * result in associating with each vertex a number T such that the corresponding one-vertex contribution is proportional to Ω^{-T} (in the limit of large Ω). This number T is called the *topological index of the diagram.* The topological indices of the six basic vertices have been tabulated in Table 6.1 (at the end of the book).

We may now extend the concept of T, defining it in such a way that the volume dependence of a complex (many-vertex) diagram is given by Ω^{-T}. The following fundamental theorem can be proven:

The topological index of a diagram is equal to the sum of the topological indices of its vertices.

We shall not give a proof of this theorem here, because it is rather lengthy, although very simple (induction).** The theorem can be verified on some examples. For instance, the diagram of Fig. 6.1 should be proportional to Ω^{-T}, where $T = 1+1+1 = 3$. Formula (6.1) shows, when written explicitly by using Table 6.1 and eq. (2.11), that this contribution to $\rho_{\mathbf{k}}(\alpha | \ldots; t)$ is indeed proportional to Ω^{-3}.

§ 8. Reduced Distribution Functions at Time *t*

We have developed in §§ 4–7 a formalism which enables us to *solve* formally the Liouville equation by means of a power series

* One should keep in mind the asymptotic formula (2.11) in evaluating the volume dependence (for large Ω) of some of these diagrams.

** The proof shows that there is one exceptional case: when vertices A and C are connected as in the letter "Z", vertex A must be given the index 1 instead of 0.

in e^2, and to *classify* the terms of this series with respect to their dependence on N and Ω. The next step is to derive from the expression for the complete distribution function $f_N(t)$ formulae for the reduced distribution functions $f_s(t)$. The latter are the only physically relevant functions, as already stressed repeatedly. This derivation is quite simple. As we already possess an expansion for $f_N(t)$, which we assume to converge, it is sufficient to integrate each term of the series over the positions and velocities of all particles except those of the set $\{s\}$. In particular, we obtain in this way the contributions to the Fourier components of the reduced distributions.

We now show that this process of reduction induces a substantial simplification in the choice of diagrams. The reason for this simplification is the following. Consider a contribution to a Fourier component of $f_s(1, \ldots, s; t)$. This will be obtained from a formula of the type (5.27) by integrating over all particles $s+1, \ldots, N$. We now write this formula schematically by singling out a given vertex:

$$\rho_{\{\mathbf{k}\}}(\mathbf{v}_1, \ldots, \mathbf{v}_s; t) = \int d\mathbf{v}_{s+1} \ldots d\mathbf{v}_N \{\ldots \langle \{\mathbf{k}^{(m)}\} | \mathscr{L}'_{jn} | \{\mathbf{k}^{(m+1)}\} \rangle \ldots\} \rho_{\{\mathbf{k}'\}}(\mathbf{v}_1 \ldots | \ldots \mathbf{v}_N; 0) \tag{8.1}$$

The dotted portions in this expression stand for the factors to the right and to the left of the singled-out matrix element in a formula of type (5.27). These factors are essentially operators which depend on the velocities of the particles whose indices appear in the diagram to the left and to the right, respectively, of the vertex considered (through the factors of type $[\sum \mathbf{k}_j \cdot \mathbf{v}_j - z]^{-1}$). The matrix element under consideration, on the other hand, is a differential operator, containing a factor ∂_{jn}, otherwise independent of the velocities. Therefore, we may write (8.1) in the general form

$$\rho_{\{\mathbf{k}\}}(\{s\}; t) = \int d\mathbf{v}_{s+1} \ldots d\mathbf{v}_N F(\{\mathbf{v}_r\}) \left(\frac{\partial}{\partial \mathbf{v}_j} - \frac{\partial}{\partial \mathbf{v}_n}\right) \times G(\{\mathbf{v}_m\}) \rho_{\{\mathbf{k}'\}}(\mathbf{v}_1 \ldots | \ldots \mathbf{v}_N; 0) \tag{8.2}$$

We now consider two possibilities. Assume first that neither j nor n belongs to the set $(1, \ldots, s)$. Then, either at least one of the variables $\mathbf{v}_j$ or $\mathbf{v}_n$ belongs to the set $\{\mathbf{v}_r\}$ on which F depends (in this case the integral is in general non-zero) or neither $\mathbf{v}_j$ nor $\mathbf{v}_n$ belongs to the set $\{\mathbf{v}_r\}$. In this case the integration signs over $\mathbf{v}_j$ and $\mathbf{v}_n$ can be written just in front of the derivatives:

$$\rho_{\{\mathbf{k}\}}(\{s\};t) = \int d\mathbf{v}'_{s+1} \ldots d\mathbf{v}_N F(\{\mathbf{v}_r\}) \int d\mathbf{v}_j d\mathbf{v}_n \left(\frac{\partial}{\partial \mathbf{v}_j} - \frac{\partial}{\partial \mathbf{v}_n}\right) \times G(\{\mathbf{v}_m\}) \rho_{\{\mathbf{k}'\}}(\mathbf{v}_1 \ldots | \ldots \mathbf{v}_N; 0) \tag{8.3}$$

The $\mathbf{v}_j$ and $\mathbf{v}_n$ integrals can be converted by Green's theorem into surface integrals over an infinitely large domain in $\mathbf{v}$-space. These surface terms vanish because of our assumption that $\rho_k(v)$ and its derivatives vanish at infinity (see § 4). In terms of diagrams, (8.3) vanishes unless one of the indices j or n (or both) occurs in the diagram to the left of the vertex we consider.

Suppose now that one at least of the indices j, n belongs to the set of "fixed" particles $(1, \ldots, s)$. Then this velocity is not integrated over and the previous argument does not apply: the corresponding expression does not, in general, vanish, unless F or G does.

This argument can now be applied to each of the vertices in the diagram, and the following theorem is obtained.

THEOREM I. *The only diagrams which give non-vanishing contributions to the reduced distribution function* $f_s(1, \ldots, s; t)$ *are those in which each vertex has one at least of the following properties*:

a) *One at least of the two indices of the vertex belongs to the set* $(1, \ldots, s)$.

b) *One at least of the indices of the vertex appears on a line to the left of that vertex.*

We can also formulate the following two corollaries:

Corollary 1. In a diagram without external lines to the left, one at least of the two first indices (at left) must belong to the set $(1, \ldots, s)$.

Corollary 2. If one of the indices of the vertex (j, n), say j, satisfies condition (a) or (b) and the other does not, the operator ∂_{jn} in the final expression can be replaced by ∂_j.

Fig. 8.1 shows some applications of these rules.

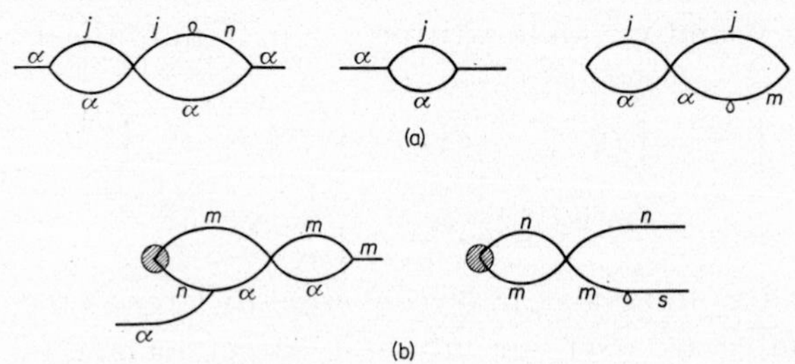

Fig. 8.1. Some examples of diagrams which give non-vanishing (a) or vanishing (b) contributions to the distribution functions. The greek particle α belongs to the set $(1, \ldots, s)$. In the vanishing contributions the vertex which is responsible for the vanishing is encircled.

The preceding rules are especially important in limiting the possible types of disconnected diagrams contributing to a reduced distribution function. If in particular a diagram has a disconnected part without external lines, one of the indices of the leftmost vertex of that part must be a fixed particle or one having already appeared to the left of that vertex. We stress the latter situation by joining with a dotted line the vertices of disconnected parts which have an index in common, and we call such diagrams "*semiconnected*".

This discussion may be summarized in the following theorem concerning the *connection* of the permitted diagrams.

THEOREM II. *The most general diagram contributing to a Fourier component of $f_s(1, \ldots, s)$ consists of at most s disconnected parts; the leftmost vertex of each of these involves a fixed particle. Each of these parts taken separately is a connected or semiconnected diagram.*

Fig. 8.2 shows some examples of the application of this theorem.

We now show that the contributions to $f_s(t)$ are finite in the limit (2.3) (i.e. $N \to \infty, \Omega \to \infty, N/\Omega = c$) if they are so initially.

In particular, we show that there exists no diagram contributing to a *reduced* distribution function which depends on N and on Ω through a combination of type $N^r c^p (r > 0)$, because in that case we would have a divergence in the limit. A basic remark is that all

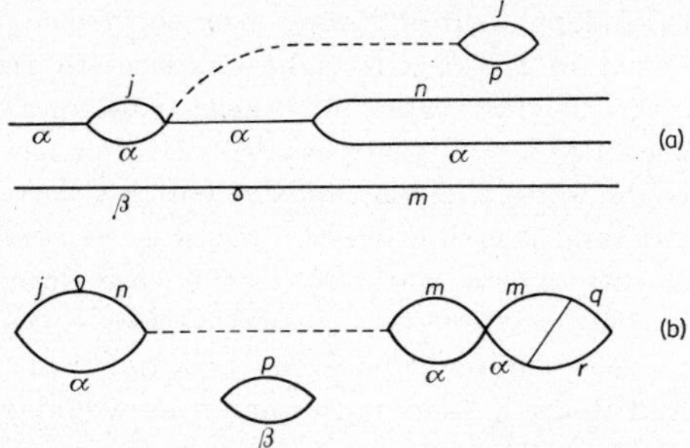

Fig. 8.2. Examples of contributions to the Fourier components $\rho_{\mathbf{k},-\mathbf{k}}(\alpha, \beta; t)$ and $\rho_0(t)$ of the two-particle reduced distribution function $f_2(\alpha, \beta; t)$.

vertices of topological index $T = 1$ have to their right at least one particle label which does not appear to their left. On the other hand, all labels to the right of the vertices $T = 0$ also appear to the left of these vertices. A glance at Table 6.1 shows this immediately.

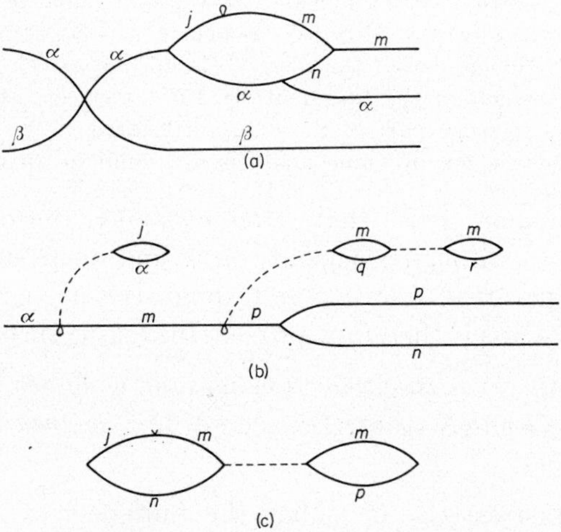

Fig. 8.3. Examples for the discussion ot the theorem on the finiteness of the reduced distribution functions.

Consider now first a *connected diagram with external lines at left* (such as the one of Fig. 8.3a) and analyze it from left to right. The particle labels of the external lines at left are necessarily *greek* labels. The diagram may begin with some vertices of index $T = 0$; these introduce no roman index, and therefore no N-factor. Hence, if the diagram consists only of such vertices, it is independent of N and of Ω, and hence independent of c.

Suppose now that in this left-to-right analysis we arrive at the first vertex of index $T = 1$: this vertex introduces a factor Ω^{-1}. On the other hand, it introduces also a new particle label. Two choices are possible for the latter: we could write a greek index: then this vertex introduces no particle summation and it is of order Ω^{-1}. Or else, we could write a roman letter, and hence introduce a factor N: in this case the vertex is of total order $N\Omega^{-1} = c$. The case of the creation vertex C requires a special remark. Indeed, we could have here the possibility of writing *two* new particle indices and choosing them both roman: we would then have a vertex of order $N^2\Omega^{-1} = Nc$, which would diverge in the limit (2.3). However, this choice violates theorem I and must be excluded.

This analysis goes on and shows that each vertex which introduces a new particle label also introduces a factor Ω^{-1}. Such diagrams depend therefore on N and on Ω either through the factor c^T (T = topological index of the diagram) and are finite in the limit (2.3), or else through the factor $c^{T-a}\Omega^{-a}$ $(a > 0)$, in which case they go to zero in the limit.

Consider now a *disconnected diagram* (see Fig. 8.3b). If all connected parts which compose it have external lines at left, the analysis is identical to the previous one. But if one part has no external lines at left, it must begin with a creation vertex C which may give trouble because it could introduce two factors of N. However, as before, this possibility is excluded by theorem I. Hence, one of the new labels must be greek, or must have appeared to the left of the fragment. In the latter case the fragment is semiconnected to some part of the diagram located to its left. Therefore the conclusions of the previous analysis are valid in all cases.

This discussion shows that, *after reduction*, the operator connecting $\rho_{\{\mathbf{k}'\}}(0)$ to $\rho_{\{\mathbf{k}\}}(t)$ depends on N and on Ω either through the finite quantity $N/\Omega = c$ or through the negligible quantity $c^{T-a}\Omega^{-a}$. We have therefore proven the basic theorem:

If the reduced distribution functions of a system are initially finite in the limit $N \to \infty$, $\Omega \to \infty$, $N/\Omega = c$, they remain so at all later times.

We must stress the fact that the finiteness of $\rho_{\{\mathbf{k}\}}(1, \ldots, s)$ at all times is a direct consequence of the fact that many diagrams contributing to the complete $\rho_{\{\mathbf{k}\}}(1, \ldots, s | \ldots N)$ have been eliminated by the process of reduction. An analysis of a diagram which contributes to $\rho_{\{\mathbf{k}\}}(1, \ldots, s| \ldots N)$ but *not* to $\rho_{\{\mathbf{k}\}}(1, \ldots, s)$, such as the one of Fig. 8.3c, immediately shows that the corresponding contribution is of the order $N^d c^r$, d being the number of *disconnected* fragments without external lines entering the diagram.

We have established that there exist no diagrams of order $N^p c^s$ which would become infinitely large in the limit (2.3). But, by using the rules developed so far, we still have the possibility of obtaining diagrams which are "too small", i.e. which vanish as Ω^{-p} in the limit. It would be desirable to have a further restriction which eliminates automatically these vanishing diagrams. The discussion of the previous theorem has shown that the occurrence of this case is due only to an inadequate choice of particle labels. The reader will easily adapt that discussion in order to prove the following theorem concerning the *labeling* of the reduced diagrams:

Theorem III. *The only reduced diagrams which are finite and non-vanishing in the limit $N \to \infty$, $\Omega \to \infty$, $N/\Omega = c$ are those which satisfy the following conditions:*

a) *Every connected part with external lines at left involves no greek label distinct from those which label these external lines.*

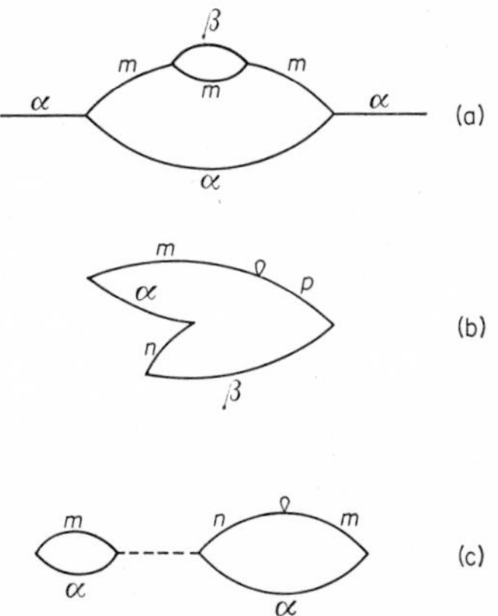

Fig. 8.4. Three diagrams which violate theorem III. (a) The connected part involves a greek label which does not label an external line at left; (b) the connected diagram involves two greek labels; (c) the second connected part involves two labels (α, m) having already appeared to the left.

b) *Every connected part without external lines at left involves either one single greek label or (if it is semiconnected to another part to its left) one single label having already appeared to its left.*

Some applications of this theorem are illustrated in Fig. 8.4.

This theorem completes the formal construction of the diagram technique in the following sense. Any diagram drawn according to the rules of §§ 6 and 8 is in one-to-one correspondence with a contribution to a reduced Fourier component. Moreover, the order of magnitude of such a diagram is given by the following rule, which combines those of § 7, B and C:

THEOREM IV. *Any diagram which satisfies the requirements of theorems I, II and III is proportional to the factor c^T, T being its topological index.*

This theorem completes the general exposition of the diagram technique and we can now go over to its application to the statistical mechanics of charged particles.

CHAPTER 2

The Short-Time Behavior of Classical Plasmas — The Vlassov Equation

§ 9. General Discussion of the Properties of Classical Plasmas

The simplest application of the general theory developed in the introduction consists in a straightforward perturbation theory, dealing with systems for which the coupling parameter e^2 is small. This type of theory would then very much resemble the elementary perturbation theory in quantum mechanics. However, it would be of no interest in statistical mechanics, as will be briefly shown now.

Consider first the case of an ordinary gas the molecules of which interact through short-range forces, which are assumed weak on the average (e^2 small). Suppose now that the contributions to some Fourier component have been ordered according to powers of e^2, say:

$$\rho_0(t) = \rho_0(0)+e^2\rho_0^{(1)}(t)+e^4\rho_0^{(2)}(t)+e^6\rho_0^{(3)}(t)+\ldots \tag{9.1}$$

All the coefficients in this series [except $\rho_0(0)$] are functions of time. This expansion is significant for short times because then all the coefficients are of roughly the same order of magnitude. Due to the smallness of e^2, the series is presumably convergent and can be broken off after a small number of terms.

However, as the time becomes longer, the structure of this series changes radically. Some terms start growing systematically and after a sufficiently long time the series (9.1) loses meaning. If we had broken off the series, say after the term in e^2, we would have neglected terms in e^4; but some contributions to the latter terms turn out to grow proportionally to t and will eventually become larger than the terms in e^2 which remain bounded in time. More details about this type of behavior will be found in Chapter 6. This short anticipation was only meant to show that the structure of the theory is completely different according to whether our interest lies in a short-time description or a long-time description.

An important question now arises: what is meant by "long" time and by "short" time? In an ordinary gas this question can be answered rather simply. In effect, two characteristic time scales appear quite naturally:

t_c: *the duration of a collision*: this is the time which a particle moving with the average velocity spends in the sphere of influence of another particle.

t_r: *the relaxation time*: this is the time which the system needs to reach thermal equilibrium; in a dilute gas it is of the order of the time between two collisions (or the inverse of the collision frequency).

In an inhomogeneous system there exists a third time scale:

t_h: *the hydrodynamic time*: defined as the time necessary for a particle to travel through the hydrodynamic length L_h (see § 3).

The crucial time scale which determines the change in behavior of the perturbation series (9.1) is t_c. For times shorter or of the order of t_c the first type of behavior prevails, whereas for times much longer than t_c the second type is relevant.

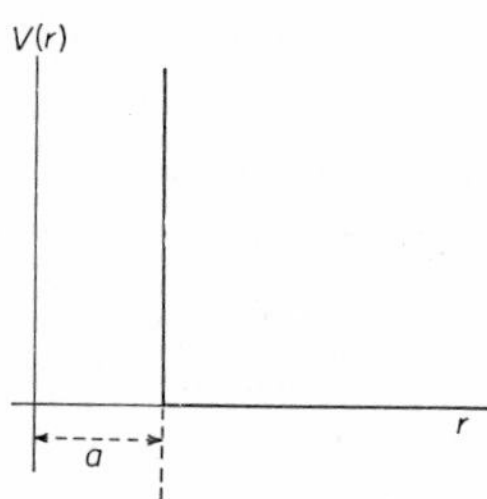

Fig. 9.1. Hard-sphere potential.

The hydrodynamic time scale is always assumed here to be the longest in inhomogeneous systems. Only under this assumption does the initial condition discussed in § 3 apply. This assumption excludes, of course, such phenomena as shock waves, in which there exist strong gradients. There exists no really satisfactory molecular theory of these phenomena at the present time, and therefore they will not be discussed here. The relaxation time and the duration of a collision, on the other hand, are intrinsic quantities, depending only on the dynamical properties and on the state of the system (density, temperature). The nature of these two time scales will show an essential difference between an ordinary gas and a plasma. These quantities can be discussed by performing a simple dimensional analysis.

Consider first the case of an ordinary gas whose molecules interact through short-range forces. We shall take the extreme model of hard spheres to describe the interactions (see Fig. 9.1).

$$V(r) = \infty \quad \text{for} \quad r < a$$
$$V(r) = 0 \quad \text{for} \quad r > a.$$

For such a system four independent characteristic parameters are available:

the radius of the molecules	a
the mass of the molecules	m
the number density	c
the temperature, in the form	$\beta = (kT)^{-1}$

With these quantities we can form one and only one independent non-dimensional constant γ (any other invariant is a function of γ)

$$\gamma = a^3 c \tag{9.2}$$

We now look for combinations of our four constants in the form $a^x m^y c^z \beta^u$ which have dimensions of time. It is easily seen that the most general time constructed with these parameters is of the form:

$$T = a\sqrt{m\beta}\,\gamma^z = (a/\bar{v})\gamma^z \tag{9.3}$$

where $\bar{v}$ is the root-mean-square velocity and z is an arbitrary number. The simplest of these times is obtained by taking $z = 0$:

$$t_c = a/\bar{v} \tag{9.4}$$

This time corresponds to the intuitive idea of a *collision time* (range of the force divided by the mean velocity). It should be noted that it is independent of the density. We now look for another time which is inversely proportional to the density, as is intuitively natural for a *relaxation time*. This is obtained by setting $z = -1$, with the result

$$t_r = c^{-1}a^{-2}\sqrt{m\beta} = (a^2 c\bar{v})^{-1} \tag{9.5}$$

which is seen to be the usual expression of the relaxation time of a hard-sphere gas. We may now assume that the gas is *dilute*; this is expressed by the non-dimensional condition:

$$\gamma \ll 1 \tag{9.6}$$

But (9.4) and (9.5) show that $t_c = \gamma t_r$, so that (9.6) implies

$$t_c \ll t_r \tag{9.7}$$

We now go over to the case of a *plasma*, in which the interaction energy is given by eq. (1.4). The most striking difference from the previous case is the fact that $V(r)$ contains no characteristic parameter having the dimensions of a length and characterizing the range of the force. Moreover, $V(r)$ decreases so slowly with distance that integrals containing factors of $V(r)$ are usually divergent. In particular, $V(r)$ has a Fourier transform only in the generalized sense:

$$V_l = \lim_{\alpha \to 0} \frac{1}{8\pi^3} \int d\mathbf{r}\, e^{-i\mathbf{l}\cdot\mathbf{r}} V(r) e^{-\alpha r} = \lim_{\alpha \to 0} \frac{1}{2\pi^2} \frac{1}{l^2 + \alpha^2} = \frac{1}{2\pi^2} \frac{1}{l^2} \tag{9.8}$$

These features are an expression of the fundamental property of the Coulomb forces: their *long range* (mathematically: infinite range).

On the other hand, the hamiltonian (1.4) contains the coupling parameter e^2. The hard-sphere interaction does not contain such a parameter, because that interaction is highly singular: it is either 0 or ∞.

We may now perform a dimensional analysis as before. We first determine the fundamental non-dimensional parameter which can be constructed with the four quantities e^2, m, c, β. This turns out to be *

* This parameter was first used by R. Milner, *Phil. Mag.*, **23**, 551 (1912); **25**, 742 (1913); see also R. Fowler and E. A. Guggenheim, *Statistical Thermodynamics*, chap. 9, Cambridge University Press, 1939; N. N. Bogoliubov, *Problems of a Dynamical Theory in Statistical Physics*, chaps. 1.4 and 2.11 (ref. see footnote p. 27); E. W. Montroll and J. C. Ward, *Phys. of Fluids*, **1**, 55 (1958).

$$\Gamma = e^2 c^{\frac{1}{3}} \beta \tag{9.9}$$

We assume for a *classical plasma*

$$\Gamma \ll 1 \tag{9.10}$$

This condition is achieved at low densities or at high temperatures. Several numerical values of the parameter Γ are given in Table 9.1.

We now construct quantities having the dimensions of time. The most general characteristic time of the form $m^x e^{2y} c^z \beta^u$ is

$$T = (m\beta)^{\frac{1}{2}} c^{-\frac{1}{3}} \Gamma^y \tag{9.11}$$

with y arbitrary. As a consequence of the absence of a characteristic range a, no one of the times contained in (9.11) can be interpreted as the duration of a collision. The simplest time, obtained from (9.11) by setting $y = 0$, is $t_0 = 1/\bar{v}c^{\frac{1}{3}}$, which is not related to the interactions (it is the time taken by an average particle to travel across the mean separation of two particles).

Table 9.1. Typical values of the parameter Γ at various temperatures and densities ($a = 1.66$).

T, °K \ c, part./cm³	10^6	10^{12}	10^{18}	10^{24}
10^2	$a \cdot 10^{-3}$	$a \cdot 10^{-1}$	$a \cdot 10$	$a \cdot 10^3$
10^4	$a \cdot 10^{-5}$	$a \cdot 10^{-3}$	$a \cdot 10^{-1}$	$a \cdot 10$
10^6	$a \cdot 10^{-7}$	$a \cdot 10^{-5}$	$a \cdot 10^{-3}$	$a \cdot 10^{-1}$
10^8	$a \cdot 10^{-9}$	$a \cdot 10^{-7}$	$a \cdot 10^{-5}$	$a \cdot 10^{-3}$

Physically speaking, this difficulty comes from the fact that the range of the Coulomb forces is so long that each particle interacts simultaneously with a large number of other particles. These interactions can therefore not be described by a succession of binary (or even multiple) collisions which take place during a finite time and are separated by periods of free motion. We must therefore look for other characteristic times related to the interactions.

A very important time contained in (9.11) is obtained by setting $y = -\frac{1}{2}$:

$$t_p = (m/e^2c)^{\frac{1}{2}} \tag{9.12}$$

This time is proportional to the square root of the mass and is independent of the velocity of the particle (or of β), and therefore immediately suggests an analogy with a harmonic oscillator. In this perspective, t_p would be the period of oscillation of a particle subject to an elastic force characterized by a spring constant equal to e^2c. It will be seen in the next chapters that this analogy with a harmonic oscillator is not an accident, but expresses a fundamental property of plasmas. At first sight it looks surprising that particles in a gas might behave like harmonic oscillators: it will be seen in Chapter 3 that the explanation lies precisely in the long range of the Coulomb forces.

We now look for a relaxation time, which should be proportional to c^{-1}. We therefore take $y = -2$ in (9.11), obtaining:

$$t_r = (e^4c)^{-1}(m/\beta^3)^{\frac{1}{2}} \tag{9.13}$$

The ratio t_p/t_r is therefore equal to

$$t_p/t_r = \Gamma^{\frac{3}{2}} \ll 1 \tag{9.14}$$

This discussion therefore leads us quite naturally to a choice of the two characteristic molecular time scales corresponding to t_c and t_r in the case of an ordinary gas. These are, correctly ordered:

$$t_r \gg t_p \tag{9.15}$$

§ 10. Choice of Diagrams for Short Times

We are now going to derive an equation which describes the behavior of the plasma over short periods of time, i.e. for $t = O(t_p) \ll t_r$. Such a problem is meaningful for a plasma in situations in which $\Gamma \ll 1$, because the relaxation time can be very long; in many experiments on very hot plasmas ("thermonuclear" plasmas) the experimentally realizable plasmas have actually a life-time which is much shorter than the relaxation time. The system has therefore no time to reach equilibrium and its behavior can be determined from an equation valid for short times.

One striking difference between a short-range potential (not necessarily a hard-sphere potential) and the Coulomb potential is the fact that the "duration of a collision" is independent of the strength of the interactions in the former case whereas for a plasma it is a function of e^2.

As was mentioned at the beginning of § 9, in the expansion (9.1) all time-independent coefficients are generally of the same order (as concerns the time dependence) for short times. We do not prove this statement here, but refer the reader to Chapter 6, where the time dependence will be studied more systematically. The discussion of the series is therefore limited to a discussion of the orders of magnitude in e^2 and c. A complete dimensional analysis of the various terms in eq. (5.26) analogous to the one performed in § 9 is rather difficult and cumbersome because of the complicated nature of the expressions. However, the simple analysis of the previous section enables us to recognize very easily the important contributions.

Two basic remarks will guide this choice. We first note that powers of the "duration of the collision", and particularly the inverse square of t_p

$$t_p^{-2} = e^2 c/m$$

necessarily depend on the charge and on the density through the combination $e^2 c$. On the other hand, the parameter of smallness Γ [see (9.9)] depends on the charge and on the density through the factor $e^2 c^{\frac{1}{3}} = e^{\frac{4}{3}}(e^2 c)^{\frac{1}{3}}$. This factor therefore contains "uncompensated" powers of e^2, i.e. factors of e^2 which are not multiplied by an equal number of factors of c.

Going back now to theorem IV of § 8, we see that the separation of the vertices of the diagrams into two classes, corresponding to $T = 1$ and to $T = 0$ (see Table 6.1), is really a subdivision into contributions proportional to $e^2 c$ and to e^2, respectively.

We now argue that when we are interested in times of the order t_p, all contributions proportional to powers of the combination $e^2 c$ must be retained. In other words, the combination

e^2c cannot be regarded as a small parameter; an infinite subseries of the form $\sum_{n=0}^{\infty}(e^2c)^n f_n$ must be extracted from (5.26) and summed completely in order to get a physically consistent result. On the other hand, terms containing uncompensated e^2 factors such as $e^{2m}(e^2c)^n$ will be smaller by m factors of Γ than the corresponding terms with $m = 0$ (when properly reduced to non-dimensional form).

To sum up, it may be stated that to a first approximation, valid for short times, *all terms proportional to arbitrary powers of* e^2c *must be retained, whereas terms containing uncompensated* e^2 *factors must be neglected.*

The previous rule might appear somewhat loose, because the quantities e^2 and e^2c are not non-dimensional quantities; one could not say that "e^2 is small, whereas e^2c is large". These rules actually ensure that no quantity of the order $t/t_p = t\sqrt{(e^2c/m)}$ is neglected. Anticipating the results of the next sections, it will be seen that the result of the summation of the subseries in e^2c will lead to functions of the type $\exp(it/t_p)$; when $t \approx t_p$ it is of course not permitted to expand such functions in powers of t/t_p and retain only a few terms.

We now go over to the choice of the diagrams of order $(e^2c)^n$. We shall calculate the one-particle reduced distribution function $f_1(\mathbf{x}_\alpha, \mathbf{v}_\alpha; t)$, which will also be denoted briefly by $f(\alpha)$. Referring to formula (2.10), it is seen that in Fourier space we need to look for all contributions to $\rho_0(\mathbf{v}_\alpha) \equiv \varphi(\alpha)$ and to $\rho_{\mathbf{k}}(\mathbf{v}_\alpha)$ which are of the desired order in e^2c. According to the general philosophy of the diagram method, we first solve a topological problem:

To find all diagrams with no external line at left and all diagrams with one external line (labeled α*) at left, such that the corresponding contributions are of order* $(e^2c)^n$, $n = 0, 1, 2, \ldots$

A. Contributions to $\varphi(\alpha)$

The diagrams of this class must have no external line at left; in particular, they must end at left with a vertex which has no lines at left. *

* Note that the latter statement is not sufficient for the absence of outgoing lines in a complex diagram! (see Fig. 6.2, third line).

The only vertex which has this structure is the destruction vertex C of Table 6.1. This vertex is of the right order (e^2c). Starting from it, and adding successive vertices to the right, a connected diagram must be built up. In order to satisfy the condition on the order of magnitude, only vertices with topological index $T = 1$ can be connected to the first one. The only such vertices which can be connected are the inhomogeneous destruction vertex D and the loop F. (It is easily seen that vertex C appearing in an intermediate state of a connected diagram always violates theorem I of § 8 on reduced diagrams.) As a result, the composite diagrams constructed in this way necessarily have two or more outgoing lines to the right, and therefore connect $\rho_0(t)$ to $\rho_{\mathbf{k},-\mathbf{k}}(0)$ or to initial Fourier components with even more non-vanishing wave-vectors. But, according to our initial condition for homogeneous systems, $\rho_{\mathbf{k},-\mathbf{k}}(0) \sim e^2$, and the latter components are proportional to even higher powers of e^2. The resulting diagrams are therefore proportional to $e^{2m}(e^2c)^n$ with $m \geqq 1$, and are therefore not of the desired order. The conclusion of this analysis is therefore: *There are no diagrams of order* $(e^2c)^n$ *contributing to* $\varphi(\alpha)$.

B. Contributions to $\rho_{\mathbf{k}}(\alpha)$

In § 3 we have discussed a special important type of initial condition for inhomogeneous systems. It was assumed that there exists a hydrodynamic length scale L_h which is much larger than any molecular length scale $L_m : L_h \gg L_m$. Although this condition will play a quite important role in the study of the long-time behavior of inhomogeneous plasmas (Chapters 7 and 9), it can be relaxed in the study of the short-time behavior in which we are presently engaged *. For the sake of generality we shall therefore derive the equation of evolution without making any assumption about the relative sizes of L_h and L_m, or equivalently, about the shape of $\rho_{\mathbf{k}}$ in Fourier space.

The crucial remark here is the fact that the decomposition (3.9) is quite general; in particular, it does not depend on the hydro-

* A general systematic theory of inhomogeneous systems has been developed by G. Severne, *Physica,* **29** (1963) (in press).

dynamic condition. More generally, a Fourier component with s non-vanishing wave-vectors is split uniquely as follows:

$$\rho_{\mathbf{k}_1,\ldots,\mathbf{k}_s}(1,\ldots,s) = \prod_{j=1}^{s} \rho_{\mathbf{k}_j}(j) + \rho_{[\mathbf{k}_1,\ldots,\mathbf{k}_s]}(1,\ldots,s); \quad \sum \mathbf{k}_j \neq 0 \qquad (10.1)$$

The first term is completely factorized into a product of s inhomogeneity factors. The second term is the sum of all possible correlation patterns of s particles: products of a binary correlation and $s-2$ inhomogeneity factors, products of a ternary correlation and $s-3$ inhomogeneity factors, etc. This separation is the exact analogue of the cluster expansion well known in equilibrium theory (see Chapter 14). For our purpose it is, however, not necessary to specify in further detail the form of the term $\rho_{[\mathbf{k}_1\ldots\mathbf{k}_s]}(1,\ldots,s)$.

Let us now go back to our discussion of the diagrams contributing to $\rho_{\mathbf{k}}(\alpha; t)$. The discussion here is very similar to the previous one. The diagram must begin at the left with one of the vertices C, D or F. These are all of the right order of magnitude, e^2c. However, connected diagrams with an external line at the left beginning with a vertex C necessarily violate theorem I of § 8, and can therefore be discarded. The most general composite diagram of the desired order consists of a succession of vertices D or F connected to the first one (see Fig. 10.1). Such a diagram has one or more external lines to the right. The latter state corresponds to a

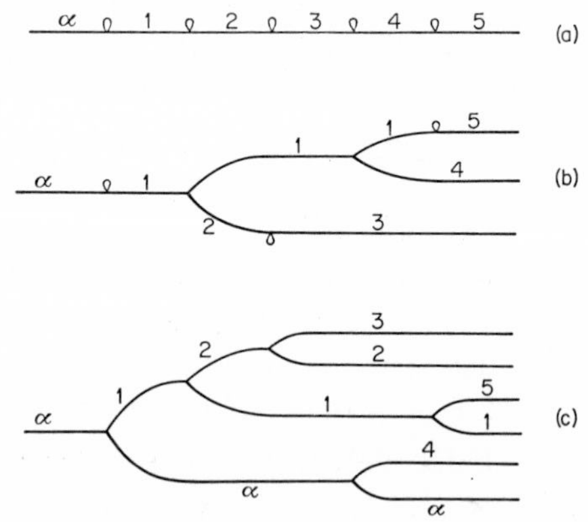

Fig. 10.1. Examples of contributions to $\rho_{\mathbf{k}}(\alpha; t)$.

Fourier component with one or more non-vanishing wave-vectors at time zero. In the case of one line, the inhomogeneity factor $\rho_{\mathbf{k}}(j; 0)$ is independent of e^2, as follows from § 3; hence the corresponding contribution of the complete diagram is of the desired order $(e^2c)^n$. If, however, there are more lines, corresponding to $\rho_{\mathbf{k}_1 \ldots \mathbf{k}_s}(1, \ldots, s; 0)$, only the first term in (10.1) is independent of e^2 and gives a contribution of the desired order. The true correlation part of the Fourier component is proportional to at least one power of e^2 (see § 3) and hence gives a negligible contribution according to our criterion.

Therefore, to sum up, the most general contribution to $\rho_{\mathbf{k}}(\alpha; t)$ of order $(e^2c)^n$ is a diagram consisting of a *connected succession of n vertices of type D or F, in any order, operating on the completely factorized part of the Fourier component at time zero.*

§ 11. Derivation of the Vlassov Equation

Once the relevant diagrams have been chosen, the next problem is their summation. More exactly, an equation will be derived, the solution of which is precisely the partial sum selected from (5.26) according to the criteria of § 10.

Consider first $\varphi(\alpha)$. There is no diagram of order $(e^2c)^n$, $n \geqq 1$. Therefore the only term remaining in (5.26) is the one for which $n = 0$, and we have

$$\varphi(\mathbf{v}_\alpha, t) = \varphi(\mathbf{v}_\alpha, 0) \tag{11.1}$$

To this order of approximation, *the velocity distribution function remains constant in time.*

Consider now the diagrams contributing to $\rho_{\mathbf{k}}(\alpha; t)$. Let us first introduce short notations for the matrix elements corresponding to the vertices D and F, taken from Table 6.1:

$$\begin{aligned} F_{jn}(\mathbf{k}) &= (8\pi^3/\Omega)m^{-1}V_k(-i\mathbf{k}\cdot\partial_{jn}) \\ D_{jn}(\mathbf{k}, \mathbf{k}') &= (8\pi^3/\Omega)m^{-1}V_{|\mathbf{k}-\mathbf{k}'|}i(\mathbf{k}'-\mathbf{k})\cdot\partial_{jn} \end{aligned} \tag{11.2}$$

We divide the diagrams into two subsets: the set (F) of all diagrams ending with a vertex F, and the set (D) of diagrams ending with a vertex D. We now split the mathematical expression corresponding to our set of diagrams in the same way, writing explicitly the matrix element of the last vertex of each subset:

$$\rho_{\mathbf{k}}(\alpha|\ldots;t)=\frac{1}{2\pi}\int_C dz\,e^{-izt}\frac{1}{i(\mathbf{k}\cdot\mathbf{v}_\alpha-z)}\rho_{\mathbf{k}}(\alpha|\ldots;0)$$
$$+\frac{1}{2\pi}\int_C dz\,e^{-izt}\frac{1}{i(\mathbf{k}\cdot\mathbf{v}_\alpha-z)}(-e^2)\sum_j F_{\alpha j}(\mathbf{k})\left\{\frac{1}{i(\mathbf{k}\cdot\mathbf{v}_j-z)}\Big[\rho_{\mathbf{k}}(j|\alpha\ldots;0)\right.$$
$$+(-e^2)\sum_n F_{jn}(\mathbf{k})\frac{1}{i(\mathbf{k}\cdot\mathbf{v}_n-z)}\rho_{\mathbf{k}}(n|\alpha,j\ldots;0)+(-e^2)\sum_{\mathbf{k}'}\sum_n D_{jn}(\mathbf{kk}')$$
$$\left.\times\frac{1}{i[\mathbf{k}'\cdot\mathbf{v}_n+(\mathbf{k}-\mathbf{k}')\cdot\mathbf{v}_j-z]}\rho_{\mathbf{k}'}(n|\alpha\ldots;0)\,\rho_{\mathbf{k}-\mathbf{k}'}(j|\ldots;0)+\ldots\Big]\right\}$$
$$+\frac{1}{2\pi}\int_C dz\,e^{-izt}\frac{1}{i(\mathbf{k}\cdot\mathbf{v}_\alpha-z)}(-e^2)\sum_{\mathbf{k}'}\sum_j D_{\alpha j}(\mathbf{kk}') \tag{11.3}$$
$$\times\left\{\frac{1}{i[\mathbf{k}'\cdot\mathbf{v}_\alpha+(\mathbf{k}-\mathbf{k}')\cdot\mathbf{v}_j-z]}\rho_{\mathbf{k}'}(\alpha|\ldots;0)\rho_{\mathbf{k}-\mathbf{k}'}(j|\ldots;0)+\ldots\right\}$$

It is easily seen from Fig. 11.1 that the diagrams obtained by clipping off the last vertex of class (F) are all the diagrams contributing to $\rho_{\mathbf{k}}(j|\ldots;t)$ to order $(e^2c)^n$. Similarly, by clipping off

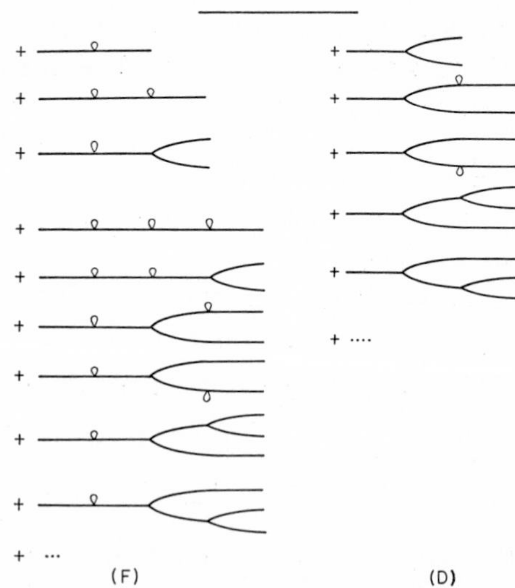

Fig. 11.1. Summation of the diagrams of order $(e^2c)^n$ contributing to $\rho_{\mathbf{k}}(\alpha;t)$ in the Vlassov approximation.

the last vertex of class (D) we obtain all diagrams contributing to $\rho_{\mathbf{k}'}(\alpha|\ldots;t)\,\rho_{\mathbf{k}-\mathbf{k}'}(j|\ldots;t)$ to the same order. (This is a consequence of the fact that only the completely factorized parts of the Fourier components are retained at time zero.) The contents of the two curly brackets should therefore become clear with the aid of Fig. 11.1.

We now differentiate the two sides of (11.3) with respect to time; this brings down a factor $-iz$ in the integrands of the r.h.s.: this factor we write as $(i\mathbf{k}\cdot\mathbf{v}_\alpha - iz) - i\mathbf{k}\cdot\mathbf{v}_\alpha$, obtaining

$$\partial_t \rho_{\mathbf{k}}(\alpha|\ldots;t) = \frac{1}{2\pi}\int_C dz\,\mathrm{e}^{-izt}\rho_{\mathbf{k}}(\alpha|\ldots;0)$$

$$+\frac{1}{2\pi}\int_C dz\,\mathrm{e}^{-izt}(-e^2)\sum_j F_{\alpha j}(\mathbf{k})\left\{\frac{1}{i(\mathbf{k}\cdot\mathbf{v}_j - z)}[\rho_{\mathbf{k}}(j|\alpha\ldots;0)+\ldots]\right\}$$

$$+\frac{1}{2\pi}\int_C dz\,\mathrm{e}^{-izt}(-e^2)\sum_j\sum_{\mathbf{k}'} D_{\alpha j}(\mathbf{k}\mathbf{k}')\left\{\frac{1}{i(\mathbf{k}'\cdot\mathbf{v}_\alpha + (\mathbf{k}-\mathbf{k}')\cdot\mathbf{v}_j - z)}\right.$$

$$\left.\times\,[\rho_{\mathbf{k}'}(\alpha|\ldots;0)\rho_{\mathbf{k}-\mathbf{k}'}(j|\ldots;0)+\ldots]\right\} - i\mathbf{k}\cdot\mathbf{v}_\alpha\rho_{\mathbf{k}}(\alpha|\ldots;t) \quad (11.4)$$

The first term on the r.h.s. is zero, as can be seen by applying the Cauchy theorem to the z-integral. In the second term, the operator $-e^2\sum_j F_{\alpha j}(\mathbf{k})$ can be drawn in front of the integral (it does not depend on z). The expression remaining to its right is then exactly $\rho_{\mathbf{k}}(j|\alpha\ldots;t)$. In the same way, in the third term $-e^2\sum_j\sum_{\mathbf{k}'} D_{\alpha j}(\mathbf{k}\mathbf{k}')$ can be written in front of the integral, and this operator acts on $\rho_{\mathbf{k}'}(a|\ldots;t)\,\rho_{\mathbf{k}-\mathbf{k}'}(j|\ldots;t)$. Substitute now eq. (11.2) into (11.4), integrate over all velocities except $\mathbf{v}_\alpha$, and take the limit $N\to\infty$, $\Omega\to\infty$, $N/\Omega = c$ [use eq. (2.11)]. The result is:

$$\partial_t\rho_{\mathbf{k}}(\alpha;t) + i\mathbf{k}\cdot\mathbf{v}_\alpha\rho_{\mathbf{k}}(\alpha;t) = 8\pi^3 e^2 c m^{-1} V_k i\mathbf{k}\cdot\partial_\alpha\int d\mathbf{v}_j\rho_{\mathbf{k}}(\mathbf{v}_j|\mathbf{v}_\alpha;t)$$

$$+8\pi^3 e^2 c m^{-1}\partial_\alpha\cdot\int d\mathbf{k}'\int d\mathbf{v}_j i(\mathbf{k}-\mathbf{k}')V_{|\mathbf{k}-\mathbf{k}'|}\rho_{\mathbf{k}'}(\mathbf{v}_\alpha;t)\rho_{\mathbf{k}-\mathbf{k}'}(\mathbf{v}_j;t) \quad (11.5)$$

Applying formula (2.24) to the first term of the r.h.s., we can write $\rho_{\mathbf{k}}(\mathbf{v}_j|\mathbf{v}_\alpha) = \rho_{\mathbf{k}}(\mathbf{v}_j)\varphi(\mathbf{v}_\alpha)$. The last step will be to transform this equation (combined with $\partial_t\varphi(\alpha) = 0$) to phase space, by multiplying both sides with $(8\pi^3)^{-1}\exp(i\mathbf{k}\cdot\mathbf{x}_\alpha)$ and integrating over $\mathbf{k}$.

Let us perform in detail the transformation of the r.h.s. Let us call $\mathbf{F}_\mathbf{k} \equiv i\mathbf{k}V_k$; it is the Fourier transform of the force $\nabla_\alpha V(|\mathbf{x}_\alpha - \mathbf{x}_j|)$. The first term is then:

$$e^2 c\partial_\alpha \cdot \left\{\varphi(\alpha)\int d\mathbf{k}\,\mathbf{F}_\mathbf{k} e^{i\mathbf{k}\cdot\mathbf{x}_\alpha}\int d\mathbf{v}_j \rho_\mathbf{k}(j)\right\}$$

$$= e^2 c[\partial_\alpha\varphi(\alpha)]\cdot\int d\mathbf{k}\int d\mathbf{k}'\int d\mathbf{v}_j \mathbf{F}_\mathbf{k}\rho_{\mathbf{k}'}(j)e^{i\mathbf{k}\cdot\mathbf{x}_\alpha}\delta(\mathbf{k}-\mathbf{k}')$$

$$= e^2 c[\partial_\alpha\varphi(\alpha)]\cdot\int d\mathbf{k}\int d\mathbf{k}'\int d\mathbf{v}_j\int d\mathbf{x}_j \mathbf{F}_\mathbf{k}\rho_{\mathbf{k}'}(j)e^{i\mathbf{k}\cdot\mathbf{x}_\alpha}e^{i(\mathbf{k}'-\mathbf{k})\cdot\mathbf{x}_j}$$

$$= e^2 c[\partial_\alpha\varphi(\alpha)]\cdot\int d\mathbf{x}_j d\mathbf{v}_j\left\{\int d\mathbf{k}\,\mathbf{F}_\mathbf{k} e^{i\mathbf{k}\cdot(\mathbf{x}_\alpha-\mathbf{x}_j)}\right\}\left\{\int d\mathbf{k}'\rho_{\mathbf{k}'}(j)e^{i\mathbf{k}'\cdot\mathbf{x}_j}\right\}$$

$$= e^2[\partial_\alpha\varphi(\alpha)]\cdot\int d\mathbf{x}_j d\mathbf{v}_j[\nabla_\alpha V(|\mathbf{x}_\alpha-\mathbf{x}_j|)][f(\mathbf{x}_j,\mathbf{v}_j)-c\varphi(\mathbf{v}_j)]$$

The last step made use of eq. (2.12). The second term of the r.h.s. of eq. (11.5) is transformed in a similar way; the result is the same as above but for the replacement of $c[\partial_\alpha\varphi(\alpha)]$ by $\partial_\alpha[f(\alpha)-c\varphi(\alpha)]$. The final result of the transformation of eq. (11.5) is:

$$\partial_t f(\alpha)+\mathbf{v}_\alpha\cdot\nabla_\alpha f(\alpha) = (e^2/m)[\partial_\alpha f(\alpha)]\cdot\nabla_\alpha\int d\mathbf{x}_j d\mathbf{v}_j V(|\mathbf{x}_\alpha-\mathbf{x}|)[f(j)-c\varphi(j)] \quad (11.6)$$

This equation was first proposed by Vlassov * and is the basis of a large amount of the existing work on microscopic plasma physics. Before discussing it further, we shall make the following remarks.

Suppose that at the initial time $\rho_\mathbf{k}(\alpha; 0)$ is very small, so that products of many factors $\rho_\mathbf{k}$ may be neglected; suppose moreover that it remains small for all further times. Then the only type of diagram to be retained is one built up only with loops F (see Fig. 10.1a). The calculations of this section are then con-

* A. A. Vlassov, *J. Exptl. Theoret. Phys. U.S.S.R.*, **8**, 291 (1938).

siderably simplified.* As a result, the last term on the r.h.s. of (11.5) is neglected, and the equation in phase space becomes:

$$\partial_t f(\alpha) + \mathbf{v}_\alpha \cdot \nabla_\alpha f(\alpha) = (e^2 c/m)[\partial_\alpha \varphi(\alpha)] \cdot \nabla_\alpha \int d\mathbf{x}_j d\mathbf{v}_j V(|\mathbf{x}_\alpha - \mathbf{x}_j|)[f(j) - c\varphi(j)] \quad (11.7)$$

Equation (11.7) is called the *linearized Vlassov equation*. Being linear, it is much simpler to handle than the complete equation (11.6), and most actual calculations are done with this simplified equation.

There is no fundamental difficulty in extending the previous calculation to an s-component plasma. One then finds a set of s equations for the s distribution functions describing such a system.

$$\partial_t f^{(\mu)}(\alpha) + \mathbf{v}_\alpha \cdot \nabla_\alpha f^{(\mu)}(\alpha) = (e^2/m_\mu)[\partial_\alpha f^{(\mu)}(\alpha)] \cdot \nabla_\alpha \sum_{\nu=1}^{s} z_\mu z_\nu \int d\mathbf{x}_j d\mathbf{v}_j V^{(\mu\nu)}(|\mathbf{x}_\alpha - \mathbf{x}_j|)[f^{(\nu)}(j) - c_\nu \varphi^{(\nu)}(j)] \quad (11.8)$$

§ 12. Some General Properties of the Vlassov Equation

Let us now investigate several properties of the Vlassov equation. We first show that it can be written in a simple and suggestive form. Consider the integral on the r.h.s. of eq. (11.6): it is a function of $\mathbf{x}_\alpha$.

Let us introduce the notation:

$$\Phi(\mathbf{x}_\alpha; t) = e \int d\mathbf{x}_j V(|\mathbf{x}_\alpha - \mathbf{x}_j|)[n(\mathbf{x}_j; t) - c] \quad (12.1)$$

where $n(\mathbf{x}; t)$ is the local density defined by eq. (1.14). This quantity Φ has the dimensions of an electrical potential. It has a simple molecular meaning: it is $(1/e)$ times the average molecular interaction energy, weighted by the local density excess of the plasma. Keeping in mind that in our model a homogeneous electron gas is electrically neutral at every point (because the charge of the electrons is canceled by the continuous background), the weighting factor in (12.1) is really minus the local electrical charge density, defined as

* See § 13.

$$\sigma(\mathbf{x}_\alpha; t) = -e[n(\mathbf{x}_\alpha; t) - c] \tag{12.2}$$

Formula (12.1) can therefore be rewritten as:

$$\Phi(\mathbf{x}_\alpha; t) = -\int d\mathbf{x}' \frac{1}{|\mathbf{x}_\alpha - \mathbf{x}'|} \sigma(\mathbf{x}'; t) \tag{12.3}$$

Written in this form, one immediately recognizes the classical expression of the potential at a point $\mathbf{x}_\alpha$ due to a charge distribution $\sigma(\mathbf{x}; t)$ (see any textbook on electricity). It should be noted at this point that the value of the potential Φ at time t is determined by the value of the charge distribution at the *same* time. There is no retardation effect here (in the electrodynamical sense), because we have assumed from the beginning that the molecular interactions are instantaneous.

Keeping in mind that the Coulomb potential $r_{\alpha j}^{-1}$ is a harmonic function, we may take the laplacian of eq. (12.3)

$$\begin{aligned} -\nabla_\alpha^2 \int d\mathbf{x}' \frac{1}{|\mathbf{x}_\alpha - \mathbf{x}'|} \sigma(\mathbf{x}'; t) &= -4\pi \int d\mathbf{x}' \delta(\mathbf{x}_\alpha - \mathbf{x}') \sigma(\mathbf{x}'; t) \\ &= -4\pi\sigma(\mathbf{x}_\alpha; t) \end{aligned}$$

Equation (11.6) is therefore equivalent to the system of equations

$$\partial_t f(\alpha) + \mathbf{v}_\alpha \cdot \nabla_\alpha f(\alpha) - (e/m)[\nabla_\alpha \Phi(\mathbf{x}_\alpha; t)] \cdot \partial_\alpha f(\alpha) = 0 \tag{12.4}$$

$$\nabla_\alpha^2 \Phi(\mathbf{x}_\alpha; t) = -4\pi\sigma(\mathbf{x}_\alpha; t) \tag{12.5}$$

Equation (12.4) has the same form as the Liouville equation for a single charged particle moving in a space- and time-dependent field $\mathbf{E} = -\nabla_\alpha \Phi(\alpha)$. This field is determined by the phenomenological Poisson equation, i.e. by the same equation which determines the field due to a continuous macroscopic charge distribution. This charge distribution itself is determined by the distribution function of the particles. The Vlassov equation is thus of the form of a *self-consistent field equation*: the change in time of the distribution function is determined by the field $\mathbf{E}$, and the latter in turn changes in the course of time because f changes. This equation is quite analogous to the well known Hartree equation of quantum mechanics. In this perspective the plasma can be viewed as a continuous medium to which the ordinary laws of macroscopic electrodynamics can be applied. This was the basic idea which guided Vlassov when he wrote his equation. It should be noted that this starting point is the same as in the Debye–Hückel theory which was so successful in interpreting the properties of electrolytes (see Chapter 12). In the latter theory the thermodynamic properties are calculated by introducing an average electric potential, which is again determined by a Poisson equation relating the charge density to the (pair) distribution function.

The fact that a plasma can behave under certain conditions as a continuous medium is quite remarkable from the molecular point of view. It is closely related to the long range of the Coulomb force, which in turn introduces collective interactions involving many particles at the same time; this therefore creates a quite characteristic cohesion of the particles, which are able to respond in phase to external perturbations. The collective behavior tends, however, to be opposed by the free thermal motion of the particles: this effect enters the Vlassov equation through the term $\mathbf{v}_\alpha \cdot \nabla_\alpha f(\alpha)$. It will be seen in Chapter 3 how this opposition determines the overall behavior of the plasma.

The rigorous molecular foundation of the Vlassov equation was first investigated by Bogoliubov,* using his theory of irreversible processes. A derivation along the same lines as his has been given recently by Guernsey.** Although the formal part of their theory uses ideas rather similar to those of our derivation, it does not show very clearly the physical basis of the equation. Rostoker and Rosenbluth† have derived the Vlassov equation by taking the so-called "fluid-limit" ($e \to 0$, $m \to 0$, $c \to \infty$ in such a way that $(e/m) =$ const, $ec =$ const.) of the Liouville equation [or the Y–B–G–B–K equation (4.1)]. This, however, does not constitute in our opinion a molecular proof. It is rather obvious that in this mathematical limit, the plasma becomes a continuous medium and that phenomenological

* N. N. Bogoliubov, *Problems of a Dynamical Theory in Statistical Physics*, Moscow, 1946; translated by E. K. Gora, in *Studies in Statistical Mechanics*, vol. 1, edited J. de Boer and G. E. Uhlenbeck, Interscience, New York, 1962.

** R. L. Guernsey, *The Kinetic Theory of Fully Ionized Gases*, Off. Nav. Res. Contract No. Nonr. 1224 (15), 1960.

† N. Rostoker and M. N. Rosenbluth, *Phys. of Fluids*, **3**, 1 (1960).

electrodynamics should apply. The problem is to show *why* under certain conditions a *real* plasma behaves like a medium.

The derivation presented here has the advantage of showing what assumptions are necessary for the validity of the Vlassov equation. Let us discuss them again very briefly.

The main assumption is the limitation of our study to *times which are short compared to the relaxation time* t_r.

This assumption has led to the choice of diagrams proportional to $(e^2c)^n$. In contrast to the case of ordinary gases, the short-time behavior is represented already in zeroth order by an infinite sum of diagrams. This remarkable fact is a consequence of the fact that the "duration of the collision" t_p depends on e^2, which in turn is due to the long range of the Coulomb force. This choice of diagrams is the feature which introduces the collective effects, can be seen very easily from the diagrams. Each e^2c vertex introduces a new particle. A diagram with n e^2c vertices therefore involves $n + 1$ particles. The behavior of the plasma in the Vlassov approximation is therefore determined by contributions involving an arbitrary number of particles.

CHAPTER 3

The Theory of Plasma Oscillations

§ 13. The Solution of the Initial-Value Problem for the Linearized Vlassov Equation

We now show that it is possible to obtain an exact closed solution to the initial-value problem of the *linearized* Vlassov equation. We specify the problem as follows:

To find a solution of the equation (11.7) *which reduces at time* 0 *to the given function*:

$$f(\mathbf{x}_\alpha, \mathbf{v}_\alpha; 0) = c\left\{\varphi(\mathbf{v}_\alpha) + \int d\mathbf{k}\, q_{\mathbf{k}}(\mathbf{v}_\alpha)e^{-i\mathbf{k}\cdot\mathbf{x}_\alpha}\right\} \tag{13.1}$$

We already know that, in the Vlassov approximation, $\varphi(\mathbf{v}_\alpha; t) = \varphi(\mathbf{v}_\alpha; 0)$. Therefore the only non-trivial problem is to find the Fourier component $\rho_{\mathbf{k}}(\mathbf{v}_\alpha; t)$. But we have already solved the Liouville equation retaining only contributions of order $(e^2c)^n$, and we have derived the Vlassov equation by time-differentiating this approximate solution of the Liouville equation. As already stated, in the case of the linearized Vlassov equation, the series (11.2) is considerably simplified, and can actually be summed in closed form. In this simple case we must suppress all matrix elements $D_{jn}(\mathbf{kk}')$. We also integrate the terms of this series over all velocities except $\mathbf{v}_\alpha$, obtaining:

$$\begin{aligned}\rho_{\mathbf{k}}(\mathbf{v}_\alpha; t) = \frac{1}{2\pi}\int_C dz\, e^{-izt} \frac{1}{i(\mathbf{k}\cdot\mathbf{v}_\alpha - z)}\Big\{\rho_{\mathbf{k}}(\mathbf{v}_\alpha; 0) + (-e^2N)\int d\mathbf{v}_j \\ \cdot F_{\alpha j}(\mathbf{k})\frac{1}{i(\mathbf{k}\cdot\mathbf{v}_j - z)}\rho_{\mathbf{k}}(\mathbf{v}_j|\mathbf{v}_\alpha; 0) + (-e^2N)\int d\mathbf{v}_j \\ \cdot F_{\alpha j}(\mathbf{k})\frac{1}{i(\mathbf{k}\cdot\mathbf{v}_j - z)}(-e^2N)\int d\mathbf{v}_n \\ \cdot F_{jn}(\mathbf{k})\frac{1}{i(\mathbf{k}\cdot\mathbf{v}_n - z)}\rho_{\mathbf{k}}(\mathbf{v}_n|\mathbf{v}_j, \mathbf{v}_\alpha; 0) + \ldots\Big\}\end{aligned} \tag{13.2}$$

In the various terms of (13.2) we can apply formula (2.24) to obtain

$$\rho_{\mathbf{k}}(\mathbf{v}_r|\mathbf{v}_1 \ldots \mathbf{v}_{r-1}; 0) = \rho_{\mathbf{k}}(\mathbf{v}_r; 0) \prod_{j=1}^{r-1} \varphi(\mathbf{v}_j) = q_{\mathbf{k}}(\mathbf{v}_r) \prod_{j=1}^{r-1} \varphi(j) \quad (13.3)$$

We rewrite formula (13.2), using (13.3) and eq. (11.2):

$$\rho_{\mathbf{k}}(\alpha; t) = \frac{1}{2\pi} \int_C dz e^{-izt} \frac{1}{i(\mathbf{k} \cdot \mathbf{v}_\alpha - z)} \Big\{ q_{\mathbf{k}}(\mathbf{v}_\alpha)$$

$$+ie^2 c m^{-1} 8\pi^3 V_k \mathbf{k} \cdot \partial_\alpha \varphi(\alpha) \int d\mathbf{v}_j \frac{q_{\mathbf{k}}(\mathbf{v}_j)}{i(\mathbf{k} \cdot \mathbf{v}_j - z)}$$

$$+ie^2 c m^{-1} 8\pi^3 V_k \mathbf{k} \cdot \partial_\alpha \varphi(\alpha) \int d\mathbf{v}_j \frac{1}{i(\mathbf{k} \cdot \mathbf{v}_j - z)} ie^2 c m^{-1} 8\pi^3 V_k \mathbf{k} \cdot \partial_j \varphi(j)$$

$$\cdot \int d\mathbf{v}_n \frac{q_{\mathbf{k}}(\mathbf{v}_n)}{i(\mathbf{k} \cdot \mathbf{v}_n - z)} + ie^2 c m^{-1} 8\pi^3 V_k \mathbf{k} \cdot \partial_\alpha \varphi(\alpha) \quad (13.4)$$

$$\cdot \int d\mathbf{v}_j \frac{1}{i(\mathbf{k} \cdot \mathbf{v}_j - z)} ie^2 c m^{-1} 8\pi^3 V_k \mathbf{k} \cdot \partial_j \varphi(j)$$

$$\cdot \int d\mathbf{v}_n \frac{1}{i(\mathbf{k} \cdot \mathbf{v}_n - z)} ie^2 c m^{-1} 8\pi^3 V_k \mathbf{k} \cdot \partial_n \varphi(n) \int d\mathbf{v}_m \frac{1}{i(\mathbf{k} \cdot \mathbf{v}_m - z)} q_{\mathbf{k}}(\mathbf{v}_m)$$

$$+ \ldots \Big\}$$

From these first four terms the structure of the entire series can already be seen. We see in particular the appearance of a repeating unit:

$$J_{\mathbf{k}}(z) = 8\pi^3 e^2 c m^{-1} V_k \int d\mathbf{v}_j \frac{1}{\mathbf{k} \cdot \mathbf{v}_j - z} \mathbf{k} \cdot \partial_j \varphi(j) \quad (13.5)$$

For reasons of clarity we rewrite the series (13.4) once more.

$$\rho_{\mathbf{k}}(\alpha; t) = \frac{1}{2\pi i} \int_C dz e^{-izt} \Big\{ \frac{1}{\mathbf{k} \cdot \mathbf{v}_\alpha - z} q_{\mathbf{k}}(\mathbf{v}_\alpha) \quad (13.6)$$

$$+ \frac{e^2 c m^{-1}}{\mathbf{k} \cdot \mathbf{v}_\alpha - z} 8\pi^3 V_k \mathbf{k} \cdot \partial_\alpha \varphi(\alpha) [1 + J_{\mathbf{k}}(z) + J_{\mathbf{k}}^2(z) + \ldots] \int d\mathbf{v}_1 \frac{q_{\mathbf{k}}(\mathbf{v}_1)}{\mathbf{k} \cdot \mathbf{v}_1 - z} \Big\}$$

The expression enclosed in square brackets is a simple geometric series, which can be summed exactly:

$$\rho_{\mathbf{k}}(\alpha; t) = \frac{1}{2\pi i}\int_C dz\, \mathrm{e}^{-izt}\left\{\frac{1}{\mathbf{k}\cdot\mathbf{v}_\alpha - z}\, q_{\mathbf{k}}(\mathbf{v}_\alpha)\right. \tag{13.7}$$

$$\left. + 8\pi^3 e^2 c\, m^{-1} V_k \mathbf{k}\cdot\partial_\alpha \varphi(\alpha)\,\frac{1}{\mathbf{k}\cdot\mathbf{v}_\alpha - z}\,\frac{1}{1 - J_{\mathbf{k}}(z)}\int d\mathbf{v}_1 \frac{1}{\mathbf{k}\cdot\mathbf{v}_1 - z}\, q_{\mathbf{k}}(\mathbf{v}_1)\right\}$$

This formula solves the problem, as it represents a closed summation of the series solution of the linearized Vlassov equation. It should be remarked that the summation can be done only if $|J_{\mathbf{k}}(z)| < 1$; but formula (13.7) is valid for all values of J, by the principle of analytical continuation. We now rewrite (13.7) explicitly, using for V_k the Fourier transform of the Coulomb potential (9.8)

$$\rho_{\mathbf{k}}(\mathbf{v}_\alpha; t) = (2\pi)^{-1}\int_C dz\, \mathrm{e}^{-izt}\rho_{\mathbf{k}}(\mathbf{v}_\alpha; z) \tag{13.8}$$

$$\rho_{\mathbf{k}}(\mathbf{v}_\alpha; z) = \frac{q_{\mathbf{k}}(\mathbf{v}_\alpha)}{i(\mathbf{k}\cdot\mathbf{v}_\alpha - z)} + \frac{\omega_p^2}{k^2}\mathbf{k}\cdot\partial_\alpha\varphi(\mathbf{v}_\alpha)\,\frac{1}{\mathbf{k}\cdot\mathbf{v}_\alpha - z}\,\frac{1}{\varepsilon_+(k; z)} \times \int d\mathbf{v}_1 \frac{q_{\mathbf{k}}(\mathbf{v}_1)}{i(\mathbf{k}\cdot\mathbf{v}_1 - z)} \tag{13.9}$$

where

$$\varepsilon_+(k; z) = 1 - \frac{\omega_p^2}{k^2}\int d\mathbf{v}_2 \frac{\mathbf{k}\cdot\partial_2\varphi(\mathbf{v}_2)}{\mathbf{k}\cdot\mathbf{v}_2 - z} \tag{13.10}$$

The parameter ω_p appearing in this formula is defined by

$$\omega_p^2 = 4\pi e^2 c/m \tag{13.11}$$

ω_p, which has the dimensions of a frequency, is called the *plasma frequency*, for reasons which will become clear in the next sections. It is seen to be (apart from a numerical factor) just the inverse of the characteristic time t_p defined by eq. (9.12).

Formula (13.9) is the starting point of the whole theory of the linearized Vlassov equation. It is equivalent to a formula first obtained by Landau * in his basic work on the theory of plasma

* L. Landau, *J. Exptl. Theoret. Phys. U.S.S.R.*, **16**, 574 (1946); translated in *J. Phys. U.S.S.R.*, **10**, 25 (1946).

oscillations. He obtained the result by solving the linearized Vlassov equation (11.7) by a Laplace transform, which is a more compact method. However, we prefer to use here the perturbation method through the resolvent formalism, and this is not only to preserve unity in this book. Indeed, this method brings out clearly the collective origin of the denominator appearing in the second term of the r.h.s. of eq. (13.9), which will be seen to play a crucial role in the statistical mechanics of plasmas.

§ 14. Individual Particle Behavior and Collective Behavior

In the theory of plasma oscillations the main interest lies in a function which is simpler than the inhomogeneity factor $\rho_{\mathbf{k}}(\alpha; t)$. This function is the *local density excess* $h_{\mathbf{k}}(t)$ defined in Table 2.2. This quantity is directly proportional to the local charge density and can thus be measured experimentally. Its Laplace transform $h_{\mathbf{k}}(z)$ is obtained from (13.9) by an integration over the velocity $\mathbf{v}_\alpha$:

$$h_{\mathbf{k}}(z) = \int d\mathbf{v}_\alpha \frac{q_{\mathbf{k}}(\mathbf{v}_\alpha)}{i(\mathbf{k}\cdot\mathbf{v}_\alpha - z)} + \frac{\omega_p^2}{k^2}\int d\mathbf{v}_\alpha \frac{\mathbf{k}\cdot\partial_\alpha\varphi(\alpha)}{\mathbf{k}\cdot\mathbf{v}_\alpha - z}\,\frac{1}{\varepsilon_+(k; z)}\int d\mathbf{v}_1 \frac{q_{\mathbf{k}}(\mathbf{v}_1)}{i(\mathbf{k}\cdot\mathbf{v}_1 - z)}$$

Using the definition (13.10) of $\varepsilon_+(k; z)$ we are left with the very compact formula:

$$h_{\mathbf{k}}(z) = \frac{1}{\varepsilon_+(k; z)}\int d\mathbf{v} \frac{q_{\mathbf{k}}(\mathbf{v})}{i(\mathbf{k}\cdot\mathbf{v} - z)} \tag{14.1}$$

The density excess itself is obtained from (14.1) by an inverse Laplace transformation

$$h_{\mathbf{k}}(t) = (2\pi)^{-1}\int_C dz\, e^{-izt} h_{\mathbf{k}}(z) \tag{14.2}$$

The integral can be evaluated by deforming the contour C and pulling it toward infinity in the direction of the negative imaginary axis, as explained in Appendix 1. The value of the integral depends then on the behavior of the analytical continuation of $h_{\mathbf{k}}(z)$ into the half-plane below the original contour C. We assume for simplicity that $h_{\mathbf{k}}(z)$ has only simple poles in this region. (Other types of singularities will be met later in this book, e.g. in Chapter 6.) Then $h_{\mathbf{k}}(t)$ can be written as follows (see Appendix 1):

$$h_{\mathbf{k}}(t) = -i \sum_j e^{-iz_j t}[\text{Res}\, h_{\mathbf{k}}(z)]_{z=z_j} \tag{14.3}$$

Hence, *the location of the poles fixes the (complex) frequencies associated with the various modes of time variation. The amplitudes of the corresponding modes are given by the residues at these poles.* In particular, we may find three types of behavior according to the location of the poles: *steady oscillations* (z_j on the real axis), *damped oscillations* ($z_j \in S_-$) and *exponentially growing oscillations* ($z_j \in S_+$).

In the present section it will be shown that a number of important properties follow merely from the form of $h_{\mathbf{k}}(z)$, eq. (14.1), without any special assumption on $\varphi(\alpha)$ or on $q_{\mathbf{k}}(\alpha)$. The function $h_{\mathbf{k}}(z)$ appears as a product of two factors: $[\varepsilon_+(k; z)]^{-1}$ and an integral involving the initial condition $q_{\mathbf{k}}(\alpha)$. This binary splitting reflects a fundamental aspect of the plasma behavior, which appears as a "leitmotiv" throughout this book.

Consider first the integral factor, which we write as

$$h^0_{\mathbf{k}}(z) = \int d\mathbf{v} \frac{q_{\mathbf{k}}(\mathbf{v})}{i(\mathbf{k}\cdot\mathbf{v} - z)} \tag{14.4}$$

We first note that the denominator $(\mathbf{k}\cdot\mathbf{v} - z)$ depends only on one of the three components of the vector $\mathbf{v}$. The integral over the two components of $\mathbf{v}$ perpendicular to $\mathbf{k}$ affects only the factor $q_{\mathbf{k}}(\mathbf{v})$. We introduce at this point an operation which will play an important role in this book, and which we call the "*barring*" *operation*. Let us call ν the component of $\mathbf{v}$ parallel to $\mathbf{k}$

$$\nu = \mathbf{k}\cdot\mathbf{v}/k \tag{14.5}$$

We associate with any function of the vector $\mathbf{v}$, $f(\mathbf{v}) \equiv f(\mathbf{v}_\perp, \nu)$, a barred function $\bar{f}(\nu)$ which depends on the single scalar variable ν:

$$\bar{f}(\nu) = \int d\mathbf{v}_\perp f(\mathbf{v}_\perp, \nu) \equiv \int d\mathbf{v}\, \delta(\nu - \mathbf{k}\cdot\mathbf{v}/k) f(\mathbf{v}) \tag{14.6}$$

In terms of the barred function $\bar{q}_{\mathbf{k}}(\nu)$, eq. (14.4) becomes

$$h^0_{\mathbf{k}}(z) = \int_{-\infty}^{\infty} d\nu \frac{\bar{q}_{\mathbf{k}}(\nu)}{i(k\nu - z)}; \qquad z \in S_+ \tag{14.7}$$

We note that if the interactions are switched off, or equivalently if the charge e is set equal to zero, $\varepsilon_+(k; z) \to 1$ [see eq. (13.10)]. Hence $h_{\mathbf{k}}^0(z)$ is exactly the (F–L)excess density produced by an initial disturbance $q_{\mathbf{k}}(\mathbf{v})$ *in a gas of non-interacting particles*. Each pole of this function contributes to $h_{\mathbf{k}}(t)$ (14.3) a term which varies with a complex frequency determined only by the initial inhomogeneity factor. Physically, the initial disturbance is carried away bodily by the particles in their free flight. In general, the various particles have different velocities. Hence one expects that the initial perturbation will be spatially dislocated and that the system will eventually reach a homogeneous state. We now show that this follows mathematically from (14.7).

In order to discuss the analytical behavior of $h_{\mathbf{k}}^0(z)$, we note that this function is expressed by (14.7) as a *Cauchy integral*. We thus use for the discussion the theory of this integral, which is given in some detail in Appendix 2. We must keep in mind that eq. (14.1) has been derived under the assumption that z lies in the upper half-plane S_+. More exactly, we need the function $h_{\mathbf{k}}(z)$ for values of z lying on the contour C of the inverse Laplace transformation (Fig. 4.1) which is originally situated in the upper half-plane. We know from the general theory that the integral of eq. (14.7) defines a function $h_{\mathbf{k}}^0(z)$ which is regular in the whole upper half-plane. However, in obtaining formula (14.3) we have deformed the contour C by pulling it into the lower half-plane. This process is legitimate only if the integrand $h_{\mathbf{k}}^0(z)$ is analytical in the region swept out by the contour during its deformation. It has been shown in Appendix 2 that the Cauchy integral in eq. (14.7) defines, for $z \in S_-$, a function which is *not* the analytical continuation of $h_{\mathbf{k}}^0(z)$ into S_-. In calculating $h_{\mathbf{k}}(t)$ by eq. (14.3) one must take in the right-hand side the residues of the proper analytical continuation of $h_{\mathbf{k}}^0(z)$ into S_-. If the function $\bar{q}_{\mathbf{k}}(\nu)$ itself possesses such a continuation, it is shown in eq. (A2.1.12) that the integral must be taken along the contour Γ shown in Fig. 14.1, or equivalently:

$$h_{\mathbf{k}}^0(z) = \frac{1}{ik}\int_{-\infty}^{\infty} d\nu \frac{\bar{q}_{\mathbf{k}}(\nu)}{\nu - z/k} + \frac{2\pi}{k}\bar{q}_{\mathbf{k}}(z/k); \qquad z \in S_- \qquad (14.8)$$

The function $h_{\mathbf{k}}^{0}(z)$ is known to be regular in S_{+}. If it had no singularities in S_{-} or on the real axis, it would be a constant (as follows from a well known theorem in analysis); as it vanishes at infinity, it would then vanish identically everywhere. If this

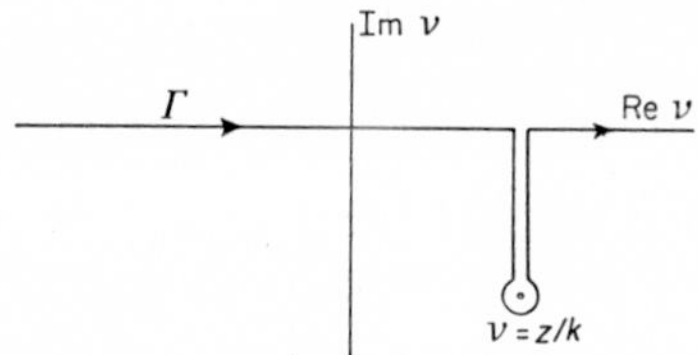

Fig. 14.1. Contour of integration Γ for the analytical continuation of the Cauchy integral (14.7).

trivial case is excluded, it follows that $h_{\mathbf{k}}^{0}(z)$ *necessarily has singularities in the lower half-plane* (*or on the real axis*). This is the main result needed here. As expected, the factor $h_{\mathbf{k}}^{0}(z)$ contributes *damped oscillations* to $h_{\mathbf{k}}(t)$. We are here in the presence of an irreversible homogenization process in which the interactions play no role. As explained above, it is produced by a dislocation of the initial perturbation carried away by particles with different velocities. For this reason it will be called an *individual particle behavior* of the plasma. The location of the poles in S_{-}, and hence the rates of damping, depend only on the initial inhomogeneity factor, $q_{\mathbf{k}}(\mathbf{v})$. These rates are therefore not a characteristic property of the plasma.

We now turn to the second factor, ε_{+}^{-1}, eq. (14.1). This factor can also have poles, located at the zeros of the function $\varepsilon_{+}(k; z)$. The poles $\bar{z}_j$ are thus solutions of the transcendental equation:

$$\varepsilon_{+}(k; \bar{z}_j) \equiv 1 - \frac{\omega_p^2}{k^2} \int_{\Gamma} d\mathbf{v} \frac{\mathbf{k} \cdot \partial \varphi(\mathbf{v})}{\mathbf{k} \cdot \mathbf{v} - \bar{z}_j} = 0 \qquad (14.9)$$

This equation is called the *dispersion equation*. It gives the frequencies $\bar{z}_j$ as functions of k. These frequencies are moreover functionals of the velocity distribution $\varphi(\mathbf{v})$, which is a constant of the motion in the Vlassov (short-time) approximation. The functions $\bar{z}_j$ are therefore *natural or intrinsic frequencies* independent of the type of initial perturbation. The physical mechanism producing this motion is much more complex than the free motion mechanism described above. The interactions play an essential role here ($\bar{z}_j$ is a function of e^2 through ω_p^2). This can be seen mathematically from the fact that an infinite perturbation series had to be summed in order to obtain the denominator $\varepsilon_+(k; z)$.

A characteristic difficulty arises from the fact that, in contrast to the previous type of poles, *the zeros of* $\varepsilon_+(k; z)$ *are not restricted to a definite region of the complex plane.* They may lie in S_-, on the real axis or even in S_+: their location is determined entirely by the shape of the function $\varphi(\mathbf{v})$. The question of finding criteria relating the properties of the velocity distribution to the stability of the plasma against initial disturbances is therefore a crucial one. It should be stressed again that the stability of the plasma depends *only* on $\varphi(\mathbf{v})$: if a plasma is stable toward one type of disturbance, it is stable against any disturbance.

In this section two possible types of complex frequencies, i.e. two types of response to an initial disturbance, have been singled out:

The complex frequencies which are functionals of the initial disturbance and which describe transient oscillations, damped by the free motion of the particles.

The natural frequencies of the plasma, which depend only on the unperturbed state of the plasma and which may represent damped, steady or growing oscillations.

These various types of frequencies will be analyzed in more detail in the next sections.

§ 15. Individual Particle Behavior

The two types of time dependence isolated in the last paragraph give additive contributions to the Fourier components of the distribution function. It has already been pointed out in the previous paragraph that the individual particle contribution to the local density excess $h_{\mathbf{k}}(t)$ will generally lead to a damped oscillation. We shall analyze this behavior in more detail by taking a specific example. Consider an initial disturbance of the following type:

$$q_{\mathbf{k}}(\mathbf{v}) = \frac{f(\mathbf{k})}{\pi^2 w_0^3} \frac{1}{[1+(\mathbf{v}-\mathbf{v}_0)^2/w_0^2]^2} \tag{15.1}$$

The form of the function $f(\mathbf{k})$ is not specified. Equation (15.1) represents a spherically symmetrical distribution around the average velocity $\mathbf{v}_0$, depending on the parameter w_0 (which characterizes the spread of velocities). The corresponding barred function (see eq. 14.6) is

$$\bar{q}_{\mathbf{k}}(\nu) = \frac{f(\mathbf{k})}{\pi w_0} \frac{1}{1+(\nu-\nu_0)^2/w_0^2} \tag{15.2}$$

This distribution is represented graphically in Fig. 15.1. The parameter w_0 is very convenient, because it permits the shape of the curve to vary from a δ-function-like peak ($w_0 \to 0$) to a flat curve ($w_0 \to \infty$).

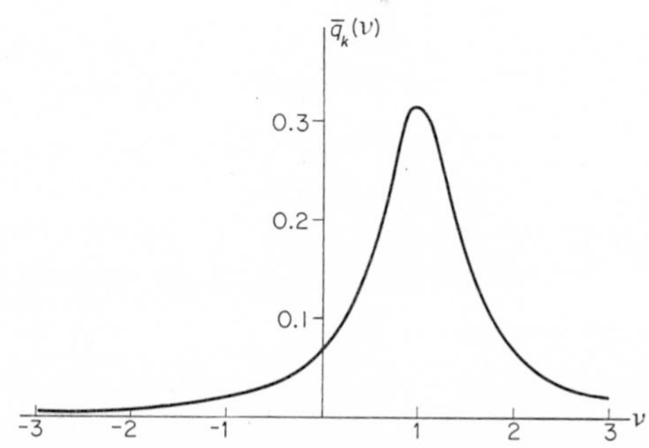

Fig. 15.1. The function $\bar{q}_k(\nu)$.

We now calculate the free-motion contribution to $h_{\mathbf{k}}(t)$ or, more simply, the function $h^0_{\mathbf{k}}(t)$ defined as the inverse Laplace transform of $h^0_{\mathbf{k}}(z)$.* We first calculate the Cauchy integral (14.7):

$$h^{0(+)}_{\mathbf{k}}(z) = i^{-1} \int d\nu \frac{\bar{q}_{\mathbf{k}}(\nu)}{k\nu - z} = \frac{f(\mathbf{k})}{i\pi w_0} \int_{-\infty}^{\infty} d\nu \frac{1}{(k\nu-z)[1+(\nu-\nu_0)^2/w_0^2]} \tag{15.3}$$

* The residue at the "individual particle" pole z_m obtained in this way differs from the exact residue appearing in eq. (14.3) by the factor $[\varepsilon_+(k; z_m)]^{-1}$. But we are interested here mainly in the frequency rather than the amplitude of this mode.

This integral is easily evaluated by completing the real axis with a semi-circle at infinity in the lower half-plane. By definition, z lies in S_+ and therefore outside the contour. Hence the integral equals the residue at the pole $\nu = \nu_0 - iw_0$, leading to the result

$$h_{\mathbf{k}}^{0(+)}(z) = i^{-1} f(\mathbf{k}) \frac{1}{\mathbf{k} \cdot \mathbf{v}_0 - ikw_0 - z} \tag{15.4}$$

This result, derived for $z \in S_+$, is taken as the definition of $h_{\mathbf{k}}^{0(+)}(z)$ for $z \in S_-$, by virtue of the principle of analytical continuation.

It should be noted that the same result (15.4) is obtained from (15.3) directly for $z \in S_-$ by replacing the path of integration (i.e. the real axis) by the contour Γ of Fig. 14.1. If, however, $z \in S_-$ and the integration is carried out along the real axis instead of the contour Γ, one gets an additional contribution from the pole $\nu = z/k$, which is now inside the contour, with the result

$$\begin{aligned} h_{\mathbf{k}}^{0(-)}(z) &= i^{-1} f(\mathbf{k}) \frac{1}{\mathbf{k} \cdot \mathbf{v}_0 + ikw_0 - z} \\ &= h_{\mathbf{k}}^{0(+)}(z) - (2\pi/k)\bar{q}_{\mathbf{k}}(z/k), \qquad z \in S_- \end{aligned}$$

Coming back to the integral (14.2) in which (15.4) is inserted instead of $h_{\mathbf{k}}(z)$, the integration is easily performed by a contour integration, with the result

$$h_{\mathbf{k}}^{0}(t) = f(\mathbf{k}) e^{-i\mathbf{k} \cdot \mathbf{v}_0 t} e^{-kw_0 t} \tag{15.5}$$

As expected, we find here a *damped oscillation*. The initial disturbance propagates with a group velocity $\mathbf{v}_0$ equal to the average velocity of the particles in the initial inhomogeneity. The rate of decay is $\tau_d = (kw_0)^{-1}$. It is thus very short for short wave-lengths (large k) and for initial disturbances with a large spread of velocities (w_0 large). By contrast, for very sharply peaked distributions, the initial disturbance is maintained for a very long time and, in the limit $w_0 \to 0$, in which $q_{\mathbf{k}}(\mathbf{v}) \to f(\mathbf{k})\delta(\mathbf{v} - \mathbf{v}_0)$, it becomes a purely harmonic wave.

It is a very suggestive exercise to perform the same calculation in another way. Instead of performing the ν-integration first and the z-integration after, we shall invert the order of integrations in eq. (14.2)

$$h_{\mathbf{k}}^{0}(t) = \int d\mathbf{v}\, q_{\mathbf{k}}(\mathbf{v}) \left\{ (2\pi i)^{-1} \int dz\, \mathrm{e}^{-izt} (\mathbf{k}\cdot\mathbf{v} - z)^{-1} \right\}$$

$$= \int d\mathbf{v}\, q_{\mathbf{k}}(\mathbf{v}) \mathrm{e}^{-i\mathbf{k}\cdot\mathbf{v}t} \tag{15.6}$$

The subsequent **v**-integration yields, of course, the same result (15.5) as before. Formula (15.6) is, however, very interesting, because it shows explicitly the role of free-particle motion. The overall motion consists of a superposition of waves exp $(-i\mathbf{k}\cdot\mathbf{v}t)$. Each of these partial waves propagates with a group velocity equal to the velocity of the individual particles, **v**, and an amplitude equal to the number of particles of velocity **v** in the initial disturbance. These waves will in general interfere destructively in the course of time, because of the spread of particle velocities. This process leads to the typical free-motion damping discussed above.

§ 16. The "Collective" Behavior

We now investigate the "natural frequencies" of the plasma. This question is an important one and can be studied from many points of view. The next chapter will be devoted to a detailed study of these problems. In this section we illustrate the main types of behavior by means of an example. As we have already stated, these frequencies are characterized by the fact that they are independent of the initial perturbation, but depend crucially on the strength of the interactions and on the properties of the unperturbed plasma.

Let us consider a velocity distribution of the following simple form:

$$\varphi(\mathbf{v}_\alpha) = \frac{1}{\pi^2 u_0^3} \frac{1}{[1 + v_\alpha^2/u_0^2]^2} \tag{16.1}$$

where u_0 is a (scalar) constant, proportional to the root-mean-square velocity. This function has the same general shape as the one represented in Fig. 15.1. The frequencies are given by the roots of eq. (14.9), which is called the "*dispersion equation*".

Substituting (16.1) into (14.9) we obtain the following equation:

$$1 - \frac{1}{\pi^2 u_0^3} \frac{\omega_p^2}{k^2} \int d\mathbf{v} \frac{-4\mathbf{k}\cdot\mathbf{v}/u_0^2}{(1+v^2/u_0^2)^3} \frac{1}{\mathbf{k}\cdot\mathbf{v}-\bar{z}} = 0 \tag{16.2}$$

Integrating over the components of $\mathbf{v}$ perpendicular to $\mathbf{k}$, and calling as usual $\nu = \mathbf{k}\cdot\mathbf{v}/k$, we obtain

$$1 - \frac{2u_0}{\pi} \frac{\omega_p^2}{k^2} \int_{-\infty}^{\infty} d\nu \frac{\nu}{(\nu^2+u_0^2)^2(\nu-\bar{z}/k)} = 0 \tag{16.3}$$

This integral can be evaluated by adding to the real axis a semi-circle in the lower half-plane. $\bar{z}$ is by definition in S_+, so that the integral is equal to $(-2\pi i)$ times the residue at the double pole $\nu = -iu_0$. The evaluation of the integral leads to the result

$$\varepsilon_+(k; \bar{z}_j) \equiv 1 - \frac{\omega_p^2}{k^2} \frac{1}{(iu_0+\bar{z}_j/k)^2} = 0 \tag{16.4}$$

This quadratic equation, derived for $\bar{z}_j \in S_+$, is to hold, by analytical continuation, for $\bar{z}_j \in S_-$ also. The dispersion equation is now readily solved, giving the two complex roots:

$$\bar{z}_{1,2} = \pm\omega_p - iku_0 \tag{16.5}$$

This very simple result represents two waves running in opposite directions, and damped with the damping constant ku_0. The real part of the frequency is here exactly equal to the constant ω_p, which is called the *plasma frequency*. In more general cases, there exists a dispersion of the frequencies (i.e. Re $\bar{z}_j$ is a function of $\mathbf{k}$), but it will be seen in the next chapter that in the limit of long wave-lengths, $\mathbf{k} \to 0$, Re $\bar{z}_j \to \omega_p$. This is the main reason for the importance of the parameter ω_p, which appears as a natural time scale in all problems involving plasmas. Let us recall the definition:

$$\omega_p = (4\pi e^2 c/m)^{\frac{1}{2}} \tag{16.6}$$

It is seen that the frequency is proportional to the charge e, which decisively shows the role of the interactions in producing collective effects. It is also proportional to $c^{\frac{1}{2}}$ and to $m^{-\frac{1}{2}}$ (which is an effect of inertia).

Let us now consider the damping constant. We notice first that the imaginary part of $\bar{z}_j$ is negative, whatever the value of k. There are no growing waves for the unperturbed distribution considered here: the distribution $\varphi(\mathbf{v})$ is called *stable*. The rate of damping is proportional to the wave-vector k. For long wave-lengths, there is almost no damping. This is again a manifestation of the collective character of the plasma oscillations. Long wave-length oscillations involve many particles moving in phase for very long times; this is only possible because of the long-range electrostatic interactions. On the other hand, very short wave-lengths are rapidly damped. There appears to be a natural limit to the collective behavior; the limit is attained when the rate of damping becomes equal to the frequency. This occurs for a wave-length corresponding to

$$k_0 = \omega_p/u_0$$

If one interprets the mean-square-root velocity u_0 in terms of temperature, this limit becomes

$$k_0 = \omega_p \beta^{\frac{1}{2}} m^{-\frac{1}{2}} = (4\pi e^2 c \beta)^{\frac{1}{2}} = \kappa \tag{16.7}$$

Here appears quite naturally the second characteristic parameter of a plasma: *the Debye length*, κ^{-1}. The Debye length is also the only molecular length entering this problem. The breakdown of the collective behavior for phenomena whose wave-length is smaller than the molecular length is quite natural. It is, of course, impossible for the particles to respond in phase when the local physical conditions vary rapidly over molecular distances. Such phenomena can therefore not persist in time and are rapidly damped.

It is interesting to comment on the mathematical origin of the collective behavior. The most remarkable fact in this connection is the impossibility of obtaining a collective description of the plasma by a straightforward perturbation calculation on the Liouville equation. In order to derive the Vlassov equation we had to sum an *infinite series* of contributions. This clearly shows that even if the strength of the interactions is very small, their effect on the behavior of the plasma is very large: the structure of the equations is radically changed. The long wave-length motions $k \to 0$ are no longer described by a factor e^{ikvt} which is almost constant, but by $e^{i\omega_p t}$ which is rapidly oscillating (ω_p is generally a very high frequency). The cause of this large effect is, of course, the very long range of the Coulomb forces.

Such radical rearrangements in the perturbation series even for small values of the perturbation parameter are a well known feature in modern physics. Phenomena such as superfluidity and superconductivity are explained theoretically by a very similar rearrangement of the perturbation series. In these phenomena the collective effects also play a crucial role, although their origin is different (they are due to quantum statistics).

§ 17. The Overall Behavior of the Plasma

We now collect the results of the previous sections of this chapter and examine the various types of behavior which can occur for different types of velocity distributions and initial perturbations. We assume that the initial perturbation is periodic in space:

$$f(\mathbf{x}_\alpha, \mathbf{v}_\alpha; 0) - c\varphi(\mathbf{v}_\alpha) = e^{i\mathbf{k}_0 \cdot \mathbf{x}_\alpha} \psi(\mathbf{v}_\alpha) \tag{17.1}$$

It follows from (13.1) that

$$q_{\mathbf{k}}(\mathbf{v}_\alpha) = \delta(\mathbf{k} - \mathbf{k}_0)\psi(\mathbf{v}_\alpha) \tag{17.2}$$

From here on, we assume that the $\mathbf{k}$-integrations are done and drop the subscript 0 in the results to simplify the notations. We assume that the velocity distribution is given by (16.1), and the initial perturbation by (15.1) [with $f(\mathbf{k}) = \delta(\mathbf{k} - \mathbf{k}_0)$]. With these simple functions, all integrations in the general solution (14.1) can be performed exactly and easily (using straightforward contour integration). The following formula is obtained for the density at time t

$$n(\mathbf{x}; t) = e^{i\mathbf{k}\cdot\mathbf{x}} \Big\{ e^{-i\mathbf{k}\cdot\mathbf{v}_0 t - k w_0 t} \Big[1 + \frac{\omega_p^2}{[\mathbf{k}\cdot\mathbf{v}_0 + i(u_0 - w_0)k]^2 - \omega_p^2} \Big] \tag{17.3}$$

$$+ \tfrac{1}{2}\omega_p e^{-k u_0 t} \Big[\frac{e^{-i\omega_p t}}{\mathbf{k}\cdot\mathbf{v}_0 + ik(u_0 - w_0) - \omega_p} - \frac{e^{i\omega_p t}}{\mathbf{k}\cdot\mathbf{v}_0 + ik(u_0 - w_0) + \omega_p} \Big] \Big\}$$

We may consider several interesting limiting cases. *

A. $w_0 \approx u_0$

In this case, the distribution of velocities in the initial disturbance is roughly the same as in the unperturbed gas (see Fig. 17.1a).

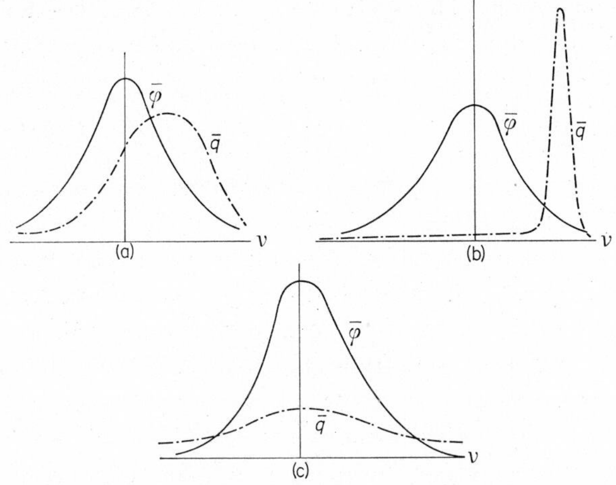

Fig. 17.1. Various types of velocity distributions (full lines) and of initial perturbations (dotted lines).

This is the type of perturbation which could occur, e.g., through a small random fluctuation. In this case, formula (17.3) shows that "individual" and "collective" processes are mixed together. After an initial period in which oscillations are excited, the disturbance dies out completely. The time of decay is given by

$$\tau^{-1} \sim kw_0 \approx ku_0$$

This is a typical dispersion phenomenon: particles which were moving in phase spread out over a wave-length after this time. The only way by which one could distinguish the two types of phenomena is their frequency.

B. $w_0 \ll u_0$

In this case the initial perturbation is sharply peaked around

* For a similar discussion, see N. G. van Kampen, *Physica*, **21**, 949 (1955).

$\mathbf{v}_0$ (see Fig. 17.1b). This type of perturbation occurs when an almost mono-energetic beam of particles is injected into the plasma. In this case, the damping of the first term is very small. After a time $(ku_0)^{-1}$, the contribution of the second term has died out; the persistent phenomenon is due to the "individual-particle" behavior. The disturbance is bodily carried away by the jet of incoming particles. The amplitude, however, is modified by the interactions with the medium.

C. $w_0 \gg u_0$

This is, in a sense, the opposite extreme. The perturbation is spread over a wide range of velocities, including velocities for which there are almost no particles in the unperturbed gas. It is now the first term in (17.3) which will be rapidly damped out. After a time $(kw_0)^{-1}$, the effect of the first term has disappeared, and we are left with a collective oscillation of frequency ω_p, which is Landau-damped. This example shows that collective effects can be called *"non-thermal" effects*. They manifest themselves isolately whenever there is an appreciable fraction of particles having velocities much larger than the thermal velocities of the medium. This peculiar non-thermal character is a very important feature of these effects and will show up, under different aspects, in many phenomena studied in this book.

CHAPTER 4

The Dispersion Equation

§ 18. The Dielectric Constant of the Plasma

Before studying in further detail the dispersion equation (14.9) we first show that the fundamental function $\varepsilon_+(k; z)$ has a simple and important phenomenological interpretation.

From the general solution (14.1) of the Vlassov equation we can immediately calculate the average potential $\Phi(\mathbf{x}_\alpha; t)$ defined by eqs. (12.2) and (12.3). In order to make contact with our representation, we take the Fourier transforms of the latter equations with respect to $\mathbf{x}_\alpha$:

$$\sigma_{\mathbf{k}}(t) = -eh_{\mathbf{k}}(t) \tag{18.1}$$

$$\Phi_{\mathbf{k}}(t) = -(4\pi/k^2)\sigma_{\mathbf{k}}(t) \tag{18.2}$$

Equation (18.2) derives from eq. (12.3) by an application of the convolution theorem for Fourier transforms. In order to obtain simpler expressions, we perform moreover a Laplace transformation with respect to time (see Appendix 1). Equation (18.2) then becomes

$$\Phi_{\mathbf{k}}(z) = -(4\pi/k^2)\sigma_{\mathbf{k}}(z) = +(4\pi e/k \quad h_{\mathbf{k}}(z) \tag{18.3}$$

Substituting the value of $h_{\mathbf{k}}(z)$ from eq. (4.1) we obtain the following expression for the average potential

$$\Phi_{\mathbf{k}}(z) = \frac{4\pi e}{k^2}\,\frac{1}{\varepsilon_+(k; z)}\int d\mathbf{v}\,\frac{q_{\mathbf{k}}(\mathbf{v})}{i(\mathbf{k}\cdot\mathbf{v}-z)} \tag{18.4}$$

This formula has the very simple form of a product of three factors. It can be understood in two ways. Its fundamental meaning is dictated by eq. (18.3): it is the product of the exact Coulomb interaction $(4\pi/k^2)$ and the charge density. But we also note that the integral appearing in (18.4) is the Laplace transform of the charge density at time t, $\sigma^0_{\mathbf{k}}(z)$, *calculated as if there were no interactions* [see eq. (14.4)]

$$\sigma_{\mathbf{k}}^{0}(z) = -e\int d\mathbf{v}\,\frac{q_{\mathbf{k}}(\mathbf{v})}{i(\mathbf{k}\cdot\mathbf{v}-z)}$$

The function $4\pi e^2/k^2\varepsilon_+(k;z)$ plays then the role of an *effective interaction between these fictitious free particles*. This effective interaction depends on z and on k. The consequence of this can be seen more easily in space–time.

Let us call $U(r;t)$ the inverse Fourier–Laplace transform of the effective interaction:

$$U(r;t) = \frac{1}{(2\pi)^4}\int_C dz\int d\mathbf{k}\,\frac{4\pi e^2}{k^2}\,\frac{1}{\varepsilon_+(k;z)}\,\mathrm{e}^{-i\mathbf{k}\cdot\mathbf{r}-izt} \tag{18.5}$$

This function depends on the distance r and on time. The potential $\Phi(\mathbf{x};t)$ calculated from (18.4) is then obtained by applying the convolution theorem:

$$\Phi(\mathbf{x};t) = -\int_0^t dt'\int d\mathbf{x}'\,U(|\mathbf{x}-\mathbf{x}'|;t-t')\sigma^0(\mathbf{x}';t') \tag{18.6}$$

In this perspective the potential is calculated from the density σ^0, evolving according to the laws of *free* motion, by means of a propagator which is no longer the Coulomb potential, as in (12.3), but the effective interaction $U(r;t)$. Moreover, this effective interaction is no longer instantaneous: eq. (18.6) has the form of a *retarded potential.*

To sum up, it is clear that the whole effect of the interactions on the motion of the particles is contained in the function $\varepsilon_+(k;z)$. One can view it as the effect of the presence of a medium of interacting particles on the interaction between two charged particles. Physically speaking, the presence of the charged test-particles *polarizes* the medium by creating a non-uniform charge distribution around each of them. Because of this polarization, the interaction between the two particles will no longer be the pure Coulomb force, but will be screened by the medium. This screening effect is described in phenomenological electrodynamics by a dielectric constant. It is therefore quite natural to call $\varepsilon_+(k;z)$ the *dielectric constant of the plasma,* although this current name is rather inadequatc, because $\varepsilon_+(k;z)$ is not constant.

It is well known from electrodynamics that two fields are needed in order to describe the electrical effects in the presence of matter: the electrical field **E** and the electrical induction **D**. In order to make Maxwell's equations complete, one needs an assumption about the relation between **E** and **D**. If the external field is constant and if the matter is homogeneous and isotropic, one assumes a linear relation

$$\mathbf{D} = \varepsilon \mathbf{E}$$

The dielectric constant ε is a characteristic parameter of the material medium. It describes the polarization produced within the matter by the external field. If we now consider rapidly varying fields, the polarization cannot follow the variation of the field because of the inertia of the molecules. There appears a time lag between the cause and the response: this phenomenon is known as a *dispersion phenomenon*. In this case the linear relation between field and induction takes the form:

$$\mathbf{D}(t) = \int_0^t dt' \, \varepsilon(t-t')\mathbf{E}(t')$$

If we Laplace transform this equation we obtain

$$\mathbf{D}(\omega) = \varepsilon(\omega)\mathbf{E}(\omega)$$

If there is moreover a spatial dispersion, the relation between field and induction will be taken to be of the form

$$\mathbf{D}(\mathbf{x}; t) = \int d\mathbf{x}' \int_0^t dt' \, \varepsilon(\mathbf{x}-\mathbf{x}'; t-t')\mathbf{E}(\mathbf{x}'; t') \tag{18.7}$$

and a Fourier–Laplace transform yields

$$\mathbf{D}_{\mathbf{k}}(\omega) = \varepsilon(\mathbf{k}; \omega)\mathbf{E}_{\mathbf{k}}(\omega) \tag{18.8}$$

It can easily be shown from the linearized Vlassov equation, by introducing an external field term, that (18.8) is the relation appropriate to our system.

The frequency-dependent dielectric constant has a simple physical interpretation only when the frequency ω is real. However, many general properties of this function can be demonstrated when the function $\varepsilon(k; \omega)$ is analytically continued into the upper half-plane of ω. This analytical continuation is precisely our function $\varepsilon_+(k; z)$. We already know the essential property of this function: $\varepsilon_+(k; z)-1$ *is a Cauchy integral, and hence is a regular function of* z *in the upper half-plane*. In other words, $\varepsilon_+(k; \omega)-1$ for *real* ω is a plus-function (see Appendix 2). This is why we write a subscript $+$ in order to stress this property.

We do not intend to go systematically into details about the macroscopic theory of the dielectric constant. A very thorough treatment is given in the textbook of Landau and Lifshitz.* The complex dielectric constant is a concept which is extensively used in the theory of wave propagation in plasmas and in the physics of the ionosphere. We refer the reader interested in more details to the specialized literature, in particular to the recent book of Ginzburg.**

It might seem artificial at first sight to describe the behavior of the plasma as in eq. (18.6) by introducing a set of fictitious free particles interacting through non-coulombian retarded forces $U(r; t)$. It should, however, be realized that this description has the advantage of being *intrinsic*, i.e. independent of the initial perturbation, whereas the description in terms of a polarized charge distribution (12.3) is of course different for each initial configuration even if the "permanent" features of the plasma do not change (i.e. for the same velocity distribution). This is why a knowledge of the effective interaction, or equivalently of the dielectric constant, contains complete information about the evolution of a plasma of given intrinsic characteristics (of course, as far as the linear Vlassov approximation is valid).

In order to gain further insight into the properties of U we shall evaluate the effective interaction for a specific example, i.e. for a particular velocity distribution of the plasma. For simplicity we consider the same situation as in § 16. With the dielectric constant ε_+ calculated in eq. (16.4), the effective interaction is given by

$$U(r; t) = \frac{4\pi}{(2\pi)^4} \int_C dz \int d\mathbf{k} \, \frac{1}{k^2} \, \frac{1}{1-[\omega_p/(iku_0+z)]^2} \, e^{i\mathbf{k}\cdot\mathbf{r}-izt} \tag{18.9}$$

Some care must be taken in the evaluation of the z-integral, because the resulting function is singular at $t = 0$. The integration is performed as follows:

$$\frac{1}{2\pi}\int_C dz \frac{1}{1-\omega_p^2/(z+iku_0)^2} e^{-izt} = \frac{1}{2\pi}\int_C dz \frac{(z+iku_0)^2}{(z+iku_0)^2-\omega_p^2} e^{-izt}$$

$$= \frac{1}{2\pi}\int_C dz \left\{1+\frac{\omega_p}{(z+iku_0)^2-\omega_p^2}\right\} e^{-izt} = \delta(t)-\omega_p \sin \omega_p t \, e^{-ku_0 t}$$

* L. D. Landau and E. M. Lifshitz, *Electrodynamics of Continuous Media*, Pergamon Press, Oxford, 1960.

** V. L. Ginzburg, *Propagation of Electromagnetic Waves through a Plasma* (in Russian), Fizmatgiz, Moscow, 1960.

The subsequent k-integration is rather easy

$$\begin{aligned} U(r;t) &= \frac{1}{(2\pi)^3}\int d\mathbf{k}\,\frac{4\pi}{k^2}\,[\delta(t)-\omega_p \sin\omega_p t\, e^{-ku_0 t}]e^{-i\mathbf{k}\cdot\mathbf{r}} \\ &= \frac{1}{r}\left\{\delta(t)+\omega_p \sin\omega_p t\left(1-\frac{2}{\pi}\tan^{-1}\frac{u_0 t}{r}\right)\right\} \end{aligned} \tag{18.10}$$

This formula clearly shows that the interacting medium introduces two effects:

a) a change of the r-dependence of the effective interaction,
b) a retardation effect.

In the limit $\omega_p \to 0$, both effects disappear. The general properties of this function will be better seen from the following figures.

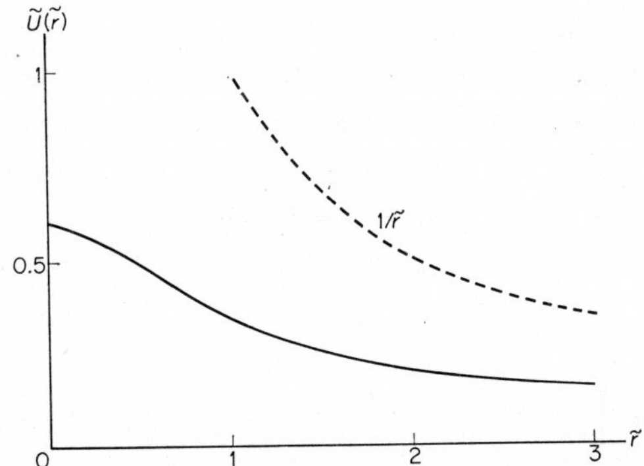

Fig. 18.1. The effective potential $U(r;t_0)$ (full line) at a given time t_0. For comparison, the Coulomb potential is represented by a dotted line. Dimensionless quantities are used: $\tilde{r} = \kappa r$, $\tilde{U} = U/\kappa^2 u_0$.

Fig. 18.1 shows the function $U(r;t_0)$ at a given time t_0. It is seen that the effective interaction, as expected, is smaller than r^{-1} at all distances. This constitutes the *screening effect* mentioned above.

It is very illuminating to calculate the integral

$$\bar{U}(r;t) = \int_0^t d\tau\, U(r;\tau) \tag{18.11}$$

This integral represents the action of a unit point charge switched on at time zero at the origin $[\sigma(\mathbf{x}) = -e\delta(\mathbf{x})]$ on a point at distance r at time t. Using the properties of the Laplace transformation this can be written as:

$$\overline{U}(r;t) = \frac{1}{(2\pi)^4}\int d\mathbf{k}\int_C dz \frac{1}{-iz}\frac{4\pi}{k^2}\frac{1}{\varepsilon_+(k;z)} e^{-i\mathbf{k}\cdot\mathbf{r}-izt} \quad (18.12)$$

The value of this function for $t \to \infty$ deserves special interest, because it is the stationary potential set up after all transient effects have been damped out. It is obtained from (18.12) by evaluating $(-2\pi i)$ times the residue of the integrand at $z = 0$. For our typical example, one easily obtains from eqs. (18.9) and (16.7)

$$\overline{U}(r;\infty) = \frac{1}{(2\pi)^3}\int d\mathbf{k}\frac{4\pi}{k^2+\omega_p^2/u_0^2} = \frac{1}{r} e^{-\kappa r} \quad (18.13)$$

Thus the persistent effect of the polarization is to replace the true Coulomb interaction by a Debye interaction. This shows the reason for the all-important role of the Debye potential in plasma physics. Fig. 18.2 shows the Debye potential as compared with the Coulomb potential, showing clearly the screening effect.

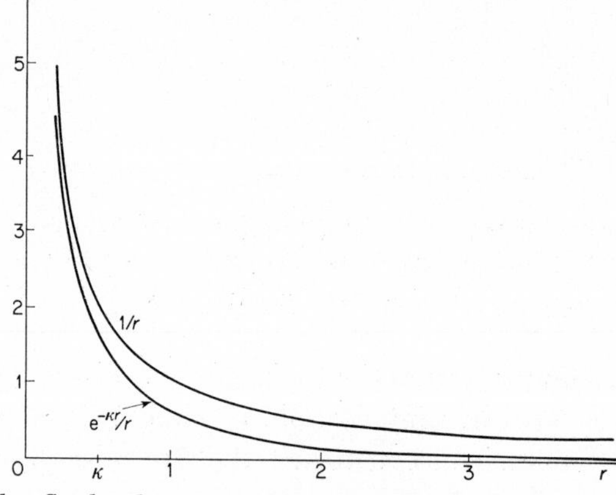

Fig. 18.2. The Coulomb potential $1/r$ compared with the Debye potential $\exp(-\kappa r)/r$.

Fig. 18.3 shows the retardation effect, $U(r_0; t)$, for fixed r_0. Here a very interesting phenomenon shows up. The influence of a point of the freely moving charge distribution on a particle at a distance r_0 through the medium is a damped periodical function of time. The effect of that point on the particle almost dies out after a time of the order r_0/u_0. The characteristic feature is the periodic nature of the effective interaction. It clearly shows the relation between plasma oscillations and effective interactions. It can be said that the effective interaction between particles proceeds through an emission of plasma oscillations at one point and their absorption by other particles at another point.

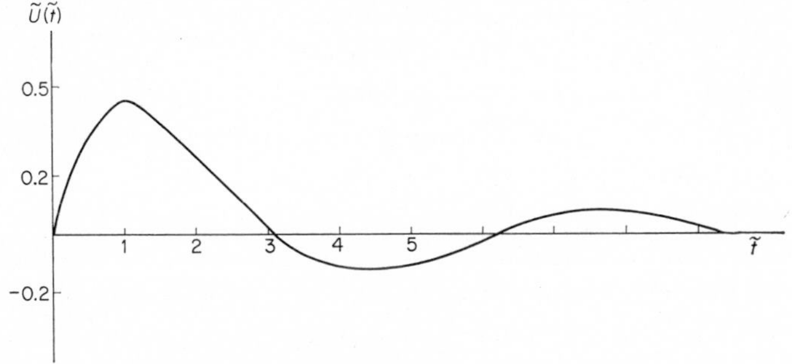

Fig. 18.3. The effective potential $U(r_0; t)$ as a function of time t at a fixed point in space. $\tilde{U} = U/\kappa^2 u_0$, $\tilde{t} = \omega_p t$.

The fact that the retardation effect is limited (i.e. that a given point influences another one only for a finite time) is a direct consequence of the Landau damping.

We now go over to a more detailed study of some properties of the dielectric constant $\varepsilon_+(k; z)$ and of the related dispersion equation.

§ 19. The Stability Criterion

A first important question arises in connection with the stability of the plasma. We have seen in § 14 that the poles of the resolvent corresponding to "individual-particle" behavior are always located on the real axis or in the lower half-plane. This restriction, however, does not apply to the "collective" poles. According to the possible types of velocity distribution function, these poles may lie anywhere in the complex plane.

It is obvious that if distributions exist for which there are poles in the upper half-plane S_+, the whole theory breaks down. The Fourier components corresponding to these unstable modes grow exponentially in time; this means in the first place that the linear approximation certainly becomes invalid after a very short time. It has usually been stated that the answer to this difficulty lies in the full, non-linear Vlassov equation (11.6) (which, of course, cannot be solved exactly). The non-linear terms of the Vlassov equation do not, however, provide the only mechanism of stabilization. Very recent work in this field has put forward several other quite different mechanisms of stabilization. These show that in unstable situations the velocity distribution $\varphi(\mathbf{v})$ evolves concurrently with $\rho_{\mathbf{k}}(\mathbf{v})$ in such a way as to stop the exponential growth of the latter (i.e. to bring the unstable zeros of the dielectric constant down to the real axis). An account of this recent work will be given in Appendix 9.

The question of finding criteria of stability was investigated first by Nyquist * and more recently by a number of authors. ** We shall essentially follow here the graphical method used by Ghertsenstein, Jackson and Penrose.

The method is based on the following well known theorem in the theory of analytical functions (its proof can be found in any textbook on complex variables):

Consider an analytical function $F(z)$ and a contour G lying inside the region in which $F(z)$ is defined. Consider the expression

$$N = \frac{1}{2\pi i} \int_G dz \frac{F'(z)}{F(z)} = \frac{1}{2\pi i} \int_G d[\ln F(z)] \tag{19.1}$$

The number N is an integer, and is equal to the difference between the number of zeros of $F(z)$ and the number of its poles inside the contour G.

We now apply this theorem to the dielectric constant $\varepsilon_+(k; z)$

* H. Nyquist, *Bell System Tech. J.*, **11**, 126 (1932).

** M. E. Ghertsenstein, *J. Exptl. Theoret. Phys. U.S.S.R.*, **23**, 669 (1952), F. Berz, *Proc. Phys. Soc. London*, **B69**, 939 (1956); O. Buneman, *Phys. Rev. Letters*, **1**, 8 (1958); P. L. Auer, *Phys. Rev. Letters*, **1**, 411 (1958); J. D. Jackson, *J. Nucl. Energy*, Part C, **1**, 171 (1960); O. Penrose, *Phys. of Fluids*, **3**, 258 (1960); G. Backus, *J. Math. Phys.*, **1**, 178 (1960).

defined by eq. (13.10). We study ε_+ as a function of z for a fixed value of k [we therefore write simply $\varepsilon_+(z)$ instead of $\varepsilon_+(k; z)$]. Using the notations $\bar{\varphi}$ and ν introduced in eqs. (14.5–6) as well as $\bar{\varphi}'(\nu) \equiv d\bar{\varphi}(\nu)/d\nu$, eq. (13.10) can be rewritten as follows:

$$\varepsilon_+(z) = 1 - \frac{\omega_p^2}{k^2} \int_\Gamma d\nu \frac{\bar{\varphi}'(\nu)}{\nu - z/k} \tag{19.2}$$

The zeros of $\varepsilon_+(z)$ are the frequencies of the plasma oscillations for that value of k. We choose the contour shown in Fig. 19.1,

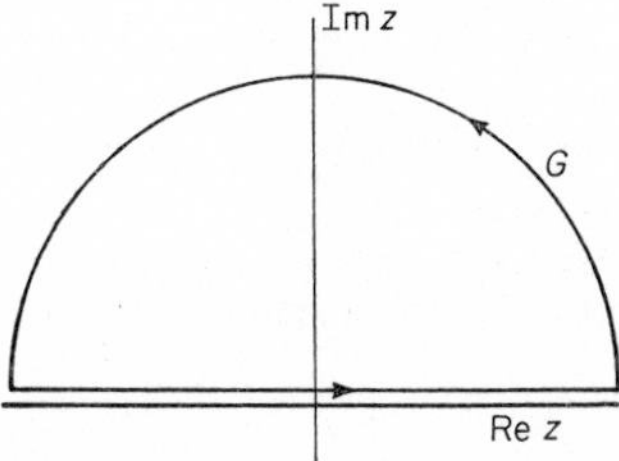

Fig. 19.1. The contour of integration G chosen for the integral (19.3).

consisting of the real axis and a semi-circle at infinity in the upper half-plane S_+. We know from the general theory of Cauchy integrals (see Appendix 2) that the function $\varepsilon_+(z)$ is regular (i.e. has no poles) for $z \in S_+$; therefore the integral

$$N_0 = \frac{1}{2\pi i} \int_G dz \frac{\varepsilon_+'(z)}{\varepsilon_+(z)} \tag{19.3}$$

equals the number of zeros of $\varepsilon_+(z)$ in the upper half-plane. The question of stability therefore amounts to finding conditions on $\bar{\varphi}(\nu)$ such that the integral (19.3) vanishes. But this integral equals $(2\pi i)^{-1}$ times the variation of $\ln \varepsilon_+(z)$, or equivalently, $(2\pi)^{-1}$ times the variation of the argument of the complex number $\varepsilon_+(z)$ when z varies along the contour of Fig. 19.1. We may therefore associate with Fig. 19.1 another graph (conveniently called the *hodograph*) such that to each point z of the contour G, Fig. 19.1, corresponds in the hodograph the affix of the complex number $\varepsilon_+(z)$. We now study the main properties of the hodograph.

From the theory of Cauchy integrals (see Appendix 2), we know that the integral appearing in (19.2) tends to zero as $1/z$ for $z \to \infty$. Therefore along the semi-circle of Fig. 19.1, $\varepsilon_+(z) = 1$. The non-trivial analysis concerns the behavior of $\varepsilon_+(z)$ on the real axis. From the Plemelj formulae (see Appendix 2), it is known that for $z = \omega$, where ω is *real*,

$$\varepsilon_+(\omega) = 1 - \frac{\omega_p^2}{k^2} \mathscr{P} \int_{-\infty}^{\infty} d\nu \frac{\bar{\varphi}'(\nu)}{\nu - \omega/k} - \pi i \frac{\omega_p^2}{k^2} \bar{\varphi}'(\omega/k), \qquad \omega \text{ real} \qquad (19.4)$$

Before going further the following remark is interesting, although it is outside the present problem. Let us separate real and imaginary parts of $\varepsilon_+(\omega)$:

$$\varepsilon_+(\omega) = \varepsilon_1(\omega) - i\varepsilon_2(\omega)$$

(the minus sign is written for later convenience). Equation (19.4) shows that the following relation holds between ε_1 and ε_2

$$\varepsilon_1(\omega) - 1 = -\frac{1}{\pi} \mathscr{P} \int_{-\infty}^{\infty} d\nu \frac{\varepsilon_2(\nu)}{\nu - \omega}, \qquad \omega \text{ real} \qquad (19.5)$$

This relation is known as the *Kramers–Kronig relation*, and is the simplest type of a so-called *dispersion relation*. It can be shown that its validity is due only to the fact that $\varepsilon_+(\omega) - 1$ is a plus-function. The importance of this formula comes from the fact that an approximate knowledge of the imaginary part of $\varepsilon_+(\omega)$ makes possible the calculation of an approximation to the real part which is consistent with all physical requirements (in particular: causality).

From eq. (19.4) it is seen that $\varepsilon_+(\omega)$ is real whenever $\bar{\varphi}'(\nu) = 0$. In other words [assuming $\bar{\varphi}(\nu)$ continuous], the hodograph crosses the real axis at all points where the distribution $\bar{\varphi}(\nu)$ has a minimum or a maximum.

We may even state somewhat more. If the zero of $\bar{\varphi}'(\nu)$ corresponds to a *maximum* of $\bar{\varphi}$, then $\bar{\varphi}'(\nu - \eta) > 0$ and $\bar{\varphi}'(\nu + \eta) < 0$ (η being a sufficiently small positive number); (19.4) then shows that when ω grows the imaginary part of $\varepsilon_+(\omega)$ goes from negative, through zero, to positive values. The hodograph therefore crosses the real axis upwards. However, in the vicinity of a minimum, the hodograph crosses the real axis downwards (see Fig. 19.2). As $\varepsilon_+(\pm\infty) = 1$, the curve is necessarily closed. It can be stated moreover that if the curve encircles the origin, it necessarily does so counterclockwise; otherwise the number N_0 would be negative. As $\varepsilon_+(z)$ is holomorphic in S_+, it can only have zeros in that region, so that the number N_0 defined by (19.1) is positive.

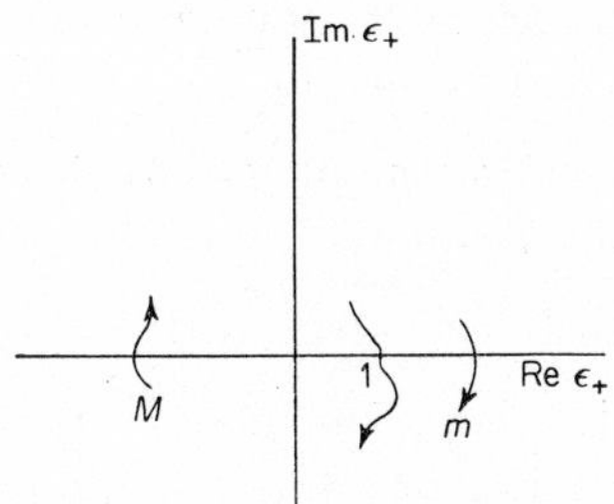

Fig. 19.2. Typical features in the behavior of the hodograph of $\varepsilon_+(\omega)$. The branch M corresponds to the neighborhood of a maximum of $\bar{\varphi}(\omega/k)$; the branch m corresponds to a minimum. $\varepsilon_+(\pm\infty) = 1$.

The problem can now be answered as follows: If the closed curve does not encircle the origin, the variation $[\arg \varepsilon_+(z)]_G$ as z goes around the contour is equal to zero and the system is stable. If, however, the hodograph encircles the origin n times, the argument of $\varepsilon_+(z)$ varies by $2\pi n$, and there will be n zeros of $\varepsilon_+(z)$ in the upper half-plane.

We now prove an important theorem (Nyquist *):

If the distribution function $\bar{\varphi}(v)$ has a single maximum, the plasma is stable.

We know that $\varepsilon_+(\pm\infty) = 1$, and therefore is real. On the other hand, $\bar{\varphi}(\infty) = 0$, so that the point at infinity can be considered as a minimum. The hodograph therefore goes downwards through the point $\varepsilon_+ = 1$. As the distribution function has only one maximum at $v = v_0$, the hodograph will cross the real axis upwards at some other point. The theorem will be proven if we can show that this second crossing point lies to the right of the origin. We will even show that it lies to the right of the first crossing point.

* See reference p. 96.

The value of $\varepsilon_+(\omega)$ for $\omega/k = v_0$ is

$$\varepsilon_+(kv_0) = 1 - \frac{\omega_p^2}{k^2} \lim_{\eta \to 0} \left\{ \int_{-\infty}^{v_0-\eta} dv \frac{\bar{\varphi}'(v)}{v - v_0} + \int_{v_0+\eta}^{+\infty} dv \frac{\bar{\varphi}'(v)}{v - v_0} \right\} \tag{19.6}$$

Now, as $\bar{\varphi}(v)$ has a single maximum,

$$\begin{cases} \bar{\varphi}'(v) > 0 & \text{for} \quad v < v_0 \\ \bar{\varphi}'(v) < 0 & \text{for} \quad v > v_0 \end{cases} \tag{19.7}$$

The bracketed expression in (19.6) is therefore obviously definite-negative and it follows that

$$\varepsilon_+(kv_0) > 1$$

which proves the theorem. The hodograph corresponding to a "single-humped" distribution function has the general shape shown in Fig. 19.3.

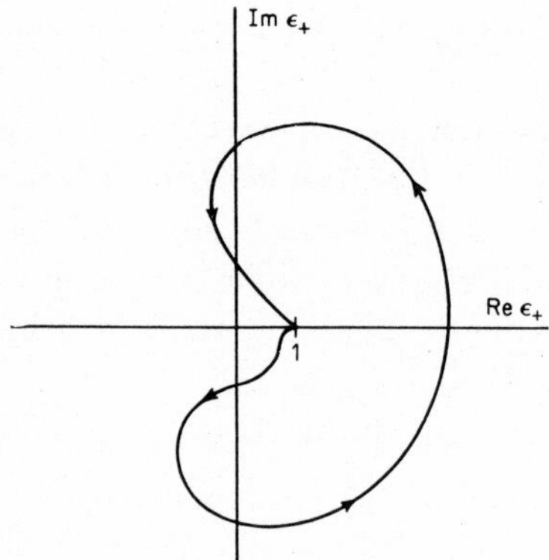

Fig. 19.3. Typical hodograph corresponding to a velocity distribution having a single maximum.

This theorem is very satisfactory from the physical point of view. In effect, the most important physical distributions are of the type required by the theorem, a typical example being the gaussian.

We now consider velocity distribution functions which have two maxima. These correspond physically to systems in which particles are essentially divided into two groups differing by their velocity. The previous type of argument shows that the hodograph will now cross the real axis at four points, one of them still being the point $+1$ (corresponding to $\omega = \pm\infty$). But the previous argument no longer holds for determining the location of the other points: there are stable situations or unstable ones. Intuitively, the unstable situations correspond to functions with very widely separated maxima. These correspond to two streams of plasma moving with widely different velocities. A simple example of an unstable plasma will be studied in some detail in § 22.

The criterion can be made more precise (Penrose*) by expressing the fact that in unstable situations the real value of $\varepsilon_+(\omega_0) < 0$, where $\omega_0 = kv_0$ corresponds to a minimum of $\bar{\varphi}$. From (19.4) we obtain

$$\begin{aligned}\operatorname{Re} \varepsilon_+(kv_0) &= 1 - \frac{\omega_p^2}{k^2} \lim_{\eta\to 0} \left\{ \int_{-\infty}^{v_0-\eta} + \int_{v_0+\eta}^{\infty} \right\} \frac{d[\bar{\varphi}(v) - \bar{\varphi}(v_0)]}{v - v_0} \\ &= 1 - \lim_{\eta\to 0} \frac{\omega_p^2}{k^2} [2\bar{\varphi}(v_0) - \bar{\varphi}(v_0 - \eta) - \bar{\varphi}(v_0 + \eta)]/\eta \\ &\quad - \frac{\omega_p^2}{k^2} \mathscr{P} \int_{-\infty}^{\infty} dv \frac{\bar{\varphi}(v) - \bar{\varphi}(v_0)}{(v - v_0)^2}\end{aligned}$$

The limit (second term on the r.h.s.) is zero. In order that $\operatorname{Re} \varepsilon_+(kv_0)$ be negative (instability), the following condition must be satisfied:

$$\mathscr{P} \int_{-\infty}^{\infty} dv \frac{\bar{\varphi}(v) - \bar{\varphi}(v_0)}{(v - v_0)^2} > \frac{k^2}{\omega_p} \tag{19.8}$$

To sum up, we can formulate the following criterion of instability: the system will display unstable oscillations of wave number k if its velocity distribution function $\bar{\varphi}(v)$ has a minimum at a value $v = v_0$ for which condition (19.8) is satisfied.

§ 20. Limiting Forms of the Dispersion Equation

A. The Long Wave-length Limit

We now look for a general solution of the dispersion equation in the limit of long wave-length oscillations, i.e. in the limit $k \to 0$. From the discussion of the simple example considered in § 16 we should expect that the real part of the collective frequency approaches a non-zero constant, whereas the damping tends to zero. This is the expression of a general property already stressed before: *long-range phenomena are almost purely collective phenomena.*

* See reference p. 96.

We shall assume that the distribution function $\bar{\varphi}(v)$ is symmetrical with respect to the origin, so that

$$\int_{-\infty}^{\infty} dv\, v\bar{\varphi}(v) = 0 \tag{20.1}$$

We then make the following ansatz: we are looking for solutions:

$$z_0(k) \equiv kx_0(k) + iky_0(k) \tag{20.2}$$

which satisfy the dispersion equation (14.9)

$$\varepsilon_+(k; z_0) = 0$$

and which behave as follows in the limit $k \to 0$:

$$\begin{cases} x_0(k) \to \infty \\ y_0(k) \to 0 \end{cases} \quad \text{for} \quad k \to 0 \tag{20.3}$$

It should be stressed that the solutions which will be found in this way are not necessarily *all* the solutions in the limit $k \to 0$; it could also happen that there are no solutions of this type [the example of § 16 is not of this type: it violates the second condition (20.3)]. Therefore, it should always be verified on the result that the conditions of the ansatz are satisfied.

We shall essentially follow here the treatment given by Jackson.* By virtue of the condition (20.3), the solution is "almost" real. We shall therefore expand the function $\varepsilon_+(z)$ in the vicinity of the real value kx_0. Keeping in mind formula (19.4), we obtain:

$$\begin{aligned} \varepsilon_+(z_0) = 1 - \frac{\omega_p^2}{k^2} \Bigg\{ & \mathscr{P}\int_{-\infty}^{\infty} dv \frac{\bar{\varphi}'(v)}{v - x_0} + \pi i \bar{\varphi}'(x_0) \\ & + iy_0 \left[\mathscr{P}\int_{-\infty}^{\infty} dv \frac{\bar{\varphi}''(v)}{v - x_0} + \pi i\, \bar{\varphi}''(x_0) \right] + \dots \Bigg\} \end{aligned} \tag{20.4}$$

Separating real and imaginary parts we find the two equations:

$$\begin{aligned} & 1 - \frac{\omega_p^2}{k^2} \left\{ \mathscr{P}\int_{-\infty}^{\infty} dv \frac{\bar{\varphi}'(v)}{v - x_0} - \pi y_0 \bar{\varphi}''(x_0) + \dots \right\} = 0 \\ & \pi\bar{\varphi}'(x_0) + y_0 \mathscr{P}\int_{-\infty}^{\infty} dv \frac{\bar{\varphi}''(v)}{v - x_0} + \dots = 0 \end{aligned} \tag{20.5}$$

* J. D. Jackson, *J. Nucl. Energy*, Part C, **1**, 171 (1960).

These equations will be solved up to the first order in y_0. It should be noted that in the first equation (20.5), the term $y_0\bar{\varphi}''(x_0)$ is actually of the second order, because in the limit $k \to 0$, $x_0 \to \infty$ and therefore $\bar{\varphi}''(x_0)$ tends exponentially to zero. We therefore have to solve the two equations:

$$1 - \frac{\omega_p^2}{k^2} \mathscr{P} \int_{-\infty}^{\infty} dv \frac{\bar{\varphi}'(v)}{v - x_0} = 0 \tag{20.6}$$

$$\pi\bar{\varphi}'(x_0) + y_0 \mathscr{P} \int_{-\infty}^{\infty} dv \frac{\bar{\varphi}''(v)}{v - x_0} = 0 \tag{20.7}$$

We first solve for y_0. In order to evaluate the term involving the second derivative in (20.7), we use a method due to Jackson. Multiply eq. (20.6) by k^2 and differentiate with respect to k. After an integration by parts we obtain:

$$2k = \omega_p^2 \frac{dx_0}{dk} \mathscr{P} \int_{-\infty}^{\infty} dv \frac{\bar{\varphi}''(v)}{v - x_0} \tag{20.8}$$

Substituting (20.8) into eq. (20.7) we immediately obtain:

$$y_0 = -\frac{\pi\omega_p}{2k} \frac{dx_0}{dk} \bar{\varphi}'(x_0) \tag{20.9}$$

We now solve eq. (20.6) for x_0 in the same approximation. We should say at this point that eq. (20.6) is the dispersion equation first derived by Vlassov * and later by Pines and Bohm** (in a slightly different form). These authors actually neglected the damping of the plasma oscillations, the existence of which was first proposed by Landau. † If the first condition (20.3) is satisfied, then $v/x_0 \ll 1$ for all values of v for which $\bar{\varphi}'(v)$ is effectively non-zero; we may therefore expand the integral as follows:

$$1 + \frac{\omega_p^2}{k^2 x_0} \mathscr{P} \int_{-\infty}^{\infty} dv \bar{\varphi}'(v) \left\{ 1 + \frac{v}{x_0} + \left(\frac{v}{x_0}\right)^2 + \left(\frac{v}{x_0}\right)^3 + \ldots \right\} = 0 \tag{20.10}$$

The successive terms are evaluated by partial integration. The resulting equation is [keeping in mind (20.1) and the fact that $\bar{\varphi}(v)$ is normalized to 1]:

$$1 - \frac{\omega_p^2}{k^2} \left[\frac{1}{x_0^2} + 3 \frac{\langle v^2 \rangle}{x_0^4} + \ldots \right] = 0 \tag{20.11}$$

* A. A. Vlassov, *J. Exptl. Theoret. Phys. U.S.S.R.*, **8**, 291 (1938).
** D. Pines and D. Bohm, *Phys. Rev.*, **85**, 338 (1952).
† L. D. Landau, *J. Phys. U.S.S.R.*, **10**, 25 (1946).

Up to order k^2, we find the solution:

$$x_0^2 = \frac{\omega_p^2}{k^2} + 3\langle v^2\rangle + \dots \tag{20.12}$$

or, coming back to the true frequency $\omega \equiv kx_0$,

$$\omega^2 = \omega_p^2 + 3k^2\langle v^2\rangle + \dots \tag{20.13}$$

Substituting (20.12) into eq. (20.9) we get the damping rate $\gamma = -ky_0$ to the first order:

$$\gamma = -\frac{\pi\omega_p^3}{2k^2}\bar{\varphi}'\left(\frac{\omega_p}{k}\right) \tag{20.14}$$

(This quantity is positive, because $\bar{\varphi}'(\omega_p/k) < 0$.)

Let us now comment on the two formulae obtained. A first remarkable feature of eq. (20.13) is the fact that in the limit $k \to 0$, ω goes to ω_p, a constant independent of k. This means physically that in this limit plasma oscillations do not propagate in space: *the disturbance remains localized in space*. In effect, if the initial disturbance is

$$f(\mathbf{x}; 0) = \int d\mathbf{k}\, e^{i\mathbf{k}\cdot\mathbf{x}} q_{\mathbf{k}}$$

and if $q_{\mathbf{k}}$ contains only very small wave-vectors, then (disregarding damped individual modes):

$$f(\mathbf{x}; t) \approx \int d\mathbf{k}\, e^{i\mathbf{k}\cdot\mathbf{x}} e^{i\omega_p t} q_{\mathbf{k}} = e^{i\omega_p t} f(\mathbf{x}; 0)$$

This behavior is in marked contrast with all usual types of oscillating phenomena, such as acoustic or optical waves. For a

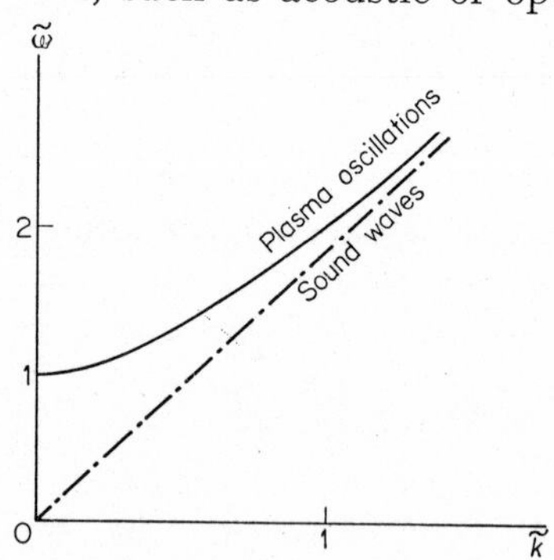

Fig. 20.1. Dispersion curves for plasma oscillations and for sound waves.

sound wave, for instance, it is well known that, for long wave-lengths, $\omega = c_0 k$. As a result the wave propagates in space: $f(x; t) = f(x \pm c_0 t; 0)$. The type of dispersion relation eq. (20.13) is therefore characteristic of plasma oscillations: it is represented schematically in Fig. 20.1.

For small but finite wave-lengths, there appears a small dispersion, described by the complete equation (20.13). This means that the waves will propagate in space and progressively die out as a consequence of destructive interference (this spatial damping is due to the non-linear relationship between ω and k). The physical origin of this dispersion is clearly the thermal motion ($\langle v^2 \rangle$ is a measure of the "temperature" of the system).

We now consider the Landau damping, eq. (20.14). As expected, for $k \to 0$, the damping in general tends rapidly to zero, because it is proportional to the derivative of the distribution function in the tail of the distribution. It is very interesting to notice that the Landau damping is very sensitive to the distribution function, whereas the dispersion relation (20.13) only depends on such overall characteristics of the distribution as the mean-square velocity or possibly some higher moments: The limiting value ω_p is even independent of the shape of the distribution: it only depends on the concentration. As a striking example of this difference consider the two distributions:

$$\bar{\varphi}_1(v) = \frac{2u_0^3}{\pi} \frac{1}{(u_0^2+v^2)^2}, \quad \bar{\varphi}_2(v) = \frac{1}{\sqrt{2\pi}\,u_0} \mathrm{e}^{-(v^2/2u_0^2)}$$

For *both* distributions,

$$\omega^2 = \omega_p^2 + 3k^2 u_0^2$$

However, the damping rates, as given by (20.14) in the limit $k \to 0$, are quite different:

$$\gamma_1 = \frac{4u_0^3}{\omega_p^2} k^3; \qquad \gamma_2 = \sqrt{\frac{\pi}{2}} \frac{\omega_p^4}{u_0^3} \frac{\mathrm{e}^{-\omega_p^2/2(ku_0)^2}}{k^3}$$

The damping goes to zero much faster in the gaussian case than for $\bar{\varphi}_1$, where it decays only as k^3.

B. The Short Wave-length Limit

It has been shown in § 16 on a simple example that the collective behavior of the plasma completely breaks down if the wave-length of the disturbance becomes smaller than the Debye radius. We show in this section that for realistic distribution functions, this breakdown takes the aspect of a divergence. We shall essentially follow here Landau's treatment * (see also Jackson **).

We are calling a "realistic" distribution $\bar{\varphi}(v)$ one which falls off exponentially as $v \to \pm\infty$. (The simple example of § 16 is not of this type.) We shall investigate the limiting form of the dispersion equation (14.9) as $k \to \infty$. Again we shall make *a priori* reasonable assumptions about the location of the pole in the v-plane, which should be checked in all particular cases. Writing again:

$$z_0(k) = kx_0(k) + iky_0(k)$$

we should expect that $kx_0(k)$ is a slowly growing function of k, whereas the damping grows rapidly as $k \to \infty$. We assume the ansatz

$$\begin{cases} x_0(k) \to 0 \\ y_0(k) \to \infty \end{cases} \quad \text{for} \quad k \to \infty \tag{20.15}$$

This means that the real part of the pole grows more slowly than k whereas the imaginary part grows faster than k. For large

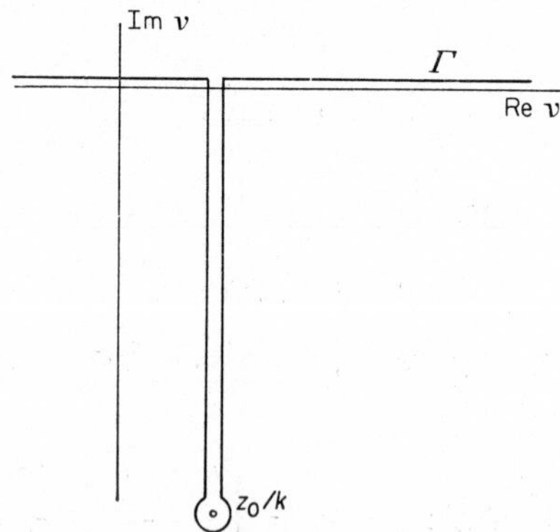

Fig. 20.2. Contour of integration Γ for large k.

* See reference p. 103.

** See reference p. 102.

values of k the pole is very near the imaginary axis and far down in the lower half-plane. The contour of integration Γ [see eq. (14.9) and Fig. 14.1] is shown schematically in Fig. 20.2.

The function $\varepsilon_+(k; z)$ can therefore be written as

$$\varepsilon_+(k; z) = 1 - \frac{\omega_p^2}{k^2}\int_{-\infty}^{\infty} dv \frac{\bar{\varphi}'(v)}{v - z_0/k} - \frac{\omega_p^2}{k^2}\int_{\gamma} dv \frac{\bar{\varphi}'(v)}{v - z_0/k} \tag{20.16}$$

We now introduce our assumption about the exponential decay of $\bar{\varphi}(v)$ (and therefore of $\bar{\varphi}'(v)$ also) for $v \to \pm\infty$. The first integral will therefore equal some finite function of z_0/k, which is multiplied by k^{-2}; this is therefore a small number. The second integral, taken along the small circle γ around the pole, equals $2\pi i$ times $\bar{\varphi}'(z_0/k)$. By virtue of (20.15), z_0/k has a large imaginary part. But if $\bar{\varphi}'(v)$ decays exponentially for real $v \to \pm\infty$, it will correspondingly *grow* exponentially for large imaginary values of v. The last term of (20.16) therefore represents the dominant contribution to the integral along Γ. Therefore, in this limit the dispersion equation becomes:

$$1 - 2\pi i \frac{\omega_p^2}{k^2} \bar{\varphi}'(z_0/k) = 0 \tag{20.17}$$

Separating real and imaginary parts, we obtain:

$$\text{Re}\left[2\pi i \frac{\omega_p^2}{k^2} \bar{\varphi}'(x_0 + iy_0)\right] = 1 \tag{20.18}$$

$$\text{Im}\left[2\pi i \frac{\omega_p^2}{k^2} \bar{\varphi}'(x_0 + iy_0)\right] = 0 \tag{20.19}$$

We therefore obtain two transcendental equations which determine x_0 and y_0. These equations depend crucially on the shape of the distribution.

We shall go further by considering the particular case of a Maxwell distribution:

$$\bar{\varphi}(v) = \frac{1}{\sqrt{2\pi}\, u_0} e^{-v^2/2u_0^2} \tag{20.20}$$

We have:

$$\bar{\varphi}'(z_0/k) = -\frac{x_0 + iy_0}{\sqrt{2\pi}\, u_0^3} \exp\left[-\frac{(x_0 + iy_0)^2}{2u_0^2}\right]$$

Neglecting x_0 in front of the exponential and x_0^2 in the argument of the exponential, the two equations (20.18–19) become:

$$(\sqrt{2\pi}\,\omega_p^2 y_0/k^2 u_0^3)e^{y_0^2/2u_0^2}\cos(x_0 y_0/u_0) = 1 \tag{20.21}$$

$$y_0 e^{y_0^2/2u_0^2}\sin(x_0 y_0/u_0) = 0 \tag{20.22}$$

As y_0 has been assumed large and negative, the only solutions of (20.22) are:

$$x_0 y_0/u_0 = m\pi \tag{20.23}$$

Among these solutions, only those for which m is odd are compatible with (20.21) [because otherwise $\cos(x_0 y_0/u_0) = 1$ and the l.h.s. of (20.21) would be negative, as $y_0 < 0$]. We shall only consider $m = 1$ (the rate of damping is independent of m, at least for small m). Equation (20.21) becomes:

$$-(\sqrt{2\pi}\,\omega_p^2 y_0/u_0^3 k^2)e^{y_0^2/2u_0^2} \tag{20.24}$$

or else

$$-(\alpha/k^2)y_0 = e^{-y_0^2/2u_0^2} \tag{20.25}$$

It is easy to see graphically that this equation has a single solution for $y_0 < 0$. This solution is a growing function of k (Fig. 20.3). It is also seen in this figure that the damping is stronger when the temperature ($\sim u_0$) increases. It is seen from eq. (20.25) that $|y_0| \sim \sqrt{(\ln k)}$ for large k, and therefore the rate of damping $\gamma = -ky_0 \sim k\sqrt{(\ln k)}$. From (20.23) it follows that $\omega = kx_0 \sim k/y_0 \sim k/\sqrt{(\ln k)}$: the frequency also goes to infinity, but more slowly than k. Thus the initial ansatz is satisfied.

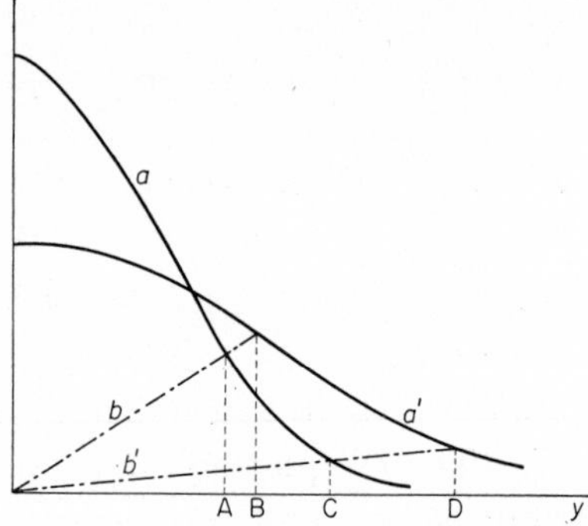

Fig. 20.3. Graphical solution of eq. (20.25). The curves a and a' represent the r.h.s. of the equation as a function of y_0. a corresponds to a low temperature (small u_0), a' to a high temperature. b and b' represent the l.h.s. b corresponds to small k, b' to large k.

§ 21. The Dielectric Constant for a Maxwellian Distribution

We collect in this paragraph several formulae which will be useful not only for the problem of plasma oscillations but also in the remainder of this book. It should be realized from the beginning that as far as the short-time behavior is concerned, the Maxwell (Gaussian) distribution does not play any privileged role: it is just one distribution among many other possible ones. In effect, we have seen that *any* distribution function of momenta (i.e. any non-negative normalized function of $\mathbf{v}$) is a stationary solution of Vlassov's equation. Only if the evolution of the plasma is followed over longer periods of time — for which the Vlassov equation is no longer valid — does the Maxwell distribution appear as the unique end-point of the evolution: the thermal equilibrium.

However, many physical problems are of the following type. The plasma is left to evolve freely over a long period of time in which we are not interested. At a given time, which we call $t = 0$, we impose an external disturbance $q_{\mathbf{k}}(\mathbf{v})$ and we follow the free evolution of the system for short positive times. In this case the system has reached thermal equilibrium in the period of time prior to $t = 0$. Hence, at time zero the velocity distribution is maxwellian (the system is "aged"). This is why the Maxwell distribution plays a special role even in the short-time study of a plasma.

Let the barred velocity distribution be

$$\bar{\varphi}(v) = (m\beta/2\pi)^{\frac{1}{2}} \exp\left(-\tfrac{1}{2}\beta m v^2\right) \tag{21.1}$$

The dielectric constant $\varepsilon_+(k; z)$, eq. (13.10), becomes in this case:

$$\varepsilon_+(k; z) = 1 + \frac{\omega_p^2}{k^2} \frac{(\beta m)^{\frac{3}{2}}}{(2\pi)^{\frac{1}{2}}} \int_\Gamma dv \frac{v \exp\left(-\tfrac{1}{2}\beta m v^2\right)}{v - z/k} \tag{21.2}$$

We shall be interested in the values of ε_+ for z real. We simplify the notation by introducing the real dimensionless variables

$$\begin{aligned} \chi &= k/\omega_p\sqrt{\beta m} = k/\kappa \\ \zeta &= z/\sqrt{2}\,\omega_p \\ y &= z\sqrt{\beta m}/\sqrt{2}\,k = \zeta/\chi \end{aligned} \tag{21.3}$$

where κ is the inverse Debye length, eq. (16.7). The integral in eq. (21.2) can now be written as follows [see eq. (19.4)]

$$\varepsilon_+(\chi; \chi y) = \frac{1}{\chi^2}\left\{\chi^2 + \pi^{-\frac{1}{2}} \mathscr{P}\int dv \frac{v e^{-v^2}}{v - y} + i\sqrt{\pi}\, y e^{-y^2}\right\} \tag{21.4}$$

We now evaluate the integral:

$$J(y) = \pi^{-\frac{1}{2}}\mathscr{P}\int_{-\infty}^{\infty} dv \frac{v\,\mathrm{e}^{-v^2}}{v-y} = 1+\pi^{-\frac{1}{2}}y\mathscr{P}\int_{-\infty}^{\infty} dv \frac{\mathrm{e}^{-v^2}}{v-y} \tag{21.5}$$
$$= 1-2y\,\mathrm{e}^{-y^2}\Psi(y)$$

where

$$\Psi(y) = -\frac{1}{2\sqrt{\pi}}\mathscr{P}\int_{-\infty}^{\infty} dt \frac{\exp(-t^2-2ty)}{t} \tag{21.6}$$

(We have first added and subtracted ye^{-v^2} from the numerator of the original integrand in $J(y)$, then made the substitution $t = v-y$). We note that

$$\Psi(0) = 0 \tag{21.7}$$

Taking the derivative of (21.6) with respect to y we find:

$$d\Psi(y)/dy = \pi^{-\frac{1}{2}}\mathscr{P}\int_{-\infty}^{\infty} dt \exp(-t^2-2ty) = \exp(y^2) \tag{21.8}$$

We now integrate (21.8) over y, keeping in mind eq. (21.7):

$$\Psi(y) = \int_0^y dt\,\mathrm{e}^{t^2} \tag{21.9}$$

$\Psi(y)$ is known as the *error function of a complex argument*. The main properties of this function can be found in Bateman's book. * A short table of $\Psi(y)$ for real y can be found in Jahnke and Emde. ** An extensive table has been published recently by Fried and Conte. †

Collecting eqs. (21.4), (21.5), (21.6) and (21.9) we obtain the following expression for $\varepsilon_+(\chi;\,\chi y)$:

$$\varepsilon_+(\chi;\,\chi y) = \chi^{-2}\{\chi^2+1-2y\mathrm{e}^{-y^2}\Psi(y)+i\pi^{\frac{1}{2}}y\mathrm{e}^{-y^2}\} \tag{21.10}$$

For later reference we give the formulae and curves for the behavior of the complex function $\varepsilon_+(\chi;\,\chi y)$ for *real* y. The function can be separated into its real and imaginary parts:

$$\varepsilon_+(\chi;\,\chi y) \equiv \varepsilon_1(\chi;\,y)-i\varepsilon_2(\chi;\,y) \tag{21.11}$$

$$\varepsilon_1(\chi;\,y) = 1+\chi^{-2}[1-2y\,\mathrm{e}^{-y^2}\Psi(y)] \tag{21.12}$$

$$\varepsilon_2(\chi;\,y) = -\pi^{\frac{1}{2}}\chi^{-2}y\mathrm{e}^{-y^2} \tag{21.13}$$

* H. Bateman (A. Erdelyi, editor), *Higher Transcendental Functions*, vol. 2, McGraw-Hill, New York, 1953.

** E. Jahnke and F. Emde, *Tables of Functions, with Formulae and Curves*, Dover Publ. Co., New York, 1945.

† B. D. Fried and S. D. Conte, *The Plasma Dispersion Function*, Academic Press, New York, 1961.

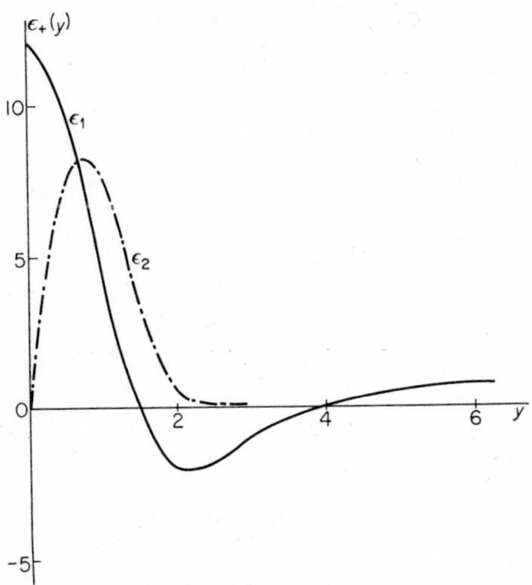

Fig. 21.1. Real and imaginary parts of the dielectric constant ($\varepsilon_+ = \varepsilon_1 + i\varepsilon_2$) of a classical gas in thermal equilibrium. (Note that the sign of ε_2 is opposite to the one introduced in the main text).

The real and the imaginary parts of the dielectric constant are shown in Fig. 21.1. Fig. 21.2 shows the hodograph of the gaussian, which is of course of the general type drawn in Fig. 19.3.

It is interesting to derive an asymptotic formula for $\varepsilon_1(\chi; y)$

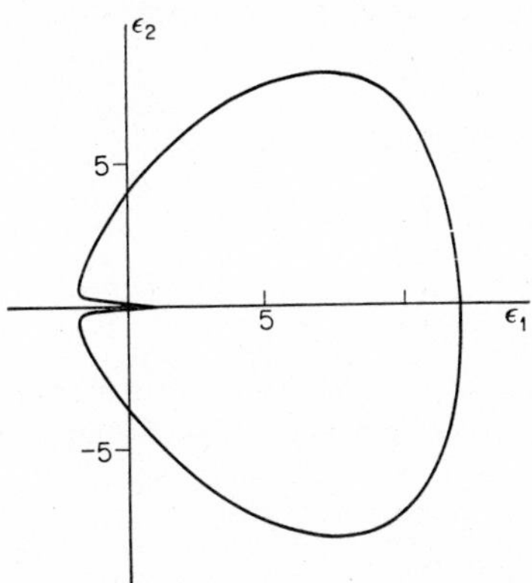

Fig. 21.2. Hodograph of the dielectric constant $\varepsilon_+ = \varepsilon_1 + i\varepsilon_2$ represented in Fig. 21.1.

valid for large values of y. We make use of the following asymptotic expansion for $\Psi(y)$ given in Bateman *

$$\Psi(y) = -\tfrac{1}{2}i\pi^{+\frac{1}{2}}+\tfrac{1}{2}e^{y^2}\left\{\sum_{m=0}^{M-1}[\Gamma(\tfrac{1}{2}+m)/\Gamma(\tfrac{1}{2})]y^{-2m-1}+O(|y^{-2M-1}|)\right\} \tag{21.14}$$

Substituting this expression (but for the first term) into (21.12) we obtain

$$\varepsilon_1(\chi; y) = 1-\frac{1}{2\chi^2 y^2}-\frac{3}{4\chi^2 y^4}-\ldots \tag{21.15}$$

We rewrite this by using ordinary variables (21.3) and calling the real frequency $z = \omega$

$$\varepsilon_1(k; \omega) = 1-\frac{\omega_p^2}{\omega^2}-\frac{3\omega_p^2 k^2}{\beta m\omega^4}-\ldots \tag{21.16}$$

This is an expansion of the real part of the dielectric constant valid for small wave-vectors and large frequencies. If this expansion is set equal to zero and the resulting equation is solved for ω one obtains at a first approximation the root found in eq. (20.13) from a more general treatment. [Note that $\langle v^2\rangle = (\beta m)^{-1}$.]

§ 22. An Example of an Unstable Distribution

It has been shown in § 19 that there exist momentum distributions which are unstable, in the sense that any small perturbation at the initial time results in oscillations of exponentially growing amplitude. In such situations, the Vlassov equation ceases to be valid. We shall study, in this section, a detailed example which shows how this breakdown arises.

We choose the following distribution (Jackson)**

$$\bar{\varphi}(v) = \frac{u_0}{2\pi}\left\{\frac{1}{(v-v_0)^2+u_0^2}+\frac{1}{(v+v_0)^2+u_0^2}\right\} \tag{22.1}$$

This is a superposition of two functions of the type considered in § 16. Fig. 22.1 shows a plot of this function, which has a minimum at $v = 0$. It has two peaks, separated by a distance $2v_0$. As was shown in § 19, this distribution can be unstable for certain wave-lengths. The function $\varepsilon_+(k; z)$ is readily calculated:

* See reference p. 110.

** J. D. Jackson, *J. Nucl. Energy*, Part C, **1**, 171 (1960).

$$\varepsilon_+(k; z) = 1 + \frac{\omega_p^2}{k^2}\frac{u_0}{\pi}\int_\Gamma dv\, \frac{1}{v - z/k}\left\{\frac{v - v_0}{[(v-v_0)^2 + u_0^2]^2} + \frac{v + v_0}{[(v+v_0)^2 + u_0^2]^2}\right\} \qquad (22.2)$$

The integrations are readily done (see § 16) and result in the following dispersion equation:

$$1 - \frac{\omega_p^2}{2k^2}\left\{\frac{1}{[(z/k) + iu_0 - v_0]^2} + \frac{1}{[(z/k) + iu_0 + v_0]^2}\right\} = 0 \qquad (22.3)$$

This is a simple algebraic equation of the fourth degree in z, which is easily solved.

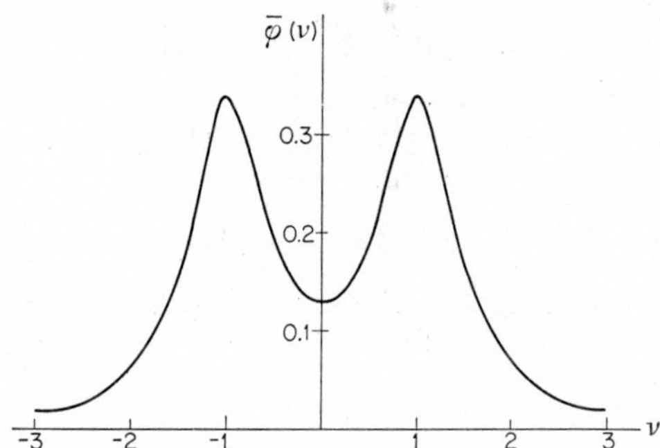

Fig. 22.1. The unstable distribution function $\bar{\varphi}(v)$, eq. (22.1).

We shall call the four roots z_+^A, z_-^A, z_+^B, z_-^B:

$$\begin{aligned} z_\pm^A &= -iku_0 + [(kv_0)^2 + \tfrac{1}{2}\omega_p^2 \pm \sqrt{\tfrac{1}{4}\omega_p + 2\omega_p(kv_0)^2}]^{\frac{1}{2}} \\ z_\pm^B &= -iku_0 - [(kv_0)^2 + \tfrac{1}{2}\omega_p^2 \pm \sqrt{\tfrac{1}{4}\omega_p + 2\omega_p^2(kv_0)^2}]^{\frac{1}{2}} \end{aligned} \qquad (22.4)$$

The discussion of stability is now reduced to the discussion of the nature of the roots. We first note that for z_+^A and for z_+^B, the bracketed expressions in (22.4) are positive. Therefore these roots are complex, but their imaginary part ($-ku_0$) is negative: they correspond to damped oscillations. Consider now the case in which

$$[(kv_0)^2 + \tfrac{1}{2}\omega_p^2]^2 < \tfrac{1}{4}\omega_p^4 + 2\omega_p^2(kv_0)^2$$

or else:

$$kv_0 < \omega_p \qquad (22.5)$$

In this case, the bracketed expression is negative, and its square root is purely imaginary. The solution z_-^B represents then a purely imaginary negative number. The only root which can lead to instability is, therefore, the root z_-^A. In order to facilitate the discussion, we shall introduce the following non-dimensional quantities:

$$\frac{z_-^A}{\omega_p} = \tilde{z}, \qquad \frac{kv_0}{\omega_p} = V \qquad \frac{ku_0}{\omega_p} = \chi \qquad (22.6)$$

Then the possibly unstable root is:

$$\bar{z} = -i\chi + [V^2 + \tfrac{1}{2} - \sqrt{2V^2 + \tfrac{1}{4}}]^{\frac{1}{2}} \tag{22.7}$$

In order to have an unstable situation, two conditions must be simultaneously satisfied:

(a) the square root must be imaginary; this requires condition (22.5) or:

$$V < 1 \tag{22.8}$$

(b) $\bar{z}/i$ must be positive; this requires

$$\chi^2 < |V^2 + \tfrac{1}{2} - \sqrt{2V^2 + \tfrac{1}{4}}| \tag{22.9}$$

This condition can be worked out and leads after some algebra to the discussion of the sign of a quadratic form. The general condition of instability is obtained in the form:

$$-\chi^2 + \tfrac{1}{2} - \sqrt{\tfrac{1}{4} - 2\chi^2} < V^2 < -\chi^2 + \tfrac{1}{2} + \sqrt{\tfrac{1}{4} - 2\chi^2} \tag{22.10}$$

[which must of course be supplemented by (22.8)]. In the case $\chi \ll 1$, these expressions can be simplified and lead to:

$$\chi < V < \sqrt{1 - 3\chi^2} \tag{22.11}$$

Let us discuss this simplified formula in some detail. It can be looked at in two ways. We may first consider excitations of *given wave-vector* k. We also fix the spread u_0, so that χ is also given. Equation (22.11) then becomes a condition on the separation of the two peaks:

$$u_0 < v_0 < \frac{1}{k}\sqrt{[\omega_p^2 - 3(ku_0)^2]} \tag{22.11a}$$

This shows that there is only a limited range of distributions which are unstable for given k. This peculiar property expresses a competition between the Landau damping $(-iku_0)$ [see eq. (16.5)] and the instability. This range of instability is wider the smaller k is. This point of view was the one taken in § 19, where we discussed stability criteria for a given k. As a concrete illustration of those criteria, we have drawn in Fig. 22.2 the hodographs of the function ε_+ for the three situations: $v_0 < u_0$, v_0 in the range (22.11a), and v_0 above the upper limit of that equation.

An alternative, maybe more interesting, interpretation of (22.11) is to consider the distribution as given (u_0, ω_p, v_0 fixed) and to look at (22.11) as giving the values of k which give rise to growing excitations.

We must now employ the exact inequalities (22.10), because we cannot be sure that the simplifying assumptions are valid throughout the range of variation. The two inequalities (22.10) can be combined into the single one:

$$|V^2 + \chi^2 - \tfrac{1}{2}| < \sqrt{\tfrac{1}{4} - 2\chi^2}$$

When worked out, this inequality provides the following condition for k:

$$k^2 < \omega_p^2(v_0^2 - u_0^2)/(v_0^2 + u_0^2)^2 \tag{22.12}$$

In other words, for a given distribution, only values of k which satisfy (22.12) lead to instabilities. It is immediately seen that if $v_0 < u_0$, the condition cannot be satisfied. For $v_0 > u_0$, (22.12) shows that there is an upper bound to the wave-vector of the unstable modes. Only long wave-

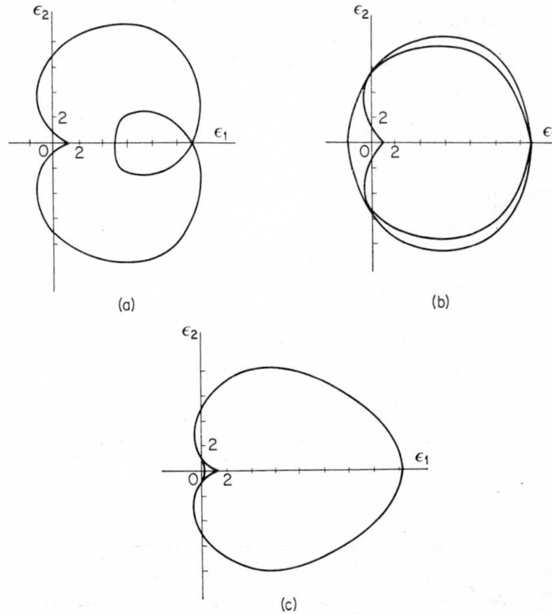

Fig. 22.2. Typical hodographs showing the passage from stability to instability and back to stability as the parameter grows (k and u_0 fixed).

(a) $v_0 < u_0$

(b) $u_0 < v_0 < k^{-1}[\omega_p^2 - 3(ku_0)^2]^{\frac{1}{2}}$

(c) $v_0 > k^{-1}[\omega_p^2 - 3(ku_0)^2]^{\frac{1}{2}}$

length disturbances can grow: short wave-lengths are Landau-damped. This remarkable property shows that the unstable growth is a purely collective phenomenon.

Fig. 22.3 shows a plot of $\bar{z}$ as a function of k for given $u_0 < v_0$, ω_p: it shows the passage from instability to stability as k grows.

Fig. 22.4 shows how the limit of instability (22.12) varies with v_0: it is seen that this limit tends to zero as the separation of the peaks $2v_0 \to \infty$.

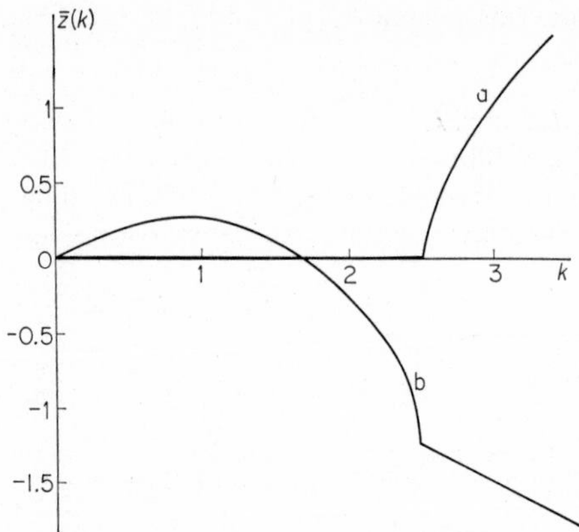

Fig. 22.3. The possibly unstable frequency $\bar{z}(k)$, showing the passage from instability to stability as the wave-vector k grows. (a): Re $\bar{z}$; (b): Im $\bar{z}$.

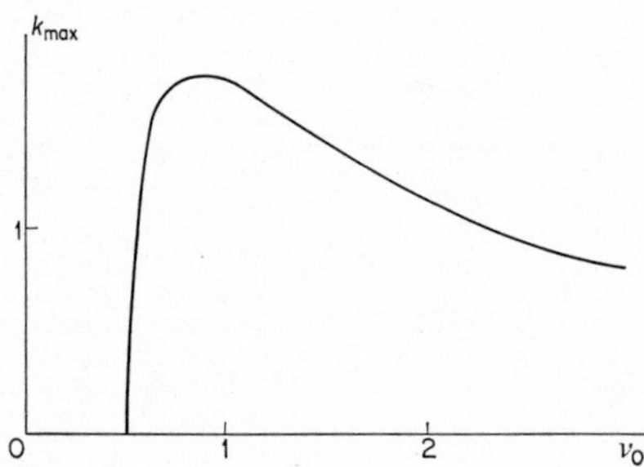

Fig. 22.4. The variation of the limit of instability with the parameter ν_0.

CHAPTER 5

The van Kampen–Case Treatment of the Vlassov Equation

§ 23. Eigenfunctions of the Vlassov Operator

We shall describe in this chapter an alternative method for solving the Vlassov equation which, although equivalent to the resolvent method discussed previously, starts from a completely different point of view. This method is due to van Kampen * and was slightly modified and simplified by Case. ** Not only is this method very interesting in itself, but it will also provide an excellent preparation for the mathematical techniques which will be used in subsequent chapters (especially Chapters 9 and 10).

Let us start from the Fourier-transformed linearized Vlassov equation [eq. (11.5) in which the last term is suppressed]:

$$\partial_t \rho_{\mathbf{k}}(\mathbf{v}; t) + i\mathbf{k} \cdot \mathbf{v} \rho_{\mathbf{k}}(\mathbf{v}; t) = (\omega_p^2/k^2) i\mathbf{k} \cdot \partial\varphi(\mathbf{v}) \int d\mathbf{v}' \rho_{\mathbf{k}}(\mathbf{v}'; t) \tag{23.1}$$

We shall try to find a solution of this equation in the form

$$\rho_{\mathbf{k}}(\mathbf{v}; t) = \chi_{\mathbf{k}, v}(\mathbf{v}) \, e^{-ikvt} \tag{23.2}$$

In particular, we shall ask for what values of v there exists a solution of this form; we are thus faced with an *eigenvalue problem.* Substituting (23.2) into (23.1) we obtain the following equation:

$$\mathbf{k} \cdot \mathbf{v} \chi_{\mathbf{k}, v}(\mathbf{v}) - \omega_p^2 k^{-2} \mathbf{k} \cdot \partial\varphi(\mathbf{v}) \int d\mathbf{v}' \chi_{\mathbf{k}, v}(\mathbf{v}') = kv \chi_{\mathbf{k}, v}(\mathbf{v}) \tag{23.3}$$

This is a typical eigenvalue equation: $\chi_{\mathbf{k}, v}(\mathbf{v})$ is the eigenfunction of the Vlassov equation corresponding to the eigenvalue kv. It is an integral equation for the function $\chi_{\mathbf{k}, v}(v_x, v_y, v_z)$. We now show that it is sufficient to solve a simpler integral equation, in one dimension; the complete function $\chi_{\mathbf{k}, v}(\mathbf{v})$ is then determined by purely algebraic relations. Define a function $\bar{\chi}_v(\nu)$ from $\chi_{\mathbf{k}, v}(\mathbf{v})$ by the "barring" operation defined in eq. (14.6)

* N. G. van Kampen, *Physica,* **21**, 949 (1955); **23**, 641 (1957).
** K. M. Case, *Ann. Phys.* **7**, 349 (1959).

$$\bar{\chi}_v(\nu) = \int d\mathbf{v}\,\delta(\nu - \mathbf{k}\cdot\mathbf{v}/k)\chi_{\mathbf{k},v}(\mathbf{v}) \tag{23.4}$$

(we drop the index $\mathbf{k}$ from $\bar{\chi}_v(\nu)$ for brevity, although $\bar{\chi}_v(\nu)$ is also a function of $\mathbf{k}$). We obtain an equation for $\bar{\chi}_v(\nu)$ by multiplying both sides of (23.3) by $\delta(\nu - \mathbf{k}\cdot\mathbf{v}/k)$ and integrating over $\mathbf{v}$

$$k\nu\bar{\chi}_v(\nu) - v_p^2 k\bar{\varphi}'(\nu)\int d\nu'\,\bar{\chi}_v(\nu') = kv\bar{\chi}_v(\nu) \tag{23.5}$$

where

$$v_p = \omega_p/k \tag{23.6}$$

Noting that

$$k\nu = \mathbf{k}\cdot\mathbf{v}$$

$$\int d\mathbf{v}'\chi_v(\mathbf{v}') = \int d\nu'\,\bar{\chi}_v(\nu')$$

we can eliminate the integral between eqs. (23.3) and (23.5); we obtain an algebraic relation between $\chi_{\mathbf{k},v}(\mathbf{v})$ and $\bar{\chi}_v(\nu)$:

$$\chi_{\mathbf{k},v}(\mathbf{v}) = \frac{\mathbf{k}\cdot\partial\varphi(\mathbf{v})}{k\bar{\varphi}'(\nu)}\bar{\chi}_v(\nu) \tag{23.7}$$

It is thus sufficient to solve the simpler equation (23.5), which we rewrite as:

$$(\nu - v)\bar{\chi}_v(\nu) = v_p^2\bar{\varphi}'(\nu)\int_{-\infty}^{\infty} d\nu'\bar{\chi}_v(\nu') \tag{23.8}$$

We now note that eq. (23.8) is homogeneous; the function is determined only up to an arbitrary factor. We can therefore require an additional normalization condition:

$$\int_{-\infty}^{\infty} d\nu\,\bar{\chi}_v(\nu) = 1 \tag{23.9}$$

Combining eqs. (23.8) and (23.9) we obtain the *inhomogeneous* equation:

$$(\nu - v)\bar{\chi}_v(\nu) = v_p^2\bar{\varphi}'(\nu) \tag{23.10}$$

We can then write the following solution directly:

$$\bar{\chi}_v^*(\nu) = v_p^2\bar{\varphi}'(\nu)[\nu - v]^{-1} \tag{23.11}$$

However, this solution is not satisfactory, because it has a singularity for $\nu = v$. In order to make it satisfactory, we need an unambiguous rule stating the contribution of the singularity when it occurs under an integral. In effect, $\bar{\chi}_v(\nu)$ is essentially a distribution function and hence an intermediate tool for calculating averages. It therefore need not be a "good function": it is sufficient to define it as a "distribution" in the sense of Schwartz * (i.e. a singular but integrable "function"). In order to derive these rules we note the following facts:

(a) $\bar{\chi}_v^*(\nu)$ has an unambiguous meaning as long as $\nu \neq v$. We can therefore replace that solution by $\mathscr{P}[v_p^2\bar{\varphi}'(\nu)/(\nu-v)]$.

(b) $\bar{\chi}_v^*(\nu)$ is not the complete solution of eq. (23.10): we can still add an arbitrary solution of the associated homogeneous equation:

$$(\nu-v)f_v(\nu) = 0$$

This equation has the solution $f_v(\nu) = \delta(\nu-v)$, which can be multiplied by an arbitrary function of v, say $\varepsilon_1(v)$.

Hence the most general solution of eq. (23.10) is

$$\bar{\chi}_v(\nu) = v_p^2\bar{\varphi}'(\nu)\mathscr{P}\frac{1}{\nu-v}+\varepsilon_1(v)\delta(\nu-v) \tag{23.12}$$

This form of $\bar{\chi}_v(\nu)$ is now an unambiguous distribution. But we still need to determine the arbitrary function $\varepsilon_1(v)$. This is done by the normalization condition (23.9)

$$v_p^2\mathscr{P}\int_{-\infty}^{\infty} d\nu\bar{\varphi}'(\nu)/(\nu-v)+\varepsilon_1(v)\int_{-\infty}^{\infty} d\nu\,\delta(\nu-v) = 1$$

from which we obtain:

$$\varepsilon_1(v) = 1+v_p^2\mathscr{P}\int_{-\infty}^{\infty} d\nu\,\bar{\varphi}'(\nu)/(v-\nu) \tag{23.13}$$

We note at this point the important fact that, through eq. (23.13), one can find an eigenfunction $\bar{\chi}_v(\nu)$ for *any* value of the frequency kv and for *any* value of the wave-vector $\mathbf{k}$. This means that strictly speaking *there exists no dispersion relation* of the form $v = v(k)$. In other words, the spectrum of eigenvalues of the

* See L. Schwartz, *Théorie des Distributions*, éd. Hermann, Paris, 1950; M. J. Lighthill, *Introduction to Fourier Analysis and Generalized Functions*, Cambridge University Press, 1958.

Vlassov equation consists of the whole real axis, from $-\infty$ to $+\infty$. We shall come back to the implications of this property in § 26.

Let us now introduce the notation

$$\varepsilon_2(v) = \pi v_p^2 \bar{\varphi}'(v) \tag{23.14}$$

It is then seen, by comparison with eq. (19.4), that the complex function

$$\varepsilon_+(v) = \varepsilon_1(v) - i\varepsilon_2(v) \tag{23.15a}$$

is identical with the dielectric constant for the real value $z = kv$ of the frequency. We note moreover that, whereas $\varepsilon_+(v) - 1$ is a "plus-function", its complex conjugate $[\varepsilon_+(v)]^* - 1$ is a "minus-function" (because $v_p^2 \bar{\varphi}'(v)$ is real) (see Appendix 2). This justifies the following notation:

$$\varepsilon_-(v) = \varepsilon_1(v) + i\varepsilon_2(v) \tag{23.15b}$$

We can now write the eigenfunctions in the following useful form

$$\bar{\chi}_v(\nu) = \varepsilon_2(\nu)\pi^{-1}\mathscr{P}(\nu - v)^{-1} + \varepsilon_1(\nu)\delta(\nu - v) \tag{23.16}$$

Keeping in mind the definitions (A2.2.2) and (A2.2.3) of the singular functions $\delta_\pm(x)$ we may also write this expression in the following equivalent forms:

$$\bar{\chi}_v(\nu) = \operatorname{Re}\,[\varepsilon_-(\nu)\delta_-(\nu - v)] \tag{23.17a}$$

$$\bar{\chi}_v(\nu) = \operatorname{Re}\,[\varepsilon_+(\nu)\delta_+(\nu - v)] \tag{23.17b}$$

§ 24. Eigenfunctions of the Adjoint Vlassov Operator

One of the most important properties of the set of eigenfunctions of a hermitian operator, such as those occurring in quantum mechanics, consists in providing a complete set of mutually orthogonal normalized functions.

This statement implies:

(a) *If the set is complete*, any sufficiently well behaved function of ν, say $\alpha(\nu)$, can be expanded in a series of eigenfunctions. In our case the spectrum is continuous, hence the series becomes an integral:

$$\alpha(\nu) = \int_{-\infty}^{\infty} dv\, \bar{\chi}_v(\nu) A(v) \tag{24.1}$$

(b) *If the functions are mutually orthogonal and are normalized,* the coefficients of the expansion are given by:

$$A(v) = \int_{-\infty}^{\infty} d\nu \, \bar{\chi}_v(\nu)\alpha(\nu) \tag{24.2}$$

However, in our case we must be very careful before making the statements (24.1–2) for the following two reasons:

A. The Vlassov operator is *not* hermitian.

B. The eigenfunctions $\bar{\chi}_v(\nu)$ are not true functions, but distributions.

As a first consequence, the eigenfunctions are *not* mutually orthogonal:

$$\int d\nu \, \bar{\chi}_v(\nu)\bar{\chi}_{v'}(\nu) \neq 0 \quad \text{for} \quad v \neq v'$$

and hence (24.2) is certainly not true [even if (24.1) is true]. We can, however, as is usual in the theory of non-hermitian operators, introduce another set of functions $\tilde{\chi}_v(\nu)$ with the following properties:

(α) There is a one-to-one correspondence between the functions $\bar{\chi}_v(\nu)$ and $\tilde{\chi}_v(\nu)$.

(β) The following orthogonality property holds:

$$\int d\nu \, \bar{\chi}_v(\nu)\tilde{\chi}_{v'}(\nu) = C_v \delta(v-v') \tag{24.3}$$

The new set of functions is defined as the set of eigenvalues of the *adjoint Vlassov equation*:

$$(\nu - v)\tilde{\chi}_v(\nu) = v_p^2 \int d\nu' \, \bar{\varphi}'(\nu')\tilde{\chi}_v(\nu') \tag{24.4}$$

This is a homogeneous equation of the same general type as (23.8) and can be solved by the same method as in § 23. We again have the freedom to impose a normalization condition; it is convenient to take it in the form

$$v_p^2 \int d\nu' \, \bar{\varphi}'(\nu')\tilde{\chi}_v(\nu') = 1 \tag{24.5}$$

The same arguments as before then yield the solution:

$$\tilde{\chi}_v(\nu) = \mathscr{P}(\nu-v)^{-1}+\tilde{\lambda}(v)\delta(\nu-v) \tag{24.6}$$

where $\tilde{\lambda}(v)$ is determined by the normalization condition (24.5) as:

$$\tilde{\lambda}(v) = \pi\varepsilon_1(v)/\varepsilon_2(v) \tag{24.7}$$

It is very easy to show that the functions $\bar{\chi}_v(\nu)$ and $\tilde{\chi}_{v'}(\nu)$ are orthogonal. Rewrite eqs. (23.8) and (24.4) in the form:

$$v\bar{\chi}_v(\nu) = \nu\bar{\chi}_v(\nu)-v_p^2\bar{\varphi}'(\nu)\int d\nu'\bar{\chi}_v(\nu')$$

$$v'\tilde{\chi}_{v'}(\nu) = \nu\tilde{\chi}_{v'}(\nu)-v_p^2\int d\nu'\,\bar{\varphi}'(\nu')\tilde{\chi}_{v'}(\nu')$$

Multiply the first one through by $\tilde{\chi}_{v'}(\nu)$ and the second by $-\bar{\chi}_v(\nu)$, integrate both over ν and add them term by term. We thus obtain:

$$(v-v')\int d\nu\,\bar{\chi}_v(\nu)\tilde{\chi}_{v'}(\nu) = 0$$

This relation is equivalent to (24.3) and therefore the orthogonality property is proven. However, this simple proof does not provide the value of the normalization constant C_v. The latter is obtained by a somewhat delicate calculation which is given in Appendix 3, with the result:

$$C_v = \pi|\varepsilon_-(v)|^2/\varepsilon_2(v) \tag{24.8}$$

§ 25. Completeness of the Set of Vlassov Eigenfunctions

At the present stage of our investigation, we have at our disposal two mutually orthogonal sets of eigenfunctions. Coming back to the problem of the expansion of an arbitrary function of ν, we can state the following.

If an arbitrary function $\alpha(\nu)$ can be expanded as

$$\alpha(\nu) = \int_{-\infty}^{\infty} dv\,A(v)\bar{\chi}_v(\nu) \tag{25.1}$$

then the coefficients are given by

$$A(v) = C_v^{-1}\int_{-\infty}^{\infty} d\nu\,\alpha(\nu)\tilde{\chi}_v(\nu) \tag{25.2}$$

We shall now examine under what conditions the set $\bar{\chi}_v(\nu)$ is complete. Mathematically this problem is stated as follows: what are the conditions under which the integral equation (25.1) has a solution $A(v)$ for any prescribed $\alpha(\nu)$?

Using formula (23.16), eq. (25.1) can be rewritten as:

$$\varepsilon_1(\nu)A(\nu)-\varepsilon_2(\nu)\pi^{-1}\mathscr{P}\int dv\, A(v)[v-\nu]^{-1} = \alpha(\nu) \tag{25.3}$$

This is a typical singular integral equation of the Cauchy type. These equations are dealt with in detail in Appendix 2, § A2.3. However, it turns out that eq. (25.3) is of an especially simple type, for the solution of which the general theory can be avoided.

It is shown in Appendix 2, § A2.2, that any integrable function $f(\nu)$ of a real variable can be split into two parts, a "plus part" and a "minus part":

$$f(\nu) = f_+(\nu)-f_-(\nu) \tag{25.4}$$

The two components are defined as follows:

$$f_\pm(\nu) = \pm\tfrac{1}{2}\int d\nu'\, \delta_\pm(\nu-\nu')f(\nu') \tag{25.5}$$

The characteristic property of a plus-function is the fact that it possesses an analytical continuation into the upper half-plane S_+, and that the latter is regular in the whole half-plane S_+. An analogous property holds for the analytical continuation of a minus-function into S_-. It follows from (25.5) that

$$i\pi^{-1}\mathscr{P}\int_{-\infty}^{\infty} d\nu'\,(\nu-\nu')^{-1}f(\nu') = f_+(\nu)+f_-(\nu) \tag{25.6}$$

Applying eqs. (25.4) and (25.6) to the function $A(\nu)$, and keeping in mind the definitions (23.15 a–b), we can rewrite eq. (25.3) as follows

$$\varepsilon_+(\nu)A_+(\nu) - \varepsilon_-(\nu)A_-(\nu) = \alpha(\nu) \tag{25.7}$$

It has been shown in § 23 that $\varepsilon_+(\nu)$ equals 1 plus a "plus-function". Hence $\varepsilon_+(\nu)A_+(\nu)$ is a plus-function, and $\varepsilon_-(\nu)A_-(\nu)$ is a minus-function. This is an enormous simplification over the general type of equation (A2.3.1).

Indeed, it is now sufficient to split $\alpha(\nu)$ into its plus and minus parts and identify separately the plus terms and the minus terms:

$$\begin{aligned} \varepsilon_+(\nu)A_+(\nu) &= \alpha_+(\nu) \\ \varepsilon_-(\nu)A_-(\nu) &= \alpha_-(\nu) \end{aligned} \tag{25.8}$$

These equations are, however, not always soluble. In order to be so, it is necessary that

$$A_+(\nu) = \alpha_+(\nu)/\varepsilon_+(\nu)$$

is a plus-function. In other words, the analytical continuation of $A_+(\nu)$ into the upper half-plane,

$$A_+(z) = \alpha_+(z)/\varepsilon_+(z), \qquad z \in S_+$$

must be regular in S_+. As $\alpha_+(z)$ is regular by the definition of a plus-function, it is thus necessary and sufficient that

$$\varepsilon_+(z) \neq 0, \qquad z \in S_+ \tag{25.9}$$

Remarkably enough, this condition is identical with the *stability condition* which has been discussed in detail in § 19. The same condition ensures the fact that

$$\varepsilon_-(z) \neq 0, \qquad z \in S_-$$

and consequently that

$$A_-(\nu) = \alpha_-(\nu)/\varepsilon_-(\nu)$$

is a minus-function. Hence we have proved the following:

THEOREM: *If the momentum distribution $\bar{\varphi}(\nu)$ is stable, the set of eigenfunctions $\bar{\chi}_v(\nu)$ is complete.* Any function $\alpha(\nu)$ can be expanded in the form (25.1), with the coefficients given by

$$A(v) = [\alpha_+(v)/\varepsilon_+(v)] - [\alpha_-(v)/\varepsilon_-(v)] \tag{25.10}$$

It is an easy matter to prove the equivalence of (25.10) and (25.2). Indeed, using (24.6–8), (25.4), (25.6) and (23.15 a–b), we find successively:

$$\begin{aligned} C_v^{-1}\int d\nu\bar{\chi}_v(\nu)\alpha(\nu) &= [\varepsilon_2(v)/\pi|\varepsilon_-(v)|^2]\int d\nu\,\alpha(\nu)[\mathscr{P}(\nu-v)^{-1} \\ &\quad +\pi[\varepsilon_1(v)/\varepsilon_2(v)]\delta(\nu-v)] \\ &= |\varepsilon_-(v)|^{-2}\{\varepsilon_2(v)[i\alpha_+(v)+i\alpha_-(v)]+\varepsilon_1(v)[\alpha_+(v)-\alpha_-(v)]\} \\ &= [\alpha_+(v)\varepsilon_-(v)-\alpha_-(v)\varepsilon_+(v)]/|\varepsilon_-(v)|^2 \end{aligned}$$

which is identical with (25.10). The completeness theorem can be expressed compactly by a *"closure relation"*

$$\int dv\, \bar{\chi}_v(\nu_1)\tilde{\chi}_v(\nu_2)C_v^{-1} = \delta(\nu_1 - \nu_2) \tag{25.11}$$

The main result of this section is the connection between stability and completeness of the eigenfunction system $\bar{\chi}_v(\nu)$. If the system is unstable, it turns out that there are "not enough" functions in the set $\bar{\chi}_v(\nu)$: one can no longer expand an arbitrary function. This is, however, not a serious drawback in itself. Case * has shown how to generalize the previous procedure in order to include unstable situations too. One essentially has to add to the continuous real spectrum a set of discrete eigenvalues.

Let us indicate very briefly how this generalization is performed. We assume, by definition, that the "plus-dielectric constant" $\varepsilon_+(z)$, eq. (19.2), has s simple complex zeros z_i in the upper half-plane:

$$\varepsilon_+(z_i) = 0; \quad z_i \in S_+; \quad i = 1, 2, \ldots, s \tag{25.12}$$

It then follows that the "minus-dielectric constant" $\varepsilon_-(z)$, i.e. the function defined by eq. (19.2) for $z \in S_-$, or equivalently the analytical continuation into S_- of $\varepsilon_-(v)$ defined by eq. (23.15b), also has s zeros in the lower half-plane:

$$\varepsilon_-(z_i) = 0; \quad z_i \in S_-; \quad i = s+1,\ s+2, \ldots, 2s \tag{25.13}$$

Actually the latter zeros are complex conjugate to the corresponding zeros of $\varepsilon_+(z)$:

$$z_{s+j} = z_j^*; \quad j = 1, \ldots, s \tag{25.14}$$

It is very easily verified that, besides the eigenfunctions $\bar{\chi}_v(\nu)$ defined by (23.16), one can construct $2s$ more eigenfunctions by the formula:

$$\bar{\chi}_i(\nu) = \frac{1}{\pi}\frac{\varepsilon_2(\nu)}{\nu - z_i}; \quad i = 1, 2, \ldots, 2s \tag{25.15}$$

These eigenfunctions are normalized:

$$\int_{-\infty}^{\infty} d\nu\, \bar{\chi}_i(\nu) = 1. \tag{25.16}$$

Indeed, this equation is just another way of writing (25.12) and (25.13).

The corresponding adjoint eigenfunctions are given by

$$\tilde{\chi}_i(\nu) = \frac{1}{\nu - z_i}; \quad i = 1, 2, \ldots, 2s \tag{25.17}$$

* See ref. p. 117.

and they are normalized as follows

$$\pi^{-1}\int d\nu\, \varepsilon_2(\nu)\tilde{\chi}_i(\nu) = 1 \tag{25.18}$$

One can easily extend the calculations of this chapter and prove the following theorems.

1) *Orthogonality relations*:

$$\int d\nu\, \bar{\chi}_v(\nu)\tilde{\chi}_i(\nu) = \int d\nu\, \bar{\chi}_i(\nu)\tilde{\chi}_v(\nu) = 0$$

$$\int d\nu\, \bar{\chi}_i(\nu)\tilde{\chi}_j(\nu) = C_i\delta_{ij} \tag{25.19}$$

$$C_i = \frac{1}{\pi}\int d\nu\, \frac{\varepsilon_2'(\nu)}{\nu - z_i} = -\varepsilon_{\pm}'(z_i)$$

where the plus sign refers to $i = 1, \ldots, s$ and the minus sign to $i = s+1, \ldots, 2s$. These equations complete eq. (24.3) in the case of unstable systems.

2) Equation (25.11) is replaced by the following more general *closure relation*:

$$\int d\nu\, \bar{\chi}_v(\nu_1)\tilde{\chi}_v(\nu_2)C_v^{-1} + \sum_{i=1}^{2s} \bar{\chi}_i(\nu_1)\tilde{\chi}_i(\nu_2)C_i^{-1} = \delta(\nu_1 - \nu_2) \tag{25.20}$$

§ 26. Application of the Eigenvalue Expansion to the Solution of the Initial-Value Problem

We now show that the initial-value problem for the Vlassov equation can be solved very simply by means of the eigenfunction expansion. To be definite, we shall calculate the density excess in a *stable* plasma

$$h_{\mathbf{k}}(t) = \int d\mathbf{v}\, \rho_{\mathbf{k}}(\mathbf{v}; t) = \int d\nu\, \bar{\rho}_{\mathbf{k}}(\nu; t)$$

assuming that initially the "barred" $\bar{\rho}_{\mathbf{k}}(\nu; 0)$ is given by

$$\bar{\rho}_{\mathbf{k}}(\nu; 0) = \bar{q}_{\mathbf{k}}(\nu) \tag{26.1}$$

We can expand $\bar{q}_{\mathbf{k}}(\nu)$ in terms of eigenfunctions in the form:

$$\bar{q}_{\mathbf{k}}(\nu) = \int_{-\infty}^{\infty} dv\, Q(v)\bar{\chi}_v(\nu) \tag{26.2}$$

with

$$Q(v) = C_v^{-1}\int_{-\infty}^{\infty} d\nu\, \bar{q}_{\mathbf{k}}(\nu)\tilde{\chi}_v(\nu) \tag{26.3}$$

Then we immediately have the general solution of the initial-value problem in the form:

$$\bar{\rho}_{\mathbf{k}}(\nu; t) = \int dv Q(v) \bar{\chi}_v(\nu) \mathrm{e}^{-ikvt} \tag{26.4}$$

(This formula is the exact analog of the eigenfunction expansion currently used in quantum mechanics.) The integration over ν is trivial because of the normalization condition (23.9). The final result is:

$$h_{\mathbf{k}}(t) = \int_{-\infty}^{\infty} dv Q(v) \mathrm{e}^{-ikvt} \tag{26.5}$$

This expansion can be further simplified if we use for $Q(v)$ the form (25.10). We then obtain:

$$h_{\mathbf{k}}(t) = \int_{-\infty}^{\infty} dv \mathrm{e}^{-ikvt} \{[\bar{q}_+(v)/\varepsilon_+(v)] - [\bar{q}_-(v)/\varepsilon_-(v)]\}$$

where $\bar{q}_+(v)$ is the plus part of $\bar{q}_{\mathbf{k}}(v)$. For positive times, we can complete the contour of integration by a half-circle at infinity in the lower half-plane. As $\bar{q}_-(v)/\varepsilon_-(v)$ is a minus-function, it has no poles in S_-, and we are left with:

$$h_{\mathbf{k}}(t) = \int_{-\infty}^{\infty} dv \mathrm{e}^{-ikvt} \bar{q}_+(v)/\varepsilon_+(v) \tag{26.6}$$

We now proceed to show the equivalence of this result with the result obtained by the resolvent method. The point of comparison is obtained from eqs. (14.1–2):

$$h_{\mathbf{k}}(t) = (2\pi)^{-1} \int_C dz \frac{\mathrm{e}^{-izt}}{\varepsilon_+(z)} \int d\nu \frac{\bar{q}_{\mathbf{k}}(\nu)}{i(k\nu - z)} \tag{26.7}$$

As we deal with *stable* systems, $\varepsilon_+(z)$ has no zeros in the upper half-plane, and the contour of integration C can approach the real axis. Then let z be given by

$$z \equiv k(v + i\eta), \qquad \eta > 0$$

Applying the limiting formula (A2.2.8) we obtain

$$\begin{aligned} h_{\mathbf{k}}(t) &= \lim_{\eta \to 0} \frac{1}{2\pi} \int_C dvk \frac{\mathrm{e}^{-ik(vt + i\eta t)}}{\varepsilon_+(v)} \int d\nu \frac{\bar{q}_{\mathbf{k}}(\nu)}{ik(\nu - v - i\eta)} \\ &= \int_C dv \mathrm{e}^{-ikvt} \frac{1}{\varepsilon_+(v)} \tfrac{1}{2} \int d\nu \delta_+(v - \nu) \bar{q}_{\mathbf{k}}(\nu) = \int_C dv \mathrm{e}^{-ikvt} \frac{\bar{q}_+(v)}{\varepsilon_+(v)} \end{aligned}$$

This proves the complete equivalence of the Landau and of the van Kampen–Case treatments. We need thus comment no more on the implications of (26.7). It is, however, very interesting to point out the different point of view of van Kampen. The most important result of van Kampen was to show that *one can, for given k, obtain a solution of the Vlassov equation oscillating with any desired frequency kv, and without any damping*. This solution is obtained by choosing an appropriate initial condition, viz. $\bar{\varphi}_{\mathbf{k}}(\nu) = \bar{\chi}_v(\nu)$. Hence, mathematically speaking, there is no dispersion relation. There is, however, no contradiction between this statement and the results of the previous chapters, which showed that plasma oscillations are always Landau-damped and that a dispersion relation exists. Indeed, the "eigen-distributions" are physically unrealizable because of their highly singular character. Any physically realizable initial distribution function is a superposition of eigenfunctions, and we then obtain the result (26.6), which is equivalent to our previous results.

Another very important result achieved by the van Kampen–Case method is to provide us with a basis of eigenfunctions which reflects a basic structure of the statistical mechanics of charged particles. It will be seen further that these eigenfunctions provide a very elegant and simple method for solving other apparently unrelated fundamental plasma-physics problems.

CHAPTER 6

The Long-Time Behavior of Classical Plasmas

§ 27. General Structure of the Perturbation Series for Long Times

In the first part of this book, we have studied in some detail the short-time behavior of a plasma. The dimensional analysis performed in § 9 provided us with a guide for the selection of diagrams in the complete perturbation series. The two main results of that discussion are the following:

(a) the existence of a smallness parameter $\Gamma = e^2 c^{\frac{1}{3}} \beta$,

(b) the existence of a short-time scale related to plasma oscillations, $t_p = (m/e^2 c)^{\frac{1}{2}}$.

Guided by these facts we have derived in § 10 a set of rules for the choice of diagrams. All contributions proportional to powers of e^2c were retained, whereas terms of the form $e^{2m}(e^2c)^n$ were neglected, as leading to corrections proportional to powers of Γ. It was essential for the validity of the result that the coefficients of these parameters be bounded in time. We have already met situations in which this condition is not satisfied. The plasma instabilities occur in systems in which some special types of initial velocity distributions lead to exponentially growing contributions, so that after a very short time the convergence of the e^2c series breaks down. However, these instabilities are a rather exceptional feature. We now show that there exists another quite universal type of convergence breakdown of the e^2c series. This type of behavior is actually responsible for the irreversible approach to equilibrium. The discussion can be done most easily on the basis of a simple example.

Consider the simple diagram of Fig. 27.1, which will be called a *cycle*. It represents a contribution to $\varphi(\mathbf{v}; t)$ coming from $\varphi(\mathbf{v}; 0)$. The essential topological feature of this diagram is that it starts on the right with a "vacuum" state (no lines) and, after a passage through an intermediate state in which particles

α and n are correlated, it comes back to a final state identical to the initial one. The cycle is the simplest member of a class of diagrams called "*diagonal fragments*". These are defined as diagrams whose initial and final states are strictly identical, and moreover whose intermediate states are all different from the initial and final ones.

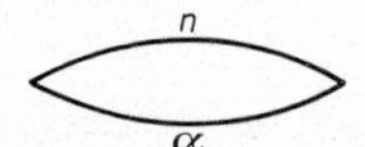

Fig. 27.1. The cycle.

Some comments are needed here in order to avoid confusion. A diagram such as the loop F in Table 6.1 is not diagonal, although it has one line on either side of the vertex: these lines do not bear the same index and therefore do not represent the same state. Also, diagram E of the same table is not diagonal: the δ-function of wave-vector conservation only requires the equality of the *sums* $(\mathbf{k}_j'+\mathbf{k}_n')$ and $(\mathbf{k}_j+\mathbf{k}_n)$. It should be kept in mind that the contribution of the cross E is actually a sum over $\mathbf{k}_j'$ and $\mathbf{k}_n'$. In this sum there is one term for which separately $\mathbf{k}_j' = \mathbf{k}_j$ and $\mathbf{k}_n' = \mathbf{k}_n$, but there is an infinity of other terms for which this equality is not true: the exceptional term contributes only N^{-1} to the total expression, and can therefore be disregarded. Such exceptional identities are not included in the class of diagonal fragments. On the contrary, in a diagram like the one of Fig. 27.2 the *separate* identities $\mathbf{k}_j = \mathbf{k}_j'''$, $\mathbf{k}_n = \mathbf{k}_n'''$ are direct consequences of the conservation law of wave-vectors [i.e. the corresponding contributions contain the factors $\delta(\mathbf{k}_j-\mathbf{k}_j''')\delta(\mathbf{k}_n-\mathbf{k}_n''')$]. This is what we call "strictly" identical states in the definition of a diagonal fragment. As a rather important consequence of this definition we note the following fact. *There is no one-vertex diagonal fragment*: the smallest possible diagonal fragment is the cycle.

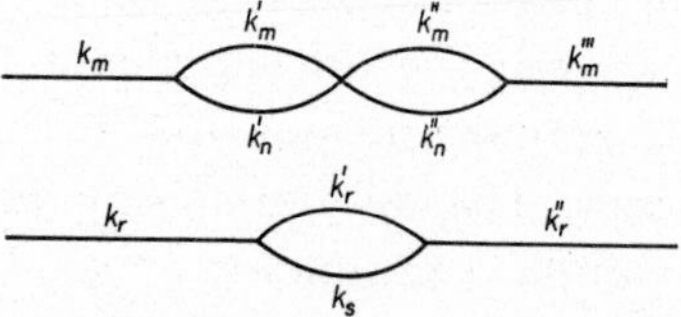

Fig. 27.2. A diagonal fragment with external lines.

Let us now go back to the cycle. The term of eq. (5.26) corresponding to this case is easily evaluated. [We use the matrix elements A and C of Table 6.1 and integrate over all particles but α; we also use eq. (2.23).]

$$\varphi(\mathbf{v}_\alpha;t)|_{\text{cycle}} = 8\pi^3 e^4 c m^{-2} \int (d\mathbf{v})^{N-1} \frac{1}{2\pi}\int_C dz\, e^{-izt} \int d\mathbf{l} \frac{1}{-iz} V_l(-i\mathbf{l}\cdot\partial_{\alpha n})$$

$$\cdot \frac{1}{i(\mathbf{l}\cdot\mathbf{g}_{\alpha n}-z)} V_l(i\mathbf{l}\cdot\partial_{\alpha n}) \frac{1}{-iz}\varphi_N(\{v\};0) \tag{27.1}$$

$$= 8\pi^3 e^4 c m^{-2} \int d\mathbf{v}_n \frac{1}{2\pi}\int_C dz\, e^{-izt} \frac{1}{-z^2} \partial_\alpha\cdot\mathbf{\Phi}(g_{\alpha n};z)\cdot\partial_{\alpha n}\varphi(\mathbf{v}_\alpha;0)\varphi(\mathbf{v}_n;0)$$

where the tensor function $\mathbf{\Phi}$ is defined as follows

$$\mathbf{\Phi}(g_{\alpha n};z) = \int d\mathbf{l} \frac{V_l^2 \mathbf{l}\mathbf{l}}{i(\mathbf{l}\cdot\mathbf{g}_{\alpha n}-z)} \tag{27.2}$$

(Remember $\mathbf{g}_{\alpha n} = \mathbf{v}_\alpha - \mathbf{v}_n$ is the relative velocity of particles α and n.) Our problem now is to evaluate the expression (27.1), and more particularly discuss its time dependence. We shall not, however, do this evaluation in detail here, but only point out the most relevant features for our present purpose.

Exactly as in the first part of this book, the resolvent method reduces the study of the time dependence to the study of the nature and the location of the singularities of an analytical function. We therefore have to study the singularities of the integrand in (27.1). It is immediately seen that this function has a *double pole at* $z = 0$. The existence of this double pole comes from *the diagonal nature of the diagram.* A diagram consisting of a succession of n diagonal fragments without external lines will have a $(n+1)$-fold pole at $z = 0$.

Consider now the function $\mathbf{\Phi}(z)$, defined by eq. (27.2). This function is a typical Cauchy integral, of the type already met and discussed in Appendix 2. One of its main properties, on which we have already insisted repeatedly, is the fact that it represents two different functions, according to whether the complex variable z lies in the upper or in the lower half-plane (S_+ or S_-); these two functions will be called $\mathbf{\Phi}^{(+)}$ and $\mathbf{\Phi}^{(-)}$ respectively. Each of them is regular in the domain of its original definition. However, $\mathbf{\Phi}^{(-)}$ is not the analytical continuation of $\mathbf{\Phi}^{(+)}$: there is a discontinuity along the whole real axis. On the other hand, the function $\mathbf{\Phi}^{(+)}$ can in general be continued analytically into the lower half-plane;

the methods for doing this are discussed in Appendix 2. The function $\mathbf{\Phi}^{(+)}(z)$ thus defined in the whole complex plane necessarily will have singularities in S_{-}. The nature of these singularities cannot be specified without further information about V_l, i.e. about the interaction potential. Let us assume, for the moment, that these singularities are *simple poles located in the lower half-plane*. It will be shown in the next section that in specific physical examples such poles do occur, although there may also exist other types of singularities. The latter, however, will be shown not to affect the main conclusions of this section. The location of these poles can be determined roughly from a dimensional argument. If the interaction force has a finite range, the function V_l will depend on some parameter having the dimension of a length, say r_0. The pole must then be located at a point $z = -i\alpha v/r_0$, where v is some average velocity and α is some finite positive constant. It is seen that the pole is roughly located at $z_c = -it_c^{-1}$, t_c being the "duration of a collision", defined in § 9. Such a pole contributes to the integral (27.1) a term depending on time through the exponential $e^{-iz_ct} \approx e^{-t/t_c}$ (see the discussion in Appendix 1 on the relation between the location of the poles and the time dependence). We already see at this point that such contributions are damped out after a time of the order of the collision time.

Consider now the contribution of the double pole at $z = 0$. The residue at this pole is obtained as usual by expanding the coefficient of z^{-2} around $z = 0$ and looking for the coefficient of z in this expansion. The contribution to the integral (27.1) is $(-2\pi i)$ times the residue. The complete evaluation of the integral (27.1) is:

$$\varphi(\alpha; t)|_{\text{cycle}} = 8\pi^3 e^4 c\, m^{-2} \int d\mathbf{v}_n \partial_\alpha \cdot \{t\mathbf{\Phi}^{(+)}(g_{\alpha n}; 0) + i\mathbf{\Phi}^{(+)\prime}(g_{\alpha n}; 0)$$
$$-i \sum_j e^{-iz_jt}\, \text{Res}_{z=z_j} \mathbf{\Phi}^{(+)}(g_{\alpha n}; z)\} \cdot \partial_{\alpha n} \varphi(\alpha; 0)\varphi(n; 0) \qquad (27.3)$$

In this formula $\mathbf{\Phi}^{(+)\prime} = \partial\mathbf{\Phi}^{(+)}/\partial z$; the summation over j is a summation over all poles located in the lower half-plane.

We have now separated three types of terms which are completely different in their time dependence:

(a) a group of *exponentially decaying terms*,
(b) *a constant term*,
(c) a term *proportional to t*.

We are now in a position to understand the statement made at the beginning of § 9 about the profound difference between short-time and long-time behavior. For very short times, of order t_c, these three types of terms are of the same order. As time passes, however, there will be a gradual transition toward a new regime. The exponential terms die out, the constant term, of course, remains bounded, and the third type of term grows systematically, eventually dominating all the other contributions. The long-time behavior is described practically by this term alone.

The different mathematical origin of the various terms points out the *universal character* of the terms (b) and (c). By this we mean that the time dependence of these terms is determined uniquely by the topological structure of the diagram. In other words, whatever the interaction potential V_l (and therefore, whatever the nature of the gas), the contribution of a cycle will always contain these two terms. Only the numerical values of the coefficients of the various powers of t depend on the interaction potential.

On the contrary, the number of complex poles, their location and the existence of other types of singularities, and thus the number and time dependence of the terms of type (a), depend crucially on the nature of the hamiltonian.

For this reason, the terms of type (b) and (c) are called the *universal* contributions of the cycle, whereas terms (a) are called *specific* contributions. The universal term proportional to the highest power of t in a given diagram [in the present case, term (c)] is called the *leading universal contribution.* *

§ 28. The Cycle Diagram in the Case of a Plasma

It has been shown in the previous section in the simplest case,

* It should be stated that in some exceptional situations it could happen that the systematic growth of the term (c) is masked by a stronger growth of terms which are essentially of type (a). This will turn out to be the case when instabilities occur. The discussion of this question is postponed to Appendix 9.

i.e. the cycle, that there exist in general two types of terms in the contributions to $\varphi(\alpha; t)$: *universal* terms, which depend on time through powers t^n (for the cycle, $n = 0$ and 1), and *specific* terms whose time dependence is fixed by the nature of the interactions. We shall investigate more closely the latter terms in the case of a plasma.

The problem consists of the evaluation of the function $\mathbf{\Phi}^{(+)}(g; z)$ defined by eq. (27.2) and the study of its singularities. The calculation is particularly convenient in a reference system (X, Y, Z) in which the Z-axis is directed along the vector $\mathbf{g}$ (see Fig. 28.1).

It is easily seen (from parity arguments) that the tensor $\mathbf{\Phi}^{(+)}$ is diagonal in this reference system, and that moreover the XX and YY components are equal. Let us then introduce the notation

$$\mathbf{\Phi}^{(+)}(z) = \Phi_1(z)(\mathbf{1}_x\mathbf{1}_x+\mathbf{1}_y\mathbf{1}_y)+\Phi_2(z)\mathbf{1}_z\mathbf{1}_z \tag{28.1}$$

Consider first $\Phi_2(z)$ for the case of a *Coulomb potential*. We introduce into eq. (27.2) the Fourier transform of the Coulomb potential, eq. (9.8). The result is, in polar coordinates:

$$\begin{aligned}\Phi_2 &= \frac{1}{4\pi^4}\int d\mathbf{l}\,\frac{l_z^2}{i(\mathbf{l}\cdot\mathbf{g}-z)}\,\frac{1}{l^4}\\ &= \frac{2\pi}{4\pi^4 i}\int_0^\pi d\theta\,\sin\theta\,\cos^2\theta\int_0^\infty dl\,l^2\,\frac{l^2}{lg\cos\theta-z}\,\frac{1}{l^4}\\ &= \frac{1}{2\pi^3 i}\int_{-1}^{1} d\xi\,\xi^2\int_0^\infty dl\,\frac{1}{lg\xi-z}\end{aligned} \tag{28.2}$$

We meet here, for the first time, a characteristic difficulty of the Coulomb interactions. The integral over l in (28.2) is logarithmically divergent at the upper limit; moreover, for $z \to 0$, it also becomes divergent at the lower limit. [It should be kept in mind that $\Phi_2(0)$ plays an especially important role, being the coefficient of t in the leading universal term of eq. (27.3).] These two divergences have different physical origins.

The divergence for $l \to \infty$ is a *short-distance divergence*. Such a behavior can be shown to be general for any potential which becomes infinite at short distances. Intuitively, it is quite natural that at very short distances, where the interactions become very large, a straight perturbation expansion in powers of e^2 should be impracticable. The large deflections produced in close collisions cannot be described by a single cycle. This difficulty is, however, not characteristic of plasmas: it occurs for all realistic gases. Its

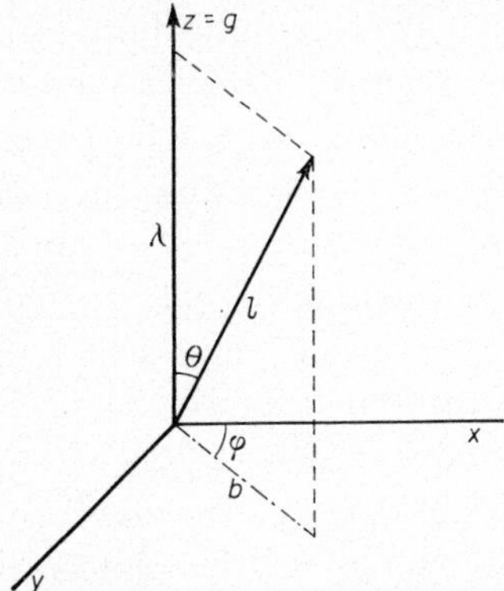

Fig. 28.1. Reference system for the evaluation of the integral (27.2).

rigorous solution involves the summation of a subseries of diagrams, representing the dominant contributions in a dilute gas, whatever the strength of the interactions (see § 33). As this problem is not specific to plasmas, we shall not treat it in full rigor here, in order not to complicate further a problem which is sufficiently difficult of itself. We shall treat this difficulty much more roughly, by cutting off the domain of integration at a value $l_{\max} = l_M$. The actual choice of this parameter is not important as far as the present section is concerned, and will be discussed later.

The divergence for $l \to 0$, i.e. for *large distances*, is a characteristic property of plasmas, which is, of course, related to the *collective effects*. The first part of this book has dealt with several aspects of these effects. Particularly, in § 18, it has been shown that the interactions in a plasma can be described in terms of an effective interaction potential, including the collective effects. This potential is essentially a screened and retarded potential whose shape is rather complicated and depends on the velocity distribution. If the transient effects are disregarded, it can be approximated rather closely by a *Debye potential*

$$V_\kappa(r) = e^{-\kappa r}/r \tag{28.3}$$

Remember the definition (16.7) of the inverse Debye length κ:

$$\kappa = (4\pi e^2 c\beta)^{\frac{1}{2}}$$

This potential arises quite naturally in all equilibrium problems concerning plasmas. Out of equilibrium it may be a good approximation in some circumstances, as a model which retains the essential collective screening. It has been customary to use this potential in most works concerning the long-time behavior of a plasma. In the present chapter, we shall also use this simple interaction in order to include roughly the collective effects. The rigorous study of the conditions under which this approximation is valid will be done later, in Chapter 11.

The use of the Debye potential (28.3) suppresses the divergences of (28.2) at $l \to 0$, although the divergence at $l \to \infty$ remains [because $V_\kappa(r) \to \infty$ as $r \to 0$]. The latter point will be taken care of by the cut-off at $l = l_M$. We now evaluate (27.2) with the use of the Debye potential (28.3). The Fourier transform of the latter is:

$$V_l = \frac{1}{2\pi^2} \frac{1}{l^2 + \kappa^2} \tag{28.4}$$

The evaluation is more conveniently done by using cylindrical coordinates (λ, b, φ) in the reference system of Fig. 28.1. We first calculate $\Phi_1(z)$. By standard procedures one obtains:

$$\begin{aligned} \Phi_1(z) &= \frac{1}{4\pi^4} \int_{-\infty}^{\infty} d\lambda \, \frac{1}{i(\lambda g - z)} \int_0^{l_M} db\, b \int_0^{2\pi} d\phi \, \frac{b^2 \sin^2 \phi}{(b^2 + \kappa^2 + \lambda^2)^2} \\ &= \frac{1}{8\pi^3 i} \int_{-\infty}^{\infty} d\lambda \left\{ -\frac{l_M^2}{\lambda^2 + \mu^2} + \ln \frac{\lambda^2 + \mu^2}{\lambda^2 + \kappa^2} \right\} \frac{1}{\lambda g - z} \end{aligned} \tag{28.5}$$

where

$$\mu = \sqrt{(\kappa^2 + l_M^2)}$$

The integration over λ can be performed by the method of residues. The first term enclosed in the curly brackets is integrated straightforwardly. The integrand of the second (logarithmic) term has a pole at z/g (which is located in S_+) and four branch-points, at $\pm i\mu$ and $\pm i\kappa$. We now complete the real axis with a half-circle at infinity in S_-; we must, however, avoid the two branch-points by making a cut. This leads to the contour of Fig. 28.2.

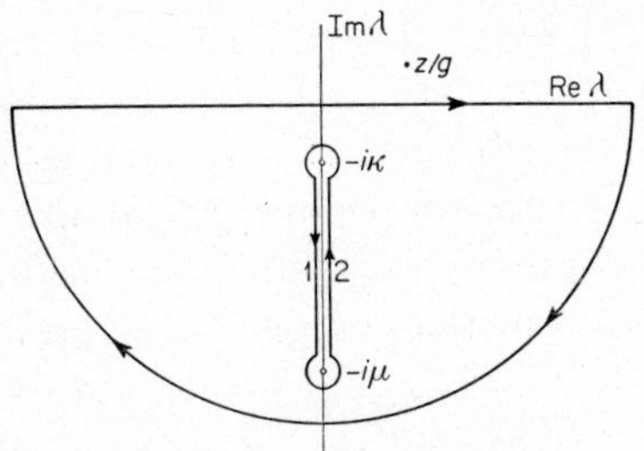

Fig. 28.2. Contour of integration in eq. (28.5).

As there is no pole within the contour, and as the large half-circle and the small circles around the branch-points do not contribute, we finally obtain

$$\int_{-\infty}^{\infty} d\lambda \frac{1}{\lambda g - z} \ln \frac{\lambda^2+\mu^2}{\lambda^2+\kappa^2} = \Big\{ \underset{(1)}{\int_{-i\mu}^{-i\kappa}} - \underset{(2)}{\int_{-i\mu}^{-i\kappa}} \Big\} d\lambda \frac{1}{\lambda g - z} \ln \frac{\lambda^2+\mu^2}{\lambda^2+\kappa^2}$$

$$= -2\pi i \int_{-i\mu}^{-i\kappa} d\lambda \frac{1}{\lambda g - z} = \frac{2\pi i}{g} \ln \frac{i\mu g + z}{i\kappa g + z}$$

Completing the calculation, we obtain the final result

$$\Phi_1(z) = \frac{l_M^2}{8\pi^2 i\mu} \frac{1}{z+i\mu g} + \frac{1}{4\pi^2 g} \ln \frac{z+i\mu g}{z+i\kappa g} \tag{28.6}$$

Similar (but simpler) calculations lead to the following result for $\Phi_2(z)$:

$$\begin{aligned} \Phi_2(z) &= \frac{1}{4\pi^4 i} \int_{-\infty}^{\infty} d\lambda \frac{\lambda^2}{\lambda g - z} \int_0^L db\, b \int_0^{2\pi} d\phi \frac{1}{(b^2+\kappa^2+\lambda^2)^2} \\ &= \frac{1}{2i} \Big\{ \frac{\kappa}{z+i\kappa g} - \frac{\mu}{z+i\mu g} \Big\} \end{aligned} \tag{28.7}$$

The function $\Phi_2(z)$ has analytical properties which agree with our qualitative analysis of § 27. It has two terms, each one having a pole in S_-, located respectively at $-i\kappa g$ and $-i\mu g$. These are the two characteristic inverse times which can be constructed with the parameters of the potential and play the role of t_c^{-1}. (It should be noted that $\mu g > \kappa g$.) The component $\Phi_1(z)$ on the other hand, while being regular in S_+ (as it should be) has an analytical continuation into S_- which, besides a pole at $z = -i\mu g$, has *two logarithmic branch-points*, at $-i\mu g$ and at $-i\kappa g$. Our previous assumption about the absence of singularities other than poles therefore breaks down for the Debye potential. Our purpose is to show now that the previous general conclusions are still valid in this case. We perform the complete z-integration in eq. (27.1), which we rewrite as

$$\varphi(\alpha; t)|_{\text{cycle}} = 8\pi^3 e^4 c\, m^{-2} \int d\mathbf{v}_n \partial_\alpha \cdot \mathbf{\Psi}(t) \cdot \partial_{\alpha n} \varphi(\alpha; 0) \varphi(n; 0) \tag{28.8}$$

with

$$\mathbf{\Psi}(t) = \frac{1}{2\pi} \int_C dz \frac{e^{-izt}}{-z^2} \mathbf{\Phi}^{(+)}(z) \equiv \Psi_1(t)[\mathbf{1}_x\mathbf{1}_x + \mathbf{1}_y\mathbf{1}_y] + \Psi_2(t)\mathbf{1}_z\mathbf{1}_z \tag{28.9}$$

We then have

$$\Psi_1(t) = \frac{1}{2\pi} \frac{1}{4\pi^2} \int_C dz \frac{e^{-izt}}{-z^2} \left\{ \frac{l_M^2}{2i\mu} \frac{1}{z+i\mu g} + \frac{1}{g} \ln \frac{z+i\mu g}{z+i\kappa g} \right\}$$

There is no difficulty in evaluating the first term by the method of residues. The second term is evaluated by taking a contour analogous to the one of Fig. 28.2 (but passing above $z = 0$). Some algebra leads to the following result:

$$\begin{aligned} \Psi_1(t) = \frac{1}{4\pi^2 g} \Bigg\{ \left[-\frac{l_M^2}{2\mu^2} + \ln \frac{\mu}{\kappa} \right] t + \Bigg[\left(\frac{l_M^2}{2\mu^2} + 1 \right) \frac{1}{\mu g} (1 - e^{-\mu g t}) \\ - \frac{1}{\kappa g} (1 - e^{-\kappa g t}) \Bigg] + t[\text{Ei}(-\kappa g t) - \text{Ei}(-\mu g t)] \Bigg\} \end{aligned} \tag{28.10}$$

where Ei (x) is the *exponential integral* defined as follows: *

$$\text{Ei}(-x) = -\int_x^\infty dt\, e^{-t}/t \tag{28.11}$$

The evaluation of the component $\Psi_2(t)$ presents no special difficulty. The result is

$$\Psi_2(t) = \frac{1}{2g} \left\{ \frac{1}{\mu g} (1 - e^{-\mu g t}) - \frac{1}{\kappa g} (1 - e^{-\kappa g t}) \right\} \tag{28.12}$$

* H. Bateman (A. Erdelyi, editor), *Higher Transcendental Functions*, vol. 2, McGraw-Hill, New York, 1953.

Let us first discuss $\Psi_1(t)$. We notice the presence of four types of terms: a term proportional to t, a constant, exponentially decaying term and terms of the form t Ei $(-\alpha t)$. The logarithmic branch-points therefore have introduced terms of the latter type. It is easy to study their behavior for long times by using the asymptotic expansion of these functions *

$$\mathrm{Ei}(-x) = -\frac{\mathrm{e}^{-x}}{x}\left[\sum_{m=0}^{M-1}\frac{m!}{(-x)^m} + O(|x|^{-M})\right] \tag{28.13}$$

Introducing this into (28.10) we obtain the following asymptotic behavior:

$$\Psi_1(t) \sim \frac{1}{4\pi^2 g}\left\{\left[-\frac{l_M^2}{2\mu^2} + \ln\frac{\mu}{\kappa}\right]t + \left(\frac{1}{\mu g} - \frac{1}{\kappa g}\right)\right.$$
$$\left. -\frac{l_M^2}{2\mu^3 g}\mathrm{e}^{-\mu g t} + \frac{1}{(\kappa g)^2}\frac{\mathrm{e}^{-\kappa g t}}{t} - \frac{1}{(\mu g)^2}\frac{\mathrm{e}^{-\mu g t}}{t} + \ldots\right\}, \qquad \kappa g t \gg 1 \tag{28.13}$$

The dominant term grows proportionally to t. The decaying terms do so quite strongly $(\mathrm{e}^{-\alpha g t}/t)$. Thus, the existence of logarithmic branch-points in S_- improves the approximation of neglecting non-universal terms in the long-time expressions. It can also be verified that the coefficient of t in (28.13) agrees with that found in eq. (27.3)

$$\frac{1}{4\pi^2 g}\left[-\frac{l_M^2}{2\mu^2} + \ln\frac{\mu}{\kappa}\right] = \Phi_1(0) \tag{28.14}$$

It is also illuminating to consider the behavior of (28.10) for small times. We then use the expansion *

$$\mathrm{Ei}(-x) = \gamma + \ln x + \sum_{n=1}^{\infty}\frac{(-x)^n}{n!\,n} \tag{28.15}$$

where γ is the Euler–Mascheroni constant. Substituting this into (28.10) and expanding the exponentials, we obtain

$$\Psi_1(t) \sim \frac{1}{4\pi^2 g}\left\{-\frac{l_M^2}{2\mu^2}\mu g t^2 + \tfrac{3}{4}t(\mu g t)^2 + \ldots\right\}, \quad \mu g t \ll 1 \tag{28.16}$$

As a result of many cancellations, it is seen that the dominant term for short times is proportional to t^2. This is a quite general

* Bateman, *loc. cit.*

feature of all contributions. The fact that all contributions are proportional to t^2 for short times is connected with the *reversible* behavior for short times: the dominant short-time contribution is invariant with respect to time inversion ($t \to -t$, $\mathbf{v} \to -\mathbf{v}$). On the contrary, the dominant contribution for long times is proportional to t and is thus not invariant with respect to time inversion. The component $\Psi_2(t)$ has the same property of beginning with a term proportional to t^2. However, it has no contribution proportional to t. Thus, in the "leading universal term" approximation, i.e. the long-time approximation, the tensor $\mathbf{\Psi}$ has only two non-vanishing (and equal) components in the reference system of Fig. 28.1. This result will be obtained by another method in Chapter 7.

The detailed calculation of this section, besides drawing attention to the convergence difficulties encountered in plasma physics, also shows on an exactly calculable example the change of structure of the perturbation series from the short-time behavior to the long-time behavior.

§ 29. Choice of Contributions for a Plasma

We are now in a position to generalize our foregoing results in order to derive specific rules for the choice of diagrams. We have seen in the case of the cycle that the terms can be divided into two groups: universal terms which are proportional to powers of t, and specific terms whose time dependence depends on the nature of the potential. The latter terms are exponentially decaying (at least asymptotically); we do not discuss them further here (for a detailed discussion, see Prigogine*). The universal terms arise from the presence of multiple poles on the real axis. Such multiple poles are necessarily due to the existence of diagonal fragments. The straightforward generalization of the previous calculations shows that a succession of n diagonal fragments contains $n+1$ identical states and therefore a $(n+1)$-fold pole. Two cases can occur:

(*a*) If the identical states are "vacuum" states, the pole is

* I. Prigogine, *Non-Equilibrium Statistical Mechanics*, Interscience, New York, 1963.

at $z = 0$. Applying the asymptotic theorem of Laplace transforms (see Appendix 1) we know that this pole contributes terms of the form:

$$t^n,\ t^{n-1}, \ldots, t^0$$

(*b*) If the repeated states involve a certain number of lines, the pole is at $z = \sum \mathbf{k}_j \cdot \mathbf{v}_j$. The same asymptotic theorem then tells us that this multiple pole contributes growing oscillations:

$$t^n \exp(-i\sum \mathbf{k}_j \cdot \mathbf{v}_j t),\ t^{n-1} \exp(-i\sum \mathbf{k}_j \cdot \mathbf{v}_j t), \ldots,\ \exp(-i\sum \mathbf{k}_j \cdot \mathbf{v}_j t)$$

Let us consider now very briefly some other types of diagrams which are not diagonal fragments. We cannot go into a full and systematic study of all possible types of diagrams here, but shall only discuss those which occur in subsequent calculations in this book. More details are found in Prigogine's monograph referred to above.

(*c*) *Pure destruction fragment.* This is defined as a connected diagram with n external lines at right and m external lines at left, with the condition $n > m$. We assume there is no diagonal fragment within the diagram. Let us discuss, for instance, the case $n = 2$, $m = 0$. The contribution begins with a factor $(-iz)^{-1}$ (corresponding to the vacuum state). The internal lines as well as the external lines at right correspond to wave-vectors which are integrated over: indeed the final expression, which is a contribution to ρ_0, does not depend on any wave-vector. The situation is the same as for the internal lines of a cycle, which correspond to a wave-vector $\mathbf{l}$ involved in an integral $\mathbf{\Phi}(z)$, eq. (27.2). Hence, this part of the diagram gives rise to poles shifted into S_-. There is only one pole at $z = 0$, giving a *constant universal contribution.*

(*d*) *Pure creation fragment.* This type of diagram is symmetrical to the previous one: it has more external lines at left than at right. Consider again an example with two external lines at left and no external lines at right: this is a contribution to $\rho_{\mathbf{k},-\mathbf{k}}(\alpha, \beta; t)$. The external lines correspond here to a factor $i(\mathbf{k} \cdot \mathbf{g}_{\alpha\beta} - z)^{-1}$, where the wave-vector $\mathbf{k}$ is no longer involved in a Cauchy integral [see e.g. eq. (7.2)]. Hence it corresponds to a pole on the real axis. If there are no diagonal fragments within the creation

Table 29.1. Relation between general shape of diagrams and universal time dependence. The hatched parts represent arbitrary connected parts with the given number of external lines. The arrows indicate the part of the diagram responsible for the poles. Linked arrows represent multiple poles.

Diagram	Poles on the real axis	Time dependence of universal terms
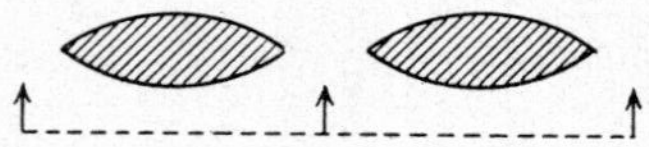	Triple pole in $z = 0$	$t^2,\ t,\ 1$
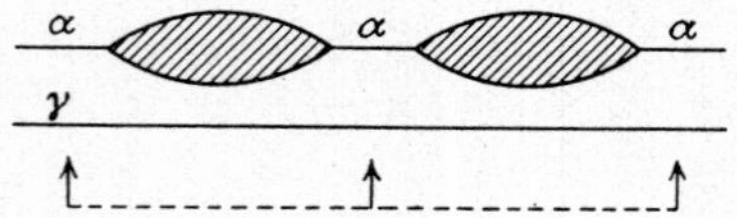 α α α γ	Triple pole in $z = \mathbf{k} \cdot \mathbf{g}_{\alpha\gamma}$	$t^2 e^{i\mathbf{k}\cdot\mathbf{g}t},\ t e^{i\mathbf{k}\cdot\mathbf{g}t},$ $e^{i\mathbf{k}\cdot\mathbf{g}t}$
	Simple pole in $z = 0$	1
	Simple pole in $z = \mathbf{k} \cdot \mathbf{g}$ Simple pole in $z = 0$	$e^{i\mathbf{k}\cdot\mathbf{g}t}$ 1
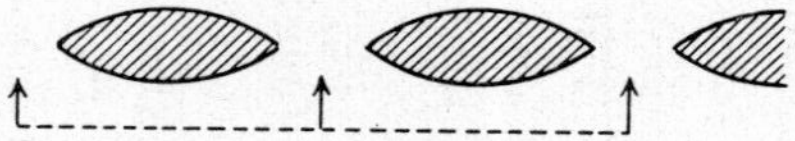	Triple pole in $z = 0$	$t^2,\ t,\ 1$
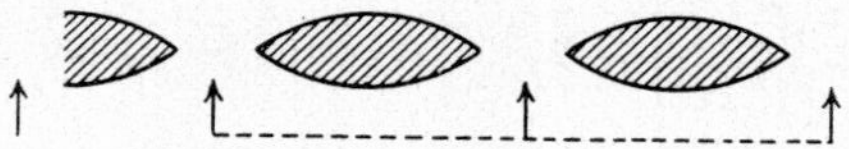	Simple pole in $z = \mathbf{k} \cdot \mathbf{g}$ Triple pole in $z = 0$	$e^{i\mathbf{k}\cdot\mathbf{g}t}$ $t^2,\ t,\ 1$

fragment, this pole is simple. The internal lines of the fragment are involved in Cauchy integrals, and therefore do not contribute real poles. But the vacuum state at right corresponds to a factor $(-iz)^{-1}$ and thus to a simple pole at the origin. The universal terms are thus proportional to $\exp(-i\mathbf{k}\cdot\mathbf{g}_{\alpha\beta}t)$ and 1.

(*e*) Creation, destruction and diagonal fragments can now be combined into complex diagrams, the detailed discussion of which is performed by using the same types of arguments as above. Some examples of the relation between the shape of the diagram and its time dependence are given in Table 29.1.

This short discussion is sufficient at the present time for the derivation of criteria for the choice of diagrams. Let us first consider again the simple case of the cycle where we had two universal terms, proportional respectively to:

$$e^4ct \quad \text{and} \quad e^4c.$$

The non-universal terms need not be discussed further, because we already know that they decay to zero for times longer than t_p. So far as the e- and c-dependence is concerned, these terms are small according to the criterion established in § 9: they contain one uncompensated power of e^2. Also, if we insert in the first one $t = t_p \sim (e^2c)^{-\frac{1}{2}}$ we find $e^3c^{\frac{1}{2}} = e^2(e^2c)^{\frac{1}{2}}$, which is also small. We thus were justified in neglecting such terms in the Vlassov approximation. But we are now interested in *long times*, of the order of the relaxation time t_r defined by eq. (9.13). Inserting that expression into e^4ct, we find that the factors e^4c are cancelled and the resulting term is no longer small, but of the order 1. On the other hand, the second universal term, e^4c, still remains small. Thus the characteristic change in structure of the perturbation series results in the fact that e^4ct becomes a quantity which must be retained to all orders, and cannot be regarded as small. Our first prescription will therefore be the following: select all diagrams of order $(e^4ct)^n$, $n = 0, 1, \ldots$, and sum completely the subseries thus obtained.

In order that this prescription be meaningful, we must, however, be sure that there are no larger terms in the perturbation series. For instance, there should be no terms of the form $(e^4c^2)^mt$. This, however, is not possible, for the following reason: in order to construct a diagonal fragment (and therefore a time factor), we need at least one creation vertex A and one destruction vertex C, Table 6.1 (or a vertex B and a vertex D, if there are outgoing lines), and thus we necessarily have a factor e^4c coming from these two vertices. In the intermediate transitions, we can only have factors e^2 or e^2c (according to the topological index of the respective vertex). Thus, if e^2 is considered small and e^2c finite (in the sense discussed in § 10), $e^2(e^2c)^nt$ is the maximum order possible for a given diagonal fragment, as far as e and c are concerned. It can, however, happen in certain exceptional circumstances (instabilities) that the *growth in time* is stronger than t (actually, exponential). These situations will be discussed in Appendix 9.

Another very important point is the following. The prescription $(e^4ct)^n$ is not complete in the case of a plasma. In order to be consistent with the rules of § 10, it must be realized that terms of order $(e^4ct)^n(e^2c)^m$, or equivalently $(e^2c)^p(e^2t)^q$, are of the same order of magnitude as the terms $(e^4ct)^n$. It will be one of the main purposes of this book to show that the summation of all the terms of this order suppresses at least one of the main theoretical difficulties of plasma physics, i.e. the long-distance divergence due to collective effects.

It must also be realized that the previous prescriptions are, of course, only a first approximation for the study of plasmas. In this approximation, the dominant contributions are retained in situations where the characteristic parameter Γ, eq. (9.9), is small. If one is interested in corrections to these results, one should consider terms of the order $e^2(e^2c)^n(e^2t)^m$, $e^4(e^2c)^n(e^2t)^m$, and so on. A brief study of the first corrections of this order will be made in Chapter 11.

Coming back to the dominant terms, the previous discussion can be summed up by stating the following prescription.

If Γ is considered as a small parameter, the dominant contributions to the distribution function are obtained by summing all the terms of order $(e^2c)^p(e^2t)^q$ *(p and q being arbitrary positive integers) in the perturbation series* (5.26).

We shall perform this double summation in two steps. In the first step, we sum over q all terms for which $p = q$. This means that we sum the diagrams of order $(e^4ct)^n$ discussed at the beginning of this section. This is done not merely for simplicity, but also because it leads to a very well known and widely used equation, i.e. the so-called *Landau equation*. This equation is especially simple and can be considered as a general framework in the study of the long-time behavior of plasmas. Moreover, it will turn out that the general equation obtained after performing the second summation over p can be cast into the same formal framework, although important mathematical differences will occur.

CHAPTER 7

The Landau ("Fokker–Planck") Approximation

§ 30. Evolution of the Velocity Distribution Function

According to the program outlined in § 29, we now extract from the perturbation series (5.26) the subseries of all terms proportional to $(e^4ct)^m$, $m = 0, 1, 2, \ldots$, and sum this partial series.

Let us explain what we mean by "sum the series" in this connection: the summation of the series will actually be reduced to the solution of a partial differential equation. In other words, it will be shown that the function represented by our selected subseries of (5.26) can also be represented as the solution (subject to the given initial condition) of a differential equation which is simpler than the original Liouville equation. The same idea has been applied in § 11 to the derivation of the Vlassov equation, although the particular case of the *linearized* Vlassov equation is one of the very few instances in which the selected subseries can be summed exactly in closed form (§ 13).

In the present approximation, we are necessarily interested in the one-particle distribution function [eq. (2.12)], which we rewrite in a more compact form in the limit (2.3):

$$f(\mathbf{x}_\alpha, \mathbf{v}_\alpha; t) = c\left\{\varphi(\mathbf{v}_\alpha; t) + \int d\mathbf{k}\, e^{i\mathbf{k}\cdot\mathbf{x}_\alpha} \rho_{\mathbf{k}}(\mathbf{v}_\alpha; t)\right\} \qquad (30.1)$$

As usual, when no confusion is possible, we occasionally drop the variable t and replace $\mathbf{v}_\alpha$ or the set $(\mathbf{x}_\alpha, \mathbf{v}_\alpha)$ by the letter α. It should be kept in mind that Fourier components representing correlations are initially proportional to $e^{2r}(e^2c)^m$, $r \geqq 1$, and are thus negligible in the present approximation where only terms of (effective) order $(e^2c)^m$ are retained.

In the present section we study the evolution of $\varphi(\alpha)$. The selection of the relevant diagrams is very simple in this case. In order to obtain a contribution of order $(e^4ct)^m$ we need m successive diagonal fragments ($= m$ factors of t). But, as stated in § 29, any diagonal fragment without external lines must begin with a creation vertex A and end with a destruction vertex C (Table 6.1); this means that every diagonal fragment contains at least the factors $e^2 \cdot e^2 c$, any other e^2 or $e^2 c$ factors coming from intermediate vertices. However, in the present case, these two factors are the only ones we need and thus there can be no intermediate vertices in this approximation. Moreover, any non-diagonal diagram would give rise to powers of e^2 which are not compensated by time factors, and must be neglected. We must moreover conform to the prescriptions of § 8 concerning reduced distribution functions. The general rule is therefore:

The most general diagram contributing to $\varphi(\alpha)$ terms of order $(e^4 ct)^m$ is a succession of m semiconnected cycles, of which at least the extreme one at left involves particle α.

We now proceed to the evaluation of the sum. Going back to eqs. (27.1) and (27.2), let us introduce the following operators:

$$8\pi^3 e^4 c\, m^{-2}\, \partial_{\alpha n} \cdot \mathbf{\Phi}^{(+)}(g_{\alpha n}; z) \cdot \partial_{\alpha n} \equiv \mathscr{C}_{\alpha n}(z)$$
$$\mathscr{C}_{\alpha n}(0) \equiv \mathscr{C}_{\alpha n} \tag{30.2}$$

A simple generalization of eq. (27.1) shows that the general term involving m semiconnected cycles contributes the following expression to $\varphi(\alpha)$ (see Fig. 30.1)

Fig. 30.1. A typical diagram contributing to $\varphi(\alpha; t)$ in the Landau approximation.

$$\varphi(\alpha; t)^{(m)} = \int d\mathbf{v}_1 \ldots \int d\mathbf{v}_m \frac{1}{2\pi} \int_C dz \frac{e^{-izt}}{(-iz)^{m+1}} [\mathscr{C}(z)]^m$$
$$\varphi(\alpha; 0)\varphi(1; 0) \ldots \varphi(m; 0) \tag{30.3}$$

It should be noted that the notation $[\mathscr{C}(z)]^m$ is actually an abbreviation for a product $\mathscr{C}_{\alpha 1}(z)\mathscr{C}_{\alpha 2}(z)\mathscr{C}_{13}(z) \ldots$ in which the indices depend on the type of semiconnection; these indices are, however, irrelevant at this point. Take $(-2\pi i)$ times the residue of the z-integral at the pole $z = 0$

$$\frac{1}{2\pi}\int_C dz\, e^{-izt}\frac{1}{(-iz)^{m+1}}[\mathscr{C}(z)]^m = \frac{t^m}{m!}[\mathscr{C}(0)]^m + i\frac{t^{m-1}}{(m-1)!}\left\{\frac{d[\mathscr{C}(z)]^m}{dz}\right\}_{z=0} + \dots \tag{30.4}$$

The terms neglected in (30.4) are of successive orders t^{m-2}, t^{m-3}, . . ., t^0 (plus, of course, the non-universal terms). According to the discussion of §§ 27 and 29, we only retain the leading term, with the result:

$$\varphi(\alpha; t)^{(m)} = \int (d\mathbf{v})^m \frac{t^m}{m!}[\mathscr{C}]^m \varphi(\alpha; 0)\varphi(1; 0) \dots \varphi(m; 0) \tag{30.5}$$

It should be pointed out that this result is very general in the sense that a succession of m semiconnected diagonal fragments of *any* type will always give rise to a leading term of the form (30.5). The only difference will lie in the form of $\mathscr{C}$ and in the number of particles involved (i.e. the number of φ-factors and $\mathbf{v}$-integrations); see appendix 4.

We must now sum the terms of type (30.5):

(1) over the number of cycles in a diagram, i.e. over m,

(2) for a given number of cycles, over all possible semiconnections.

In order to take care of the second point, the abbreviation $[\mathscr{C}]^m$ must be expanded by writing indices. Fig. 30.2, drawn for the case $m = 3$, will help in understanding the classification of all possible semiconnections for a given m.

The result is then obviously:

$$\varphi(\alpha; t) = \sum_{m=0}^{\infty}\frac{t^m}{m!}\left\{\int d\mathbf{v}_1 \mathscr{C}_{\alpha 1}\int d\mathbf{v}_2[\mathscr{C}_{\alpha 2}+\mathscr{C}_{12}]\int d\mathbf{v}_3[\mathscr{C}_{\alpha 3}+\mathscr{C}_{13}+\mathscr{C}_{23}] \dots \int d\mathbf{v}_m\left[\mathscr{C}_{\alpha m}+\sum_{r=1}^{m-1}\mathscr{C}_{rm}\right]\right\}\varphi(\alpha; 0)\varphi(1; 0)\dots\varphi(m; 0) \tag{30.6}$$

This is the formal sum we are interested in, representing the velocity distribution for $t \gg t_p$ in terms of $\varphi(\alpha; 0)$. We now derive an equation by time differentiation of this expression:

$$\partial_t \varphi(\alpha; t) = \sum_{m=1}^{\infty} \frac{t^{m-1}}{(m-1)!} \left\{ \int d\mathbf{v}_1 \mathscr{C}_{\alpha 1} \right.$$

$$\left. \ldots \int d\mathbf{v}_m \left[\mathscr{C}_{\alpha m} + \sum_{r=1}^{m-1} \mathscr{C}_{rm} \right] \right\} \varphi(\alpha; 0) \ldots \varphi(m; 0)$$

$$= \int d\mathbf{v}_1 \mathscr{C}_{\alpha 1} \left\{ \sum_{s=0}^{\infty} \frac{t^s}{s!} \left[\int d\mathbf{v}_2 [\mathscr{C}_{\alpha 2} + \mathscr{C}_{12}] \right. \right. \tag{30.7}$$

$$\ldots \int d\mathbf{v}_{s+1} \left[\mathscr{C}_{\alpha, s+1} + \mathscr{C}_{1, s+1} + \sum_{r=2}^{s} \mathscr{C}_{r, s+1} \right] \varphi(\alpha; 0) \varphi(1; 0) \ldots \varphi(s+1; 0)$$

$$= \int d\mathbf{v}_n \mathscr{C}_{\alpha n} \left\{ \sum_{s=0}^{\infty} \frac{t^s}{s!} \int d\mathbf{v}_1 [\mathscr{C}_{\alpha 1} + \mathscr{C}_{n1}] \right.$$

$$\ldots \int d\mathbf{v}_s [\mathscr{C}_{\alpha, s} + \mathscr{C}_{n, s} + \sum_{r=1}^{s-1} \mathscr{C}_{r, s}] \varphi(\alpha; 0) \varphi(n; 0) \varphi(1; 0) \ldots \varphi(s; 0)$$

In going from the first line to the second, the summation index has been changed from m to $s = m - 1$; in the next step,

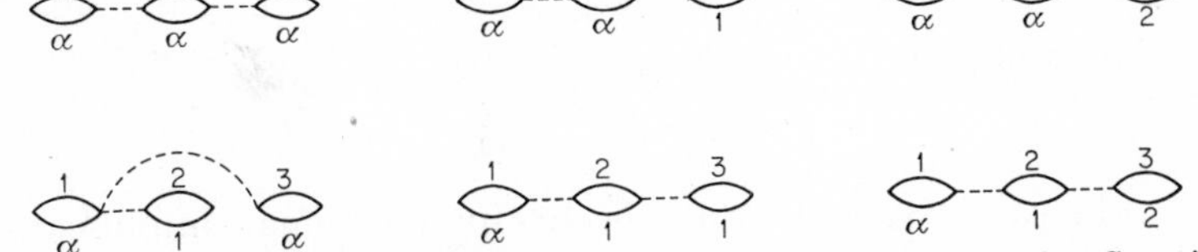

Fig. 30.2. All possible semiconnections of three successive cycles. Starting from the left, the first cycle necessarily involves α and a free particle, 1. The second cycle may have in common with the first either particle α (first line) or particle 1 (second line). The third cycle may share with the two previous ones either particle α (first column) or particle 1 (second column) or particle 2 (third column).

the free particle indices (which are mere integration variables) have been renamed as follows: $1 \to n,\ 2 \to 1, \ldots, s+1 \to s$. It is now a matter of somewhat tedious but straightforward algebra to show that the expression standing to the right of $\mathscr{C}_{\alpha n}$ in eq. (30.7) is identical with the product of two factors $\varphi(\alpha; t)\varphi(n; t)$, both expressed in the form (30.6).*

* The following obvious property is used in the proof: two operators $\mathscr{C}_{12}\mathscr{C}_{34}$ having no common particles commute. This enables one, for instance, to symmetrize such a product: $\mathscr{C}_{12}\mathscr{C}_{34} = \frac{1}{2}(\mathscr{C}_{12}\mathscr{C}_{34} + \mathscr{C}_{34}\mathscr{C}_{12})$.

The final expression then becomes:

$$\partial_t \varphi(\alpha; t) = \int d\mathbf{v}_n \mathscr{C}_{\alpha n} \varphi(\alpha; t) \varphi(n; t) \tag{30.8}$$

Before going over to writing out eq. (30.8) in an explicit form, we would like to note here that the last step in the previous proof is actually a proof of the important theorem of "propagation of the molecular chaos". This theorem states that if the two-body velocity distribution is factorized at the initial time: $\varphi_2(\alpha, n) = \varphi(\alpha)\varphi(n)$, then it remains so at all later times. It is significant that the proof is valid for *long* times.

We now write the equation of evolution (30.8) explicitly by evaluating the operator $\mathscr{C}_{\alpha n}$, or, equivalently [see eq. (30.2)], the function $\mathbf{\Phi}^{(+)}(g_{\alpha n}; 0)$, eq. (27.2). But $\mathbf{\Phi}^{(+)}(0)$ is the limiting value of a Cauchy integral for z approaching the real axis from above. This value is given by the Plemelj formulae (see Appendix 2):

$$\mathbf{\Phi}^{(+)}(0) = \int d\mathbf{l} |V_l|^2 \mathbf{l}\mathbf{l} \pi \delta_+(-\mathbf{l} \cdot \mathbf{g}_{\alpha n}) = \int d\mathbf{l} |V_l|^2 \mathbf{l}\mathbf{l} \pi \delta_-(\mathbf{l} \cdot \mathbf{g}_{\alpha n}) \tag{30.9}$$

where the functions $\delta_+(x)$ and $\delta_-(x)$ are defined as follows (see Appendix 2)

$$\delta_\pm(x) = \delta(x) \pm \frac{i}{\pi} \mathscr{P} \frac{1}{x}. \tag{30.10}$$

In the case of a cycle, there occurs an additional simplification. The δ part of the δ_- function is an even function of $\mathbf{l}$, whereas the principal part of $(\mathbf{l} \cdot \mathbf{g})^{-1}$ is odd. As the remainder of the integrand in (30.9) is even, only the δ part of δ_- contributes to the integral. Combining (30.9), (30.2) and (30.8), the final form of the equation of evolution of $\varphi(\alpha)$ is

$$\partial_t \varphi(\alpha) = \frac{8\pi^4 e^4 c}{m^2} \int d\mathbf{v}_n \int d\mathbf{l}\, V_l^2 \mathbf{l} \cdot \partial_\alpha \delta(\mathbf{l} \cdot \mathbf{g}_{\alpha n}) \mathbf{l} \cdot \partial_{\alpha n} \varphi(\alpha) \varphi(n) \tag{30.11}$$

This is the so-called "*Landau equation*" for a homogeneous plasma. Its discussion will be deferred to § 33 and to Chapter 8.

Let us mention the generalization of this equation to an s-component plasma:

$$\begin{aligned} \partial_t \varphi^{(\mu)}(\alpha) = \sum_{\nu=1}^{s} (8\pi^4 e^4 c_\nu / m_\mu) \int d\mathbf{v}_n \int d\mathbf{l}\, V^2 \mathbf{l} \cdot \partial_\alpha \delta(\mathbf{l} \cdot \mathbf{g}_{\alpha n}) \\ \mathbf{l} \cdot (m_\mu^{-1} \partial_\alpha - m_\nu^{-1} \partial_n) \varphi^{(\mu)}(\alpha) \varphi^{(\nu)}(n) \end{aligned} \tag{30.11'}$$

§ 31. The Concept of Pseudo-Diagonal Fragments

We now go over to the study of the inhomogeneity factor $\rho_{\mathbf{k}}(\mathbf{v}_\alpha; t)$. It has already been stressed repeatedly (§§ 3 and 10) that the choice of diagrams for inhomogeneous Fourier components is always more complex than for homogeneous components. The reason is the occurrence of an additional large characteristic length L_h (or characteristic time t_h) related to the spatial variation of the inhomogeneity. It appears that in the long-time behavior this additional difficulty plays a quite important role.

We always assume that the initial condition is the one discussed in detail in § 3. In the present section it will be shown that the universal terms of a certain type of diagram, which would normally be bounded in time, are actually growing systematically for times much shorter than t_h (although much longer than t_p).

The simplest example of such a diagram is shown in Fig. 31.1a.

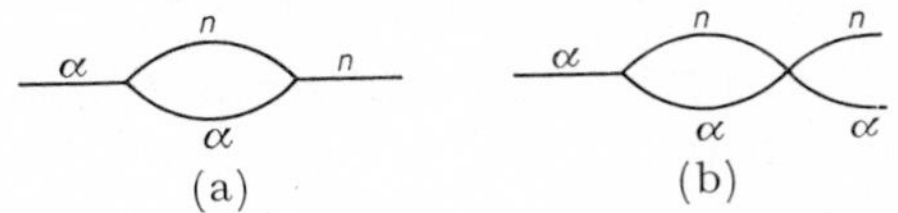

Fig. 31.1. The simplest pseudo-diagonal diagrams.

This diagram will be discussed in detail here. Using Table 6.1 and formula (5.26), it is seen that its contribution is:

$$\rho_{\mathbf{k}}(\alpha; t)|_a = \frac{8\pi^3 e^4 c}{2\pi m^2} \int d\mathbf{l} \int d\mathbf{v}_n \int_C dz\, e^{-izt} \frac{1}{i(\mathbf{k}\cdot\mathbf{v}_\alpha - z)} iV_{|\mathbf{k}-\mathbf{l}|}(\mathbf{k}-\mathbf{l}) \cdot \partial_{\alpha n}$$

$$\frac{1}{i(\mathbf{k}\cdot\mathbf{v}_n + \mathbf{l}\cdot\mathbf{g}_{\alpha n} - z)} V_l i\mathbf{l}\cdot\partial_{\alpha n} \frac{1}{i(\mathbf{k}\cdot\mathbf{v}_n - z)} \rho_{\mathbf{k}}(n; 0)\varphi(\alpha; 0) \qquad (31.1)$$

As it stands, the integrand of this expression has no double poles and consequently it has no terms proportional to powers of t.

We now consider in more detail the consequences of the separation between t_h and all molecular time-scales. We are interested in the derivation of an equation valid for times $t \sim t_r$. This automatically implies $t \ll t_h$. We thus need an approximation of (31.1) valid in the situations defined by the double inequality

$$t_p \ll t \ll t_h \qquad (31.2)$$

A simple inspection of (31.1) shows that the two characteristic times t_p and t_h are easily identified. Two types of wave-vectors appear in (31.1). The vector $\mathbf{k}$ must be considered as very small, because $\rho_{\mathbf{k}}(\alpha; 0)$ is a sharply peaked function around $\mathbf{k} = 0$ (see § 3). The effective length of $\mathbf{k}$ is L_h^{-1}. On the other hand, the vector $\mathbf{l}$ has a large effective length: indeed, V_l, being the Fourier transform of a short-range potential (Debye potential), has a large spread in Fourier space. The effective order of magnitude of $\mathbf{l}$ is therefore κ, the inverse Debye length. As a result, $\mathbf{k} \cdot \mathbf{v}$ can be roughly identified with t_h^{-1}, whereas $\mathbf{l} \cdot \mathbf{v}$ is a measure of t_p^{-1}. The two conditions (31.2) can therefore be replaced by

$$\begin{aligned} \mathbf{l} \cdot \mathbf{v}t &\gg 1 \qquad \text{(a)} \\ \mathbf{k} \cdot \mathbf{v}t &\ll 1 \qquad \text{(b)} \end{aligned} \tag{31.3}$$

This rule is mathematically rather rough, because $\mathbf{k}$, $\mathbf{v}$ and $\mathbf{l}$ are all integration variables; however, the correctness of the results can be verified by a more careful (but longer) evaluation of the integrals.

As a result, $\mathbf{k} \cdot \mathbf{v}$ appears as a small quantity and the integrand in (31.1) can be expanded in a power series in this quantity:

$$\begin{aligned} \rho_{\mathbf{k}}(\alpha; t)|_a = \frac{8\pi^3 e^4 c}{2\pi m^2} \int d\mathbf{l} \int d\mathbf{v}_n \int_C dz\, e^{-izt} \frac{1}{-iz} \left(1 + \frac{\mathbf{k} \cdot \mathbf{v}_\alpha}{z} + \ldots\right) \\ \left(V_l - \underline{\mathbf{k} \cdot \frac{\partial V_l}{\partial \mathbf{l}}} + \ldots\right) i(\underline{\mathbf{k}} - \mathbf{l}) \cdot \partial_{\alpha n} \frac{1}{i(\mathbf{l} \cdot \mathbf{g}_{\alpha n} - z)} \left(1 + \underline{\frac{\mathbf{k} \cdot \mathbf{v}_n}{\mathbf{l} \cdot \mathbf{g}_{\alpha n} - z}} + \ldots\right) \\ V_l i \mathbf{l} \cdot \partial_{\alpha n} \frac{1}{-iz} \left(1 + \frac{\mathbf{k} \cdot \mathbf{v}_n}{z} + \ldots\right) \rho_{\mathbf{k}}(n; 0) \varphi(\alpha; 0) \end{aligned} \tag{31.4}$$

It is now seen that in this approximation, a multiple pole has appeared at $z = 0$, giving rise to powers of t. Before proceeding with the evaluation of this integral we note that the underlined terms can be eliminated. They are of relative order k/l, or t_p/t_h, and are therefore very small. The evaluation of residues for the remaining terms yields:

$$\begin{aligned} \rho_{\mathbf{k}}(\alpha; t)|_a = 8\pi^4 e^4 c m^{-2} \int d\mathbf{l} \int d\mathbf{v}_n V_l^2 \{ & t\, \mathbf{l} \cdot \partial_{\alpha n} \delta_-(\mathbf{l} \cdot \mathbf{g}_{\alpha n})\, \mathbf{l} \cdot \partial_{\alpha n} \\ & - \tfrac{1}{2} t^2\, \mathbf{l} \cdot \partial_{\alpha n} \delta_-(\mathbf{l} \cdot \mathbf{g}_{\alpha n})\, \mathbf{l} \cdot \partial_{\alpha n} (i\mathbf{k} \cdot \mathbf{v}_n) \\ & - \tfrac{1}{2} t^2\, (i\mathbf{k} \cdot \mathbf{v}_\alpha) \mathbf{l} \cdot \partial_{\alpha n} \delta_-(\mathbf{l} \cdot \mathbf{g}_{\alpha n})\, \mathbf{l} \cdot \partial_{\alpha n} \\ & + O[t(\mathbf{k} \cdot \mathbf{v}t)^2] + O(t_p/t) \}\, \rho_{\mathbf{k}}(n; 0) \varphi(\alpha; 0) \end{aligned} \tag{31.5}$$

The terms of order $t(\mathbf{k}\cdot\mathbf{v}t)^2$ come from the retention of higher powers of $\mathbf{k}\cdot\mathbf{v}$ in the expansion; the terms of order t_p/t are the non-leading universal terms [analogous to $\Phi'(0)$ in eq. (27.3)]. Formula (31.5) clearly shows that by our procedure we have obtained an expansion in powers of $\mathbf{k}\cdot\mathbf{v}t \sim t/t_h$, although it is simultaneously an asymptotic (long-time) expansion with respect to t/t_p. This is precisely the type of expression needed in the "hydrodynamic" approximation defined by (31.2). We note the important fact that the zeroth order term in eq. (31.5) is nothing other than the homogeneous Landau operator [eq. (30.11)] acting on $\rho_{\mathbf{k}}(n;0)\varphi(\alpha;0)$.

The previous method is easily generalized to other diagrams of the same type. For instance, the diagram drawn in Fig. 31.1b gives the following exact contribution

$$\rho_{\mathbf{k}}(\alpha;t)|_b = \frac{8\pi^3e^4c}{2\pi m^2}\int d\mathbf{v}_n\int d\mathbf{l}\int d\mathbf{k}'\int_C dz\, \mathrm{e}^{-izt}\frac{1}{i(\mathbf{k}\cdot\mathbf{v}_\alpha-z)}V_l(i\mathbf{l}\cdot\partial_{\alpha n})$$
$$\frac{1}{i(\mathbf{k}\cdot\mathbf{v}_\alpha-\mathbf{l}\cdot\mathbf{g}_{\alpha n}-z)}V_{|\mathbf{k}'-\mathbf{l}|}i(\mathbf{k}'-\mathbf{l})\cdot\partial_{\alpha n} \tag{31.6}$$
$$\frac{1}{i[\mathbf{k}'\cdot\mathbf{v}_n+(\mathbf{k}-\mathbf{k}')\cdot\mathbf{v}_\alpha-z]}\rho_{\mathbf{k}-\mathbf{k}'}(\alpha;0)\rho_{\mathbf{k}'}(n;0)$$

Use was made here of eq. (3.14). But in order to be consistent with the use of that equation, both $\mathbf{k}'$ and $\mathbf{k}-\mathbf{k}'$, or equivalently both $\mathbf{k}$ and $\mathbf{k}'$, must be considered as small (of order L_h^{-1}). The same method as before can be applied to eq. (31.6), with the result:

$$\rho_{\mathbf{k}}(\alpha;t)|_b = 8\pi^3e^4cm^{-2}\int d\mathbf{v}_n\int d\mathbf{l}\int d\mathbf{k}'V_l^2$$
$$\cdot\{t\,\mathbf{l}\cdot\partial_{\alpha n}\delta_-(\mathbf{l}\cdot\mathbf{g}_{\alpha n})\,\mathbf{l}\cdot\partial_{\alpha n} \tag{31.7}$$
$$-\tfrac{1}{2}t^2\,\mathbf{l}\cdot\partial_{\alpha n}\,\delta_-(\mathbf{l}\cdot\mathbf{g}_{\alpha n})\,\mathbf{l}\cdot\partial_{\alpha n}[i\mathbf{k}'\cdot\mathbf{v}_n+i(\mathbf{k}-\mathbf{k}')\cdot\mathbf{v}_\alpha]$$
$$-\tfrac{1}{2}t^2\,(i\mathbf{k}\cdot\mathbf{v}_\alpha)\,\mathbf{l}\cdot\partial_{\alpha n}\,\delta_-(\mathbf{l}\cdot\mathbf{g}_{\alpha n})\mathbf{l}\cdot\partial_{\alpha n}\}\rho_{\mathbf{k}'}(\alpha;0)\rho_{\mathbf{k}-\mathbf{k}'}(n;0)$$

Both diagrams have the following common topological structure: they have a central body which would be a cycle but for the presence of external lines. The latter correspond to wave-vectors in the hydrodynamic range, i.e. of order L_h^{-1}. Because of this feature the expressions can be expanded in powers of $\mathbf{k}\cdot\mathbf{v}t$.

The main property of the diagrams of Fig. 31.1 is that they have the same dependence on e^2 and on c and the same leading universal time dependence as the cycle. We now generalize these diagrams in the following way.

Starting with an arbitrary *homogeneous* diagonal fragment (i.e. a diagonal fragment without external lines) a certain type of inhomogeneous diagram can be constructed by drawing at least one external line to the right and at least one external line to the left. The external lines added to the right may start at any vertex which is not of type E (Table 6.1). Those added to the left must start at a destruction vertex C and be labelled by the fixed index appearing at that vertex [see theorem IIIb of § 8]. Moreover, these added external lines must represent wave-vectors in the hydrodynamic range ($k \sim L_h^{-1}$). A diagram constructed according to these rules is called a *pseudo-diagonal fragment*. The initial homogeneous diagonal fragment is called the *prototype* associated with the pseudo-diagonal fragment. Fig. 31.2 shows an example of such a construction.

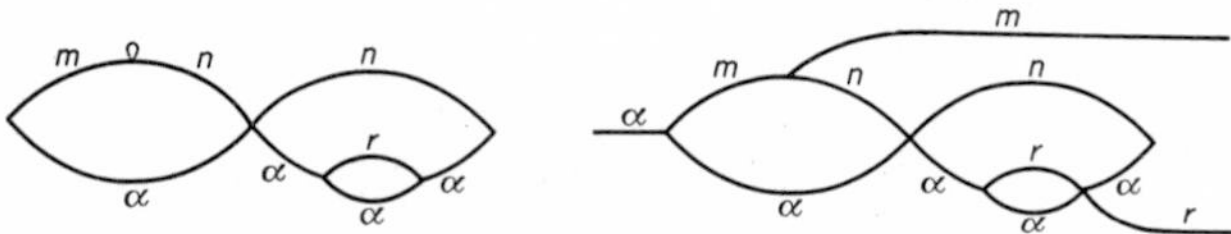

Fig. 31.2. A prototype and a pseudo-diagonal fragment derived from it.

The following comments justify the rules of construction of pseudo-diagonal fragments. Adding a line to the right transforms a vertex A into B, a vertex B into E and a vertex F into D. In these transformations the topological index is left invariant. Also, the number of fixed and of free particles is unchanged; hence the dependence on c of the resulting diagram is the same as that of the prototype.

On the other hand, adding a line to the left transforms a vertex D into E and a vertex F into B: here the topological index changes from 1 to 0. Also, being an external line at left, the new line must bear a Greek label, hence a summation over particles is lost. Consequently, diagrams constructed by adding external lines to the left of vertices D or F are proportional to one less factor c than the original prototypes. They cannot be classified in the same group as the latter. But adding a line to the left of vertex C transforms it into D, with conservation of the topological index 1. We have seen, moreover, in § 8 that one of the two particles involved at a vertex C is fixed. Hence, adding a line bearing the fixed index to the left does not change the number of particle summations: the resulting diagram has the same dependence on e^2 and c as the prototype.

The following statement results from the previous rules and the theorems of § 8. *A pseudo-diagonal fragment arising from a connected prototype necessarily has a single external line at left.* (Indeed, a connected prototype can have only one vertex C.) Note that there also exist pseudo-diagonal fragments which have several external lines at left; these are, however, generated by disconnected prototypes (see e.g. Fig. 27.2).

With each prototype one can associate a class of diagrams containing the inhomogeneous diagonal fragment and the set of all possible pseudo-diagonal fragments constructed from the prototype. This class is called *the class of pseudo-diagonal fragments generated by the given prototype.* Fig. 31.3 shows the class of pseudo-diagonal fragments generated by a prototype. The importance of this concept lies in the following theorem whose general proof can be inferred from the examples studied above.

For times satisfying the conditions (31.2), *each member of a class of pseudo-diagonal fragments has the same dependence on e^2 and c and the same leading universal time dependence as the prototype; the latter is corrected by a series in powers of* $\mathbf{k} \cdot \mathbf{v}t \sim t/t_h$.

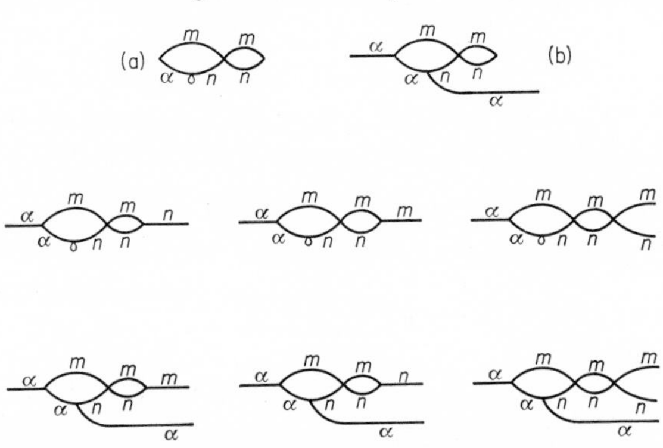

Fig. 31.3. The members of the class of pseudo-diagonal fragments generated by the prototype (a). Diagram (b) is the diagonal fragment of the class.

Pseudo-diagonal fragments are therefore non-diagonal fragments which, under the condition (31.2), behave like diagonal ones. In the same way, one can prove the following generalization.

A diagram consisting of a succession of m pseudo-diagonal

fragments has a time dependence of the type

$$\frac{t^m}{m!}(1+\alpha\mathbf{k}\cdot\mathbf{v}t+\beta(\mathbf{k}\cdot\mathbf{v}t)^2+\ldots)$$

An example of such a diagram is given in Fig. 31.4.

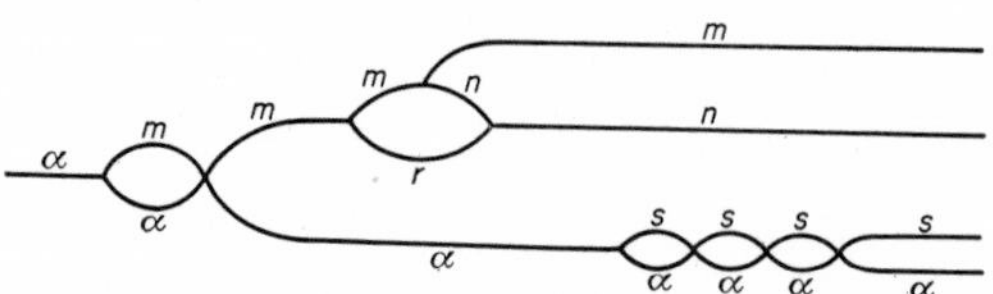

Fig. 31.4. A succession of pseudo-diagonal fragments. This diagram has the following time dependence: $(t^3/3!)(1+\mathbf{k}\cdot\mathbf{v}t+\ldots)$.

§ 32. The Evolution of $\rho_k(\alpha; t)$ and of $f(\alpha; t)$

We are now going to derive an equation of evolution for $\rho_{\mathbf{k}}(\alpha; t)$. Taking into account the results of § 31, we are looking for contributions of order $(e^4ct)^n$ and of order $(e^4ct)^n(\mathbf{k}\cdot\mathbf{v}t)$. In this manner the first two terms of an expansion in powers of $\mathbf{k}\cdot\mathbf{v}t$ are obtained. The choice of the corresponding diagrams is a simple matter. The most general is *a succession of connected pseudo-cycles, and of semiconnected homogeneous cycles*. The whole diagram must have an outgoing α-line at left. (The pseudo-cycles are of course the diagrams belonging to the set of pseudo-diagonal fragments generated by a cycle. They are shown on the second line of Fig. 32.2.) Each diagram must be evaluated up to the first order in $\mathbf{k}\cdot\mathbf{v}t$.

We first examine the general structure of a term in the series, e.g. the one corresponding to the diagram of Fig. 32.1. A simple

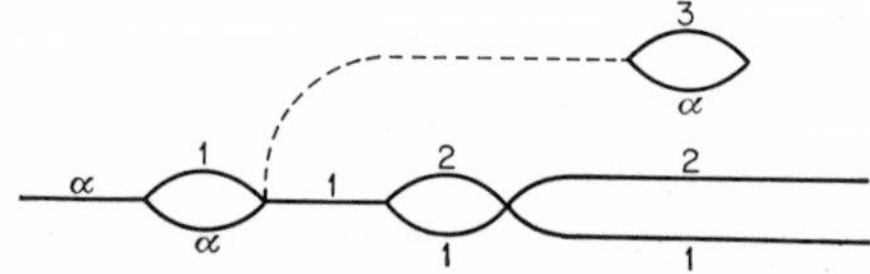

Fig. 32.1. A typical diagram consisting of two pseudo-cycles and a semi-connected homogeneous cycle.

generalization of the calculation leading to (31.5) or to (31.7) shows that the contribution of this diagram is

$$\begin{aligned}
\int d\mathbf{v}_1 d\mathbf{v}_2 d\mathbf{v}_3 \int d\mathbf{k}' \{ & (t^3/3!)\mathscr{C}_{\alpha 1}\mathscr{C}_{12}\mathscr{C}_{\alpha 3} \\
& -(t^4/4!)[i\mathbf{k}\cdot\mathbf{v}_\alpha \mathscr{C}_{\alpha 1}\mathscr{C}_{12}\mathscr{C}_{\alpha 3} \\
& +\mathscr{C}_{\alpha 1} i\mathbf{k}\cdot\mathbf{v}_1 \mathscr{C}_{12}\mathscr{C}_{\alpha 3} \\
& +\mathscr{C}_{\alpha 1}\mathscr{C}_{12}[i\mathbf{k}'\cdot\mathbf{v}_1 + i(\mathbf{k}-\mathbf{k}')\cdot\mathbf{v}_2]\mathscr{C}_{\alpha 3} \\
& +\mathscr{C}_{\alpha 1}\mathscr{C}_{12}\mathscr{C}_{\alpha 3}[i\mathbf{k}'\cdot\mathbf{v}_1 + i(\mathbf{k}-\mathbf{k}')\cdot\mathbf{v}_2]]\} \\
& \cdot \rho_{\mathbf{k}'}(1; 0)\rho_{\mathbf{k}-\mathbf{k}'}(2; 0)\varphi(\alpha, 0)\varphi(3, 0)
\end{aligned} \tag{32.1}$$

The first term is the zeroth-order term, whereas the terms enclosed in square brackets are of first order in $\mathbf{k}\cdot\mathbf{v}t$. They are obtained from the former by inserting a factor $(-i\sum \mathbf{k}_j\cdot\mathbf{v}_j t/4)$ in front, between and behind each operator $\mathscr{C}_{nm}$.

The summation is most easily done diagrammatically: this method is very intuitive and avoids writing cumbersome expressions. In Fig. 32.2 we have drawn the first terms of the

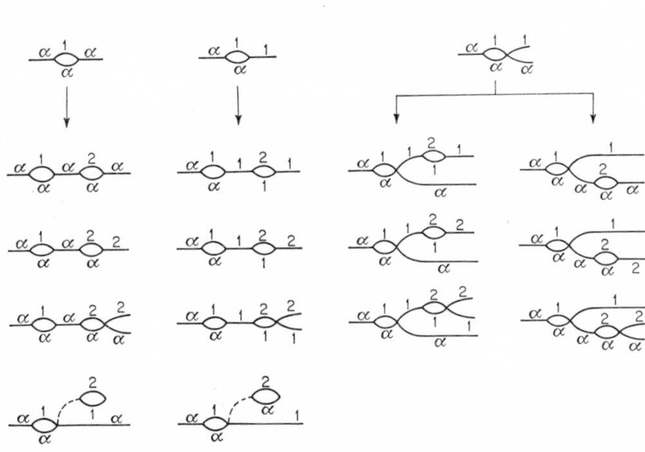

Fig. 32.2. The recurrent construction of the terms in the series of pseudo-cycles. The three pseudo-cycles of the second line are called (from left to right) (a), (b), (c) in the main text.

series, classifying them according to the number of their cycles and pseudo-cycles. A recurrence law is immediately apparent from this figure. In order to obtain all diagrams with $n+1$

pseudo-cycles, one takes each diagram in the group with n pseudo-cycles and one successively connects to each of its external lines at right each of the three pseudo-cycles. Moreover, a homogeneous cycle is semiconnected to each diagram containing at least one pseudo-cycle of type (a) or (b). This recurrent structure ensures the following properties. Clipping off the extreme left cycle in the diagrams beginning with the diagonal cycle (a) leaves a series consisting of disconnected diagrams with two components (see Fig. 32.3): a connected succession of pseudo-cycles with homogeneous cycles semiconnected to the main branch, and a string of semiconnected homogeneous cycles beginning with particle *1*. The number of cycles and pseudo-cycles in these components can vary from 0 to ∞. By virtue of the results of § 30 a series of disconnected diagrams represents a *product*, which in the present case is $\rho_{\mathbf{k}}(\alpha; t)\varphi(1; t)$. The same operation performed on

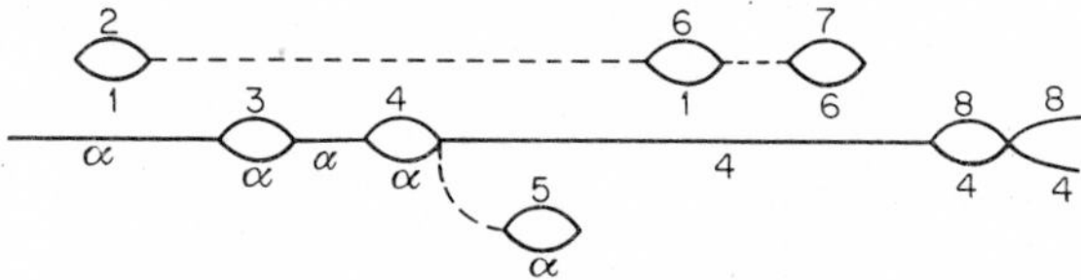

Fig. 32.3. A typical diagram of the series obtained by clipping off the last cycle in the series for $\rho_{\mathbf{k}}(\alpha; t)$.

the diagrams beginning with (b) reproduces the series for $\rho_{\mathbf{k}}(1; t)\varphi(\alpha; t)$ whereas under the same operation the diagrams beginning with (c) give rise to $\rho_{\mathbf{k}'}(\alpha; t)\rho_{\mathbf{k}-\mathbf{k}'}(1; t)$.

A first rough derivation of the kinetic equation will be given now. Keeping in mind the derivation of § 30, it is seen that a straightforward generalization of the procedure used there is the following. Take the time derivative of $\rho_{\mathbf{k}}(\alpha; t)$. This derivative

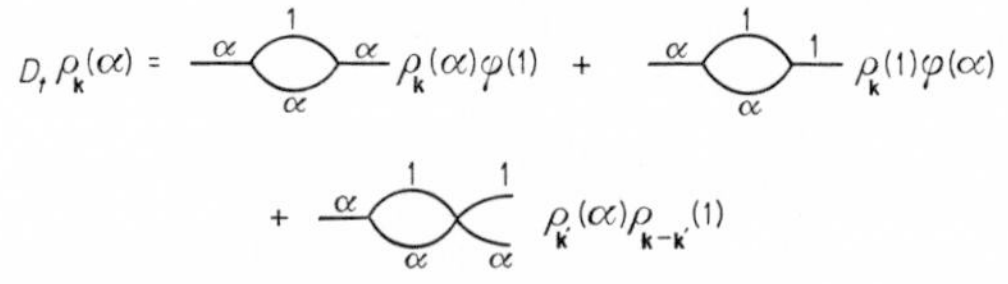

$$D_t \rho_{\mathbf{k}}(\alpha) = \ldots \rho_{\mathbf{k}}(\alpha)\varphi(1) + \ldots \rho_{\mathbf{k}}(1)\varphi(\alpha) + \ldots \rho_{\mathbf{k}'}(\alpha)\rho_{\mathbf{k}-\mathbf{k}'}(1)$$

Fig. 32.4. Diagrammatic representation of the kinetic equation for $\rho_{\mathbf{k}}(\alpha; t)$. The symbol D_t is a "total" derivative: $D_t \equiv (\partial_t + i\mathbf{k}\cdot\mathbf{v})$.

is equal to the operator corresponding to the last diagonal (or pseudo-diagonal) fragment, acting on the series obtained by clipping off this fragment in all diagrams. To this must be added a term $-i\mathbf{k} \cdot \mathbf{v}_\alpha \rho_\mathbf{k}(\alpha; t)$ coming from the free propagation line (as in § 11). Diagrammatically this is shown in Fig. 32.4. But in the hydrodynamic approximation, the three pseudo-cycles represent the same operator as the homogeneous cycle. The external lines simply indicate the functions on which this operator acts. Therefore, Fig. 32.4 represents the following equation:

$$\begin{aligned} \partial_t \rho_\mathbf{k}(\alpha; t) + i\mathbf{k} \cdot \mathbf{v}_\alpha \rho_\mathbf{k}(\alpha; t) &= \int d\mathbf{v}_1 \mathscr{C}_{\alpha 1} \\ \cdot \Big\{ \rho_\mathbf{k}(\alpha; t)\varphi(1; t) + \rho_\mathbf{k}(1; t)\varphi(\alpha; t) &+ \int d\mathbf{k}' \rho_{\mathbf{k}'}(\alpha; t) \rho_{\mathbf{k}-\mathbf{k}'}(1; t) \Big\} \end{aligned} \tag{32.2}$$

From this equation we may easily derive a kinetic equation for the one-particle distribution function:

$$f(\alpha) = c \Big\{ \varphi(\alpha) + \int d\mathbf{k}\, e^{i\mathbf{k}\cdot\mathbf{x}_\alpha} \rho_\mathbf{k}(\alpha) \Big\} \tag{32.3}$$

Taking the time derivative of eq. (32.3) and substituting for $\partial_t \varphi(\alpha)$ from (30.8) and for $\partial_t \rho_\mathbf{k}(\alpha)$ from (32.2) we obtain

$$\begin{aligned} \partial_t f(\alpha) = \int d\mathbf{v}_1 \mathscr{C}_{\alpha 1} \Big\{ \varphi(\alpha)\varphi(1) \\ + \int d\mathbf{k}\, e^{i\mathbf{k}\cdot\mathbf{x}_\alpha} \Big[\rho_\mathbf{k}(\alpha)\varphi(1) + \rho_\mathbf{k}(1)\varphi(\alpha) + \int d\mathbf{k}' \rho_{\mathbf{k}'}(\alpha) \rho_{\mathbf{k}-\mathbf{k}'}(1) \Big] \Big\} \end{aligned} \tag{32.4}$$

The $\mathbf{k}$-integral can be transformed as follows

$$\begin{aligned} \int d\mathbf{k}\, e^{i\mathbf{k}\cdot\mathbf{x}_\alpha}[\ldots] = \int d\mathbf{x}_1 \delta(\mathbf{x}_\alpha - \mathbf{x}_1) \int d\mathbf{k} \Big[e^{i\mathbf{k}\cdot\mathbf{x}_\alpha} \rho_\mathbf{k}(\alpha)\varphi(1) \\ + e^{i\mathbf{k}\cdot\mathbf{x}_1} \rho_\mathbf{k}(1)\varphi(\alpha) + \int d\mathbf{k}'\, e^{i\mathbf{k}'\cdot\mathbf{x}_\alpha + i(\mathbf{k}-\mathbf{k}')\cdot\mathbf{x}_1} \rho_{\mathbf{k}'}(\alpha) \rho_{\mathbf{k}-\mathbf{k}'}(1) \Big] \end{aligned} \tag{32.5}$$

Combining eqs. (32.5), (32.4), (32.3) as well as (30.2) and (30.9), we finally obtain the result

$$\begin{aligned} \partial_t f(\mathbf{x}_\alpha, \mathbf{v}_\alpha; t) &+ \mathbf{v}_\alpha \cdot \nabla_\alpha f(\mathbf{x}_\alpha, \mathbf{v}_\alpha; t) \\ &= (8\pi^4 e^4/m^2) \int d\mathbf{v}_1 \int d\mathbf{x}_1 \int d\mathbf{l}\, V_l^2 \mathbf{l} \cdot \partial_\alpha \delta(\mathbf{l} \cdot \mathbf{g}_{\alpha 1}) \mathbf{l} \cdot \partial_{\alpha 1} \\ &\quad f(\mathbf{x}_\alpha, \mathbf{v}_\alpha; t) f(\mathbf{x}_1, \mathbf{v}_1; t) \delta(\mathbf{x}_\alpha - \mathbf{x}_1) \end{aligned} \tag{32.6}$$

This method of derivation gives a quick answer in all cases. However, it has the disadvantage of not providing a separation of the orders of magnitude in $\mathbf{k}\cdot\mathbf{v}t$. We shall therefore now give a complete, although lengthier, derivation, coming out with a result which can be guessed directly from (32.6). In treating similar problems later, we shall take advantage of the justification given here of the simple diagrammatic method used above and shall not repeat these cumbersome calculations.

We introduce the following notations:

$\rho_{\mathbf{k}}^{(0)}(\alpha; t)$ is the sum of all contributions to $\rho_{\mathbf{k}}(\alpha; t)$ consisting of an operator independent of $\mathbf{k}\cdot\mathbf{v}t$ acting on $\rho_{\mathbf{k}}(\alpha; 0)$;

$\rho_{\mathbf{k}}^{(1)}(\alpha; t)$ is the sum of all contributions consisting of an operator of first order in $\mathbf{k}\cdot\mathbf{v}t$ acting on $\rho_{\mathbf{k}}(\alpha; 0)$.

The same splitting is defined for other inhomogeneous Fourier components. Clearly, $\varphi(\alpha) \equiv \varphi^{(0)}(\alpha)$.

For instance, in eq. (32.1) the first term on the r.h.s. is a contribution to $\rho_{\mathbf{k}}^{(0)}(\alpha)$ whereas the remaining terms are contributions to $\rho_{\mathbf{k}}^{(1)}(\alpha)$.

We also define a corresponding decomposition for $f(\mathbf{x}_\alpha, \mathbf{v}_\alpha; t)$:

$$f^{(0)}(\alpha) = c\left\{\varphi(\alpha) + \int d\mathbf{k}\, \rho_{\mathbf{k}}^{(0)}(\alpha) e^{i\mathbf{k}\cdot\mathbf{x}_\alpha}\right\} \tag{32.7}$$

$$f^{(1)}(\alpha) = c\int d\mathbf{k}\, \rho_{\mathbf{k}}^{(1)}(\alpha) e^{i\mathbf{k}\cdot\mathbf{x}_\alpha} \tag{32.8}$$

Formula (32.1) shows that the functions $\rho_{\mathbf{k}}^{(0)}(\alpha)$ and $\rho_{\mathbf{k}}^{(1)}(\alpha)$ are both expressed as power series in t and in e^4c. Let us introduce the notation

$$\begin{aligned} \rho_{\mathbf{k}}^{(0)}(\alpha) &= \sum_{n=0}^{\infty} (t^n/n!)[\rho_{\mathbf{k}}^{(0)}(\alpha)]_n \\ \rho_{\mathbf{k}}^{(1)}(\alpha) &= \sum_{n=0}^{\infty} [t^{n+1}/(n+1)!][\rho_{\mathbf{k}}^{(1)}(\alpha)]_n \end{aligned} \tag{32.9}$$

The summation index n equals the number of factors e^4c in each term. In the case of $\rho_{\mathbf{k}}^{(1)}(\alpha)$ the exponent of t is one unit higher than the number of e^4c-factors, because of the extra factor $i\mathbf{k}\cdot\mathbf{v}t$ [see eq. (32.1)].

The following definition is self-explanatory:

$$\begin{aligned} [\rho_{\mathbf{k}}^{(p)}(\alpha)\varphi(j)]_n = {}& [\rho_{\mathbf{k}}^{(p)}(\alpha)]_n[\varphi(j)]_0 + [\rho_{\mathbf{k}}^{(p)}(\alpha)]_{n-1}[\varphi(j)]_1 + \cdots \\ & + [\rho_{\mathbf{k}}^{(p)}(\alpha)]_0[\varphi(j)]_n, \qquad p = 0,1 \end{aligned} \tag{32.10}$$

A similar definition holds for $[\rho_{\mathbf{k}}^{(p)}\rho_{\mathbf{k}'}^{(p')}]_n$.

We first calculate $\rho_{\mathbf{k}}^{(0)}(\alpha; t)$. This term is obtained by making the following correspondence between diagrams and contributions:

$$\text{cycle or pseudo-cycle} \to \mathscr{C}_{\alpha n}$$

$$\text{external line or line connecting two pseudo-cycles} \to 1.$$

[see first term of eq. (32.1)].

In the following expressions (32.11–14) we omit writing an integral over $\mathbf{v}_1$ and an integral over $\mathbf{k}'$ in all terms involving these variables.

Making use of the previously quoted properties, we can write

$$\rho_{\mathbf{k}}^{(0)}(\alpha; t) = \sum_{n=0}^{\infty} \frac{t^n}{n!} \mathscr{C}_{\alpha 1} \{[\rho_{\mathbf{k}}^{(0)}(\alpha)\varphi(1)]_{n-1} + [\rho_{\mathbf{k}}^{(0)}(1)\varphi(\alpha)]_{n-1} + [\rho_{\mathbf{k}'}^{(0)}(\alpha)\rho_{\mathbf{k}-\mathbf{k}'}^{(0)}(1)]_{n-1}\} \tag{32.11}$$

The situation is somewhat more complicated for $\rho_{\mathbf{k}}^{(1)}(\alpha)$. It is convenient to write separately the term for which the factor $i\mathbf{k}\cdot\mathbf{v}t$ is in front [see second term in (32.1)] and group the other terms together:

$$\begin{aligned}\rho_{\mathbf{k}}^{(1)}(\alpha; t) = \sum_{n=0}^{\infty} \frac{t^{n+1}}{(n+1)!} (-i\mathbf{k}\cdot\mathbf{v}_\alpha)\mathscr{C}_{\alpha 1}\{[\rho_{\mathbf{k}}^{(0)}(\alpha)\varphi(1)]_{n-1} + [\rho_{\mathbf{k}}^{(0)}(1)\varphi(\alpha)]_{n-1} \\ + [\rho_{\mathbf{k}'}^{(0)}(\alpha)\rho_{\mathbf{k}-\mathbf{k}'}^{(0)}(1)]_{n-1}\} \\ + \sum_{n=0}^{\infty} \frac{t^{n+1}}{(n+1)!} \mathscr{C}_{\alpha 1}\{[\rho_{\mathbf{k}}^{(1)}(\alpha)\varphi(1)]_{n-1} + [\rho_{\mathbf{k}}^{(1)}(1)\varphi(\alpha)]_{n-1} \\ + [\rho_{\mathbf{k}'}^{(1)}(\alpha)\rho_{\mathbf{k}-\mathbf{k}'}^{(0)}(1)]_{n-1} + [\rho_{\mathbf{k}'}^{(0)}(\alpha)\rho_{\mathbf{k}-\mathbf{k}'}^{(1)}(1)]_{n-1}\}\end{aligned} \tag{32.12}$$

The derivation of an equation of evolution is now an easy matter. Taking the time derivative of eq. (32.11) and using (32.9), we get

$$\begin{aligned}\partial_t \rho_{\mathbf{k}}^{(0)}(\alpha; t) &= \sum_{n=1}^{\infty} \frac{t^{n-1}}{(n-1)!} \mathscr{C}_{\alpha 1}\{[\rho_{\mathbf{k}}^{(0)}(\alpha)\varphi(1)]_{n-1} + \ldots\} \\ &= \mathscr{C}_{\alpha 1}\{\rho_{\mathbf{k}}^{(0)}(\alpha; t)\varphi(1; t) + \rho_{\mathbf{k}}^{(0)}(1; t)\varphi(\alpha; t) \\ &\quad + \rho_{\mathbf{k}'}^{(0}(\alpha; t)\rho_{\mathbf{k}-\mathbf{k}'}^{(0)}(1; t)\}\end{aligned} \tag{32.13}$$

In a similar way:

$$\begin{aligned}\partial_t \rho_{\mathbf{k}}^{(1)}(\alpha; t) &= -i\mathbf{k}\cdot\mathbf{v}_\alpha \sum_{n=1}^{\infty} \frac{t^n}{n!} \mathscr{C}_{\alpha 1}\{[\rho_{\mathbf{k}}^{(0)}(\alpha)\varphi(1)]_{n-1} + \ldots\} \\ &\quad + \mathscr{C}_{\alpha 1} \sum_{n=1}^{\infty} \frac{t^n}{n!} \{[\rho_{\mathbf{k}}^{(1)}(\alpha)\varphi(1)]_{n-1} + \ldots\} \\ &= -i\mathbf{k}\cdot\mathbf{v}_\alpha \rho_{\mathbf{k}}^{(0)}(\alpha; t) + \mathscr{C}_{\alpha 1}\{\rho_{\mathbf{k}}^{(1)}(\alpha; t)\varphi(1; t) + \rho_{\mathbf{k}}^{(1)}(1; t)\varphi(\alpha; t) \\ &\quad + \rho_{\mathbf{k}'}^{(1)}(\alpha; t)\rho_{\mathbf{k}-\mathbf{k}'}^{(0)}(1; t) + \rho_{\mathbf{k}'}^{(0)}(\alpha; t)\rho_{\mathbf{k}-\mathbf{k}'}^{(1)}(1; t)\}\end{aligned} \tag{32.14}$$

We now derive kinetic equations for $f^{(0)}(\alpha)$ and $f^{(1)}(\alpha)$, as we did for (32.6). Starting from eq. (32.13) we obtain:

$$\partial_t f^{(0)}(\mathbf{x}_\alpha, \mathbf{v}_\alpha; t) = (8\pi^4 e^4/m^2) \int d\mathbf{v}_1 \int d\mathbf{x}_1 \int d\mathbf{l} \\ V_l^2 \mathbf{l} \cdot \partial_\alpha \delta(\mathbf{l} \cdot \mathbf{g}_{\alpha 1}) \mathbf{l} \cdot \partial_{\alpha 1} f^{(0)}(\mathbf{x}_\alpha, \mathbf{v}_\alpha; t) f^{(0)}(\mathbf{x}_1, \mathbf{v}_1; t) \cdot \delta(\mathbf{x}_\alpha - \mathbf{x}_1) \tag{32.15}$$

A similar calculation starting with (32.14) yields:

$$\partial_t f^{(1)}(\mathbf{x}_\alpha, \mathbf{v}_\alpha; t) + \mathbf{v}_\alpha \cdot \nabla_\alpha f^{(0)}(\mathbf{x}_\alpha, \mathbf{v}_\alpha; t) \\ = (8\pi^4 e^4/m^2) \int d\mathbf{v}_1 \int d\mathbf{x}_1 \int d\mathbf{l}\, V_l^2 \mathbf{l} \cdot \partial_\alpha \delta(\mathbf{l} \cdot \mathbf{g}_{\alpha 1}) \mathbf{l} \cdot \partial_{\alpha 1} \tag{32.16} \\ [f^{(0)}(\mathbf{x}_\alpha, \mathbf{v}_\alpha; t) f^{(1)}(\mathbf{x}_1, \mathbf{v}_1; t) + f^{(1)}(\mathbf{x}_\alpha, \mathbf{v}_\alpha; t) f^{(0)}(\mathbf{x}_1, \mathbf{v}_1; t)] \delta(\mathbf{x}_\alpha - \mathbf{x}_1)$$

Comparing eq. (32.6), obtained by a very simple diagrammatic method, with the two equations (32.15–16), derived by a more elaborate procedure, it is seen that the latter equations can be obtained from the former by the following prescription:

Substitute $f = f^{(0)} + f^{(1)}$ into eq. (32.6)

Assume

$$\begin{cases} f^{(1)} \ll f^{(0)} \\ \mathbf{v} \cdot \nabla f^{(0)} \text{ is of the same order as } f^{(1)} \end{cases}$$

Identify terms of the same order in eq. (32.6).

The process defined in this way will be called the *Chapman–Enskog* expansion of eq. (32.6). To sum up, assuming the hydrodynamic condition (31.2), we are able to derive eq. (32.6) very simply, with the additional condition of restricting it to the Chapman–Enskog approximation. In further work [see eq. (42.2)] we shall derive more general inhomogeneous equations by the simple method which yields non-linear expressions of type (32.6). The latter can then always be shown to be valid with restriction to the approximation above. The physical nature of the Chapman–Enskog approximation will be discussed further in § 33.

§ 33. The Boltzmann Equation and its Relation to the Landau Equation

In §§ 30 and 32 we derived equations of evolution for $\varphi(\alpha)$ and for $f(\alpha)$, pertaining respectively to a homogeneous and to a non-homogeneous plasma. These equations have very similar right-hand sides, consisting of a unique operator, $\mathscr{C}_{\alpha n}$, acting on a certain product or sum of products of two distribution functions φ or f, and integrated over the argument of one of the latter. This right-hand side is called the *collision term* of the equation, and $\mathscr{C}_{\alpha n}$ is called the *collision operator* in the Landau approximation. The reason for this name will appear shortly. The left-hand sides of the equations contain the time rate of change of f or φ and, in eqs. (32.6) and (32.16), an additional term, $\mathbf{v}_\alpha \cdot \nabla_\alpha f^{(0)}(\alpha)$, which is called the *flow term*. Let us first discuss the equation for homogeneous systems, (30.11), which is simpler.

The ordinary kinetic theory of gases is based on a fundamental equation of evolution which is well known as the *Boltzmann equation*. In homogeneous systems, the customary derivation of this equation is based on the idea that the rate of change of the distribution function is due to collisions between the molecules. These are regarded as *random instantaneous* encounters of a group of molecules, the effect of which is to produce a marked deflection of the original straight paths. The amount of this deflection depends on the mechanical parameters characterizing the relative motion of the collision partners, and can, in principle, be evaluated from elementary dynamics. In a very dilute gas ($c \to 0$), the collisions are very rare; moreover, among the few collisions which take place the overwhelming majority involve only two molecules. The rate of change of $\varphi(\mathbf{v}; t)$ is then due to two competing causes: an increase, due to molecules which initially had an arbitrary velocity $\mathbf{v}'$ and after the collision acquire a velocity $\mathbf{v}$ (this is the "inverse collision" process), and a decrease, due to the fact that a collision changes the initial velocity $\mathbf{v}$ into a different one. The result of the calculation is *

$$\partial_t \varphi(\mathbf{v}_\alpha; t) = 2\pi \int d\mathbf{v}_1 \int db\, b g_{\alpha 1} [\varphi(\mathbf{v}'_\alpha; t)\varphi(\mathbf{v}'_1; t) - \varphi(\mathbf{v}_\alpha; t)\varphi(\mathbf{v}_1; t)] \tag{33.1}$$

* See, for instance, S. Chapman and T. Cowling, *The Mathematical Theory of Non-Uniform Gases,* Cambridge University Press, 1939; R. C. Tolman, *The Principles of Statistical Mechanics,* Oxford University Press, 1955; J. O. Hirschfelder, C. F. Curtiss and R. B. Bird, *Molecular Theory of Gases and Liquids,* Wiley, New York, 1951; or any other book on the kinetic theory of gases.

where b is the "collision parameter", i.e. the distance between the two initial trajectories. The velocities $\mathbf{v}'_\alpha$ and $\mathbf{v}'_1$ of the particles involved in the inverse collision restituting $\mathbf{v}_\alpha$ and $\mathbf{v}_1$ are regarded as functions of b and of $\mathbf{g}_{\alpha 1}$. Let us comment on the main properties of this equation.

A. (33.1) *is an irreversible equation.* This means that the distribution function $\varphi(\alpha)$ tends, in some sense monotonously, toward a unique final distribution, $\varphi^0(\alpha)$, which represents the thermal equilibrium. More precisely, consider the following functional of φ: $(-\ln \varphi)$; Boltzmann has shown that whatever the initial value of φ, the average of this functional is a monotonously increasing function of time which can be identified with the macroscopic entropy density. The constant asymptotic value of the entropy is achieved when φ has the unique form φ^0. This is the famous Boltzmann H-theorem.

B. *The collisions are "instantaneous" events.* More precisely the only feature of the collision entering the Boltzmann equation is the total deflection of the particles. This asymptotic trajectory is determined by the interaction law; however, it represents the motion of the two particles in a state in which the interaction has completely ceased. The motion of the particles *during* their actual interaction is disregarded as being very short on the time scale in which we are interested here.

C. (33.1) *describes two-body collisions.* This has been assumed from the beginning in the elementary derivation. Mathematically this is reflected in the *non-linear* character of the equation. On the r.h.s. appears a product of *two* distribution functions, just as many as there are particles involved in the collision process.

D. (33.1) *represents collisions of arbitrary strength.* In the elementary derivation, no assumption has been made concerning the strength of the collision. Mathematically this is expressed by the fact that $\mathbf{v}$ and $\mathbf{v}'$ (which is essentially the velocity after the collision) differ by some finite amount. This is best seen by noting that (33.1) can be written in the form (30.8) by replacing the operator $\mathscr{C}_{\alpha 1}$ by the "Boltzmann operator"

$$\mathscr{C}^B_{\alpha n} = 2\pi \int db\, b g_{\alpha n}\{\exp[(\mathbf{v}'_\alpha - \mathbf{v}_\alpha)\cdot \partial_\alpha + (\mathbf{v}'_1 - \mathbf{v}_1)\cdot \partial_1] - 1\} \qquad (33.2)$$

which is a *finite displacement* operator acting on the velocities $\mathbf{v}_\alpha$ and $\mathbf{v}_1$.

It will be shown in this book that none of the kinetic equations appropriate for ionized gases possesses *all* these four properties. However, most of them possess the two properties A and B. Any kinetic equation possessing properties A and B will be called a *Boltzmann-like equation.* It will be seen in Chapter 13 that such equations play a quite important role in the theory of transport coefficients.

Let us now come back to the Landau equation, (30.11). It will be shown in § 34 that this equation has the property A of irreversibility required by the Boltzmann H-theorem. It obviously also possesses property C. Property B is a result of the asymptotic treatment of the diagrams. The neglect of the non-universal terms means precisely that we disregard the specific details of the trajectory during the short duration of the effective interaction. The leading universal term on the other hand describes precisely what is left of the motion after a long time, i.e. the asymptotic trajectory. But eq. (30.11) does not possess property D. Instead of being a finite displacement operator, $\mathscr{C}_{\alpha 1}$ is a second-order *differential* operator. This immediately suggests that $\mathscr{C}_{\alpha 1}$ describes "soft" collisions in which the deflections are very small. This idea is the basis of the first derivation of eq. (30.11), due to Landau. * He started with the Boltzmann equation (33.1) and expanded everything in powers of the deflection $(\mathbf{v}'_\alpha - \mathbf{v}_\alpha)$. The lowest non-vanishing term in this expansion is precisely (30.11). The Landau equation thus appears as a first approximation to the Boltzmann equation, valid when the collisions are soft, or equivalently when the interactions are weak.

In our formalism, the concept of weak coupling appears rather naturally. If one selects from the general perturbation series (5.26) the contributions which for a given power of t are proportional to the lowest power in e^2, one again obtains the cycles,

* L. D. Landau, *Phys. Z. Sowjetunion,* **10**, 154 (1936).

and thus the Landau approximation. This was the guiding idea of Brout and Prigogine,[*] who derived a "master equation" [i.e. an equation for $\varphi_N(\mathbf{v}_1, \ldots, \mathbf{v}_N; t)$] from which the Landau equation can be derived by integration over all variables but one (Prigogine and Balescu[**]).

In order to obtain the Boltzmann equation (33.1) within the present formalism, one in a sense uses the Landau procedure backwards. More precisely, for a given dependence on t, one allows for an arbitrary power of e^2, which is considered now as a finite parameter to be retained to all orders, but one expresses the low concentration of the gas by taking only the smallest power of c. This leads to a double summation of all diagrams of the order $(e^{2m}ct)^n$. Graphically, this choice amounts to replacing the cycle by the sum of all diagrams called "*chains*" (Fig. 33.1). These represent

Fig. 33.1. The sum of chains leading to the Boltzmann equation.

repeated interactions between the same two particles, resulting, so to speak, in stronger and stronger collisions. Prigogine and Henin and Résibois[†] have shown that the summation of the chains leads precisely to the Boltzmann equation.

Consider now inhomogeneous systems. The Boltzmann equation derived from elementary considerations is[††]

$$\partial_t f(\alpha) + \mathbf{v}_\alpha \cdot \nabla_\alpha f(\alpha) = \int d\mathbf{x}_1 d\mathbf{v}_1 \mathscr{C}^B_{\alpha 1} f(\alpha) f(1) \delta(\mathbf{x}_\alpha - \mathbf{x}_1) \tag{33.3}$$

where $\mathscr{C}^B_{\alpha 1}$ was defined by (33.2). This equation has the following properties, corresponding to properties A–D above:

A. *Equation* (33.3) *is a superposition of a reversible and an irreversible process.* The rate of change of f is due to two terms: a reversible

[*] R. Brout and I. Prigogine, *Physica*, **22**, 621 (1956).

[**] I. Prigogine and R. Balescu, *Physica*, **23**, 555 (1957); I. Prigogine and R. Balescu, *Physica*, **25**, 302 (1959).

[†] I. Prigogine and F. Henin, *Physica*, **24**, 214 (1958); P. Résibois, *Physica*, **25**, 725 (1959); see also I. Prigogine, *Non-Equilibrium Statistical Mechanics*, Interscience, New York, 1963.

[††] See S. Chapman and T. Cowling, R. C. Tolman, etc.; refs. p. 163.

flow term, which is the same as that in the Liouville equation for a free particle, and a collision term similar to the one of eq. (33.1).

B, C and D are also properties of eq. (33.3).

E. *Equation* (33.3) *expresses a certain type of randomness in the collision process* (*molecular chaos*). This is a very important point which has been discussed at length by all authors who have worked on the rigorous foundation of the Boltzmann equation. * The difficulty lies in the occurrence of the factors $f(\alpha)f(1)$ in the collision term instead of the factor $f_2(\alpha, 1)$, which would be more appropriate at first sight for expressing the joint probability of finding molecules α and *1* with their respective coordinates. On the other hand, this point is essential, because it *closes* the equation. If it were not realized one would get a hierarchy of equations expressing f_1 in terms of f_2, f_2 in terms of f_3, etc. The chaos property replaces this infinite system of linear equations by a *single non-linear* equation for f_1.

It should be stressed that the real difficulty only appears in the inhomogeneous case. In eq. (33.1), involving $\varphi(\alpha)$, the factorization must always be assumed at the initial time by virtue of the assumption of the decay of molecular correlations at large distances (see § 2). The propagation of chaos was shown in § 30 to be a fairly direct consequence of the structure of the diagrams. The factorization of space-dependent distribution functions is, however, a much deeper problem. It has been shown in § 3 that the initial factorization of $\rho_{\mathbf{k}\mathbf{k}'}(\alpha, n)$ [which induces the factorization of $f_2(\alpha, n)$] is only valid if a stronger assumption is made concerning the differences between hydrodynamic and molecular lengths. The Boltzmann equation cannot be expected to be valid outside of the "hydrodynamic regime" defined by $L_h \gg L_m$.

* L. Boltzmann, *Vorlesungen über Gastheorie*, Barth, Leipzig, 1895; P. and T. Ehrenfest, *Enzykl. Math. Wiss.*, vol. IV, 4, Teubner Verlag, Leipzig, 1911; J. G. Kirkwood, *J. Chem. Phys.*, **14**, 180 (1946); M. Kac, *Proc. 3rd Berkeley Sympos. Math. Statistics and Probability*, Univ. of California Press, Berkeley, Cal., 1956; M. Green, *J. Chem. Phys.*, **25**, 836 (1956); R. Brout, *Physica*, **22**, 509 (1956); H. Grad, *Handbuch der Physik*, **12**, p. 205, Springer Verlag, Berlin, 1958.

F. *Equation* (33.3) *is a local equation.* This property is expressed by the function $\delta(\mathbf{x}_\alpha - \mathbf{x}_1)$, which implies that the two distribution functions are calculated at the same point in space. One would expect in general a less strict localization; the δ-function would be replaced by a function with a finite width, of the order of the range of the force. This would mean that one had allowed for variations of the physical properties over molecular distances, and that one takes into account the probabilities of finding the two collision partners at their actual positions. The strict localization appearing in eq. (33.3) is another direct consequence of the "hydrodynamic condition" $L_h \gg L_m$, which ensures the constancy of $f(\mathbf{x}, \mathbf{v})$ over distances of molecular order.

Exactly as in the homogeneous case, one can derive eq. (32.6) from eq. (33.3) by the method of Landau. This equation is very similar, but not identical, to the equations (32.15–16). This is due to the fact that, as they stand, eqs. (33.3) and (32.6) are not yet quite explicit. In view of the discussion concerning property E above, these two equations are only valid in the "hydrodynamic region". This means that one can apply to them a certain particular type of series expansion, well known as the *Chapman–Enskog* expansion. * The idea of this expansion is as follows. Because of the smallness of the gradients, the distribution function is assumed to evolve locally in zeroth approximation as if the gradients (i.e. the flow term in the Boltzmann equation) were absent. In the next approximation, the flow term is introduced, acting on $f^{(0)}$, and the collision term is linearized around $f^{(0)}$. The subsequent approximations are obvious. It is clear that this expansion leads precisely to eqs. (32.15–16). *The discussion of* § 32 *thus amounts to a proof of the Boltzmann equation in the Chapman–Enskog approximation.* But the latter approximation is exactly equivalent to the hydrodynamic condition and it follows from our discussion that the Boltzmann equation can only be expected to be valid within the first few orders of approximation of the Chapman–Enskog expansion. In the pages which follow we will sometimes use eq. (32.6) instead of (32.15–16)

* See S. Chapman and T. Cowling, *loc. cit.*

for compactness. In that case it is always implied that this notation is merely an abbreviation for the Chapman–Enskog expansion.

The discussion of this section presented the Landau equation as a weak-force approximation to the Boltzmann equation. This is actually the principle of most of its elementary derivations. Another idea, due to Chandrasekhar, for the derivation of the Landau equation essentially makes use of the same principle of weak interactions. It is more appropriate to discuss it in a later paragraph (38). In this book we have taken a different view which is more appropriate in the case of plasmas. We are regarding the Landau equation not as an approximation to the Boltzmann equation, but as an approximation to a diametrically different equation, which is obtained by retaining all powers of $(e^2c)^n(e^2t)^m$. The latter equation will be derived in Chapter 9 and will be seen to describe rigorously the *collective interactions* in the plasma. These have been shown in the past sections to be the characteristic and fundamental property of gases of charged particles.

CHAPTER 8

Properties of the Landau Equation

§ 34. Irreversibility and the *H*-Theorem

The most important property of the Landau equation, as opposed to the short-time Vlassov equation, is its irreversible character. This means physically that starting with an arbitrary distribution function $f(\mathbf{x}, \mathbf{v}; 0)$ at time 0, the system evolves in such a way as to reach after a sufficiently long time (of order t_r or t_h) a unique homogeneous state described by the Maxwell equilibrium distribution:

$$f(\mathbf{x}, \mathbf{v}; t \to \infty) \to c\varphi^0(\mathbf{v}) = c(m\beta/2\pi)^{\frac{3}{2}} \exp\left(-\tfrac{1}{2}\beta m v^2\right) \quad (34.1)$$

where $\beta = 1/kT$, k is the Boltzmann constant and T the absolute temperature. In the discussion of this property, homogeneous and inhomogeneous systems will be treated separately, beginning with the former.

For homogeneous systems, it will be shown successively that:

(a) the equilibrium distribution is a stationary solution of eq. (30.11);

(b) there exists a function — interpreted as the entropy density — which increases monotonously and attains a stationary value only when φ reaches equilibrium. (This is the proof of unicity of the equilibrium state.)

The first point is very simply proven when the equation stands in the form (30.11). Substituting $\varphi^0(\alpha)$ for $\varphi(\alpha)$, the left-hand side of that equation becomes:

$$\frac{8\pi^4 e^4 c}{m^2}\left(\frac{m\beta}{2\pi}\right)^3 \int d\mathbf{v}_n \int d\mathbf{l}\, V_l^2 \mathbf{l}\cdot\partial_\alpha \delta(\mathbf{l}\cdot\mathbf{g}_{\alpha n})\mathbf{l}\cdot\partial_{\alpha n} \exp\left[-\tfrac{1}{2}\beta m(v_\alpha^2+v_n^2)\right]$$

$$= -\tfrac{1}{2}\pi m^2 \beta^4 e^4 c \int d\mathbf{v}_n \int d\mathbf{l}\, V_l^2 \mathbf{l}\cdot\partial_\alpha \delta(\mathbf{l}\cdot\mathbf{g}_{\alpha n})\mathbf{l}\cdot\mathbf{g}_{\alpha n}$$

$$\exp\left[-\tfrac{1}{2}\beta m(v_\alpha^2+v_n^2)\right] = 0 \quad (34.2)$$

The last step follows from the presence of the combination $\mathbf{l}\cdot\mathbf{g}\delta(\mathbf{l}\cdot\mathbf{g})$.

The next stage consists in introducing the function

$$s(t) = -\int d\mathbf{v}_\alpha c\varphi(\mathbf{v}_\alpha; t) \ln c\varphi(\mathbf{v}_\alpha; t) \tag{34.3}$$

The rate of change in time is easily calculated by using eq. (30.11)

$$\begin{aligned}
\partial_t s &= -c\int d\mathbf{v}_\alpha[\ln c\varphi(\alpha)+1]\partial_t\varphi(\alpha) \\
&= -A\int d\mathbf{v}_\alpha\int d\mathbf{v}_n\int d\mathbf{l}[\ln c\varphi(\alpha)+1]V_l^2\mathbf{l}\cdot\partial_{\alpha n}\delta(\mathbf{l}\cdot\mathbf{g}_{\alpha n})\mathbf{l}\cdot\partial_{\alpha n}\varphi(\alpha)\varphi(n) \\
&= -A\int d\mathbf{v}_\alpha\int d\mathbf{v}_n\int d\mathbf{l}[\ln c\varphi(n)+1]V_l^2\mathbf{l}\cdot\partial_{\alpha n}\delta(\mathbf{l}\cdot\mathbf{g}_{\alpha n})\mathbf{l}\cdot\partial_{\alpha n}\varphi(\alpha)\varphi(n) \\
&= -A\int d\mathbf{v}_\alpha\int d\mathbf{v}_n\int d\mathbf{l}\,\tfrac{1}{2}V_l^2[\ln c\varphi(\alpha) \\
&\qquad + \ln c\varphi(n)+2]\mathbf{l}\cdot\partial_{\alpha n}\delta(\mathbf{l}\cdot\mathbf{g}_{\alpha n})\mathbf{l}\cdot\partial_{\alpha n}\varphi(\alpha)\varphi(n) \\
&= \tfrac{1}{2}A\int d\mathbf{v}_\alpha\int d\mathbf{v}_n\int d\mathbf{l}[\varphi(\alpha)\varphi(n)]^{-1}V_l^2\{\mathbf{l}\cdot\partial_{\alpha n}\varphi(\alpha)\varphi(n)\}^2\delta(\mathbf{l}\cdot\mathbf{g}_{\alpha n}) \geqq 0
\end{aligned} \tag{34.4}$$

In going from the second to the third step, use has been made of the fact that particles α and n play the same role in the collision term; the fourth step results from an obvious symmetrization, and the last step is the result of a partial integration. As $\varphi(\alpha)$ is by definition a non-negative function, the whole expression is evidently non-negative. The function $s(t)$ can only increase in the course of time. It attains its maximum value only when $\varphi(\alpha) = \varphi^0(\alpha)$. In that case it is shown, as in eq. (34.2), that $\partial_t s = 0$. The function $s(t)$ therefore has all the characteristics which are needed for the thermodynamic entropy density.

We now discuss the inhomogeneous plasma. Here the situation is more complex, because there exist two opposing mechanisms of evolution: the collisions and the mechanical flow. The calculation will be performed in two stages, corresponding to the zeroth and first Chapman–Enskog approximations. We first discuss the stationary solution.

The zeroth-order equation (32.15) is very similar to the homogeneous equation (30.11). However, the space-dependence of $f(\alpha)$ and the localization factor $\delta(\mathbf{x}_\alpha - \mathbf{x}_n)$ allow for a richer variety of solutions. Actually, it is seen that the following function, called (rather improperly) "*local equilibrium*", is a stationary solution of eq. (32.15)

$$f^0(\mathbf{x}_\alpha, \mathbf{v}_\alpha) = [m\beta(\mathbf{x}_\alpha)/2\pi]^{\frac{3}{2}} c(\mathbf{x}_\alpha) \exp\{-\tfrac{1}{2}m\beta(\mathbf{x}_\alpha)[\mathbf{v}-\mathbf{u}(\mathbf{x}_\alpha)]^2\} \quad (34.5)$$

The proof is analogous to the one considered in homogeneous situations. Consider now the next approximation, eq. (32.16). It is immediately seen that f^0 is no longer a stationary solution. On substituting $f = f^{(0)}+f^{(1)} = f^0$, or equivalently $f^{(0)} = f^0$, $f^{(1)} = 0$, the collision term vanishes, but the flow term does not. The only stationary distribution is attained when β, c and $\mathbf{u}$ in eq. (34.5) become constant in space and time. This process of uniformization takes place in a time of order t_h. To sum up, the long-time evolution process in an inhomogeneous system is the following. Starting from an arbitrary initial state, the distribution function approaches in a time of order t_r a state close to the local equilibrium state (34.5). This state cannot, however, be attained, because of the opposing action of the flow mechanism. The latter acts with a time scale t_h and has a homogenizing effect. Because of the difference in the two time scales, the net effect is a slow evolution from a state very close to an instantaneous (time-dependent) local equilibrium toward the true thermodynamic equilibrium.

The evolution of the entropy also reflects this double process. It is easier and more compact for the present discussion to use the short formulation (32.6) (with the implicit restriction to the Chapman–Enskog approximation). We define a *local* density of entropy as:

$$s(\mathbf{x}_\alpha; t) = -\int d\mathbf{v}_\alpha f(\alpha) \ln f(\alpha) \quad (34.6)$$

Then a calculation, completely similar to the previous one, leads to the result

$$\partial_t s(\mathbf{x}_\alpha; t) - \boldsymbol{\nabla}_\alpha \cdot \int d\mathbf{v}_\alpha \mathbf{v}_\alpha f(\alpha)[\ln f(\alpha)+1] \quad (34.7)$$
$$= \tfrac{1}{2}A\int d\mathbf{v}_\alpha d\mathbf{v}_n d\mathbf{x}_n d\mathbf{l}\, \delta(\mathbf{x}_\alpha - \mathbf{x}_n)[f(\alpha)f(n)]^{-1} V_l^2 [\mathbf{l}\cdot\partial_{\alpha n} f(\alpha)f(n)]^2 \delta(\mathbf{l}\cdot\mathbf{g}_{\alpha n})$$

The change in time of the entropy density at one point is described as a sum of two contributions:

(a) The divergence of a vector function which is interpreted as an *entropy flow* $\mathbf{\Phi}_s$: this quantity describes the transfer of entropy from one region of space to another through the flow of matter and energy.

(b) A term in the nature of a *source* which describes local creation of entropy due to irreversible processes (i.e. collisions).

The divergence of $\mathbf{\Phi}_s$ has no definite sign, meaning that entropy may decrease or increase locally as a result of the flow; the source, however, is always positive, so that irreversible processes can only increase, never decrease the amount of entropy at a given point. This behavior is nothing other than an expression of the second law of thermodynamics. If we consider a system, which is not necessarily isolated, the rate of change of the entropy can be split up as follows *

$$dS/dt = d_e S/dt + d_i S/dt \tag{34.8}$$

where $d_e S/dt$ represents the change of entropy due to exchanges between the system and the external world (it is the amount of entropy which crosses the boundary of the system per unit time), whereas $d_i S/dt$ is the amount of entropy produced or absorbed per unit time within the system. The second law of thermodynamics asserts that

$$d_i S/dt \geqq 0 \tag{34.9}$$

The second law can also be expressed in local form. * Taking for the system an infinitesimal volume of fluid, and defining an entropy density $s(\mathbf{x}; t)$ such that the amount of entropy in the chosen infinitesimal volume is $S = s(\mathbf{x};\, t)dx\,dy\,dz$, eq. (34.8) is readily transformed into:

$$\partial_t s + \mathbf{\nabla} \cdot \mathbf{\Phi}_s = \sigma \tag{34.10}$$

whereas (34.9) becomes:

$$\sigma \geqq 0 \tag{34.11}$$

* I. Prigogine, *Introduction to the Thermodynamics of Irreversible Processes*, 2nd. ed., Interscience, New York, 1961; S. R. de Groot and P. Mazur, *Non-Equilibrium Thermodynamics*, North Holland Publ. Co., Amsterdam, 1962.

The quantity σ is called the *local entropy production* and is always positive by virtue of the second law of thermodynamics. Equation (34.7) is precisely of the form (34.10), and amounts to a proof of the irreversibility in situations for which the Landau equation is valid.

§ 35. Explicit Form of the Landau Equation

The Landau equation (30.11) or (32.6) can be transformed into a variety of forms which are adapted to various purposes. In the present section we perform one of these transformations, which brings the equation into the form originally obtained by Landau. * This form of the equation is a good starting point for the computation of transport coefficients. As these transformations only involve the collision operator, we shall only discuss the homogeneous equation in the present section, the results being trivially extended to the inhomogeneous case.

Let us write eq. (30.11) in the form:

$$\partial_t \varphi(\mathbf{v}; t) = A \int d\mathbf{v}' \, \partial_r C_{rs} (\partial_s - \partial'_s) \varphi \varphi' \tag{35.1}$$

We use here slightly modified notations. Instead of $\mathbf{v}_\alpha$, $\mathbf{v}_n$, we now use $\mathbf{v}$ and $\mathbf{v}'$. r and s are not particle indices, but denote cartesian components ($r = v_x, v_y, v_z$); ∂_r are components of the gradient: $\partial_r = (\partial/\partial v_x, \partial/\partial v_y, \partial/\partial v_z)$. Repeated indices are summed over. This notation avoids heavy accumulations of subscripts. The tensor C_{rs} is defined as:

$$C_{rs} = 4\pi^4 \int d\mathbf{l} \, V_l^2 l_r l_s \delta(\mathbf{l} \cdot \mathbf{g}) \tag{35.2}$$

A is a constant defined as

$$A = 2e^4 c/m^2 \tag{35.3}$$

Our purpose here is to perform explicitly the $\mathbf{l}$-integration. In order to do this, we choose a coordinate system such that its z-axis is directed along the vector $\mathbf{g}$. The components of the tensor C_{rs} then become

* L. D. Landau, *Phys. Z. Sowjetunion,* **10**, 154 (1936).

$$C_{rs} = 4\pi^4 \int_0^{l_M} dl\, l^4 V_l^2 \int_0^{\pi} d\theta \sin\theta \int_0^{2\pi} d\phi\, \delta(lg\cos\theta)$$

$$\begin{Bmatrix} \sin\theta\cos\phi \\ \sin\theta\sin\phi \\ \cos\theta \end{Bmatrix} \begin{Bmatrix} \sin\theta\cos\phi \\ \sin\theta\sin\phi \\ \cos\theta \end{Bmatrix} \quad (35.4)$$

It is immediately seen by integration over ϕ that the components (xy, xz, yz, yx, zx, zy) are zero. The zz component also vanishes because of the δ-function, and the xx and yy components are equal. In this reference system the tensor has the form

$$\mathbf{C} = \begin{pmatrix} C & 0 & 0 \\ 0 & C & 0 \\ 0 & 0 & 0 \end{pmatrix}$$

The component C is easily evaluated:

$$C = 4\pi^5 \int_0^{l_M} dl\, l^4\, V_l^2 \int_{-1}^{1} d\xi\, \delta(lg\,\xi)\,(1-\xi^2) = \frac{\pi}{g} B$$

where B is defined as

$$B = 4\pi^4 \int_0^{l_M} dl\, l^3\, V_l^2 \quad (35.5)$$

The tensor $\mathbf{C}$ in this reference system can be written as:

$$\mathbf{C} = [\mathbf{I} - \mathbf{1}_z \mathbf{1}_z] \frac{\pi}{g} B \quad (35.6)$$

where $\mathbf{I}$ is the unit tensor and $\mathbf{1}_z$ a unit vector along the z-axis. As in further operations there appears an integration over $\mathbf{v}'$, the z-axis cannot be considered as a fixed direction, and the tensor (35.6) must be rewritten in a form which is independent of the reference system. This is easily done, noting that $\mathbf{1}_z = \mathbf{g}/g$. Combining the result with eqs. (35.1) and (35.3) we obtain:

$$\partial_t \varphi = \frac{2\pi e^4 cB}{m^2} \int d\mathbf{v}' \cdot \partial_r [g^{-3}(g^2 \delta_{rs} - g_r g_s)] (\partial_s - \partial'_s) \varphi\varphi' \quad (35.7)$$

which is exactly Landau's original form of the equation. The same equation has been obtained by Bogoliubov, * using his general theory of irreversibility.

A remarkable feature of the Landau equation is the fact that the exact form of the interactions affects the equation only through the numerical value of the constant B. This situation is in marked contrast with the Boltzmann equation (33.1), where the nature of the interactions affects [through $\mathbf{v}'(b, g)$] the functional form of the equation itself. It can be said that systems described by the Landau equation obey a *principle of corresponding states*, in the sense that suitably scaled quantities obey universal laws, whatever the interactions. ** The case of the plasmas treated in this approximation does not, however, follow this law. The reason is found in the fact that the Landau equation applies to plasmas only if a Debye potential whose Fourier transform is cut off at $l = l_M$ is used in the calculation of the tensor C_{rs} (see § 28). This potential is not of a purely dynamical origin (like the Coulomb potential) but must be regarded as a first approximation to an effective potential of the type discussed in § 18. This potential includes collective polarization and screening effects, which depend on parameters characterizing the state of the system (density, temperature, etc.) and will alter the universal temperature or density dependence predicted by the law of corresponding states.

In the case of the Debye potential, the integral B, eq. (35.5), is very easily evaluated [see eq. (28.4)]:

$$B = \int_0^{l_M} dl \frac{l^3}{(l^2 + \kappa^2)^2} = \frac{1}{2} \left[- \frac{l_M^2}{l_M^2 + \kappa^2} + \ln \frac{l_M^2 + \kappa^2}{\kappa^2} \right] \tag{35.8}$$

The result (35.8) agrees, of course, with the calculation of the tensor $\mathbf{\Psi}$ performed in § 28. Equation (35.8) substituted into (35.6) yields the dominant part of $\mathbf{\Psi}$, eq. (28.10), divided by t.

We now discuss a reasonable value for the choice of the cut-off l_M, although there is a certain arbitrariness in this choice. As the

* N. N. Bogoliubov, *Problems of a Dynamical Theory in Statistical Physics*, translated by E. K. Gora, in *Studies in Statistical Mechanics*, vol. 1, edited by J. de Boer and G. E. Uhlenbeck, Interscience, New York, 1962.

** J. O. Hirschfelder, C. F. Curtiss and R. B. Bird, *Molecular Theory of Gases and Liquids*, Wiley, New York, 1951.

physical reason for the divergence at the upper limit is the strong repulsion at short distances, we have to eliminate very close encounters, which produce large deflections. We may set a limit of weak deflections at 90°. It can be shown from a simple analysis of the dynamics of a collision * that such a deflection occurs when the potential energy at a distance equal to the impact parameter b_c equals twice the original kinetic energy:

$$2\frac{mv^2}{2} = \frac{\mathrm{e}^{-\kappa b_c}}{b_c} e^2 \tag{35.9}$$

As presumably $b_c^{-1} \gg \kappa$, the exponential factor approximately equals 1. We then choose for l_M the inverse of the critical impact parameter b_c^{-1}, which is given from (35.9) by

$$l_M \approx mv^2/e^2$$

Taking moreover an average value of this cut-off in terms of the temperature, we obtain

$$l_M = 2/\beta e^2 \tag{35.10}$$

With this choice, it is seen that

$$\kappa/l_M = 4\sqrt{\pi}\, e^3 \beta^{\frac{3}{2}} c^{\frac{1}{2}} = 4\sqrt{\pi}\, \Gamma^{\frac{3}{2}} \tag{35.11}$$

where Γ is the characteristic dimensionless parameter of the plasma, eq. (9.9). The ratio κ/l_M is thus indeed very small in the range of validity of the Landau equation. We may take advantage of this smallness in order to simplify somewhat expression (35.8). Expanding the right-hand side and neglecting for consistency all terms of order κ/l_M, we are left with

$$B = [\ln (l_M/\kappa) - \tfrac{1}{2}] \tag{35.12}$$

This is the final value we adopt. It is interesting to note that the first term in B is identical with the value of B calculated straightforwardly with a pure Coulomb potential, by cutting off the **l**-integral at both ends:

$$B_{\text{coul}} = \int_{\kappa}^{l_M} dl\, \frac{l^3}{l^4} = \ln \frac{l_M}{\kappa} \tag{35.13}$$

* L. Spitzer, Jr., *Physics of Fully Ionized Gases*, Interscience, New York, 1956.

Because of the smallness of Γ, the logarithm is the dominant term in eq. (35.12). Consequently, the use of a Debye potential or the use of a Coulomb potential cut-off at $l = \kappa$ gives practically the same result, and either of them can be used in calculations. However, the use of a Debye potential has a better theoretical foundation (see Chapter 11).

§ 36. The Hydrodynamic Equations

Equation (32.6), whose r.h.s. is transformed as in § 35, is a good starting point for the derivation of the equations of hydrodynamics. This derivation will be sketched in the present section. The kinetic equation is written as:

$$\partial_t f + \mathbf{v} \cdot \nabla f = a \int d\mathbf{v}'\, \partial_r G_{rs} (\partial_s - \partial'_s) f f' \tag{36.1}$$

with $f = f(\mathbf{x}, \mathbf{v}; t)$, $f' = f(\mathbf{x}' = \mathbf{x}, \mathbf{v}'; t)$

$$a = 2\pi e^4 B / m^2 \tag{36.2}$$

$$G_{rs} = g^{-3} (g^2 \delta_{rs} - g_r g_s) \tag{36.3}$$

We now wish to derive equations of evolution for the mass density $\rho(\mathbf{x}; t) \equiv mn(\mathbf{x}; t)$, local velocity $\mathbf{u}(\mathbf{x}; t)$ and local kinetic energy density $E(\mathbf{x}; t)$. These functions have been defined in formulae (1.14), (1.15) and the first term on the r.h.s. of (1.16). Let us call $J(\mathbf{v})$ any one of the quantities m, v_x, v_y, v_z, $\frac{1}{2}mv^2$. The hydrodynamic equations are obtained by multiplying both sides of (36.1) by $J(\mathbf{v})$ and integrating over $\mathbf{v}$:

$$\partial_t \int d\mathbf{v}\, J f + \nabla \cdot \int d\mathbf{v}\, J \mathbf{v} f = a \int d\mathbf{v}\, d\mathbf{v}'\, J \partial_r G_{rs} (\partial_s - \partial'_s) f f' \tag{36.4}$$

Before proceeding with the transformations, we first show that the right-hand side of (36.4) is zero. This important theorem justifies the denomination of "*collision invariants*" given to the five quantities grouped under the symbol J. Their meaning will be discussed below.

The property is obvious for $J = m = \text{const.}$, because the $\mathbf{v}$-integral affects an expression of the type $\partial_r F_r$, and F_r is zero for infinite $\mathbf{v}$. Take now $J = v_n$:

$$\int d\mathbf{v}\, d\mathbf{v}'\, v_n \partial_r G_{rs}(\partial_s - \partial'_s) f f'$$
$$= \int d\mathbf{v}\, d\mathbf{v}'\, f f' (\partial_s - \partial'_s) G_{ns} = -4 \int d\mathbf{v}\, d\mathbf{v}'\, f f' g^{-3}(v_n - v'_n) = 0$$

The first step arises from two partial integrations, the second from a differentiation of the tensor G_{ns}. The last integral obviously vanishes by symmetry, because the functions f and f' are identical functions of $\mathbf{v}$ and $\mathbf{v}'$ respectively, and g (absolute value!) is symmetrical with respect to a permutation of $\mathbf{v}$ and $\mathbf{v}'$.

Taking now $J = v^2$ we similarly obtain

$$\int d\mathbf{v}\, d\mathbf{v}'\, v^2 \partial_r G_{rs}(\partial_s - \partial'_s) f f' = 2 \int d\mathbf{v}\, d\mathbf{v}'\, f f' (\partial_s - \partial'_s) v_r G_{rs}$$
$$= 4 \int d\mathbf{v}\, d\mathbf{v}'\, f f' g^{-3}(g^2 - 2v_r g_r) = -4 \int d\mathbf{v}\, d\mathbf{v}'\, f f' g^{-3}(v_r + v'_r)(v_r - v'_r) = 0$$

In the last step we used the identity

$$g^2 - 2v_r g_r = (v_r - v'_r - 2v_r) g_r = -(v_r + v'_r) g_r \tag{36.5}$$

The equations of hydrodynamics are now easily derived. Taking $J = m$, and using (1.14) and (1.15), eq. (36.4) becomes

$$\partial_t \rho + \boldsymbol{\nabla} \cdot (\rho \mathbf{u}) = 0 \tag{36.6}$$

which is the *continuity equation*. Taking $J = v_r$, we obtain after some easy transformations, using (36.6) *

$$\rho(\partial_t + u_k \nabla_k) u_i = -\nabla_k P_{ik} \tag{36.7}$$

where the *pressure tensor* P_{ik} is defined as

$$P_{ik} = m \int d\mathbf{v}\, (v_i - u_i)(v_k - u_k) f \tag{36.8}$$

According to the form of the pressure tensor, eq. (36.7) reduces to Euler's equation $[P_{ik} = p\delta_{ik})$ or to the Navier–Stokes equation $[P_{ik} = p\delta_{ik} - \eta(\nabla_k u_i + \nabla_i u_k - \frac{2}{3}\nabla_j u_j \delta_{ik})]$. It can be shown that the zeroth Chapman–Enskog approximation leads

* See, e.g., A. Sommerfeld, *Vorlesungen über Theoretische Physik*, B.V.: *Thermodynamik und Statistik*, Dietrich Verlag, Wiesbaden, 1952; or any of the books on the kinetic theory of gases referred to in § 33.

to Euler's equation, while the first approximation leads to the Navier–Stokes equation. *

The derivation of the energy equation requires somewhat more algebra, * but is also very simple. Introducing a density of internal energy per unit mass, defined as the average kinetic energy with respect to the motion of the center of mass:

$$q = \frac{1}{2n} \int d\mathbf{v} (\mathbf{v}-\mathbf{u})^2 f \tag{36.9}$$

the energy equation is:

$$\begin{aligned} \rho(\partial_t + \mathbf{u}\cdot\boldsymbol{\nabla})q + \boldsymbol{\nabla}\cdot\mathbf{Q} = -\rho\boldsymbol{\nabla}\cdot\mathbf{u} - (P_{ik}-p\delta_{ik})[\nabla_k u_i \\ +\nabla_i u_k - \tfrac{2}{3}\delta_{ik}(\nabla_j u_j)] \end{aligned} \tag{36.10}$$

The left-hand side contains the heat flux

$$\mathbf{Q} = \frac{1}{2n} \int d\mathbf{v} (\mathbf{v}-\mathbf{u})(\mathbf{v}-\mathbf{u})^2 f \tag{36.11}$$

whereas the r.h.s. describes the dissipation of internal energy through compression and viscosity.

From here on the usual methods of the kinetic theory of gases can be applied in order to develop a complete plasma hydrodynamics. This will not be done in detail in this book, as it has already been fully treated in other textbooks on plasma physics.** We will, however, come back to the problem of transport coefficients from a more general point of view in Chapter 13.

Before closing this section, we should like to make some comments on the collision invariants introduced above. It is sometimes stated loosely that the vanishing of the r.h.s. of eq. (36.4) implies the conservation of momentum and of energy in the collision process. This statement is misleading. There are many instances in which that term does *not* vanish, although the dynamical laws of conservation of momentum and energy are perfectly satisfied. It is clearly seen from our derivation that the essential condition for the r.h.s. of eq. (36.4) to vanish is that f and f' are identical functions of $\mathbf{v}$ and $\mathbf{v}'$ respectively. This is achieved in our case by the fact that all particles are identical and moreover that f and f' are calculated at the same

* A. Sommerfeld, *loc. cit.*

** See, e.g., J. L. Delcroix, *Introduction to the Theory of Ionized Gases*, Interscience, New York, 1960; J. G. Linhart, *Plasma Physics*, North Holland Publ. Co., Amsterdam, 1961.

point in space (presence of the factor $\delta(\mathbf{x}-\mathbf{x}')$). The invariance under consideration is thus a consequence of the local property of the Landau equation (Property F, § 33), which is itself a consequence of the hydrodynamic condition $L_h \gg L_m$. Its physical meaning is as follows: all the momentum and energy lost by one particle is gained by its collision partner *at the same point* in space. There can be no local change in momentum or energy at any point. If, however, we consider a mixture of gases (e.g. ions and electrons) or particles whose state can be distinguished (see next section), the r.h.s. of eq. (36.4) will in general not vanish. The average momentum and energy of one group of particles will of course change at each point, as a result of the transfer to the other type of particles. However, the dynamical conservation laws are not violated!

§ 37. The Fokker–Planck Equation

In order to investigate further the properties of the Landau equation (35.7), we shall study an idealized situation which is called the *brownian motion* problem or the *test-particle* problem. In this situation the particles of the gas no longer all play the same role. We fix our attention on one particle (or equivalently, a small group of mutually non-interacting particles) moving inside the bulk of a plasma which is in thermal equilibrium. The extra particle (called the test particle) is distinct from the medium because its distribution function is not maxwellian. Through interactions with the medium, this particle will eventually reach equilibrium; the problem consists in the study of the mechanism by which equilibrium is attained.

Strictly speaking, the particles of the medium do not remain in equilibrium, and also the final equilibrium state is different from the initial one. However, at the end of the process the energy excess of the test particle is distributed among N particles of the medium; therefore, the change in density and temperature in the final state as compared to the initial one is of order N^{-1}, and is thus negligible: the particles of the medium can safely be assumed to have a stationary equilibrium distribution.

We also assume for simplicity that the distribution of the test particle is homogeneous. Although this model is very idealized, it gives a good insight into the nature of the phenomena under consideration.

We now derive the kinetic equation for the distribution of the test particle. Starting from (35.7) we introduce our basic brownian motion assumption:

$$\varphi' = (m\beta/2\pi)^{\frac{3}{2}} \exp\left(-\tfrac{1}{2}\beta m v'^2\right) \tag{37.1}$$

where $\beta = (kT)^{-1}$, as usual. In order to simplify the notations, it is convenient to measure the velocities in units of the thermal velocity, and to measure the time in units of the relaxation time. We therefore make the following change of variables:

$$\begin{aligned} \mathbf{w} &\equiv (\tfrac{1}{2}\beta m)^{\frac{1}{2}}\mathbf{v}, \quad \mathbf{u} \equiv (\tfrac{1}{2}\beta m)^{\frac{1}{2}}\mathbf{v}' \\ \mathbf{q} &\equiv (\tfrac{1}{2}\beta m)^{\frac{1}{2}}\mathbf{g} = \mathbf{w}-\mathbf{u} \\ \partial &\equiv \partial/\partial\mathbf{w}, \quad \partial' \equiv \partial/\partial\mathbf{u} \\ \tau &\equiv (\sqrt{2m}/\pi\beta^{\frac{3}{2}} e^4 cB)t \end{aligned} \tag{37.2}$$

The variables $\mathbf{w}$, $\mathbf{u}$, $\mathbf{q}$, τ are dimensionless. Substituting (37.1–2) into eq. (35.7), the latter becomes:

$$\partial_\tau \varphi = \pi^{-\frac{3}{2}} \int d\mathbf{u}\, \partial_r G_{rs}(\partial_s - \partial'_s) \mathrm{e}^{-u^2} \varphi \tag{37.3}$$

with

$$G_{rs} = q^{-3}(q^2 \delta_{rs} - q_r q_s) \tag{37.4}$$

As the distribution function φ' is now prescribed, it is seen that eq. (37.3) is a *linear partial differential equation*. Compared to eq. (35.7), which is a non-linear integro-differential equation, one immediately realizes the enormous simplification achieved.

Before proceeding, we may add a short remark. The fact that eq. (37.3) is linear does not imply that the distribution φ is close to equilibrium. The linearization has been achieved here only by the assumptions that the medium has a stationary distribution and that the test particles do not interact among themselves but only with the medium. Another type of linearization of (35.7) would be obtained if it were assumed that the distribution of *all* the particles was close to equilibrium. Then one writes:

$$\varphi = \pi^{-\frac{3}{2}} \mathrm{e}^{-w^2} + \psi, \quad \varphi' = \pi^{-\frac{3}{2}} \mathrm{e}^{-u^2} + \psi', \quad |\psi| \ll \varphi, \quad |\psi'| \ll \varphi'$$

and one obtains

$$\partial_t \psi = \pi^{-\frac{3}{2}} \int d\mathbf{u}\, \partial_r G_{rs}(\partial_s - \partial'_s)[\mathrm{e}^{-w^2}\psi' + \mathrm{e}^{-u^2}\psi] \tag{37.5}$$

In spite of their apparent similarity, eq. (37.5) is essentially different from (37.3). It is a linear integro-differential equation similar to the first Chapman–Enskog approximation [see eq. (32.16)], and describes a completely different physical situation.

We now show that eq. (37.3) can be written in a completely explicit form. The terms can first be rearranged as follows:

$$\partial_\tau \varphi = \partial_r\{-F_r + \tfrac{1}{2}\partial_s T_{rs}\}\varphi \tag{37.6}$$

where

$$\begin{aligned} F_r(w) &= \pi^{-\frac{3}{2}} \int d\mathbf{u}\, e^{-u^2}(-2u_s + \partial_s) G_{rs} \\ &= -4\pi^{-\frac{3}{2}} \int d\mathbf{u}\, u_s e^{-u^2} G_{rs} \end{aligned} \tag{37.7}$$

(one notes that as G_{rs} depends only on $\mathbf{q}$, $\partial_s G_{rs} = -\partial_s' G_{rs}$; one then integrates by parts to obtain the second line), and

$$T_{rs}(w) = 2\pi^{-\frac{3}{2}} \int d\mathbf{u}\, e^{-u^2} G_{rs} \tag{37.8}$$

Let us now evaluate the integral F_r. We use polar coordinates in the particular reference system shown in Fig. 37.1: the Z-axis is taken along the fixed vector $\mathbf{w}$. We have also drawn the vector

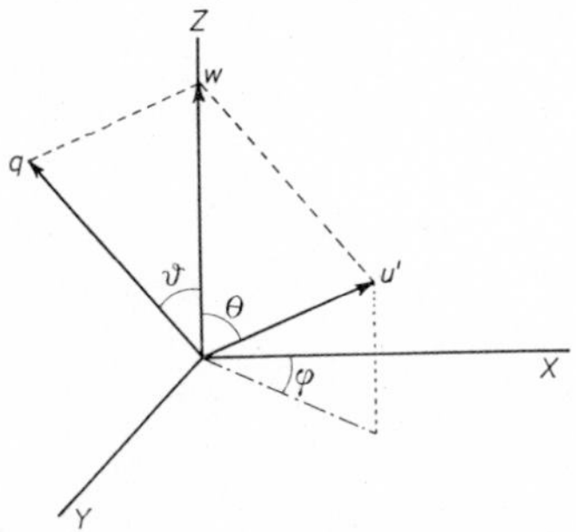

Fig. 37.1. Coordinate system for the integration over $\mathbf{u}$ in eqs. (37.7–8).

$\mathbf{q} = \mathbf{w} - \mathbf{u}$. By its definition, it follows that this vector lies in the same plane as $\mathbf{w}$ and $\mathbf{u}$. Substituting (37.4) into (37.7), we obtain the following formula for the three components of $\mathbf{F}$:

$$F_r = -4\pi^{-\frac{3}{2}} \int_0^\infty du\, u^2 \int_{-1}^{1} d\cos\theta \int_0^{2\pi} d\phi\, e^{-u^2} q^{-3}$$

$$\left[q^2 u \begin{Bmatrix} \sin\theta\cos\phi \\ \sin\theta\sin\phi \\ \cos\theta \end{Bmatrix} - (\mathbf{q}\cdot\mathbf{u}) q \begin{Bmatrix} \sin\vartheta\cos\phi \\ \sin\vartheta\sin\phi \\ \cos\vartheta \end{Bmatrix} \right]$$

It is obvious that through the ϕ-integration, the X and Y components vanish, and only F_Z is non-zero:

$$F_Z = -8\pi^{-\frac{1}{2}} \int_0^\infty du\, u\, \mathrm{e}^{-u^2} \int_{-1}^{1} d\cos\theta\; q^{-3}[q^2 u \cos\theta - (\mathbf{q}\cdot\mathbf{u}) q \cos\vartheta] \tag{37.9}$$

We now note that, from the definition (37.2) of $\mathbf{q}$, we have the following identities:

$$\begin{aligned} \mathbf{q}\cdot\mathbf{u} &= wu\cos\theta - u^2 \\ q\cos\vartheta &= \mathbf{q}\cdot\mathbf{w}/w = w - u\cos\theta \end{aligned} \tag{37.10}$$

Therefore (37.9) reduces to

$$F_Z = -8\pi^{-\frac{1}{2}} w \int_0^\infty du\, u^4 \mathrm{e}^{-u^2} \int_{-1}^{1} d\cos\theta\; q^{-3} \sin^2\theta \tag{37.11}$$

This integration is easily performed by using the following trick (due to Chandrasekhar): one expresses θ in terms of q, w and u and takes q as an independent integration variable instead of $\cos\theta$:

$$\begin{aligned} q^2 &= w^2 + u^2 - 2wu\cos\theta \\ \cos\theta &= (w^2+u^2-q^2)/2wu \\ d\cos\theta &= -q\, dq/wu \end{aligned} \tag{37.12}$$

Upon substitution in (37.11) one finds

$$F_Z = -2\pi^{-\frac{1}{2}} w^{-2} \int_0^\infty du\, \mathrm{e}^{-u^2} u\, J(w,u) \tag{37.13}$$

where

$$\begin{aligned} J(w,u) &= \int_{|w-u|}^{w+u} dq\, \{-(w^2-u^2)^2 q^{-2} + 2(w^2+u^2) - q^2\} \\ &= \begin{cases} (16/3)u^3 & \text{for } u < w \\ (16/3)w^3 & \text{for } u > w \end{cases} \end{aligned} \tag{37.14}$$

and thus

$$F_Z = -(32/3\sqrt{\pi}) w^{-2} \left\{ \int_0^w du\, u^4 \mathrm{e}^{-u^2} + w^3 \int_w^\infty du\, u\, \mathrm{e}^{-u^2} \right\} \tag{37.15}$$

The integrals occurring here can easily be related to the *error function* $\Phi(w)$:

$$\Phi(w) = 2\pi^{-\frac{1}{2}} \int_0^w dt\, \mathrm{e}^{-t^2} \tag{37.16}$$

We also write the gaussian as:

$$\Phi'(w) = d\Phi(w)/dw = 2\pi^{-\frac{1}{2}} e^{-w^2} \tag{37.17}$$

It is easily shown that

$$\begin{aligned} 2\pi^{-\frac{1}{2}} \int_w^\infty dt\, t\, e^{-t^2} &= \tfrac{1}{2}\Phi'(w) \\ 2\pi^{-\frac{1}{2}} \int_0^w dt\, t^4 e^{-t^2} &= \tfrac{1}{4}[3\Phi(w) - 3w\Phi'(w) - 2w^3\Phi'(w)] \end{aligned} \tag{37.18}$$

The final result of the integration is thus:

$$F_Z = -4w^{-2}[\Phi(w) - w\Phi'(w)] \tag{37.19}$$

This result can now be expressed in a general coordinate system, as follows:

$$F_s = F_Z(w_s/w) = -4w^{-3}[\Phi(w) - w\Phi'(w)]w_s \tag{37.20}$$

We now evaluate the components of the tensor T_{rs}, eq. (37.8). Using again the reference system of Fig. 37.1, we find that the tensor is diagonal in that system, and its components are:

$$T_{XX} = T_{YY} = 4\pi^{-\frac{1}{2}} \int du\, u^2 e^{-u^2} \int d\cos\theta \;\; q^{-1}\{1 - \tfrac{1}{2}\sin^2\vartheta\} \tag{37.21}$$

$$T_{ZZ} = 4\pi^{-\frac{1}{2}} \int du\, u^2 e^{-u^2} \int d\cos\theta \;\; q^{-1}\sin^2\vartheta \tag{37.22}$$

We now note, from Fig. 37.1, the following trigonometrical relation:

$$(\sin\vartheta)/u = (\sin\theta)/q \tag{37.23}$$

from which follows, upon substitution into (37.22) and comparison with (37.11):

$$T_{ZZ} = -(2w)^{-1}F_Z \tag{37.24}$$

Moreover:

$$T_{XX} = -\tfrac{1}{2}T_{ZZ} + 4\pi^{-\frac{1}{2}} \int du\, u^2 e^{-u^2} \int d\cos\theta \;\; q^{-1}$$

The last integral is readily evaluated by the same method as above, with the result

$$\begin{aligned} T_{XX} &= -w^{-3}[\Phi(w) - w\Phi'(w)] + 2w^{-1}\Phi(w) \\ &= w^{-3}[(2w^2 - 1)\Phi(w) + w\Phi'(w)] \end{aligned} \tag{37.25}$$

These results can be expressed in a covariant form as follows:

$$\begin{aligned} T_{rs} &= T_{ZZ}\, w_r w_s/w^2 + T_{XX}(\delta_{rs} - w_r w_s/w^2) \\ &= T_1 \delta_{rs} + T_2 w_r w_s/w^2 \end{aligned} \tag{37.26}$$

It is convenient to introduce the following functions

$$\begin{aligned} \mathscr{G}(w) &= (2w^2)^{-1}[\Phi(w) - w\Phi'(w)] \\ \mathscr{H}(w) &= (2w^2)^{-1}[(2w^2-1)\Phi(w) + w\Phi'(w)] \\ \mathscr{E}(w) &= 3\mathscr{G}(w) - \Phi(w) = (2w^2)^{-1}[(3-2w^2)\Phi(w) - 3w\Phi'(w)] \end{aligned} \tag{37.27}$$

The coefficients of eq. (37.6) then become:

$$\begin{aligned} F_r &= -8w^{-1}\mathscr{G}(w) w_r \\ T_{rs} &= 2w^{-1}\mathscr{H}(w)\delta_{rs} + 2w^{-3}\mathscr{E}(w) w_r w_s \end{aligned} \tag{37.28}$$

The functions $\mathscr{G}(w)$ and $\mathscr{H}(w)$ have been tabulated by Chandrasekhar. *

§ 38. Connection with the Theory of Brownian Motion

Equation (37.6), whatever the form of the coefficients F_r and T_{rs},** is identical with the general *Fokker–Planck equation,* which is the basic starting point of the semi-phenomenological theory of brownian motion. Detailed discussions of this type of equation can be found in papers on this subject. † It is a generalized *diffusion equation in velocity space.* The connection with the theory of brownian motion is very intuitive. Brownian motion is the motion of a heavy particle in a medium of light particles. The heavy particle receives a large number of very small random impulses from the light ones. Its trajectory is infinitesimally changed by each individual collision; however, the cumulative effect of the collisions is effective in driving the system toward equilibrium. It was Chandrasekhar's idea to treat the motion of a test particle in a plasma (or of a test star in a globular cluster)

* S. Chandrasekhar, *Principles of Stellar Dynamics,* University Press, Chicago, Ill., 1942.

** See, however, a restriction at the end of this paragraph.

† See mainly: S. Chandrasekhar, *Rev. Mod. Phys.,* **15**, 1 (1943). This and other interesting papers in this field are collected in the monograph *Noise and Stochastic Processes* (N. Wax, editor), Dover Publ. Co., New York, 1954.

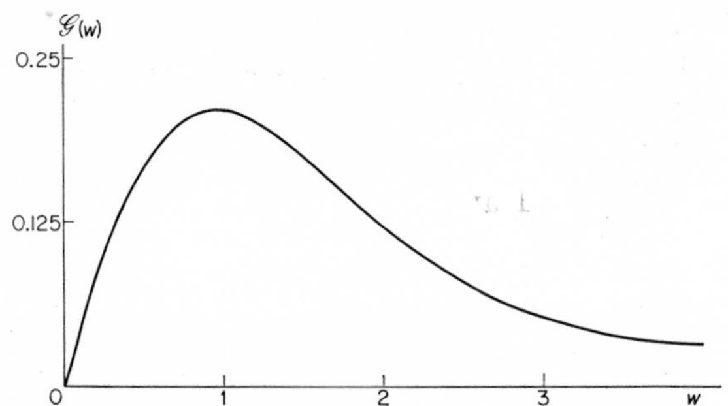

Fig. 38.1. The function $\mathscr{G}(w)$ defined in eq. (37.27).

For small values of w, $\eta(w)$ is approximately constant, as can be seen by expanding the function $\mathscr{G}(w)$:

$$\mathscr{G}(w) = (2/3\sqrt{\pi})w + \ldots; \ \eta(w) = (16/3\sqrt{\pi}) + \ldots; \ (\text{small } w) \quad (38.6)$$

For larger speeds, $\mathscr{G}(w)$ reaches a maximum (at $w = 1$) and then decreases down to zero. The asymptotic behavior is:

$$\mathscr{G}(w) = \tfrac{1}{2}w^{-2} + \ldots; \quad \eta(w) = 4w^{-3} + \ldots \ (\text{large } w) \quad (38.7)$$

This decrease has important consequences on the behavior of a plasma in an electric field, and will be discussed again in Chapter 13.

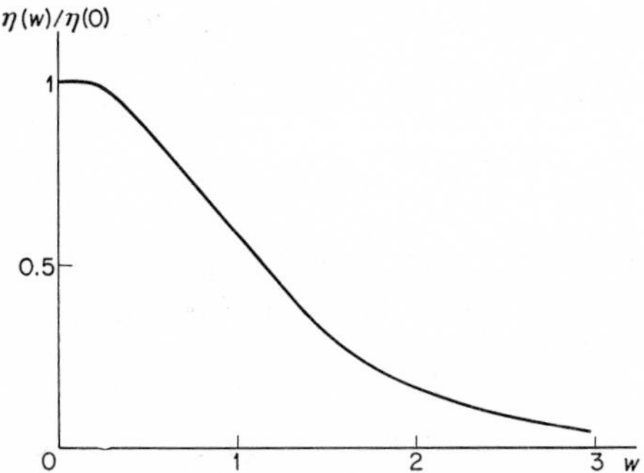

Fig. 38.2. The friction coefficient for a test particle moving through a classical electron gas at thermal equilibrium (Landau approximation).

The functions $\mathscr{H}(w)$ and $\mathscr{E}(w)$ characterizing the components of the tensor T_{rs} are shown in Fig. 38.3. It should be noted that strictly speaking not every equation of the form (37.6) can be called a Fokker–Planck equation. It must satisfy the condition

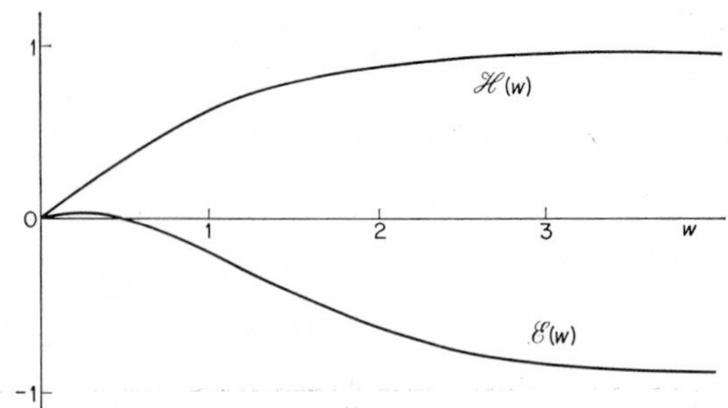

Fig. 38.3. The functions $\mathscr{H}(w)$ and $\mathscr{E}(w)$ defined in eq. (37.27).

that it admits the equilibrium distribution as a stationary solution. Introduce the following vector:

$$-F_r - T_{rs} w_s + \tfrac{1}{2} \partial_s T_{rs} \equiv J_r \tag{38.8}$$

Substituting (37.1) (with **u** changed into **w**) into eq. (37.6), it is seen that the stationarity condition implies the following relation:

$$\partial_r J_r - 2w_r J_r \equiv 0 \tag{38.9}$$

This relation between the dynamical friction and the diffusion tensor defines the most general Fokker–Planck equation. This condition was first pointed out (in a less explicit form) by Chandrasekhar. *

In the present case, it is easily seen that it is satisfied, with

$$J_r \equiv 0 \tag{38.10}$$

One can even verify from the definitions (37.27–28) that the following stronger relations hold in the present case:

$$\begin{aligned} \partial_s T_{rs} &= F_r \\ w_s T_{rs} &= -\tfrac{1}{2} F_r \end{aligned} \tag{38.11}$$

Although the brownian motion in the cycle approximation is the simplest conceivable type of long-time evolution, it is seen that this process is already considerably more complicated than the situations studied in the classical phenomenological theory of brownian motion. In the latter, one assumes that:

* S. Chandrasekhar, *Rev. Mod. Phys.*, **15,** 1 (1943).

(*a*) the friction coefficient is constant,

(*b*) the diffusion tensor equals a constant times the unit tensor.

In order to satisfy the condition (38.9) one has:

$$\begin{aligned} F_r^0 &= -\eta w_r \\ T_{rs}^0 &= \eta \delta_{rs} \end{aligned} \qquad \eta > 0 \tag{38.12}$$

With these assumptions the problem can be solved exactly and leads to a gaussian Markov process. * The brownian motion of a test particle in a plasma departs significantly from this simple model. Such a particle experiences non-linear friction, and non-linear and anisotropic diffusion.

* S. Chandrasekhar, *Rev. Mod. Phys.*, **15**, 1 (1943).

CHAPTER 9

The Ring Approximation

§ 39. The Ring Diagrams

We now go back to the discussion of § 29, which concerned the criterion of choice of contributions for the long-time evolution of a plasma. It was shown there that if the dimensionless parameter Γ is small, the dominant contributions to the distribution functions of a plasma are of the form $(e^2c)^p(e^2t)^q$, or equivalently $[e^2t(e^2c)^m]^n$. The Landau equation is an approximation to this description found by taking $p = q$ (or $m = 1$) and summing over q (or n). We are now going to complete the long-time study by performing the second step, consisting of the summation over p (or m). We are now familiar with the general idea of the diagram method (Chapters 2, 6, 7), and its formal application to the present case is rather straightforward.

We have first to choose the diagrams, using topological arguments. In §§ 39–41 we consider homogeneous systems, and therefore the evolution of $\varphi(\alpha)$. A diagram of the desired order is obtained by constructing a succession of n diagonal fragments (in order to have a factor t^n), each of which is of order $e^2(e^2c)^m t$. Each fragment must therefore have m vertices of index 1 and one vertex of index 0. We have already noted that a diagram without external lines necessarily has a creation vertex A (index 0) at right and a destruction vertex C (index 1) at left. We must therefore connect $(m - 1)$ vertices of index 1 between these two terminals. Both vertices C and D increase the number of lines to the right; if one of these is connected to the first destruction vertex C, one must connect to its right a vertex which decreases the number of lines again (in order to go back eventually to zero lines). But all vertices of the latter type have an index 0 and are therefore not appropriate. Only loops F can be connected between A and C without requiring the introduction of an extra uncompensated e^2 factor. The most general diagonal fragment of the required type is drawn in Fig. 39.1a and is called a *ring*.

A closer analysis shows that one could imagine another type of diagram contributing to the same order. Its possibility is based on the remark of p. 47, footnote. This type of diagram is drawn in Fig. 39.1b and could be called a "concave ring". It contains the exceptional "Z" configuration

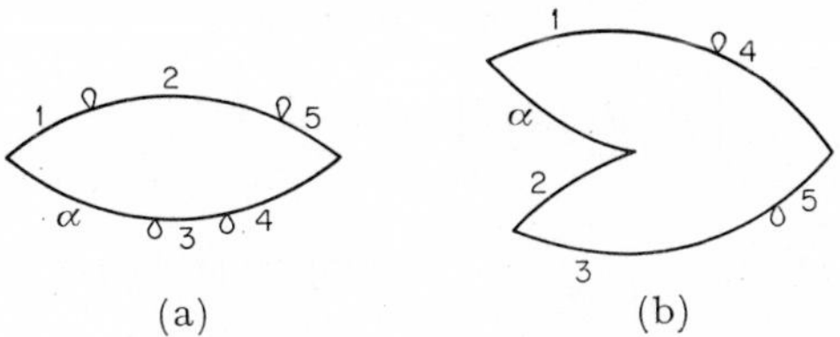

Fig. 39.1. Rings. The concave rings (b) do not contribute to reduced distribution functions, because they violate theorem I of § 8. See also Fig. 8.4.

in which vertex A has an index 1. However, a concave ring can never contribute to a reduced distribution function. In order to do so, one of its A-vertices must involve a fixed particle or be semiconnected to a previous fragment. The other vertices cannot involve fixed or semiconnected particles, otherwise the diagram would be of order $1/N$. But this diagram obviously violates rule (b) of § 8 and therefore its contribution vanishes.

We now set up rules for the construction of the mathematical expression corresponding to a ring diagram. Consider for instance the simple ring of Fig. 39.2. Applying the general rules of § 6 and the formulae of Table 6.1, one easily obtains:

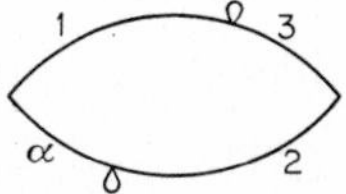

Fig. 39.2. A typical ring R_1^1.

$$\varphi(\alpha;t)\big|_1^1 = (2\pi)^{-1}\int_C dz\, e^{-izt}\int d\mathbf{v}_1 \ldots d\mathbf{v}_{N-1} \sum_1\sum_2\sum_3\sum_{\mathbf{k}_\alpha}\sum_{\mathbf{k}_1}\sum_{\mathbf{k}_2}\sum_{\mathbf{k}_3}\frac{1}{-iz}$$

$$\left\{\frac{8\pi^3}{\Omega}(-e^2m^{-1}V_{k_\alpha})(i\mathbf{k}_\alpha\cdot\boldsymbol{\partial}_{\alpha 1})\delta_{\mathbf{k}_\alpha+\mathbf{k}_1}\right\}$$

$$\frac{1}{i(\mathbf{k}_\alpha\cdot\mathbf{g}_{\alpha 1}-z)}\left\{\frac{8\pi^3}{\Omega}(-e^2m^{-1}V_{k_\alpha})(-i\mathbf{k}_\alpha\cdot\boldsymbol{\partial}_{\alpha 2})\delta_{\mathbf{k}_\alpha-\mathbf{k}_2}\right\}$$

$$\frac{1}{i(\mathbf{k}_2\cdot\mathbf{g}_{21}-z)}\left\{\frac{8\pi^3}{\Omega}(-e^2m^{-1}V_{k_1})(-i\mathbf{k}_1\cdot\boldsymbol{\partial}_{13})\delta_{\mathbf{k}_1-{}_3\mathbf{k}}\right\}$$

$$\frac{1}{i(\mathbf{k}_2 \cdot \mathbf{g}_{23} - z)} \left\{ \frac{8\pi^3}{\Omega} (-e^2 m^{-1} V_{k_3})(-i\mathbf{k}_3 \cdot \partial_{32}) \delta_{\mathbf{k}_3+\mathbf{k}_2} \right\}$$

$$\frac{1}{-iz} \rho_0(|\alpha, 1, \ldots, N-1; 0) \tag{39.1}$$

$$= \frac{1}{2\pi} \int dz\, e^{-izt} \frac{1}{-z^2} \int d\mathbf{v}_1 d\mathbf{v}_2 d\mathbf{v}_3 \int d\mathbf{l} \{-8\pi^3 e^2 c m^{-1} V_l i\mathbf{l} \cdot \partial_\alpha\} \frac{1}{i(\mathbf{l} \cdot \mathbf{g}_{\alpha 1} - z)}$$

$$\{8\pi^3 e^2 c m^{-1} V_l i\mathbf{l} \cdot \partial_\alpha\} \frac{1}{i(\mathbf{l} \cdot \mathbf{g}_{21} - z)} \{-8\pi^3 e^2 c m^{-1} V_l i\mathbf{l} \cdot \partial_1\} \frac{1}{i(\mathbf{l} \cdot \mathbf{g}_{23} - z)}$$

$$\cdot \{-e^2 m^{-1} V_l i\mathbf{l} \cdot \partial_{32}\} \varphi(\alpha; 0) \varphi(1; 0) \varphi(2; 0) \varphi(3; 0)$$

In going from the first step to the second, the following operations have been performed:

(*a*) The summations over $\mathbf{k}_1, \mathbf{k}_2, \mathbf{k}_3$ have eliminated all Kronecker deltas; the only remaining wave-vector is $\mathbf{k}_\alpha$, which is renamed $\mathbf{l}$.

(*b*) The integrations over $\mathbf{v}_4, \ldots, \mathbf{v}_{N-1}$ (the particles not involved in the ring) have transformed $\rho_0(|\alpha, 1, \ldots, N-1)$ into $\rho_0(\alpha, 1, 2, 3) = \varphi(\alpha)\varphi(1)\varphi(2)\varphi(3)$.

(*c*) The summation over $\mathbf{l}$ multiplied by $8\pi^3/\Omega$ goes over into an integral.

(*d*) The three summations over particles *1, 2, 3* combined with the three remaining factors $8\pi^3/\Omega$ give three factors $8\pi^3 c$.

(*e*) Corollary 2 of § 8 has been applied.

Equation (39.1) is exact. This expression is of the same general form as (27.1), being the inverse Laplace transform of z^{-2} times a Cauchy integral. It therefore has the same analytical properties, i.e. it is regular in the upper half-plane, has a double pole in $z = 0$ and other singularities in the lower half-plane. The arguments of § 27 can therefore be applied to the present case and one could show that the dominant term for long times is obtained by retaining only the term proportional to t in a formula corresponding to eq. (27.3). However, two difficulties appear at this stage.

A) *Choice of the potential.* It was shown in § 28 that the use of a Coulomb potential leads to divergent integrals in the calculation of the cycle. It is easily seen that the integrals in the higher order rings calculated with the Coulomb potential are even more strongly divergent. We shall not be interested in the behavior for large **l** (i.e. short distances). We therefore retain the upper cut-off l_M which was introduced in § 28. One of our main purposes is to show that the divergence for small values of **l** is suppressed as a result of the summation of the rings. In order to give a meaning to the divergent integrals, the best way is to use for V_l an approximation to the Coulomb potential which depends on a parameter α. The simplest to take is

$$V_l = \frac{1}{2\pi^2} \frac{1}{l^2 + \alpha^2} \tag{39.2}$$

This is of the same form as the Debye potential and thus leads to convergent expressions (see § 28). It should be stressed, however, that the parameter α is a purely mathematical trick without physical meaning. It allows one to handle divergent expressions in a proper way. At the end of the calculation (i.e. after the summation of the rings), α is allowed to go to zero, thus recovering a result valid (if it converges) for the true Coulomb potential. This way of handling divergent expressions is mathematically well known as the Borel summation procedure.

B) *Stability.* With the previous choice of potential, each ring is a convergent expression which has the properties described in § 27. However, the summation of the rings involves an *infinite* series. Even if each term of the series has these analytical properties and even if the series converges, it might be that the resulting sum-function has completely different analytical properties. We have already met in Chapters 3 and 4 situations in which the summation induces a radical change in the analytical properties of the distribution functions, in particular a shift of the poles into the upper half-plane, leading to an exponential instability. The same phenomenon can arise here, because the series occurring in the present problem have a striking (but not accidental) similarity with the examples mentioned. We shall, however, postpone the discussion of the stability to Appendix 9. We assume at this stage that the system is stable, so that the retention of the leading universal term in each ring is a valid approximation after a sufficiently long time.

The coefficient of t is obtained in a way similar to that leading to eq. (27.3). Expressing the limiting value of a Cauchy integral for $z \to 0$ from above, eq. (39.1) becomes

$$\varphi(\alpha; t)|_{1}^{1} = t \int d\mathbf{l} \int_{123} d_\alpha \delta_-^{\alpha 1}(-d_\alpha)\delta_-^{21} d_1 \delta_-^{23} d_{32} \cdot \varphi(\alpha; 0)\varphi(1; 0)\varphi(2; 0)\varphi(3; 0) \tag{39.3}$$

The following abbreviations have been introduced:

$$\begin{aligned} d_j &\equiv -8\pi^3 e^2 c m^{-1} V_l i\mathbf{l} \cdot \partial_j \\ d_{jm} &\equiv -e^2 m^{-1} V_l i\mathbf{l} \cdot \partial_{jm} \qquad j, m = \alpha, 1, 2, \ldots \\ \delta_-^{jm} &\equiv \pi\delta_-(\mathbf{l} \cdot \mathbf{g}_{jm}) \\ \int_{123} &\equiv \int d\mathbf{v}_1 d\mathbf{v}_2 d\mathbf{v}_3 \end{aligned} \tag{39.4}$$

We note the following symmetry properties:

$$\begin{aligned} -d_j &= (d_j)^* \\ d_{mj} &= -d_{jm} = (d_{jm})^* \\ \delta_-^{mj} &= (\delta_-^{jm})^* = \delta_+^{jm} \end{aligned} \tag{39.5}$$

The star denotes the complex conjugate.

From the study of this example, general rules for the construction of an arbitrary ring can immediately be inferred. We fix a unique orientation for the rings by adopting the following convention: *the rings involving the fixed particle* α *are drawn in such a way that the line labeled* α *lies on the lower part of the ring.* The rules are then:

A) The expression $d_\alpha \delta_-^{\alpha 1}$ corresponds to the extreme left vertex.

B) The symbol d_j is written for each loop F, j being the label of the line going out at the *left* of the vertex. These factors are given the sign + or − according to whether the loop is situated, respectively, on the upper or on the lower side of the ring.

C) Factors δ_-^{is} are intercalated between the factors d_j, i being the label of the lower line and s the index of the upper line in the state situated to the *right* of the vertex d_j.

D) A factor d_{si} corresponds to the last vertex on the right, s being the label of the upper line and i that of the lower line in the preceding state.

E′) The operator thus obtained acts on a product of momentum distribution functions at time 0: $\varphi(\alpha; 0)\varphi(1; 0) \ldots \varphi(n; 0)$; $\alpha, \ldots, n$ are the indices of all the particles involved in the ring.

F) The whole expression is integrated over the velocities $\mathbf{v}_1, \ldots, \mathbf{v}_n$ and over the wave-vector $\mathbf{l}$.

G′) The formula is multiplied by t.

We are now able to write down immediately the contribution of an arbitrary ring. Suppose that we have summed all rings (this will be done explicitly in the next section): the result will be a sum of terms of the type (39.3) involving the initial distribution functions $\varphi(j; 0)$; $j = \alpha, 1, 2, \ldots$ (it is a functional of the initial condition). We denote it by:

$$\varphi(\alpha; t)\Big|_{\text{Rings}} = t\mathscr{R}[\varphi(j; 0)] \tag{39.6}$$

$\mathscr{R}[\varphi(j, 0)]$ thus represents the sum of all one-ring diagrams over all possible types of rings. A knowledge of this expression enables one to write immediately the equation of evolution of $\varphi(\alpha; t)$:

$$\partial_t \varphi(\alpha; t) = \mathscr{R}[\varphi(j; t)] \tag{39.7}$$

This simple relation between one-ring diagrams and the kinetic equation whose solution is the sum of all semiconnected rings is proven in exactly the same way as the passage from eq. (30.5) to eq. (30.8), although the explicit algebra is more involved. A proof of this relation is given in Appendix 4. The right-hand side of eq. (39.7) therefore appears as a sum of "collision terms". We shall derive the kinetic equation directly by performing explicitly the sum of all "ring-like" collision terms, briefly called "rings". For the construction of each individual ring, the previous rules are very slightly modified:

Perform the operations A–D;

E) The operator thus obtained acts on a product of distribution functions *at time* t: $\varphi(\alpha; t)\varphi(1; t) \ldots \varphi(n; t)$; (The letter t will be omitted occasionally).

Perform operation F;
Suppress operation G′.
We are now ready for the summation of the rings.

§ 40. Summation of the Rings

In order to sum all distinct rings without repetition or omission, we establish a method which generates all these diagrams in a systematic way. Suppose we have found all the possible rings having n F-vertices (loops): there are N_n distinct members in this set. Then the complete set of rings with $n + 1$ loops is obtained as follows:

a) To each member of the set (n) add a loop on the upper line to the left of all its loops.

b) To each member add a loop on the lower line to the left of all its loops.

We thus obtain $N_{n+1} = 2N_n$ rings with $n + 1$ loops. We state that these are *all* possible distinct rings with $n + 1$ loops. In

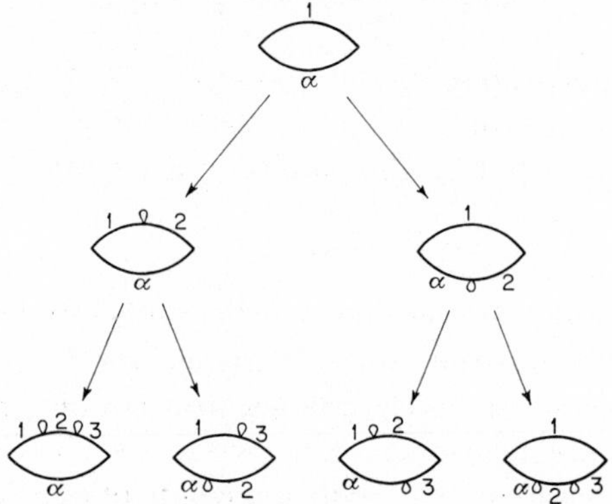

Fig. 40.1. The systematic generation of the ring diagrams.

effect, given an n-loop ring, the diagrams it generates by operations a) and b) are distinct: in a) one adds an interaction with a free particle, whereas in b) one adds an interaction with the fixed particle α (keep in mind our convention concerning the orientation of the rings). Moreover, adding a loop in any other position than the ones prescribed by operations a) and b) results in a diagram already present in the set.

This method of construction of the diagrams also provides a convenient method for their summation. We group the diagrams according to their number of loops. The first three classes of rings generated by the previous method are shown in Fig. 40.1. Applying the rules of § 39, the sum of the corresponding expressions is:

$$
\begin{aligned}
\partial_t \varphi(\alpha) = & \int d\mathbf{1} \int_1 d_\alpha \delta_-^{\alpha 1} d_{1\alpha} \varphi(\alpha)\varphi(1) \\
& + \int d\mathbf{1} \int_1 \int_2 d_\alpha \delta_-^{\alpha 1} d_1 \delta_-^{\alpha 2} d_{2\alpha} \varphi(\alpha)\varphi(1)\varphi(2) \\
& + \int d\mathbf{1} \int_1 \int_2 d_\alpha \delta_-^{\alpha 1} (-d_\alpha) \delta_-^{21} d_{12} \varphi(\alpha)\varphi(1)\varphi(2) \\
& + \int d\mathbf{1} \int_{123} d_\alpha \delta_-^{\alpha 1} d_1 \delta_-^{\alpha 2} d_2 \delta_-^{\alpha 3} d_{3\alpha} \varphi(\alpha)\varphi(1)\varphi(2)\varphi(3) \\
& + \int d\mathbf{1} \int_{123} d_\alpha \delta_-^{\alpha 1} (-d_\alpha) \delta_-^{21} d_1 \delta_-^{23} d_{32} \varphi(\alpha)\varphi(1)\varphi(2)\varphi(3) \\
& + \int d\mathbf{1} \int_{123} d_\alpha \delta_-^{\alpha 1} d_1 \delta_-^{\alpha 2} (-d_\alpha) \delta_-^{32} d_{23} \varphi(\alpha)\varphi(1)\varphi(2)\varphi(3) \\
& + \int d\mathbf{1} \int_{123} d_\alpha \delta_-^{\alpha 1} (-d_\alpha) \delta_-^{21} (-d_2) \delta_-^{31} d_{13} \varphi(\alpha)\varphi(1)\varphi(2)\varphi(3) \\
& + \dots
\end{aligned}
\tag{40.1}
$$

We now note that all the terms in (40.1) begin with the expression $\int d\mathbf{1}\, d_\alpha$. We can therefore rewrite the sum in the form:

$$
\partial_t \varphi(\alpha) = \int d\mathbf{1}\, d_\alpha F(\alpha) \tag{40.2}
$$

where the object $F(\alpha)$ is defined by the following sum derived from (40.1) [with use of the symmetry properties (39.5)]:

$$
\begin{aligned}
F(\alpha) = & \int_1 \delta_-^{\alpha 1} d_{1\alpha} \varphi(\alpha)\varphi(1) \\
& + \int_1 \delta_-^{\alpha 1} d_1 \varphi(1) \int_2 \delta_-^{\alpha 2} d_{2\alpha} \varphi(\alpha)\varphi(2) \\
& + \int_1 \delta_-^{\alpha 1} d_\alpha^* \varphi(\alpha) \int_2 \delta_-^{12*} d_{21}^* \varphi(2)\varphi(1) \\
& + \int_1 \delta_-^{\alpha 1} d_1 \varphi(1) \int_2 \delta_-^{\alpha 2} d_2 \varphi(2) \int_3 \delta_-^{\alpha 3} d_{3\alpha} \varphi(\alpha)\varphi(3) \\
& + \int_1 \delta_-^{\alpha 1} d_1 \varphi(1) \int_2 \delta_-^{\alpha 2} d_\alpha^* \varphi(\alpha) \int_3 \delta_-^{23*} d_{32}^* \varphi(3)\varphi(2)
\end{aligned}
\tag{40.3}
$$

$$+\int_1 \delta_-^{\alpha 1} d_\alpha^* \varphi(\alpha) \int_2 \delta_-^{21} d_1 \varphi(1) \int_3 \delta_-^{23} d_{32} \varphi(3)\varphi(2)$$

$$+\int_1 \delta_-^{\alpha 1} d_\alpha^* \varphi(\alpha) \int_2 \delta_-^{21} d_2^* \varphi(2) \int_3 \delta_-^{13*} d_{31}^* \varphi(3)\varphi(1)$$

$+ \ldots$

It is now seen that all the terms written to the left begin with $\int_1 \delta_-^{\alpha 1} d_1 \varphi(1)$ followed by a series which is identical with the series for $F(\alpha)$. The terms written to the right begin with $\int_1 \delta_-^{\alpha 1} d_\alpha^* \varphi(\alpha)$ followed by a series which is the complex conjugate of $F(1)$. Equation (40.3) can thus be rewritten as

$$F(\alpha) = \int_1 \delta_-^{\alpha 1} d_{1\alpha} \varphi(\alpha)\varphi(1) + \int_1 \delta_-^{\alpha 1} d_1 \varphi(1) F(\alpha) + \int_1 \delta_-^{\alpha 1} d_\alpha^* \varphi(\alpha) F^*(1) \tag{40.4}$$

In the second term of the right-hand side, $F(\alpha)$ is independent of $\mathbf{v}_1$; it can thus be written in front of the integral. Similarly $d_\alpha^* \varphi(\alpha)$ can be drawn in front of the integral in the third term. We rewrite eq. (40.4) in the form

$$\left[1 - \int_1 \delta_-^{\alpha 1} d_1 \varphi(1)\right] F(\alpha) = \int_1 \delta_-^{\alpha 1} d_{1\alpha} \varphi(\alpha)\varphi(1) + d_\alpha^* \varphi(\alpha) \int_1 \delta_-^{\alpha 1} F^*(1) \tag{40.5}$$

We now introduce the following more compact notations for the coefficients in this equation

$$\varepsilon(\alpha) = 1 - \int_1 \delta_-^{\alpha 1} d_1 \varphi(1) = 1 + 8\pi^4 e^2 c i V_l m^{-1} \int d\mathbf{v}_1 \delta_-(\mathbf{l} \cdot \mathbf{g}_{\alpha 1}) \mathbf{l} \cdot \partial_1 \varphi(1) \tag{40.6}$$

$$q(\alpha) = \int_1 \delta_-^{\alpha 1} d_{1\alpha} \varphi(\alpha)\varphi(1) = \pi i e^2 m^{-1} V_l \int d\mathbf{v}_1 \delta_-(\mathbf{l} \cdot \mathbf{g}_{\alpha 1}) \mathbf{l} \cdot \partial_{\alpha 1} \varphi(\alpha)\varphi(1) \tag{40.7}$$

$$d(\alpha) = i d_\alpha^* \varphi(\alpha) = 8\pi^3 e^2 c m^{-1} V_l \mathbf{l} \cdot \partial_\alpha \varphi(\alpha) \tag{40.8}$$

For later convenience we separate real and imaginary parts of ε and q (d is real):

$$\varepsilon(\alpha) = \varepsilon_1(\alpha) + i\varepsilon_2(\alpha)$$

$$\varepsilon_1(\alpha) = 1+8\pi^3 e^2 c m^{-1} V_l \mathscr{P}\int d\mathbf{v}_1 \frac{\mathbf{l}\cdot\partial_1\varphi(1)}{\mathbf{l}\cdot\mathbf{v}_\alpha-\mathbf{l}\cdot\mathbf{v}_1} \tag{40.6a}$$

$$\varepsilon_2(\alpha) = 8\pi^4 e^2 c m^{-1} V_l \int d\mathbf{v}_1 \delta(\mathbf{l}\cdot\mathbf{v}_\alpha-\mathbf{l}\cdot\mathbf{v}_1)\mathbf{l}\cdot\partial_1\varphi(1)$$

Comparing these equations with eq. (19.4) we immediately see that the function $\varepsilon(\alpha)$ is nothing other than the complex conjugate of (or else, the minus-function corresponding to) the dielectric constant $\varepsilon_+(\mathbf{l}; z)$ for the real value $z = \mathbf{l}\cdot\mathbf{v}_\alpha$ of the frequency [see also (23.15b)]:

$$\varepsilon(\alpha) \equiv \varepsilon_-(\mathbf{l}; \mathbf{l}\cdot\mathbf{v}_\alpha)$$

The implications of this fact will be discussed in Chapter 11.

$$q(\alpha) = q_1(\alpha)+iq_2(\alpha)$$

$$q_1(\alpha) = (e^2/m)V_l \mathscr{P}\int d\mathbf{v}_1 \frac{\mathbf{l}\cdot\partial_{\alpha 1}\varphi(\alpha)\varphi(1)}{\mathbf{l}\cdot\mathbf{v}_\alpha-\mathbf{l}\cdot\mathbf{v}_1} \tag{40.7a}$$

$$q_2(\alpha) = (\pi e^2/m)V_l \int d\mathbf{v}_1 \delta(\mathbf{l}\cdot\mathbf{v}_\alpha-\mathbf{l}\cdot\mathbf{v}_1)\mathbf{l}\cdot\partial_{\alpha 1}\varphi(\alpha)\varphi(1)$$

With these new notations, eqs. (40.2) and (40.5) become:

$$\partial_t\varphi(\alpha) = -8\pi^3 i e^2 c m^{-1}\int d\mathbf{l}\, V_l \mathbf{l}\cdot\partial_\alpha F(\alpha) \tag{40.9}$$

$$\varepsilon(\alpha)F(\alpha) = q(\alpha)+i\pi d(\alpha)\int d\mathbf{v}_1 \delta_-(\mathbf{l}\cdot\mathbf{v}_\alpha-\mathbf{l}\cdot\mathbf{v}_1)F^*(1) \tag{40.10}$$

In this way the problem of the summation of the rings has been reduced to the solution of an integral equation for the auxiliary function $F(\alpha)$. This equation will be solved exactly in the next section.

§ 41. Solution of the Integral Equation for Plasmas. (Imaginary Part of *F*)

Equation (40.10) plays a basic role in the theory of the long-time behavior of plasmas; it will be met again in many problems in plasma physics. It governs not only the first approximation which we consider here, but it can be shown that the higher order equations are also expressed in terms of the function F. F is a rather complicated mathematical object: besides being an explicit function of $\mathbf{l}$ and $\mathbf{v}_\alpha$, it is moreover a *functional of the distribution function* $\varphi(\mathbf{v}_\alpha; t)$. It is, therefore, an implicit function of time. F is a complex number, which can be separated into a real and an imaginary part:

$$F(\alpha) = F_1(\alpha)+iF_2(\alpha) \tag{41.1}$$

The integral equation (40.10) for the complex function F is thus equivalent to a system of two integral equations for the real functions F_1 and F_2. Because of the occurrence of F^* on the right-hand side of (40.10), these two equations are not independent. Before proceeding to their solution, two basic remarks will define the problem in a more precise way.

A) We first note that *only the imaginary part of F gives a non-vanishing contribution to the kinetic equation* (40.9). In effect, $\varphi(\alpha)$ is a distribution function and therefore a *real* quantity. The real part of F, substituted into (40.9), would however give an imaginary contribution, which is unphysical. Mathematically, this consistency requirement can be verified when the complete solution is known. We shall, however, defer the complete solution until Chapter 10 and will accept here this property on the basis of its physical meaning.

B) *Reduction of eq.* (40.10). The function $F(\alpha)$ depends, explicitly and through $\varphi(\alpha)$, on the *vector* variable $\mathbf{v}_\alpha$, or equivalently on the *three scalar variables* $v_{\alpha x}$, $v_{\alpha y}$, $v_{\alpha z}$. The same property is shared by the functions $q(\alpha)$ and $d(\alpha)$. The kernel of the integral equation, $\delta_-(\mathbf{l}\cdot\mathbf{v}_\alpha-\mathbf{l}\cdot\mathbf{v}_1)$, depends on $\mathbf{v}_\alpha$ and on $\mathbf{v}_1$ only through the scalar products $\mathbf{l}\cdot\mathbf{v}_\alpha$ and $\mathbf{l}\cdot\mathbf{v}_1$, i.e. on the projection of the velocity variables on the vector $\mathbf{l}$ (which enters the equation as a parameter only). As can be seen from the definition (40.6), the function $\varepsilon(\alpha)$ also depends on the single component $\mathbf{l}\cdot\mathbf{v}_\alpha/l$ of the velocity.

Let us again introduce the notation ν_α, which was used in Chapters 3, 4 and 5 [see eq. (14.5)]:

$$\nu_\alpha = \mathbf{l}\cdot\mathbf{v}_\alpha/l \tag{41.2}$$

We take advantage of this peculiar structure in order to simplify the equation. We define barred quantities by integration over the components of $\mathbf{v}_\alpha$ perpendicular to $\mathbf{1}$, as was done in (14.6):

$$\bar{f}(\nu_\alpha) = \int d\mathbf{v}_\alpha \, \delta(\nu_\alpha - \mathbf{1}\cdot\mathbf{v}_\alpha/l) f(\mathbf{v}_\alpha) \tag{41.3}$$

We now multiply both sides of eq. (40.10) by $\delta(\nu_\alpha - \mathbf{1}\cdot\mathbf{v}_\alpha/l)$ and integrate over $\mathbf{v}_\alpha$. As a result of the previous discussion,

$$\int d\mathbf{v}_\alpha \, \delta(\nu_\alpha - \mathbf{1}\cdot\mathbf{v}_\alpha/l)\varepsilon(\mathbf{1}\cdot\mathbf{v}_\alpha/l) F(\mathbf{v}_\alpha) = \varepsilon(\nu_\alpha)\bar{F}(\nu_\alpha) \tag{41.4}$$

$$\begin{aligned} &\int d\mathbf{v}_\alpha \, \delta(\nu_\alpha - \mathbf{1}\cdot\mathbf{v}_\alpha/l) d(\mathbf{v}_\alpha) \int d\mathbf{v}_1 \, \delta_-(\mathbf{1}\cdot\mathbf{v}_\alpha - \mathbf{1}\cdot\mathbf{v}_1) F^*(\mathbf{v}_1) \\ &= \bar{d}(\nu_\alpha) \int d\mathbf{v}_1 \, \delta_-(l\nu_\alpha - \mathbf{1}\cdot\mathbf{v}_1) F^*(\mathbf{v}_1) = \bar{d}(\nu_\alpha) \int d\nu_1 \, \delta_-(l\nu_\alpha - l\nu_1) \bar{F}^*(\nu_1) \\ &= l^{-1}\bar{d}(\nu_\alpha) \int d\nu_1 \, \delta_-(\nu_\alpha - \nu_1) \bar{F}^*(\nu_1) \end{aligned} \tag{41.5}$$

In deriving (41.5) we have used the obvious property

$$\begin{aligned} \int d\mathbf{v}_1 \, \delta_-(l\nu_\alpha - \mathbf{1}\cdot\mathbf{v}_1) F(\mathbf{v}_1) &= \int d\nu_1 \, \delta_-(l\nu_\alpha - l\nu_1) \int d\mathbf{v}_1 \, \delta(\mathbf{1}\cdot\mathbf{v}_1/l - \nu_1) F(\mathbf{v}_1) \\ = \int d\nu_1 \, \delta_-(l\nu_\alpha - l\nu_1) \bar{F}(\nu_1) &= l^{-1} \int d\nu_1 \, \delta_-(\nu_\alpha - \nu_1) \bar{F}(\nu_1) \end{aligned} \tag{41.6}$$

Moreover, the following important fact can be simply proved from the definitions (40.6) and (40.8):

$$(\pi/l)\bar{d}(\nu_\alpha) = \varepsilon_2(\nu_\alpha) \tag{41.7}$$

We thus obtain an equation for the barred function:

$$\varepsilon(\nu_\alpha)\bar{F}(\nu_\alpha) = \bar{q}(\nu_\alpha) + i\varepsilon_2(\nu_\alpha) \int d\nu_1 \, \delta_-(\nu_\alpha - \nu_1) \bar{F}^*(\nu_1) \tag{41.8}$$

We now note that $\bar{F}$ is related to F by a simple algebraic equation. In effect, comparing eqs. (40.10) and (41.8), and using eq. (41.6), the integral occurring on the right-hand sides of the former equations can be eliminated, and a relation between F and $\bar{F}$ is obtained in the form:

$$F = \pi \frac{d}{l\varepsilon_2} \bar{F} + \frac{1}{\varepsilon}\left[q - \frac{\pi d}{l\varepsilon_2}\bar{q}\right] \tag{41.9}$$

It is convenient for future reference to split this into real and imaginary parts. Note that from the definitions (40.7a) and (41.3):

$$\bar{q}_2(\nu_\alpha) = 0 \tag{41.10}$$

Equation (41.9) is thus equivalent to the two equations:

$$F_1 = \frac{\pi d}{l\varepsilon_2}\bar{F}_1 + \frac{1}{\varepsilon\varepsilon^*}\left[\varepsilon_1 q_1 + \varepsilon_2 q_2 - \frac{\varepsilon_1}{\varepsilon_2}\frac{\pi d}{l}\bar{q}_1\right] \tag{41.9a}$$

$$F_2 = \frac{\pi d}{l\varepsilon_2}\bar{F}_2 + \frac{1}{\varepsilon\varepsilon^*}\left[\varepsilon_1 q_2 - \varepsilon_2 q_1 + \frac{\pi d}{l}\bar{q}_1\right] \tag{41.9b}$$

The considerations developed above have thus shown that in order to calculate F, it is sufficient to solve the simpler equation (41.8). Further, a knowledge of F_2 requires only a knowledge of $\bar{F}_2$.

C) *Solution of the reduced equation* (41.8). Separating real and imaginary parts in eq. (41.8), and keeping in mind eq. (41.10), we obtain the two equations:

$$\varepsilon_1 \bar{F}_2 + \varepsilon_2 \mathscr{P} \bar{F}_2 = 0 \tag{41.11}$$

$$\varepsilon_1 \bar{F}_1 - \varepsilon_2 \mathscr{P} \bar{F}_1 = \bar{q}_1 + 2\varepsilon_2 \bar{F}_2 \tag{41.12}$$

We have introduced here the operator $\mathscr{P}$ defined as:

$$\mathscr{P}f \equiv \frac{1}{\pi}\mathscr{P}\int_{-\infty}^{\infty} d\nu_1 \frac{f(\nu_1)}{\nu_\alpha - \nu_1} \tag{41.13}$$

Equation (41.11) has three remarkable properties:

α) *It does not involve* $\bar{F}_1$. This decoupling of the two equations is a considerable simplification. It is known that single integral equations of this type can be solved in closed form, whereas, in general, systems of such equations cannot.* This simplification is due to property (41.7).

β) *It is a homogeneous equation* (in the mathematical sense: it contains no term independent of $\bar{F}_1$). This property is due to eq. (41.10).

γ) *Its index is zero.* The index of a singular Cauchy integral equation is defined in Appendix 2 as:

* N. J. Muskhelishvili, *Singular Integral Equations,* Noordhoff Publ., Groningen, 1953.

$$\chi = \frac{1}{2\pi}\left[\arg\frac{\varepsilon_1 + i\varepsilon_2}{\varepsilon_1 - i\varepsilon_2}\right]_{C'} \tag{41.14}$$

where $[\ ]_{C'}$ denotes the variation of the argument when ν varies from $-\infty$ to $+\infty$ on the real axis (R.A) and back to $-\infty$ along a half-circle at infinity in the upper half-plane. But according to the discussion of § 19, this can also be written as:

$$\chi = (N_0 - N_P) - (N_0^* - N_P^*) \tag{41.15}$$

where N_0 and N_P are respectively the number of zeros and of poles in S_+ of the analytical continuation of $\varepsilon(\nu)$ into S_+ [i.e. of $\varepsilon_-(z)$ for $z \in S_+$]; N_0^* and N_P^* are the corresponding numbers for $\varepsilon_+(z)$. But *if the system is stable* we know that $\varepsilon_+(z)$ has neither poles nor zeros in the upper half-plane. The corresponding minus-function, $\varepsilon_-(z)$, has both zeros and poles in S_+, but it can be shown that $N_0 - N_P = 0$ [see Appendix 2, § A2.3,I.] Hence, *for stable systems*, $\chi = 0$. The case of unstable plasmas will be dealt with in Appendix 9.

Property α) allows us to solve eq. (41.11) independently of (41.12). Property β) implies the existence of the trivial solution $\bar{F}_2 = 0$. Property γ) ensures the unicity of this solution (see Appendix 2, § A2.3,III). The problem is thus solved: the *only* solution of eq. (41.11) is

$$\bar{F}_2 = 0 \tag{41.16}$$

As F_2 depends only on $\bar{F}_2$, we do not need to solve eq. (41.12). We now go backwards, substituting $\bar{F}_2$ into eq. (41.9b) with the result

$$F_2 = \frac{1}{\varepsilon\varepsilon^*}\left[\varepsilon_1 q_2 - \varepsilon_2 q_1 + \frac{\pi d}{l}\bar{q}_1\right] \tag{41.17}$$

The bracket can be transformed very easily by expanding the quantities according to their definitions (40.6a, 40.7a, 40.8, 41.3, 41.13) $[\bar{f}'(\nu) \equiv d\bar{f}/d\nu;\ f'(\mathbf{v}) \equiv l^{-1}\mathbf{1}\cdot\partial f/\partial\mathbf{v}]$:

$$\begin{aligned}
F_2 = {} & \frac{1}{\varepsilon\varepsilon^*}\frac{\pi e^2}{m}V_l[\varphi'(\alpha)\bar{\varphi}(\alpha) - \varphi(\alpha)\bar{\varphi}'(\alpha)] \\
& + \frac{1}{\varepsilon\varepsilon^*}\frac{8\pi^3 e^2 c}{m}\frac{\pi e^2}{m}V_l^2\{\mathscr{P}\bar{\varphi}'[\varphi'(\alpha)\bar{\varphi}(\alpha) - \varphi(\alpha)\bar{\varphi}'(\alpha)] \\
& - \bar{\varphi}'(\alpha)[\varphi'(\alpha)\mathscr{P}\bar{\varphi} - \varphi(\alpha)\mathscr{P}\bar{\varphi}'] \\
& + \varphi'(\alpha)[\bar{\varphi}'(\alpha)\mathscr{P}\bar{\varphi} - \bar{\varphi}(\alpha)\mathscr{P}\bar{\varphi}']\}
\end{aligned}$$

The terms enclosed in curly brackets cancel pairwise, and we are left with:

$$F_2 = q_2/\varepsilon\varepsilon^* \tag{41.18}$$

Substituting the result into eq. (40.9) we finally obtain the kinetic equation:

$$\partial_t \varphi(\alpha) = \frac{8\pi^4 e^4 c}{m^2} \int d\mathbf{l} \int d\mathbf{v}_1 \mathbf{l} \cdot \partial_\alpha \left| \frac{V_l}{\varepsilon(\alpha)} \right|^2 \delta(\mathbf{l} \cdot \mathbf{g}_{\alpha 1}) \mathbf{l} \cdot \partial_{\alpha 1} \varphi(\alpha) \varphi(1) \tag{41.19}$$

$$\varepsilon(\alpha) = 1 + 2\pi^3 i \omega_p^2 V_l^2 \int d\mathbf{v}_1 \delta_-(\mathbf{l} \cdot \mathbf{g}_{\alpha 1}) \mathbf{l} \cdot \partial_1 \varphi(1) \tag{41.20}$$

ω_p being the plasma frequency defined in eq. (13.11). The properties of this equation will be discussed in Chapter 11.

Equation (41.19) was first derived by the author in 1960. * In his classical book, Bogoliubov** has considered the case of a plasma within his general theory. He obtained an integral equation essentially equivalent to eq. (40.10). However, his equation was written in ordinary space (instead of Fourier space). He did not solve this equation. Lenard † took over Bogoliubov's equation and solved it by a procedure which is very similar to the one used in this section. Another derivation, very similar to Bogoliubov's, was given by Guernsey. †† Finally, Rostoker and Rosenbluth ‡ gave a derivation of a particular case of this equation, i.e. the brownian motion approximation. Another attempt at treating the brownian motion problem in the same direction is Tchen's. ‡‡ It is not our purpose to compare all these various derivations in detail here. However, Bogoliubov's (and thus Lenard's) method will be discussed further in Chapter 11.

* R. Balescu, *Phys. of Fluids*, **3**, 52 (1960); see also *Physica*, **25**, 324 (1959).

** N. N. Bogoliubov *Problems of a Dynamical Theory in Statistical Physics*, translated by E. K. Gora, in *Studies in Statistical Mechanics*, vol. 1, ed. by J. de Boer and G. E. Uhlenbeck, Interscience, New York, 1962.

† A. Lenard, *Ann. Phys.*, **3**, 390 (1960).

†† R. L. Guernsey, *The Kinetic Theory of Fully Ionized Gases*, Off. Nav. Res., Contract No. Nonr. 1224 (15), July, 1960.

‡ N. Rostoker and M. N. Rosenbluth, *Phys. of Fluids*, **3**, 1 (1960).

‡‡ C. M. Tchen, *Phys. Rev.*, **114**, 394 (1959).

Let us mention the generalization of this equation to an s-component plasma:

$$\partial_t \varphi^{(\mu)}(\alpha) = \sum_{\nu=1}^{s} \frac{8\pi^4 e^4 z_\mu^2 z_\nu^2 c_\nu}{m_\mu} \int d\mathbf{l} \int d\mathbf{v}_1 \mathbf{l} \cdot \partial_\alpha \left| \frac{V_l}{\varepsilon^{(\mu)}(\alpha)} \right|^2 \delta(\mathbf{l} \cdot \mathbf{g}_{\alpha 1}) \\ \mathbf{l} \cdot (m_\mu^{-1} \partial_\alpha - m_\nu^{-1} \partial_1) \varphi^{(\mu)}(\alpha) \varphi^{(\nu)}(1) \tag{41.19'}$$

$$\varepsilon^{(\mu)}(\alpha) = 1 + \sum_{\sigma=1}^{s} 8\pi^4 e^2 z_\sigma z_\mu c_\sigma i V_l m_\sigma^{-1} \int d\mathbf{v}_2 \, \delta_-(\mathbf{l} \cdot \mathbf{g}_{\alpha 2}) \mathbf{l} \cdot \partial_2 \varphi^{(\sigma)}(2) \tag{41.20'}$$

§ 42. Inhomogeneous Systems

The treatment of inhomogeneous plasmas, more precisely of $\rho_{\mathbf{k}}(\alpha; t)$ in the "ring" approximation, is very easily performed with the techniques we possess at this point. It is based on a combination of the methods of §§ 31, 32, 40 and 41. We have to assume again the "*hydrodynamic condition*" (31.2). The choice of diagrams is then based on the rule of retaining only powers of $(e^2c)^n(e^2t)^m[1+\mathbf{k} \cdot \mathbf{v}t+ \ldots]$. There will be two possible types of diagrams. The first type is obtained by setting $m = 0$. It is then seen that the diagrams reduce to the ones which were studied in Chapter 2 and give rise to the Vlassov equation. The diagrams with $m \neq 0$ on the other hand contain m diagonal or pseudo-diagonal fragments which are generated by the homogeneous rings. Each ring generates a class of pseudo-diagonal fragments as shown in Fig. 42.1.

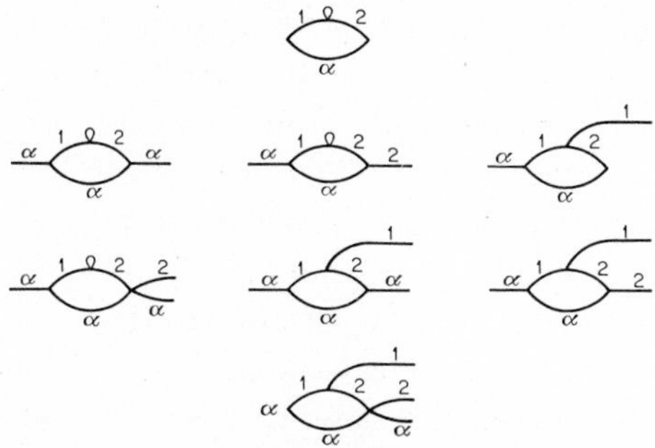

Fig. 42.1. The class of pseudo-diagonal fragments generated by a homogeneous ring.

The summation of these diagrams is based on the following remark which is a consequence of the results of § 32. The sum of the contributions of a complete class of pseudo-diagonal diagrams to $f(\alpha)$ gives rise to a collision operator which is identical to the homogeneous collision operator describing the evolution of $\varphi(\alpha)$. This operator acts on a product of functions $f(\alpha)f(1)\ldots f(n)$ evaluated at the *same* space-point $\mathbf{x}_\alpha$. This remark is an obvious generalization of the result (32.6) and it suggests the following method of summation: as the first step, the contributions of each class of pseudo-diagonal fragments are summed separately; in the second step the contributions of all classes are added.

The first step is very easily achieved. A direct generalization of the calculations of § 32 to the class shown in Fig. 42.1 shows that, to zeroth order in $\mathbf{k}\cdot\mathbf{v}t$:

$$\begin{aligned}\partial_t f(\alpha)\big|_0^1 &= \partial_t\varphi(\alpha)+\int d\mathbf{k}\, e^{i\mathbf{k}\cdot\mathbf{x}_\alpha}\,\partial_t\rho_{\mathbf{k}}(\alpha)\\ &=\int d\mathbf{l}\int d\mathbf{v}_1\, d\mathbf{v}_2\, d_\alpha\,\delta_-^{\alpha 1}\, d_1\,\delta_-^{\alpha 2}\, d_{2\alpha}\Big\{\Big[\varphi(\alpha)+\int d\mathbf{k}\,\rho_{\mathbf{k}}^{(0)}(\alpha)e^{i\mathbf{k}\cdot\mathbf{x}_\alpha}\Big]\\ &\cdot\Big[\varphi(1)+\int d\mathbf{k}\,\rho_{\mathbf{k}}^{(0)}(1)e^{i\mathbf{k}\cdot\mathbf{x}_\alpha}\Big]\Big[\varphi(2)+\int d\mathbf{k}\,\rho_{\mathbf{k}}^{(0)}(2)e^{i\mathbf{k}\cdot\mathbf{x}_\alpha}\Big]\Big\} \qquad (42.1)\\ &=\int d\mathbf{l}\int d\mathbf{v}_1\, d\mathbf{v}_2\, d_\alpha\,\delta_-^{\alpha 1}\, d_1\,\delta_-^{\alpha 2}\, d_{2\alpha}\, f^{(0)}(\alpha)f^{(0)}(1)f^{(0)}(2)\end{aligned}$$

The contributions including higher orders in $\mathbf{k}\cdot\mathbf{v}t$ are obtained by replacing on the r.h.s. of (42.1) the product of factors $f^{(0)}$ by a product of factors f. Each f is then written as $f=f^{(0)}+f^{(1)}+\ldots$, and the product is expanded up to the desired order. In order to save space, all orders are treated uniformly by dropping the superscripts $^{(0)}$ in (42.1), with the convention of expanding the final result according to the Chapman–Enskog procedure.

The contribution of a class of pseudo-diagonal fragments to $\partial_t f(\alpha)$ can thus be written automatically by means of the set of rules A–F of § 39, replacing rule E, however, by the following:

E″) The operator thus obtained acts on a product of distribution functions at time $t: f(\mathbf{x}_\alpha, \mathbf{v}_\alpha)f(\mathbf{x}_\alpha, \mathbf{v}_1)\ldots f(\mathbf{x}_\alpha, \mathbf{v}_n)$. The notation $f(\mathbf{x}_\alpha, \mathbf{v}_r)$ means the value of $f(\mathbf{x}_r, \mathbf{v}_r)$ at the point $\mathbf{x}_r=\mathbf{x}_\alpha$.

From here on, the summation of the second step goes on exactly as in §§ 40 and 41: the summation of the inhomogeneous rings has been reduced to the summation of the homogeneous rings. We can thus give the final result (including the Vlassov term) directly:

$$\partial_t f(\mathbf{x}_\alpha, \mathbf{v}_\alpha; t) + \mathbf{v}_\alpha \cdot \nabla_\alpha f(\alpha) - e[\nabla_\alpha \Phi(\mathbf{x}_\alpha; t)] \cdot \partial_\alpha f(\alpha)$$

$$= 8\pi^4 e^4 m^{-2} \int d\mathbf{l} \int d\mathbf{v}_1 \mathbf{l} \cdot \partial_\alpha \tag{42.2}$$

$$\times \frac{V_l^2}{|\varepsilon(\mathbf{x}_\alpha, \mathbf{v}_\alpha)|^2} \delta(\mathbf{l} \cdot \mathbf{g}_{\alpha 1}) \mathbf{l} \cdot \partial_{\alpha 1} f(\mathbf{x}_\alpha, \mathbf{v}_\alpha; t) f(\mathbf{x}_\alpha, \mathbf{v}_1; t)$$

where

$$\varepsilon(\mathbf{x}_\alpha, \mathbf{v}_\alpha) = 1 + 2\pi^3 i c^{-1} \omega_p^2 V_l \int d\mathbf{v}_1 \delta_-(\mathbf{l} \cdot \mathbf{g}_{\alpha 1}) \mathbf{l} \cdot \partial_1 f(\mathbf{x}_\alpha, \mathbf{v}_1; t) \tag{42.3}$$

and $\Phi(\mathbf{x}_\alpha; t)$ is defined by eq. (12.1). It is prescribed that this equation is only valid within the Chapman–Enskog approximation (see § 33).

CHAPTER 10

Binary Correlations in the Ring Approximation

§ 43. Preliminary Discussion of the Diagrams

The problem we are concerned with in the present chapter is the study of the correlation function $G(\mathbf{x}_\alpha, \mathbf{x}_\gamma, \mathbf{v}_\alpha, \mathbf{v}_\gamma; t)$ defined in eq. (2.19). Only homogeneous systems will be discussed here. The extension to non-uniform systems follows the same lines as in § 42 and will not be treated explicitly. A short account of this extension is given in a paper of the author and de Gottal.*

The correlation function of a homogeneous and isotropic system is given by eq. (2.21) combined with eq. (2.22):

$$G(r_{\alpha\gamma}, \mathbf{v}_\alpha, \mathbf{v}_\gamma; t) = c^2 \int d\mathbf{k}\, \rho_{\mathbf{k},-\mathbf{k}}(\mathbf{v}_\alpha, \mathbf{v}_\gamma; t)\, e^{i\mathbf{k}\cdot\mathbf{r}_{\alpha\gamma}} \tag{43.1}$$

where $\mathbf{r}_{\alpha\gamma} \equiv \mathbf{x}_\alpha - \mathbf{x}_\gamma$. We will use in such situations the shorter notation

$$\rho_{\mathbf{k},-\mathbf{k}}(\alpha, \gamma) = \rho_k(\alpha, \gamma) \quad \text{(isotropic system)} \tag{43.2}$$

This notation does not lead to confusion with the Fourier transform of the inhomogeneity factor $\rho_{\mathbf{k}}(\alpha)$, which has a single particle variable and depends on the *vector* index $\mathbf{k}$. **

As a result of the discussion of § 3 (see Table 2.1), the correlation function is assumed to be initially of order e^2. In a plasma, this actually means that it is proportional to one *uncompensated* e^2-factor; it may however contain an arbitrary number of factors e^2c. The collective screening effect has indeed always to be retained to all orders, as results from our previous investigation. Thus the dominant contribution to the correlation function has one more e^2-factor than the dominant contribution to $\varphi(\alpha)$. Comparing

* R. Balescu and Ph. de Gottal, *Acad. Roy. Belg. Bull. Classe Sci.*, **47**, 245 (1961).

** The notation (43.2) should be regarded as a mere shorthand. It does *not* imply that $\rho_k(\alpha, \gamma)$ depends only on the absolute value of $\mathbf{k}$. (The correlation can depend also on $\mathbf{k}\cdot\mathbf{v}_\alpha$ and $\mathbf{k}\cdot\mathbf{v}_\gamma$.) However, the corresponding *density correlation function* g_k [see table 2.2] depends only on the absolute value k of the vector $\mathbf{k}$.

this with the criteria discussed in § 29, we see that we have to select all diagrams proportional to $e^2(e^2c)^n(e^2t)^m$.

It seems, therefore, that we have here a problem of higher order than the problem of Chapter 9. However, the *relative* order of magnitude of $\rho_k(\alpha, \gamma; t)$ with respect to $\rho_k(\alpha, \gamma; 0)$ is $(e^2c)^n(e^2t)^m$. We shall therefore say that contributions of order $e^2(e^2c)^n(e^2t)^m$ to the two-body *correlation function* also belong to the ring approximation [whereas contributions of this same order to $\varphi(\alpha)$ belong to the next higher order approximation]. It will be seen that this classification is perfectly justified from the mathematical and physical points of view.

We are now faced with the problem of selecting all diagrams with two external lines at left which are of the desired order.

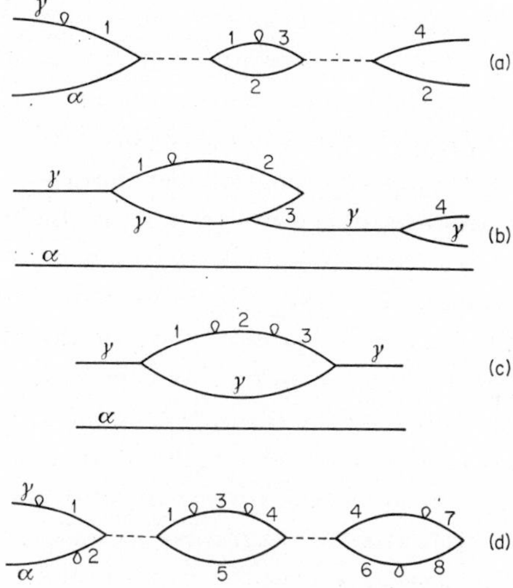

Fig. 43.1. Typical contributions to $\rho_k(\alpha, \gamma; t)$. Diagrams (a) and (b) contain destruction vertices; (c) is a diagonal fragment with external lines; (d) contains a succession of rings followed by a creation fragment.

We cannot, of course, enter here into the details of a complete theory of the diagrams with external lines, which is more delicate than the theory of diagonal and pseudo-diagonal fragments needed in the previous problems. We refer the reader for further details to Prigogine's monograph.* We shall treat here briefly the cases of interest for understanding the forthcoming chapters. In Fig. 43.1 are drawn some typical diagrams contributing to $\rho_k(\alpha, \gamma; t)$.

The diagrams (a) and (b) have the characteristic feature of beginning on the right with a destruction fragment (see § 29). It is, however, easily seen that such diagrams cannot contribute to $\rho_k(\alpha, \gamma)$ in the present approximation. Two cases are possible. In the first case the destruction fragment at right brings the system from a state in which a number of particles are correlated into a state in which they are uncorrelated [diagram (a)]. These diagrams act on an initial correlation which is of order e^2 [diagram (a)] or higher (see Table 2.1). On the other hand, they must contain at least one creation vertex A which restores two external lines at left: this vertex introduces one more e^2-factor. These diagrams thus necessarily have at least two e^2-factors and are too small. In the second case, the destruction fragment at right brings the system into a state of binary correlation, and from then on only diagonal fragments [diagram (b)] appear. Such a diagram acts on an initial state describing ternary or higher correlations: this state is proportional to at least two powers of e^2 and is thus necessarily too small. Hence the diagrams contributing to $\rho_k(\alpha, \gamma)$ in the ring approximation cannot contain destruction regions.

Let us now go over to diagrams of type (c): these consist of *successions of diagonal fragments with external lines*. The initial correlation is then of order e^2; if the diagonal fragments are rings, these diagrams are of the correct order of magnitude. It has been shown in § 29 that such diagrams, because of their multiple poles on the real axis, give rise to contributions of the type

$$e^2(e^2c)^n(e^2t)^m e^{-i\mathbf{k}\cdot\mathbf{g}_{\alpha\gamma}t} + \text{slower growing oscillations} \\ + \text{steady oscillations} + \text{decaying terms (non-universal).}$$

Hence the leading universal term gives rise to a contribution of the type

$$\rho_k'(\alpha, \gamma) = t^m e^{-i\mathbf{k}\cdot\mathbf{g}_{\alpha\gamma}t} f(k) \tag{43.3}$$

* I. Prigogine, *Non-Equilibrium Statistical Mechanics*, Interscience, New York, 1963.

These terms are perfectly defined contributions to $\rho_k(\alpha, \gamma)$. We shall, however, now restrict the field of our interest. Our purpose in studying correlation functions is not really the study of, say, density or velocity correlations in a plasma. Such quantities would not be easily measurable, because we are assuming that the intermolecular forces are the only source of correlations in the plasma (see § 3). This implies that the range of correlations is very short, of the order of a Debye length.

Let us add at this point some comments which should clarify the concept of *"range of correlation"* as we use it here. In a formula such as (43.3), the function $f(k)$ is essentially a functional of the interaction potential V_k. As a result of the summation of the diagrams this interaction potential is replaced by an effective potential (see Chapter 9) which is, roughly speaking, close to the Debye potential. Its extension in k-space is thus of order κ. Let $F(r)$ be the Fourier transform of $f(k)$:

$$F(r) = \int d\mathbf{k} f(k)\, e^{i\mathbf{k}\cdot\mathbf{r}} \tag{43.4}$$

It follows from the properties of the Fourier integral that the range of $F(r)$ in space is of order

$$\langle r \rangle \simeq \kappa^{-1} \tag{43.5}$$

The contribution of ρ'_k to the correlation function $G(r; t)$ is given by the Fourier transform of (43.3):

$$G'(r) = t^m \int d\mathbf{k}\, e^{i\mathbf{k}\cdot(\mathbf{r}-\mathbf{g}t)} f(k) = t^m F(|\mathbf{r}-\mathbf{g}t|) \tag{43.6}$$

The average extension of this function in r-space is then obtained from

$$\langle r - gt \rangle \simeq \kappa^{-1}$$

If we are interested in times of the order of the relaxation time t_r, the average value of gt equals the mean free path L_f, which is much longer than κ^{-1} in plasmas of small Γ (see § 9). Hence:

$$\langle r \rangle \simeq \kappa^{-1} + L_f \simeq L_f \tag{43.7}$$

The term ρ'_k thus describes *long-range correlations*; these correlations of range L_f are due to the appearance of the exponential in eq. (43.3). The physical process producing these correlations is the following: two particles which suffer a collision at time t must be correlated before the collision for an average time t_r (which is roughly the time between two collisions) in order to arrive together at their collision place. In other words, the exponential in eq. (43.3) describes the propagation of correlations through free flow.

It will be shown below that there also exist contributions to ρ_k which are of the same form as (43.3) but without an exponential. Equation (43.5) shows that such contributions have a spatial range of order κ^{-1}, i.e. they are *short-range correlations*. The physical cause of the latter is the existence of interactions.

Our next purpose will be to show that the long-range correlations do not contribute to any of the usual macroscopic quantities. They give rise, so to speak, to transient effects which average out to zero in a very short time (of order t_p), although ρ_k' itself has a life-time of order t_r. On the contrary, the short-range correlations give a definite contribution to the thermodynamic functions and are physically much more interesting.

In the remainder of this book, whenever we speak of "range of correlation" we shall mean the short range described above and related to the range of the effective interactions.

Correlation functions are also interesting from a point of view other than the direct measurement of density correlations. In the statistical definition of most local macroscopic observables, or more precisely of the functions occurring in irreversible thermodynamics, such as the pressure tensor, the heat flow, the local internal energy, etc., there occurs a term describing the contribution of the intermolecular forces to these quantities [see for instance eq. (1.16)]. This term is always of the form of the average of a quantity related to $V(r)$ weighted by a two-particle correlation function. Thus, by measuring for instance transport coefficients, one could get information about the molecular correlation function. In this book we shall be exclusively interested in this aspect of the correlation function. We shall therefore disregard all contributions to $\rho_k(\alpha, \gamma)$ which do not effectively contribute to the value of thermodynamic functions for times longer than t_p.

It is now easy to show that a contribution of the type (43.3) does not contribute to thermodynamic quantities for times longer than t_p. The argument is very similar to the one used in § 15 to show the connection between the real "individual-particle pole" and the "free-flow damping" occurring in the Vlassov equation. Let us discuss as an example the second term on the r.h.s. of eq. (1.16). Using formula (2.18), it is first seen that the contribution of $\varphi(\alpha)\varphi(\gamma)$ vanishes (if properly combined with the contribution of the oppositely charged medium). We are then left with:

$$
\begin{aligned}
E_p &= \tfrac{1}{2}e^2c \int d\mathbf{x}_\alpha d\mathbf{x}_\gamma d\mathbf{v}_\alpha d\mathbf{v}_\gamma V(|\mathbf{x}_\alpha - \mathbf{x}_\gamma|)\delta(\mathbf{x}_\alpha - \mathbf{x}_\gamma) \int d\mathbf{k}\, \rho_k(\alpha, \gamma)\, \mathrm{e}^{i\mathbf{k}\cdot\mathbf{r}_{\alpha\gamma}} \\
&= \tfrac{1}{2}e^2c \int d\mathbf{r}\, d\mathbf{v}_\alpha d\mathbf{v}_\gamma V(r) \int d\mathbf{k}\, \rho_k(\alpha, \gamma; t)\, \mathrm{e}^{i\mathbf{k}\cdot\mathbf{r}} \\
&= \tfrac{1}{2}e^2c \int d\mathbf{r}\, d\mathbf{v}_\alpha d\mathbf{v}_\gamma \int d\mathbf{l}\, \mathrm{e}^{i\mathbf{l}\cdot\mathbf{r}} V_l \int d\mathbf{k}\, \mathrm{e}^{i\mathbf{k}\cdot\mathbf{r}} \rho_k(\alpha, \gamma; t) \\
&= \tfrac{1}{2}e^2c \int d\mathbf{v}_\alpha d\mathbf{v}_\gamma \int d\mathbf{k}\, V_k \rho_k(\alpha, \gamma; t)
\end{aligned} \tag{43.8}
$$

Substituting the contribution (43.3) we obtain

$$
E'_p = \tfrac{1}{2}e^2ct^m \int d\mathbf{v}_\alpha d\mathbf{v}_\gamma \int d\mathbf{k}\, V_k f(k; \alpha, \gamma)\, \mathrm{e}^{-i\mathbf{k}\cdot\mathbf{g}_{\alpha\gamma}t} \tag{43.9}
$$

The behavior in time of the k-integral depends on the shape of the function $V_k \rho_k$. As stated above, ρ_k has an extension of order κ in k-space; V_k on the other hand decreases very slowly. Therefore, after a short time [of order $t_p \approx (\kappa v)^{-1}$] which is characteristic of the correlations, the integral vanishes by destructive interference of the waves $\mathrm{e}^{-i\mathbf{k}\cdot\mathbf{g}_{\alpha\gamma}t}$. *We therefore disregard all contributions of successive diagonal fragments with external lines.*

Consider now the last type of diagram of Fig. 43.1, consisting of a creation fragment acting on a succession of diagonal fragments without external lines. According to our topological rules of § 29, the contribution of this type of diagram to the resolvent has a simple pole at $z = \mathbf{k}\cdot\mathbf{g}_{\alpha\gamma}$ and a pole of order $n+1$ at $z = 0$ (n = number of diagonal fragments). The universal terms of this diagram therefore have the following time dependence:

$$
\mathrm{e}^{-i\mathbf{k}\cdot\mathbf{g}_{\alpha\gamma}t}; \quad t^n, \quad t^{n-1}, \ldots, t^0
$$

We only retain the term in t^n, which is dominant in this approximation. It is then easy to construct diagrams of this type which are of the right order of magnitude. The diagonal fragments must of course be rings: they give a contribution of order $(e^2c)^n(e^2t)^m$. The creation fragment necessarily involves the creation vertex A (Table 6.1), which is of order e^2; it can thus only contain additional e^2c-vertices. The only such vertices which can be connected to A and to one another without introducing external lines to the right are the loops F. Diagram (d) of Fig. 43.1 is thus the only permissible type of diagram in the present approximation.

To sum up, the "universal" long-time contributions to $\rho_k(\alpha, \gamma; t)$ can be separated into two parts:

$$\rho_k(\alpha, \gamma; t) = \rho_k'(\alpha, \gamma; t) + \rho_k''(\alpha, \gamma; t) \tag{43.10}$$

$\rho_k'(\alpha, \gamma)$ contains the contributions of type (c) and $\rho_k''(\alpha, \gamma)$ those of type (d). If, however, we are interested in calculating thermodynamic functions, we can completely disregard $\rho_k'(\alpha, \gamma)$. This is moreover justified, as can be seen from the diagrams, by the fact that the value of $\rho_k''(\alpha, \gamma)$ depends only on φ and *not* on ρ_k'. It can be shown also that $\rho_k'(\alpha, \gamma) \to 0$ as the system tends to equilibrium,* so that the limiting value of $\rho_k''(\alpha, \gamma)$ equals the complete equilibrium correlation function. *From now on we simply suppress the double prime, and identify* $\rho_k(\alpha, \gamma; t)$ *and* $\rho_k''(\alpha, \gamma; t)$.

§ 44. Summation of the Creation Diagrams

We now go over to the summation of the correlation diagrams. We perform this summation in two steps. In a first step we sum all the contributions ending with a given creation fragment. We adopt the convention of drawing "*modified lines*" defined in Fig. 44.1: a diagram in which such a line occurs is a shorthand

Fig. 44.1. The modified line.

for the sum of all the diagrams obtained by replacing the modified line by a simple line, one F-vertex, two F-vertices, etc. . . . With this convention the grouping of terms in the first step of the summation is shown in Fig. 44.2.

The contribution of the first term in that figure is obtained in the same way as formula (39.1). (It is just the ring of Fig. 39.2 in which the leftmost vertex has been clipped off.) Using the abbreviations (39.4) we obtain:

* I. Prigogine, *loc. cit.*

$$(2\pi)^{-1}\int_C dz\, e^{-izt}\int d\mathbf{v}_1 \ldots d\mathbf{v}_{N-2}[i(\mathbf{k}\cdot\hat{\mathbf{g}}_{\alpha\gamma}-z)]^{-1}(-d_\alpha)[i(\mathbf{k}\cdot\hat{\mathbf{g}}_{1\gamma}-z)]^{-1}$$

$$\cdot\, d_\gamma[i(\mathbf{k}\cdot\hat{\mathbf{g}}_{12}-z)]^{-1}d_{21}[-iz]^{-1}\rho_0(0) \tag{44.1}$$

$$= (2\pi)^{-1}\int dv\int_C dz\, e^{-izt}[i(\mathbf{k}\cdot\hat{\mathbf{g}}_{\alpha\gamma}-z)]^{-1}C_+(z)[-iz]^{-1}\rho_0(0)$$

where $C_+(z)$ is defined by this equation as an operator acting on everything to the right. $C_+(z)\rho_0(0)$ is a Cauchy integral, and thus a regular function for $z \in S_+$. We now introduce the following

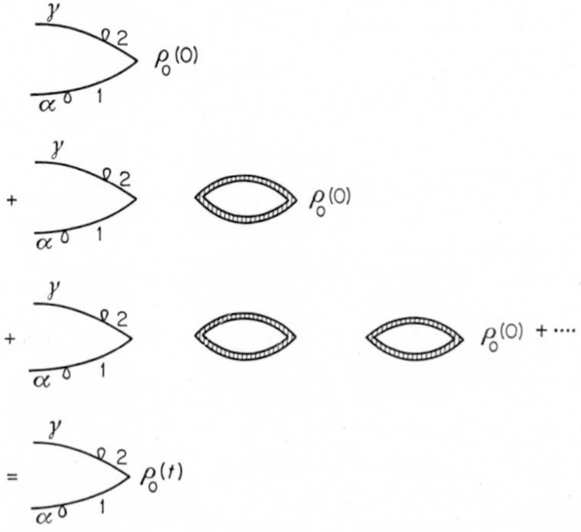

Fig. 44.2. The first step in the summation of plasma correlation diagrams.

notation for the contribution to $\rho_0(t)$ from the sum of all rings acting on $\rho_0(0)$ (i.e. the second diagram of Fig. 44.2 without the creation fragment)

$$\rho_0(t)\Big|_{\substack{\text{sum of}\\\text{single rings}}} = (2\pi)^{-1}\int dz\, e^{-izt}\langle 0|\{\Re(z)\}_{\text{rings}}|0\rangle\rho_0(0)$$

$$\equiv (2\pi)^{-1}\int dz\, e^{-izt}[-iz]^{-1}\Re(z)[-iz]^{-1}\rho_0(0) \tag{44.2}$$

In terms of the operator $\Re(z)$, the sum shown in Fig. 44.2 is obviously

$$\rho_k^{(1)}(\alpha, \gamma; t) = (2\pi)^{-1}\int dv \int_C dz\, e^{-izt}[i(\mathbf{k}\cdot\mathbf{g}_{\alpha\gamma}-z)]^{-1}C_+(z)\{[-iz]^{-1} + [-iz]^{-1}\mathfrak{R}(z)[-iz]^{-1} + [-iz]^{-1}\mathfrak{R}(z)[-iz]^{-1}\mathfrak{R}(z)[-iz]^{-1}+\ldots\}\rho_0(0) \tag{44.3}$$

The leading universal term is obtained from (44.3) by taking the residue at $z = 0$ of the successive terms and retaining in each of them the highest power of t. Some care must be taken in interpreting the function $[i(\mathbf{k}\cdot\mathbf{g}_{\alpha\gamma}-z)]^{-1}$ for $z = 0$: one cannot set brutally $z = 0$, because subsequent operations (in particular integrations) on $\rho_k(\alpha, \gamma)$ would become ambiguous. The interpretation of such a function is clearly:

$$\lim_{\substack{z\to 0\\ z\in S_+}} [i(\mathbf{k}\cdot\mathbf{g}_{\alpha\gamma}-z)]^{-1} = \pi\delta_-(\mathbf{k}\cdot\mathbf{g}_{\alpha\gamma})$$

because, by definition of the contour C in (44.3), z lies in the upper half-plane.*

Formula (44.3) then becomes

$$\rho_k^{(1)}(\alpha, \gamma; t) = \delta_-^{\alpha\gamma}\int dv\, C_+(0)\{1+t\mathfrak{R}(0)+\tfrac{1}{2}t^2\mathfrak{R}(0)\mathfrak{R}(0)+\ldots\}\rho_0(0)$$

But the bracketed operator acting on $\rho_0(0)$ is nothing other than $\rho_0(t)$ in the ring approximation, i.e. in the approximation of Chapter 9:

$$\rho_k^{(1)}(\alpha, \gamma; t) = \delta_-^{\alpha\gamma}\int dv\, C_+(0)\rho_0(t) \tag{44.4}$$

This expression is, more explicitly:

$$\rho_k^{(1)}(\alpha, \gamma; t) = \delta_-^{\alpha\gamma}\int_1\int_2 (-d_\alpha)\delta_-^{1\gamma}\, d\gamma\, \delta_-^{12} d_{21}\varphi(\alpha)\varphi(\gamma)\varphi(1)\varphi(2) \tag{44.5}$$

* In this way $\rho_k(\alpha, \gamma)$ is a singular but integrable function of $\mathbf{k}$ (more precisely, a "distribution" in the sense of Schwartz). This actually meets all mathematical requirements on $\rho_k(\alpha, \gamma)$. In effect this object has no physical meaning by itself; it is only a mathematical intermediate. Quantities which have physical meaning are constructed by multiplying $\rho_k(\alpha, \gamma)$ by some "good" function and integrating over $\mathbf{k}$ [see e.g. (43.9)]. For such functions to have a physical meaning, the distribution property of $\rho_k(\alpha, \gamma)$ is sufficient.

The second step in the summation involves a summation over all possible terminal creation fragments. Comparing formula (44.5) with eq. (39.3), one sees that the form of the contribution of each term in the sum is very closely related to the corresponding ring. To write down the contribution of an individual diagram, one applies the rules, A–D, E, F of § 39, and then:

H) suppress the symbol d_α in front of the contribution; suppress the integrations over $\mathbf{1}$ and $\mathbf{v}_1$; make the substitutions $\mathbf{1} \to \mathbf{k}$ and $1 \to \gamma$.

In performing the summation of these expressions, one groups the terms according to the same pattern as in § 40: this grouping is shown in Fig. 44.3.

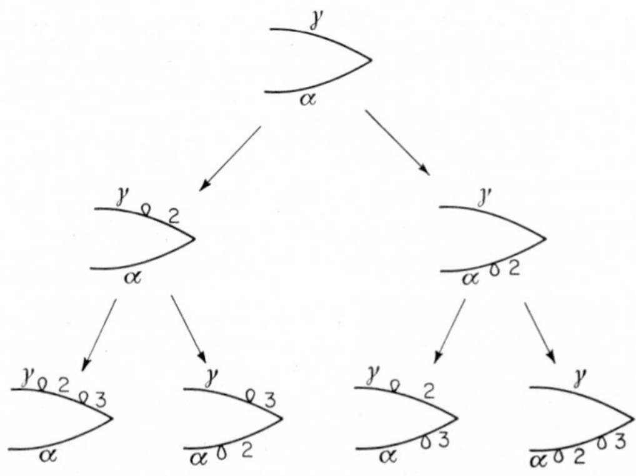

Fig. 44.3. The systematic generation of plasma creation fragments.

We need not write the corresponding formulae again. We just go back to eq. (40.1) and perform operation H on each term of the right-hand side. We then immediately get the result

$$\rho_k(\alpha, \gamma; t) = \delta_-^{\alpha\gamma}\{d_{\gamma\alpha}\varphi(\gamma)\varphi(\alpha) + d_\gamma\varphi(\gamma)F(\alpha) + d_\alpha^*\varphi(\alpha)F^*(\gamma)\} \quad (44.6)$$

where $F(\alpha)$ is the function defined by eq. (40.3) or (40.5). This formula completes the formal part of our summation problem. The summation of the creation fragments has been reduced to the solution of the integral equation which also governs the ring problem, i.e. eq. (40.10). There is, however, a difference. We have seen in § 41 that the requirement of a real $\varphi(\alpha)$ implies that a knowledge of the imaginary part F_2 of F is sufficient for determining the kinetic equation (40.9). In the present case however, $\rho_k(\alpha, \gamma)$ is not necessarily a real function. The real quantity in this problem is $G(\alpha, \gamma)$, which is the Fourier transform of $\rho_k(\alpha, \gamma)$. Therefore we also need the real part of F for a complete determination of $\rho_k(\alpha, \gamma)$. The next section is devoted to this problem.

§ 45. Solution of the Integral Equation for Plasmas. (Real part of *F*)

The first steps in the solution proceed as in § 41. The equation (40.10) for F is reduced to eq. (41.8) for $\bar{F}$, i.e. to the system of equations (41.11) and (41.12). As these equations are decoupled, eq. (41.11) is solved in the first place, yielding the solution $\bar{F}_2 = 0$. This is substituted into (41.12) with the result:

$$\varepsilon_1 \bar{F}_1 - \varepsilon_2 \mathscr{P} \bar{F}_1 = \bar{q}_1 \tag{45.1}$$

This equation is of a type very similar to eq. (41.11). It is again a singular integral equation of Cauchy type and of zero index. The latter property ensures the unicity of the solution. This equation is however inhomogeneous, and its solution will consequently be non-zero.

The close analogy of this equation with eq. (25.3) should first be pointed out. We could solve eq. (45.1) by using an argument very similar to the one used in § 25. This was done in our original paper.* We shall show here that a very simple and elegant method of solution is provided by the use of the van Kampen–Case eigenfunctions of the Vlassov equation.

Keeping in mind the definitions (24.6) and (24.7), we immediately see that eq. (45.1) can be written as follows:

$$\pi^{-1}\varepsilon_2(\nu) \int_{-\infty}^{\infty} d\nu' \tilde{\chi}_\nu(\nu') \bar{F}(\nu') = \bar{q}(\nu) \tag{45.2}$$

Let us write $\bar{F}(\nu')$ as a superposition of the eigenfunctions $\tilde{\chi}_v(\nu)$ defined by (23.16):

* R. Balescu and H. S. Taylor, *Phys. of Fluids*, **4**, 85 (1961).

$$\bar{F}(\nu') = \int_{-\infty}^{\infty} d\nu'' \,\bar{\chi}_{\nu''}(\nu') f(\nu'') \tag{45.3}$$

Substituting this into (45.2), the left-hand side is very easily transformed, using the orthogonality property (24.3) together with (24.8):

$$\pi^{-1}\varepsilon_2(\nu)\int_{-\infty}^{\infty} d\nu' d\nu'' \,\tilde{\chi}_{\nu}(\nu')\,\bar{\chi}_{\nu''}(\nu') f(\nu'')$$

$$= \pi^{-1}\varepsilon_2(\nu)\, C_\nu \int_{-\infty}^{\infty} d\nu'' \,\delta(\nu-\nu'') f(\nu'') = |\varepsilon(\nu)|^2 f(\nu)$$

Therefore eq. (45.2) yields

$$f(\nu) = \bar{q}(\nu)/|\varepsilon(\nu)|^2$$

Hence the solution is

$$\bar{F}(\nu) = \int_{-\infty}^{\infty} d\nu' \,\bar{\chi}_{\nu'}(\nu)\bar{q}(\nu')/|\varepsilon(\nu')|^2 \tag{45.4}$$

or, written more explicitly:

$$\bar{F}(\nu) = \frac{\bar{q}(\nu)}{\varepsilon(\nu)} + i\varepsilon_2(\nu)\int d\nu_1 \delta_-(\nu-\nu_1)\frac{\bar{q}(\nu_1)}{|\varepsilon(\nu_1)|^2} \tag{45.5}$$

This is the solution of eq. (45.1). Applying eq. (41.9) and remembering that $\bar{F}_2(\nu) = 0$, one finds, after some manipulations:

$$F(\mathbf{v}_\alpha) = \frac{q(\mathbf{v}_\alpha)}{\varepsilon(\nu_\alpha)} + \frac{i\pi}{k} d(\mathbf{v}_\alpha)\int d\nu_1 \delta_-(\nu_\alpha-\nu_1)\frac{\bar{q}(\nu_1)}{|\varepsilon(\nu_1)|^2} \tag{45.6}$$

Let us stress that eq. (45.6) is only valid if the distribution of velocities $\varphi(\mathbf{v})$ is *stable*. Otherwise, as shown in § 25, the functions $\bar{\chi}_{\nu'}(\nu)$ do not form a complete system; hence $\bar{F}(\nu)$ cannot be written in the form (45.3) and the method fails (see Appendix 9).

The solution (45.6) has been obtained by Balescu and Taylor.* When substituted into eq. (44.6) it yields the correlation function $\rho_k(\alpha,\gamma)$. As expected from the discussion of the previous paragraph, this equation expresses $\rho_k(\alpha, \gamma)$ as a functional of $\varphi(\alpha)$ and $\varphi(\gamma)$. This expression is exact within the frame of the ring approximation

* *loc. cit.*

It is one of the very few non-trivial instances in which such a functional relationship is known explicitly. The importance of this fact is clearly the following. In a calculation of thermodynamic quantities, such as transport coefficients, which involve a contribution from the one-particle distribution and one from the correlation function, it is not necessary to solve two separate equations. It is sufficient to solve the kinetic equation for the one-particle function. Knowing the solution, $\rho_k(\alpha, \gamma)$ or $G(r_{\alpha\gamma})$ is determined by a quadrature through eq. (45.6).

CHAPTER 11

The General Description of a Plasma in the Ring Approximation

§ 46. General Properties of the Kinetic Equation

We are now going to discuss the general properties of plasmas within the approximation established in Chapters 9 and 10, which can be denoted by the general term of the *"ring approximation"*. In order to simplify the discussion we shall mainly concentrate on homogeneous systems, and begin by considering eq. (41.19), which we rewrite here:

$$\partial_t \varphi(\alpha) = 8\pi^4 e^4 c m^{-2} \int d\mathbf{l} \int d\mathbf{v}_1 \, \mathbf{l} \cdot \partial_\alpha \frac{V_l^2}{|\varepsilon(\alpha)|^2} \delta(\mathbf{l} \cdot \mathbf{g}_{\alpha 1}) \mathbf{l} \cdot \partial_{\alpha 1} \varphi(\alpha) \varphi(1) \tag{46.1}$$

$$\varepsilon(\alpha) = 1 + 2\pi^3 i \omega_p^2 V_l \int d\mathbf{v}_2 \delta_-(\mathbf{l} \cdot \mathbf{g}_{\alpha 2}) \mathbf{l} \cdot \partial_2 \varphi(2) \tag{46.2}$$

We shall first show that it has all the desirable properties of a kinetic equation.*

A) (46.1) *is an irreversible equation.* Following the argument of § 34, we now show that:

(*a*) the Maxwell equilibrium distribution is a stationary solution of eq. (46.1),

(*b*) the entropy can only increase in time and reaches a maximum value at equilibrium (H-theorem).

The first property is ensured by the occurrence of the δ-function, exactly as in the Landau equation [see (34.2)].

The proof of the H-theorem is just as simple as for the Landau equation [see eq. (34.4)]. Let the entropy density be:

$$s(t) = -\int d\mathbf{v}_\alpha c \varphi(\alpha) \ln c\varphi(\alpha) \tag{46.3}$$

* A similar discussion was given by A. Lenard, *Ann. Phys.* **3**, 390 (1960).

Then

$$\partial_t s = -c\int d\mathbf{v}_\alpha[\ln c\varphi(\alpha)+1]\partial_t\varphi(\alpha)$$

$$= \tfrac{1}{2}A\int d\mathbf{v}_\alpha d\mathbf{v}_1\int d\mathbf{l}[\ln c\varphi(\alpha)+\ln c\varphi(\mathit{1})+2)]\mathbf{l}\cdot\partial_{\alpha 1}$$

$$\frac{V_l^2}{|\varepsilon(\alpha)|^2}\,\delta(\mathbf{l}\cdot\mathbf{g}_{\alpha 1})\mathbf{l}\cdot\partial_{\alpha 1}\varphi(\alpha)\varphi(\mathit{1})$$

$$= \tfrac{1}{2}A\int d\mathbf{v}_\alpha d\mathbf{v}_1\int d\mathbf{l}\,\frac{V_l^2}{|\varepsilon(\alpha)|^2}\,\delta(\mathbf{l}\cdot\mathbf{g}_{\alpha 1})\,\frac{[\mathbf{l}\cdot\partial_{\alpha 1}\varphi(\alpha)\varphi(\mathit{1})]^2}{\varphi(\alpha)\varphi(\mathit{1})}\geqq 0$$

The corresponding inhomogeneous equation, (42.2), is a superposition of a reversible flow term and of an irreversible collision term.

B) The hydrodynamic condition induces for eq. (42.2) the same general properties as for the inhomogeneous Landau equation (32.6), i.e. (*a*) molecular chaos, and (*b*) local character.

The latter is especially valuable in providing hydrodynamic conservation laws (see § 36). Let us show that the three classical conservations (density, velocity, energy) are satisfied. The proof is trivial for mass conservation. For the change of velocity due to collisions we have the following expression:

$$(\partial_t\mathbf{u})_{\text{coll}} = B\int d\mathbf{v}_\alpha d\mathbf{v}_1\int d\mathbf{l}\;\mathbf{v}_\alpha\mathbf{l}\cdot\partial_\alpha\frac{V_l^2}{|\varepsilon(\alpha)|^2}\,\delta(\mathbf{l}\cdot\mathbf{g}_{\alpha 1})\mathbf{l}\cdot\partial_{\alpha 1}f(\alpha)f(\mathit{1})$$

$$= B\int d\mathbf{v}_\alpha d\mathbf{v}_1\int d\mathbf{l}\,\mathbf{l}\,\frac{V_l^2}{|\varepsilon(l;\mathbf{l}\cdot\mathbf{v}_\alpha)|^2}\,\delta(\mathbf{l}\cdot\mathbf{v}_\alpha-\mathbf{l}\cdot\mathbf{v}_1)\mathbf{l}\cdot\partial_{\alpha 1}f(\alpha)f(\mathit{1})$$

This expression changes sign when the indices α and $\mathit{1}$ are interchanged in the integrand; therefore it can only be equal to zero.

The conservation of energy is proved as follows:

$$\partial(\tfrac{1}{2}v_\alpha^2) = \tfrac{1}{2}B\int d\mathbf{v}_\alpha d\mathbf{v}_1\int d\mathbf{l}\;v_\alpha^2\mathbf{l}\cdot\partial_\alpha V_l^2|\varepsilon(\alpha)|^{-2}\delta(\mathbf{l}\cdot\mathbf{g}_{\alpha 1})\mathbf{l}\cdot\partial_{\alpha 1}f(\alpha)f(\mathit{1})$$

$$= -B\int d\mathbf{v}_\alpha d\mathbf{v}_1\int d\mathbf{l}\,\mathbf{l}\cdot\mathbf{v}_\alpha V_l^2|\varepsilon(\alpha)|^{-2}\delta(\mathbf{l}\cdot\mathbf{g}_{\alpha 1})\mathbf{l}\cdot\partial_{\alpha 1}f(\alpha)f(\mathit{1})$$

$$= B\int d\mathbf{v}_\alpha d\mathbf{v}_1\int d\mathbf{l}\mathbf{l}\cdot\mathbf{v}_1 V_l^2|\varepsilon(\alpha)|^{-2}\delta(\mathbf{l}\cdot\mathbf{g}_{\alpha 1})\mathbf{l}\cdot\partial_{\alpha 1}f(\alpha)f(\mathit{1})$$

$$= -\tfrac{1}{2}B\int d\mathbf{v}_\alpha d\mathbf{v}_1\int d\mathbf{l}\,\mathbf{l}\cdot\mathbf{g}_{\alpha 1}V_l^2|\varepsilon(\alpha)|^{-2}\delta(\mathbf{l}\cdot\mathbf{g}_{\alpha 1})\mathbf{l}\cdot\partial_{\alpha 1}f(\alpha)f(\mathit{1})$$

From the existence of these conservation laws it follows that the equations of hydrodynamics are of the same form as in the Landau approximation, i.e. (36.6), (36.7), (36.10). However, the appropriate values of the pressure tensor and of the heat flow will of course be different, because they are evaluated with a different collision operator.

The previous discussion shows that the ring equation has all the necessary properties for being called a kinetic equation. We now consider the features which make it different from all usual known types of kinetic equations.

We have already noted that *a kinetic equation is necessarily a non-linear equation*. As it describes interactions between a certain number of particles, it must contain in its collision term a product of an equal number of distribution functions. We have seen, for instance, that the Landau equation is quadratically non-linear, because it describes two-body collisions.

Equation (46.1) has a striking formal similarity with the Landau equation (30.11). In particular, the same two factors $\varphi(\alpha)\varphi(1)$ occur in the collision term, which has the same general form as in (30.11), except for the occurrence of the denominator $|\varepsilon(\alpha)|^2$. This difference is however considerable because $\varepsilon(\alpha)$ *is itself a functional of* $\varphi(\alpha)$. Therefore, eq. (46.1) is "much more strongly non-linear" than any known kinetic equation. Physically, the reason is clear. Equation (46.1) describes a combination of many-body collisions. This is evident from the diagrams: a ring with n vertices represents an n-particle collision. It is also obvious from the mathematical rules (especially rule E) for the construction of the rings. Equation (46.1) thus contains the sum of two-body, three-body, four-body . . . collisions. The effect of the summation is the occurrence of φ in the denominator. The ring equation therefore describes *"collective collisions"*.

It appears worth stressing that the ring equation is one of the very few many-body processes which can be summed exactly. This may appear strange if one thinks, for instance, that even the three-body problem cannot be treated exactly. The reason for this possibility is the fact that an n-body ring is not the most general diagram involving n particles. There is an infinite number of diagrams involving a given number of particles.

In the classical three-body problem one is interested in summing *all* diagrams involving three particles. In the present problem, one isolates from each individual n-body problem a subclass of diagrams which is dominant within the approximation and then one sums over n. It so happens that the chosen diagrams are sufficiently simple to be exactly summable.

§ 47. Connection between Correlation Function and Kinetic Equation

We shall now show that there is a close connection between the kinetic equation and the correlation function, and that this justifies our joint discussion of these two concepts.

The main observation in this connection is made by comparing formulae (40.4) and (44.6). One then immediately sees the meaning of the fundamental function $F(\mathbf{l}, \mathbf{v}_\alpha)$:

$$F(\mathbf{l}, \mathbf{v}_\alpha) = \int d\mathbf{v}_1 \rho_l(\mathbf{v}_\alpha, \mathbf{v}_1) \tag{47.1}$$

The function F is the integral of the (Fourier) binary correlation function over one velocity. Chapters 9 and 10 have amply shown the central role played in the theory of plasmas by this function, or by the even simpler function:

$$\bar{F}(\mathbf{l}, \nu_\alpha) = \int d\mathbf{v}_1 d\mathbf{v}_2 \delta(\nu_\alpha - \mathbf{l} \cdot \mathbf{v}_1/l) \rho_l(\mathbf{v}_1, \mathbf{v}_2) \tag{47.2}$$

The beauty of the ring theory lies in the fact that the kinetic equation as well as the complete correlation function is determined by purely algebraic operations from this single function.

One can push the analysis even further. $F(\mathbf{l}, \mathbf{v})$ is a complex function. We have already pointed out in § 41 that only the imaginary part of F contributes to the kinetic equation. We may now calculate another central quantity of statistical mechanics, the *density correlation function,* defined as the integral of the complete two-body correlation function over both velocities (see Table 2.2). Its Fourier transform is

$$g_l = \int d\mathbf{v}_\alpha d\mathbf{v}_1 \rho_l(\mathbf{v}_\alpha, \mathbf{v}_1) = \int d\nu_\alpha \bar{F}(\nu_\alpha) = \int d\nu_\alpha \bar{F}_1(\nu_\alpha) \tag{47.3}$$

The last result follows from the fact (§ 41) that $\bar{F}_2(\nu_\alpha) \equiv 0$. The explicit ν_α-integration is trivial if one starts with (45.4):

$$\int d\nu \, \bar{F}_1(\nu) = \int d\nu \, d\nu' \, \bar{\chi}_{\nu'}(\nu) \frac{\bar{q}(\nu')}{|\varepsilon(\nu')|^2}$$

from which it follows immediately, using the normalization of $\bar{\chi}_{\nu'}(\nu)$ [eq. (23.9)], that

$$g_l = \int d\nu \frac{\bar{q}(\nu)}{|\varepsilon(\nu)|^2} \qquad (47.4)$$

This is a nice compact formula expressing the radial correlation function in terms of the dielectric constant.

The discussion above points out the following interesting property of the ring approximation.

The imaginary part of the correlation function $\rho_l(\mathbf{v}_1, \mathbf{v}_2)$ *determines completely the kinetic equation, and hence the dissipative properties of the plasma,* such as the relaxation time.

The real part of the correlation function $\rho_l(\mathbf{v}_1, \mathbf{v}_2)$ *determines completely the density correlation function, and hence the "permanent" properties of the plasma,* such as the average potential energy.

It is interesting to rewrite the kinetic equation in the form (40.9) by using our present interpretation (47.1) of the function F, and to perform a Fourier transformation. Using the relations

$$\nabla V(r) = \int d\mathbf{l} \; i\mathbf{l} V_l e^{i\mathbf{l}\cdot\mathbf{r}}$$

$$G(r, \mathbf{v}_1, \mathbf{v}_2) = c \int d\mathbf{k} \, e^{i\mathbf{k}\cdot\mathbf{r}} \rho_k(\mathbf{v}_1, \mathbf{v}_2)$$

we immediately obtain the kinetic equation in the form:

$$\partial_t \varphi(\alpha) = e^2 \int d\mathbf{v}_1 \int d\mathbf{r} [\nabla V(r)] \cdot \partial_\alpha G(r, \mathbf{v}_\alpha, \mathbf{v}_1) \qquad (47.5)$$

This equation has the same form as one of the "Y–B–G–B–K" hierarchy, eq. (4.1). However, whereas the latter is an identity having no more physical content than the Liouville equation, eq. (47.5) has a precise meaning because G is known as a functional of φ, and the equation is closed.

At this point it is very easy to make the connection with Bogoliubov's theory.* The latter theory proceeds exactly in the inverse direction from the one described here. It takes eq. (47.5) as a starting point with an as yet unspecified G. In order that (47.5) be a kinetic equation, Bogoliubov argues that G must be a time-independent functional of φ:

$$G(r, \mathbf{v}_1, \mathbf{v}_2; t) = G(r, \mathbf{v}_1, \mathbf{v}_2; \varphi) \tag{47.6}$$

In other words, the time dependence of G is completely determined once $\varphi(t)$ is known. Bogoliubov then looks for a solution of the Y–B–G–B–K equation for f_2 which is of the form (47.6). Such a solution is certainly not general, since — as we know — G is determined by the set of all reduced distributions at time zero. "Nevertheless", he says, "one may expect that for the kind of general initial distributions which are physically admissible the correlation functions will rapidly (i.e. $t \sim t_p$) approach the expressions (47.6) . . ." He then develops a method for the approximate calculation of this solution, by using an expansion procedure together with an initial condition expressing the decay of correlations at large distance.

In view of our discussion of § 43 we have been able to show that Bogoliubov's assumption that $\rho_l(\mathbf{v}_\alpha, \mathbf{v}_\beta)$ approaches a functional of φ after a time of order t_p is *not* correct. Indeed, we have seen that

$$\rho_k(\mathbf{v}_1, \mathbf{v}_2; t) = \rho_k'(\mathbf{v}_1, \mathbf{v}_2; t) + \rho_k''(\mathbf{v}_1, \mathbf{v}_2; \varphi)$$

Only ρ_k'' is a functional of φ. Moreover, we have seen that ρ_k' tends to zero not in a time t_p but after the much longer time t_r, when the system has reached equilibrium. In Chapter 10 we have actually studied only the part ρ_k'' of the correlation function, and formula (47.1) should really read

$$F(\mathbf{l}, \mathbf{v}) = \int d\mathbf{v}_1 \rho_l''(\mathbf{v}, \mathbf{v}_1) \tag{47.7}$$

* N. N. Bogoliubov, *Problems of a Dynamical Theory in Statistical Physics,* translated by E. K. Gora, in *Studies in Statistical Mechanics,* vol. 1, ed. by J. de Boer and G. E. Uhlenbeck, Interscience, New York, 1962.

However, as was shown in § 43, because of the oscillating character of ρ'_l, the following property holds:

$$\int d\mathbf{l} \int d\mathbf{v}_1 \, i\mathbf{l} V_l \rho_l(\mathbf{v}, \mathbf{v}_1) \underset{t \sim t_p}{\rightarrow} \int d\mathbf{l} \, d\mathbf{v}_1 \, i\mathbf{l} V_l \rho''_l(\mathbf{v}, \mathbf{v}_1)$$

$$= \int d\mathbf{l} \, i\mathbf{l} V_l F(\mathbf{1}, \mathbf{v})$$

In other words, the oscillations of ρ'_l are damped out by destructive interference after a few plasma oscillations. This is why the Bogoliubov theory (developed in the present case by Lenard * and by Guernsey **) although starting with a too strong assumption leads nevertheless to the correct asymptotic kinetic equation. The Bogoliubov theory will be discussed further in § 50.

§ 48. Connection between the Ring Equation and the Landau Equation

We have already pointed out the strong formal analogy between the ring equation (46.1) and the Landau equation (30.11). One can say formally that the ring equation is a Landau equation for a system whose molecules interact through the effective potential $U_l(\nu_\alpha)$:

$$U_l(\nu_\alpha) = \frac{1}{2\pi^2} \frac{1}{l^2 \varepsilon(l; \nu_\alpha)} \tag{48.1}$$

This effective potential is identical with the one encountered in § 18 in the Vlassov approximation. It is determined by the dielectric constant at the real frequency $z = \mathbf{k} \cdot \mathbf{v}_\alpha$. This potential is thus velocity-dependent.† It has been shown in § 18 that this potential contains the collective effects and in particular a screening effect. It is precisely this collective screening which removes the original divergence at small values of l (large distances). In the

* A. Lenard, *Ann. Phys.*, **3**, 390 (1960).

** R. L. Guernsey, *The Kinetic Theory of Fully Ionized Gases,* Off. Nav. Res. Contract No. Nonr. 1224 (15), July, 1960.

† It can be verified independently that if one calculates the cycle diagram with a velocity-dependent potential one gets precisely an equation identical with (46.1). In particular, the factor U_l stands to the right of the first derivative ∂_α.

Landau treatment the effective potential was taken as the Debye potential. This is now seen to be inaccurate: one has to take account of a *dynamical screening*, which is a fairly complicated velocity-dependent phenomenon. Whereas for small velocities the cloud around a particular electron has very nearly the spherical Debye shape, it is progressively deformed and for very large velocities it degenerates into a wake of plasma oscillations behind the particle. All these effects are contained in the dielectric constant $\varepsilon(\alpha)$.

There is, however, a difference between the effective potential appearing in the ring equation and the one discussed in § 18. In the Vlassov approximation, which is a short-time approximation, the effective potential is independent of time, because $\varphi(\alpha)$ is a constant of the motion. In the long-time study of the ring approximation the distribution function $\varphi(\mathbf{v}_\alpha; t)$ changes in time. $U_l(v_\alpha)$ is therefore a typical *self-consistent potential* which, from this point of view, plays a role similar to the average potential $\Phi(\mathbf{r}; t)$ appearing in the Vlassov equation (12.1): the potential U_l determines the evolution of $\varphi(t)$, but as $\varphi(t)$ changes in time, U_l is correspondingly modified. However, the relation between field and distribution function is subtler for $U_l(\mathbf{r})$ than for $\Phi(\mathbf{r})$.

Notwithstanding the previous discussion, the Landau equation with a Debye potential is still a very good approximation in many realistic situations. We will now consider this point, following closely an argument due to Lenard.* Let us write the ring equation (46.1) in the form [see (35.1)]

$$\partial_t \varphi(\mathbf{v}; t) = A \int d\mathbf{v}' \, \partial_r R_{rs}(\partial_s - \partial_s') \varphi\varphi' \tag{48.2}$$

where the tensor R_{rs} is given by

$$R_{rs}(\mathbf{v}, \mathbf{v}') = \frac{1}{4\pi^4} \int d\mathbf{l} \, \frac{l_r l_s}{l^4} \, \frac{\delta(\mathbf{l} \cdot \mathbf{g})}{|1 + l^{-2} \Psi(\mathbf{l} \cdot \mathbf{v}/l)|^2} \tag{48.3}$$

where

$$\Psi = (i\omega_p^2/\pi) \int dv_1 \, \delta_-(\mathbf{l} \cdot \mathbf{v}_\alpha/l - v_1) \bar{\varphi}'(v_1) \equiv \Psi_1 + i\Psi_2 \tag{48.4}$$

* A. Lenard, *loc. cit.*

In this way the dependence of the dielectric constant on the magnitude l has been separated out explicitly; Ψ depends only on the orientation of the vector $\mathbf{l}$. Taking a reference system whose z-axis is directed along $\mathbf{g}$, the integral (48.3) becomes

$$R_{rs}(\mathbf{v}, \mathbf{v}') = \frac{1}{4\pi^4} \int_{-1}^{1} d \cos \theta \int_0^{2\pi} d\varphi \begin{Bmatrix} \sin \theta \cos \phi \\ \sin \theta \sin \phi \\ \cos \theta \end{Bmatrix} \begin{Bmatrix} \sin \theta \cos \phi \\ \sin \theta \sin \phi \\ \cos \theta \end{Bmatrix}$$

$$\cdot \int_0^{l_M} dl \frac{l^4 \delta(lg \cos \theta)}{|l^2 + \Psi(\cos \theta, \phi)|^2} \tag{48.5}$$

$$= \frac{1}{4\pi^4 g} \int_0^{2\pi} d\phi \begin{Bmatrix} \cos \phi \\ \sin \phi \\ 0 \end{Bmatrix} \begin{Bmatrix} \cos \phi \\ \sin \phi \\ 0 \end{Bmatrix} J$$

where

$$J = \int_0^{l_M} dl \frac{l^3}{[l^2 + \Psi_1(0, \phi)]^2 + \Psi_2^2(0, \phi)} \tag{48.6}$$

This integral is fairly elementary (see also § 49):

$$J = \tfrac{1}{2} \ln \frac{(\Psi_1 + l_M^2)^2 + \Psi_2^2}{\Psi_1^2 + \Psi_2^2} - \frac{\Psi_1}{\Psi_2} \operatorname{arc\,tg} \frac{l_M^2 \Psi_2}{l_M^2 \Psi_1 + \Psi_1^2 + \Psi_2^2} - n\pi \frac{\Psi_1}{\Psi_2} \tag{48.7}$$

We need not discuss here the value of the phase factor $n\pi$ (see § 49). At present we remark that the complicated expression (48.7) can be written in the following compact form

$$J = \frac{\operatorname{Im} [\Psi \ln (1 + l_M^2/\Psi)]}{\Psi_2} \tag{48.8}$$

Lenard gives the following dimensional argument. The quantity Ψ can be written as follows

$$\Psi = \beta e^2 c \psi$$

where the function $\psi(0, \phi, v) \equiv \psi(\phi, v_\perp)$ is dimensionless. One may assume that the value of ψ is close to 1 for most velocities in the thermal range. For large velocities it tends toward zero, but the latter presumably do not contribute much to the equation.

On the other hand, the logarithm in (48.8) can be written as

$$\ln\left(1+\frac{l_M^2}{\Psi}\right) = \ln\left(1+\frac{4}{\beta^2 e^4}\,\frac{1}{\beta e^2 c\,\psi}\right) = \ln\left(1+\frac{4}{\Gamma^3 \psi}\right)$$

where we have used the value (35.10) for the upper cut-off l_M. We have assumed from the beginning that $\Gamma \ll 1$; if moreover $\psi \approx 1$, we can approximate the logarithm by

$$\ln\left(1+\frac{4}{\Gamma^3 \psi}\right) \approx \ln\frac{4}{\Gamma^3 \psi} \approx \ln\frac{4}{\Gamma^3} - \ln \psi \tag{48.9}$$

Whenever Lenard's argument applies, the second term can be neglected. Then

$$J \simeq \ln (4/\Gamma^3)$$

The result is substituted into (48.5), and the integral is evaluated and transformed into an invariant form:

$$R_{rs} = \frac{1}{4\pi^3}\left(\ln\frac{l_M}{\kappa}\right)\frac{g^2 \delta_{rs} - g_r g_s}{g^3} = C_{rs} \tag{48.10}$$

where C_{rs} is the corresponding tensor in the Landau approximation, evaluated with a Coulomb potential cut-off at the inverse Debye length κ [see eqs. (35.7) and (35.13)].

This argument shows that the Landau equation is indeed a good approximation to the ring equation in many interesting situations. However, one must be careful in two ways.

(*a*) The previous argument is presumably rather sensitive to the shape of the distribution function, and particularly to the shape of the tail of the distribution. For a strongly non-maxwellian distribution it might well fail, although one can give no general *a priori* argument.

(*b*) One must stress the logarithmic dependence on the parameter Γ. One sees from (48.9) that the error committed above is very roughly of order $|\ln \psi|/|\ln \Gamma|$. One should stress that even under thermonuclear conditions, when Γ is very small, $|\ln \Gamma|$ may not always be negligible (example: $c = 10^{17}$, $T = 10^6\,°\mathrm{K}$, $\Gamma = 7.7\times10^{-4}$, $|\ln \Gamma| = 6.8$). In colder laboratory plasmas this contribution could easily become non-negligible!

§ 49. Brownian Motion in the Ring Approximation

The ring equation, because of its peculiar non-linearity, is of a very difficult mathematical type and we shall presumably have to wait for a long time before its most interesting features, arising precisely from the non-linearity, are solved. (Not even the simplest kinetic equation, i.e. Landau's equation, has been solved completely yet!)

However, even a radically oversimplified model, such as the brownian motion, can show very interesting features. This model has been studied in particular by Rostoker and Rosenbluth * and we refer the reader to their paper for details left out here. The model has been defined in § 37: all the particles except one are assumed to be in equilibrium. Introducing the reduced variables $\mathbf{w}$, $\mathbf{u}$, τ defined in (37.2), as well as

$$\boldsymbol{\lambda} = \mathbf{l}/\kappa \tag{49.1}$$

the brownian motion ring equation can be written as

$$\partial_\tau \varphi = \pi^{-\frac{5}{2}} B^{-1} \int d\mathbf{u}\; \partial_r Q_{rs} (\partial_s - \partial'_s) e^{-u^2} \varphi \tag{49.2}$$

where

$$Q_{rs} = \frac{1}{4\pi^4} \int d\boldsymbol{\lambda}\, \frac{\lambda_r \lambda_s \delta(\boldsymbol{\lambda}\cdot\mathbf{g})}{\lambda^4 |\varepsilon^0(\lambda;\, \boldsymbol{\lambda}\cdot\mathbf{w}/\lambda)|^2} \tag{49.3}$$

ε^0 is the equilibrium dielectric constant, which has been evaluated in detail in § 21.

$$\varepsilon^0(\lambda;\, \boldsymbol{\lambda}\cdot\mathbf{w}/\lambda) = 1 + \lambda^{-2} m(\boldsymbol{\lambda}\cdot\mathbf{w}/\lambda) \tag{49.4}$$

* N. Rostoker and M. N. Rosenbluth, *Phys. of Fluids,* **3**, 1 (1960).

$$m \equiv m_1 + im_2 = 1 - 2\frac{\boldsymbol{\lambda}\cdot\mathbf{w}}{\lambda}\,\mathrm{e}^{-(\boldsymbol{\lambda}\cdot\mathbf{w}/\lambda)^2}\Psi\left(\frac{\boldsymbol{\lambda}\cdot\mathbf{w}}{\lambda}\right) + i\sqrt{\pi}\frac{\boldsymbol{\lambda}\cdot\mathbf{w}}{\lambda}\,\mathrm{e}^{-(\boldsymbol{\lambda}\cdot\mathbf{w}/\lambda)^2} \tag{49.5}$$

Equation (49.2) is of exactly the same form as eq. (37.3), and can therefore immediately be written in the form of a Fokker–Planck equation (37.6):

$$\partial_\tau \varphi = \partial_r\{-F_r + \tfrac{1}{2}\partial_s T_{rs}\}\varphi \tag{49.6}$$

where

$$F_r(w) = \pi^{-\frac{5}{2}}B^{-1}\int d\mathbf{u}\,\mathrm{e}^{-u^2}(-2u_s+\partial_s)Q_{rs} \tag{49.7}$$

$$T_{rs}(w) = 2\pi^{-\frac{5}{2}}B^{-1}\int d\mathbf{u}\,\mathrm{e}^{-u^2}Q_{rs} \tag{49.8}$$

Hence the brownian motion in the ring approximation obeys the same equation as in the Landau approximation. The only difference lies in the functional dependence of the friction and diffusion coefficients on the velocity. We shall now evaluate these coefficients.

Through the explicit form of the tensor Q_{rs}, it is immediately seen that

$$F_r = -w_s T_{rs} + \tfrac{1}{2}\partial_s T_{rs} \tag{49.9}$$

This is identical with condition (38.10), which ensures the equilibrium solution. It is thus sufficient to calculate the components of the tensor T_{rs}. Substituting (49.3) into (49.8), the integration over $\mathbf{u}$ is easily carried out with the result:

$$T_{rs} = \frac{2\pi^{-\frac{3}{2}}}{4\pi^4 B}\int d\boldsymbol{\lambda}\,\frac{\lambda_r\lambda_s \mathrm{e}^{(-\boldsymbol{\lambda}\cdot\mathbf{w}/\lambda)^2}}{\lambda^5|\varepsilon^0|^2} \tag{49.10}$$

In a reference system with its Z-axis along $\mathbf{w}$, the tensor is diagonal, with components

$$T_{ZZ} = \frac{1}{\pi^4\sqrt{\pi}B}\int_0^{A}\frac{d\lambda}{\lambda}\int_{-1}^{1} d\xi\,\xi^2\,\frac{\mathrm{e}^{-w^2\xi^2}}{|\varepsilon^0(\lambda, w, \xi)|^2} \tag{49.11}$$

$$T_{XX} = T_{YY} = \frac{1}{2\pi^4\sqrt{\pi}B}\int_0^{\Lambda}\frac{d\lambda}{\lambda}\int_{-1}^{1}d\xi\,(1-\xi^2)\frac{e^{-w^2\xi^2}}{|\varepsilon^0(\lambda, w, \xi)|^2} \tag{49.12}$$

where

$$\Lambda = \frac{l_M}{\kappa} = \frac{1}{4\sqrt{\pi}\,\Gamma^{\frac{3}{2}}} \tag{49.13}$$

Let us calculate the component T_{ZZ} in detail. Using (49.4–5) and introducing the new variables $\nu = w\xi$ and $\mu = \lambda^2$, we get:

$$\begin{aligned} T_{ZZ} &= \frac{1}{\pi^4\sqrt{\pi}B}\frac{1}{w^3}\int_0^{\Lambda}d\lambda\,\lambda^3\int_{-w}^{w}d\nu\frac{\nu^2 e^{-\nu^2}}{[\lambda^2+m_1(\nu)]^2+[m_2(\nu)]^2} \\ &= \frac{1}{\pi^5 B}\frac{1}{w^3}\int_0^{\Lambda^2}d\mu\,\mu\int_0^{w}d\nu\frac{\nu m_2(\nu)}{[\mu+m_1(\nu)]^2+[m_2(\nu)]^2} \end{aligned} \tag{49.14}$$

It is convenient to perform the μ-integration first. This integral is of the same type as (48.6), but we shall do it in more detail here.

$$\begin{aligned} J &= \int_0^{\Lambda^2}d\mu\frac{\mu}{(\mu+m_1)^2+m_2^2} \\ &= \tfrac{1}{2}\ln\frac{(\Lambda^2+m_1)^2+m_2^2}{m_1^2+m_2^2} - \frac{m_1}{m_2}\tan^{-1}x\Big|_{m_1/m_2}^{(\Lambda^2+m_1)/m_2} \end{aligned} \tag{49.15}$$

The evaluation of the $\tan^{-1}$ term requires some care. Using the well known formula:

$$\tan^{-1}x = \frac{i}{2}\ln\frac{i+x}{i-x}$$

the $\tan^{-1}$ in the second term of (49.15) can be written in the form:

$$\begin{aligned} \phi = \tan^{-1}x\Big|_a^b &= \frac{i}{2}\ln\frac{(i+b)(i-a)}{(i-b)(i+a)} = -\frac{i}{2}\ln\left(-\frac{i+A}{i-A}\right) \\ &= \frac{\pi}{2} - \operatorname{arctg} A \end{aligned} \tag{49.16}$$

where arctg means the principal value of $\tan^{-1}$ (the principal value is comprised between $-\frac{1}{2}\pi$ and $+\frac{1}{2}\pi$). A is defined as:

$$A = \frac{\Lambda^2 m_1+m_1^2+m_2^2}{\Lambda^2 m_2} \tag{49.17}$$

We can now transform the formula further by using the trigonometric relation:

$$\tan \phi = \tan (\pi/2 - \text{arctg}\, A) = 1/A$$

and thus

$$\phi = \text{arctg}\,(1/A) + k\pi \tag{49.18}$$

We must now determine the value of k:

if $A > 0$, (49.16) shows that $\phi < (\pi/2)$, and can be interpreted as a principal value of the $\tan^{-1}$; one must take $k = 0$;

if $A < 0$, $\phi > (\pi/2)$, and one must add π to the principal value of the arctg in (49.18) in order to make the latter formula equivalent to (49.16).

Collecting all the results we now obtain

$$J = \tfrac{1}{2} \ln \frac{(\Lambda^2 + m_1)^2 + m_2^2}{m_1^2 + m_2^2} - \frac{m_1}{m_2} \text{arctg} \frac{\Lambda^2 m_2}{\Lambda^2 m_1 + m_1^2 + m_2^2} - \frac{m_1}{m_2} k\pi$$

$$k = 0 \quad \text{for} \quad \Lambda^2 m_1(\nu) + |m(\nu)|^2 > 0$$
$$k = 1 \quad \text{for} \quad \Lambda^2 m_1(\nu) + |m(\nu)|^2 < 0$$

This result is exact, but complicated. A rough estimate can be obtained by using a large value of Λ. One can then replace the two first terms by the dominant term and obtain

$$\begin{gathered} J \approx 2 \ln \Lambda - k\pi [m_1(\nu)/m_2(\nu)] \\ k = 0 \quad \text{for} \quad m_1(\nu) > 0 \\ k = 1 \quad \text{for} \quad m_1(\nu) < 0 \end{gathered} \tag{49.19}$$

A glance at Fig. 21.1 shows that $m_1(\nu)$ effectively changes sign when ν varies from 0 to ∞. For small values of ν it is positive, whereas for large values it is negative. Let w_c be the critical value at which $m_1(\nu)$ vanishes:

$$m_1(w_c) = 0 \tag{49.20}$$

Substituting (49.19) into (49.14) we obtain the following approximate form:

$$T_{ZZ} = \frac{2 \ln \Lambda}{\pi^5 B} \frac{1}{w^3} \int_0^w d\nu\, \nu\, m_2(\nu) - \frac{1}{\pi^4 B} \frac{1}{w^3} \theta(w - w_c) \int_{w_c}^w d\nu\, \nu\, m_1(\nu) \tag{49.21}$$

where $\theta(x)$ is the Heaviside function [see (4.18)].

The second integral is still complicated. However, for sufficiently large values of w one could use the asymptotic expansion for $\varepsilon_1(k; w)$ given in eq. (21.16). The dominant term is then:

$$T_{zz} = \tfrac{1}{2}w^{-3}[\Phi(w) - w\Phi'(w)] + \frac{1}{2\pi^4 B} w^{-3}\theta(w-w_c)\ln\frac{w}{w_c} \tag{49.22}$$

Comparing this equation with (37.24) and (37.19) we see that the first term is identical with the Landau value. The second term is the most interesting. It occurs abruptly when the velocity of the particle exceeds a critical value which is slightly larger than the thermal velocity. This coincides with the change in character of the polarization cloud around the particle. For high velocities the particle emits plasma oscillations and this process gives rise to an extra loss of energy of the test particle.

The other coefficients of the Fokker–Planck equation can now be evaluated along the same lines. We shall only quote the final result, which we write in the same form as (37.28):

$$F_r = -\left\{8w^{-1}\mathcal{G}(w) - \frac{1}{2\pi^4 B}\theta(w-w_c)\frac{1}{w^3}\left(\ln\frac{w}{w_c} + \frac{1}{w_c^2} - \frac{3}{2w^2}\right)\right\} w_r \tag{49.23}$$

$$T_{rs} = \left\{2w^{-1}\mathcal{H}(w) + \frac{1}{2\pi^4 B}\theta(w-w_c)\frac{1}{w^3}\left(\frac{w^2}{w_c^2} - 1 - \tfrac{1}{2}\ln\frac{w}{w_c}\right)\right\}\delta_{rs} \tag{49.24}$$

$$+ \left\{2w^{-3}\mathcal{E}(w) + \frac{1}{2\pi^4 B}\theta(w-w_c)\frac{1}{w^5}\left(-\frac{w^2}{w_c^2} + 1 + \tfrac{3}{2}\ln\frac{w}{w_c}\right)\right\} w_r w_s$$

In this (rough) approximation, each coefficient is a sum of:

a) a "binary collision term", identical with the Landau contribution; here the collective effects are "static": their only effect is to provide the Debye cut-off;

b) a "plasma oscillation term". This is a purely collective term, representing the exchange of energy between an individual particle and the plasma as a whole. This is a high-speed effect which is therefore rather sensitive to the tail of the electron distribution. Although this effect is smaller than the binary collision term, it is not always negligible, especially in not too hot plasmas (in which $\Gamma \ll 1$ but $|\ln \Gamma| \sim 1$). Fig. 49.1 shows a graph of the two contributions to the friction coefficient.

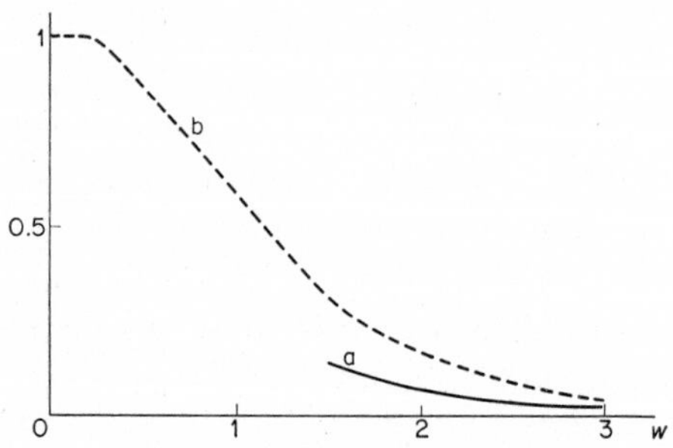

Fig. 49.1. The collective contribution $2\pi^4 B\eta_{\text{coll}}/\eta(0)$ (a) compared to the friction coefficient in the Landau approximation (b) (see also Fig. 38.2).

§ 50. Limitations of the Ring Approximation

We shall now discuss qualitatively the region of validity of the ring approximation. The limitations of this approximation are due to two distinct types of causes, which we review successively.

A. Instabilities

We have already pointed out repeatedly how crucial the stability condition is in the derivation of all our results. This is especially clear in § 45: formula (45.6) is only valid if the functions $\bar{\chi}_v(\nu)$ are a complete set, and this is only true if the system is stable. This aspect has also been noted by Lenard.* However, the difficulty can be overcome rather easily within the ring approximation: the treatment of § 45 can be modified in order to take account of the zeros of $\varepsilon_+(z)$ in the upper half-plane. Recent progress has been achieved in this field. An account of this work is given in Appendix 9.

* A. Lenard, *Bull. Am. Phys. Soc.*, **6**, 189 (1961).

B. Higher order approximations

The higher order terms also play a role in stable situations when one considers the case in which Γ is no longer very small (i.e. lower temperatures or higher densities). The principles guiding the choice of terms here are rather obvious; however some care must be exercised.

We shall discuss as an example the contributions to $\varphi(\mathbf{v}; t)$ up to the next order in Γ. Our general principles clearly indicate that in this case we have to retain one uncompensated e^2 factor, together with all contributions of order e^2t and e^2c. We thus need all the terms of order $e^2[(e^2c)^n e^2 t]^m$. All these contributions necessarily contain m semiconnected diagonal fragments evaluated in the leading universal approximation (to account for the factor t^m, see § 27). According to the nature of these fragments and of the rest of the diagram four cases occur:

1) *$m-1$ rings and one diagonal fragment of order $e^4(e^2c)^n t$,* all evaluated in the leading universal approximation. The new diagram is any diagonal fragment involving two vertices of topological index 0, and any number of vertices of index 1. One must sum over all these diagrams. They are shown in Fig. 50.1.

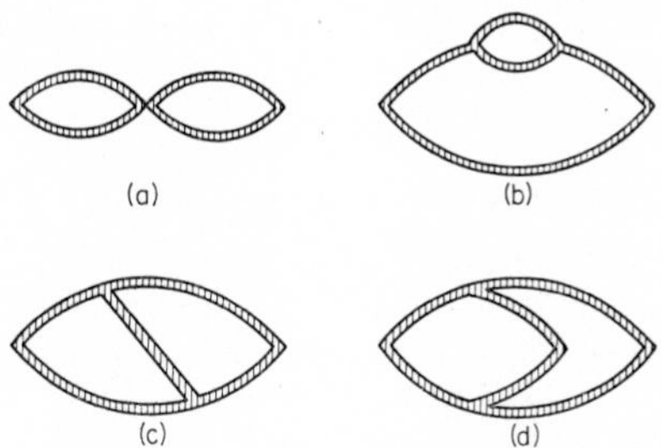

Fig. 50.1. Diagonal fragments of order $e^4(e^2c)^n t$. The modified lines have been defined in Fig. 44.1.

2) *A succession of m rings and one destruction fragment to the right,* all evaluated in the leading universal approximation. Such a diagram is shown in Fig. 50.2. The extra e^2 factor comes from $\rho_k(\alpha, \gamma; 0)$.

Fig. 50.2. Rings followed by a plasma destruction fragment.

3) *m "leading universal" rings and one "non-leading universal" ring.* Besides the term of order $(e^2c)^n e^2t$, any ring also has contributions of order $e^2(e^2c)^n$, as was shown in detail for the cycle in eq. (27.3). These terms are definitely not Boltzmann-like. They take account of the fact that the collision process (or the effective collision process in a plasma) is not instantaneous. They necessarily appear whenever the time scales t_r and t_p are not very well separated: in the present case this means that Γ is not very small. Such terms are not included in Bogoliubov's theory, which should be revised in order to take account of such contributions. This difficulty has been pointed out in a similar context by Guernsey.* In the present formalism these terms are easily isolated and can in principle be calculated straightforwardly.

4) For still higher orders, when the time scales t_r and t_p move closer together (large Γ), one must begin to examine more closely the *non-universal terms*. One can then no longer neglect the exponentially decaying terms. In that case the equations become very complicated; the evolution of the system is no longer markovian.

The meaning of all these corrections is clear. They reflect two types of facts which characterize the evolution of a plasma when the ring approximation ceases to be valid:

(*a*) *An increasing importance of individual particle behavior.* The evolution is no longer determined by a dielectric constant. The latter is characteristic of the ring approximation, in which collective effects are, explicitly or implicitly (through a Debye potential), essential for understanding the behavior of the plasma. The new diagrams of Fig. 50.1 contain the cross-vertex E, which

* R. L. Guernsey, *Relaxation Time for the Two-Particle Correlation Function in a Plasma,* Boeing Scient. Res. Lab. D1—82—0083, December, 1960.

is typical of strong two-body collisions (see Fig. 33.1) as well as more complicated three-body and four-body correlations during the collision.

(*b*) *An increasing admixture of mechanical (non-dissipative) behavior.* This is introduced by the non-dominant universal and the non-universal contributions to the diagrams. Although these terms complicate the equations very seriously, it will be shown in Chapter 13 that they do not contribute to certain important properties, such as the transport coefficients.

CHAPTER 12

The Equilibrium State

§ 51. Correlations in Equilibrium

In §§ 34 and 46 an H-theorem has been proved which shows that the velocity distribution tends toward the equilibrium value

$$\varphi^0(\mathbf{v}) = (m\beta/2\pi)^{\frac{3}{2}} e^{-(\beta m/2)v^2} \tag{51.1}$$

We now prove a generalized version of the H-theorem, by showing that not only the velocity distribution but also the correlation function tends toward its correct value in equilibrium.

According to the general properties of the correlation function $\rho_k(\alpha, \gamma)$, the instantaneous value of this function is expressed in terms of $\varphi(\mathbf{v})$ evaluated at the same time. But we have shown that $\varphi \to \varphi^0$ as $t \to \infty$; therefore $\rho_k^0(\alpha, \gamma)$ is given by formulae (44.6) and (45.6), evaluated for $\varphi = \varphi^0$. In the formulae of the present chapter, a superscript ^{0} always denotes a function evaluated at equilibrium.

One can evaluate $F^0(\mathbf{v}_\alpha)$ [and hence $\rho_k^0(\alpha, \gamma)$] directly from (45.6); however, one then finds integrals whose direct evaluation is difficult. An alternative simpler procedure consists of solving eq. (40.10) under equilibrium conditions. The coefficients of this equation are then:

$$\begin{aligned}
q^0(\mathbf{v}_\alpha) &= -e^2\beta V_k \varphi^0(\mathbf{v}_\alpha) \\
d^0(\mathbf{v}_\alpha) &= -8\pi^3 e^2 c\beta V_k k v_\alpha \varphi^0(\mathbf{v}_\alpha) \\
\varepsilon^0(v_\alpha) &= 1 - 8\pi^4 i e^2 c\beta V_k \int dv_1 \delta_-(v_\alpha - v_1) v_1 \bar{\varphi}^0(v_1)
\end{aligned} \tag{51.2}$$

These formulae immediately show that $F^0(\mathbf{v}_\alpha)$ is *real*: indeed, eq. (41.18) shows that F_2 is proportional to q_2 and (51.2) shows that $q_2^0 = 0$, hence

$$F_2^0(\mathbf{v}_\alpha) = 0 \tag{51.3}$$

[One should note that the vanishing of F_2^0 causes $\varphi^0(\mathbf{v}_\alpha)$ to be a stationary solution of the ring equation, (41.19).] We also note that upon substitution of the Coulomb potential $V_k = (2\pi^2 k^2)^{-1}$, the inverse Debye length, eq. (16.7), appears quite naturally:

$$8\pi^3 e^2 c\beta V_k = 4\pi e^2 c\beta/k^2 = \kappa^2/k^2 \tag{51.4}$$

Collecting all these results, the integral equation (40.10) becomes, in equilibrium:

$$\begin{aligned}&\left[1-(\kappa^2/k^2)\pi i\int dv_1\,\delta_-(v_\alpha-v_1)v_1\bar{\varphi}^0(v_1)\right]F^0(\mathbf{v}_\alpha)\\ &= -(e^2\beta/2\pi^2k^2)\varphi^0(\mathbf{v}_\alpha)-(\kappa^2/k^2)v_\alpha\varphi^0(\mathbf{v}_\alpha)\pi i\int dv_1\,\delta_-(v_\alpha-v_1)\bar{F}^0(v_1)\end{aligned} \tag{51.5}$$

The structure of this equation suggests that the solution is of the form:

$$F^0(\mathbf{v}_\alpha) = \varphi^0(\mathbf{v}_\alpha)Q^0(k) \tag{51.6}$$

where $Q^0(k)$ is a real function of k, independent of $\mathbf{v}_\alpha$. If a solution of this form can be found, we know that it is unique [because the index of the singular integral equation (51.5) is zero]. Substituting (51.6) into (51.5) we obtain

$$\begin{aligned}&\left[1-(\kappa^2/k^2)\pi i\int dv_1\,\delta_-(v_\alpha-v_1)v_1\bar{\varphi}^0(v_1)\right]\varphi^0(\mathbf{v}_\alpha)Q^0(k)\\ &= -(e^2\beta/2\pi^2k^2)\varphi^0(\mathbf{v}_\alpha)-(\kappa^2/k^2)v_\alpha\varphi^0(\mathbf{v}_\alpha)\pi i\int dv_1\,\delta_-(v_\alpha-v_1)\bar{\varphi}^0(v_1)Q^0(k)\end{aligned}$$

or:

$$\left[1+(\kappa^2/k^2)\int dv_1\,\pi i\,\delta_-(v_\alpha-v_1)(v_\alpha-v_1)\bar{\varphi}^0(v_1)\right]Q^0(k) = -e^2\beta/2\pi^2k^2 \tag{51.7}$$

We now use the simple identity [see Appendix 2, eq. (A2.2.3)]:

$$i\pi\delta_-(x)x = \left[\pi i\delta(x)+\mathscr{P}\frac{1}{x}\right]x = 1 \tag{51.8}$$

(This identity must, of course, be interpreted as a relation among Schwartz distributions, i.e. in connection with an integration.) Keeping in mind that $\bar{\varphi}^0(v)$ is normalized to one, eq. (51.7) immediately gives

$$Q^0(k) = -\frac{e^2\beta}{2\pi^2}\frac{1}{k^2+\kappa^2} \tag{51.9}$$

Collecting now the results (51.9), (51.6), (44.6) and (39.4), we obtain the final result

$$\rho_k^0(\alpha, \gamma) = -\frac{e^2\beta}{2\pi^2}\frac{1}{k^2+\kappa^2}\varphi^0(\alpha)\varphi^0(\gamma) \tag{51.10}$$

This is proportional to the Fourier transform of the well known Debye potential.

The binary correlation function is then given by a Fourier transformation of (51.10):

$$G^0(\alpha, \gamma) = -c^2\varphi^0(\alpha)\varphi^0(\gamma)\beta e^2\, \mathrm{e}^{-\kappa r}/r \tag{51.11}$$

It is interesting to show how the density correlation function g_k^0 (i.e. the integral of $\rho_k^0(\mathbf{v}_\alpha, \mathbf{v}_\gamma)$ over both velocities) can be calculated directly from eq. (47.4) by making use of the Kramers–Kronig relation (19.5). The method used here is due to Ichimaru.* Substituting (51.2) and (51.4) into eq. (47.5) we obtain

$$g_k^0 = -\frac{e^2\beta}{2\pi^2k^2}\int d\nu \frac{\bar{\varphi}^0(\nu)}{|\varepsilon^0(\nu)|^2}$$

But, using (51.1) and (40.6a)

$$\varepsilon_2^0(\nu) = -4\pi^2e^2c\beta k^{-2}\nu\bar{\varphi}^0(\nu) = -\pi(\kappa^2/k^2)\nu\bar{\varphi}^0(\nu)$$

Hence

$$g_k^0 = \frac{1}{8\pi^4c}\int_{-\infty}^{\infty} d\nu \frac{\varepsilon_2^0(\nu)}{\nu|\varepsilon^0(\nu)|^2} = \frac{1}{8\pi^4c}\int_{-\infty}^{\infty} d\nu \frac{1}{\nu}\,\mathrm{Im}\frac{1}{\varepsilon_+^0(\nu)} \tag{51.12}$$

But $\varepsilon_+^0(\nu)$ has no zeros in the upper half-plane because the equilibrium is a stable distribution; hence $\{[1/\varepsilon_+^0(\nu)]-1\}$ is a plus-function which vanishes at infinity. One can therefore write a generalized Kramers–Kronig formula in the form:

$$\mathrm{Re}\left\{\frac{1}{\varepsilon_+^0(\omega)}-1\right\} = \frac{1}{\pi}\mathscr{P}\int_{-\infty}^{\infty} d\nu \frac{1}{\nu-\omega}\,\mathrm{Im}\frac{1}{\varepsilon_+^0(\nu)} \tag{51.13}$$

This equation, for $\omega = 0$, gives the value of the integral appearing in (51.12). [Note that the principal part reduces to the ordinary integral for $\omega = 0$, because $\mathrm{Im}\,\varepsilon_+^0(0) = 0$.] Therefore

$$g_k^0 = \frac{1}{8\pi^4c}\pi\,\mathrm{Re}\left\{\frac{1}{\varepsilon_+^0(0)}-1\right\}$$

$$= \frac{1}{8\pi^3c}\left\{\frac{1}{1+(\kappa^2/k^2)}-1\right\} = -\frac{e^2\beta}{2\pi^2}\frac{1}{k^2+\kappa^2}$$

which agrees with (51.10) integrated over $\mathbf{v}_\alpha$ and $\mathbf{v}_\gamma$.

* S. Ichimaru, *Theory of Fluctuations in a Plasma*, Report AFCRL-62-340, Office of Aerospace Res., U.S. Airforce, 1962.

§ 52. Summation of Creation Diagrams in Equilibrium

It is of interest to show that the equilibrium correlation function (51.11) can also be derived by another method which is more closely related to the traditional methods based on the Ursell–Mayer cluster expansion.* Instead of first summing the creation diagrams for an arbitrary velocity distribution and then setting $\varphi = \varphi^0$ in the result, as was done in § 51, we may also first set $\varphi = \varphi^0$ in each creation diagram, and then sum the series. This will be done now.

In order to perform the summation of all equilibrium creation diagrams, we exploit the relation which exists between diagrams with n vertices and diagrams with $n+1$ vertices (see § 40, rules a and b). The summation will be performed by using a different type of grouping. In the first step, all the diagrams having a given number of vertices are summed. In the second step, one sums over the number of vertices. The summation of the first step is contained in the following theorem:

Let $S_n(\alpha, \gamma)$ represent the sum of all equilibrium creation diagrams with n vertices. Then

$$S_n(\alpha, \gamma) = (-e^2 \beta V_k)^n (8\pi^3 c)^{n-1} \varphi^0(\alpha) \varphi^0(\gamma), \quad n = 1, 2, 3, \ldots \tag{52.1}$$

The theorem will be proved by induction. Suppose (52.1) is true for some n; it is shown to be true also for $n+1$. According to Fig. 44.3, the sum of the creation diagrams with $n+1$ vertices equals the sum of all creation diagrams with n vertices and one extra vertex on the upper line at left plus the sum of creation diagrams with n vertices and one extra vertex on the lower line at left. This is shown diagrammatically in Fig. 52.1.

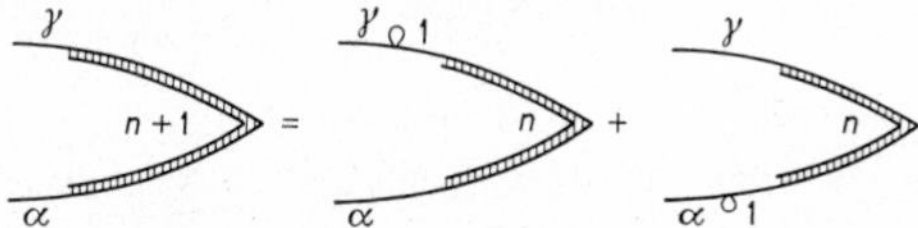

Fig. 52.1. The relation between the sum of creation fragments with $n+1$ vertices and the sum of creation fragments with n vertices.

Keeping in mind the rules of §§ 39 and 44 for writing the creation diagrams, we note that in order to obtain a diagram

* See § 54.

with one more vertex on the upper line from one with n vertices, we change in the latter the index γ into 1 (a dummy index), multiply by $\delta_-^{\alpha\gamma} d_\gamma \varphi^0(\gamma)$ and integrate over 1; a similar rule holds for adding a vertex on the lower line. Hence, Fig. 52.1 expresses the following formula:

$$S_{n+1}(\alpha, \gamma) = \int_1 \delta_-^{\alpha\gamma} d_\gamma \varphi^0(\gamma) S_n(\alpha, 1) + \int_1 \delta_-^{\alpha\gamma}(-d_\alpha)\varphi^0(\alpha) S_n(1, \gamma) \tag{52.2}$$

With the assumption (52.1) and the definitions (39.4) the formula becomes:

$$\begin{aligned} S_{n+1}(\alpha, \gamma) &= -8\pi^3 e^2 c V_k (-e^2\beta V_k)^n (8\pi^3 c)^{n-1} \int d\mathbf{v}_1 \pi \delta_-(\mathbf{k}\cdot\mathbf{g}_{\alpha\gamma}) m^{-1} \\ &\qquad \{[i\mathbf{k}\cdot\partial_\gamma \varphi^0(\gamma)]\varphi^0(\alpha)\varphi^0(1) - [i\mathbf{k}\cdot\partial_\alpha \varphi^0(\alpha)]\varphi^0(1)\varphi^0(\gamma)\} \\ &= (-e^2\beta V_k)^{n+1}(8\pi^3 c)^n [\pi i \delta_-(\mathbf{k}\cdot\mathbf{g}_{\alpha\gamma})\mathbf{k}\cdot\mathbf{g}_{\alpha\gamma}]\varphi^0(\alpha)\varphi^0(\gamma) \\ &= (-e^2\beta V_k)^{n+1}(8\pi^3 c)^n \varphi^0(\alpha)\varphi^0(\gamma) \end{aligned} \tag{52.3}$$

In the last step we made use of the identity (51.8). We have thus shown that if formula (52.1) holds for n vertices it is true for $n+1$ vertices too. We now show that it is true for $n=1$, i.e. for the single top diagram of Fig. 44.3. According to the rules of § 44, this diagram contributes the expression

$$\begin{aligned} \delta_-^{\alpha\gamma} d_{\gamma\alpha} \varphi^0(\alpha)\varphi^0(\gamma) &= -\pi\delta_-(\mathbf{k}\cdot\mathbf{g}_{\alpha\gamma}) e^2 m^{-1} V_k i\mathbf{k}\cdot\partial_{\gamma\alpha}\varphi^0(\alpha)\varphi^0(\gamma) \\ = -e^2\beta V_k[\pi i \delta_-(\mathbf{k}\cdot\mathbf{g}_{\alpha\gamma})\mathbf{k}\cdot\mathbf{g}_{\alpha\gamma}]\varphi^0(\alpha)\varphi^0(\gamma) &= -e^2\beta V_k \varphi^0(\alpha)\varphi^0(\gamma) \end{aligned}$$

where formula (51.8) was again used. The previous formula is the value of eq. (52.1) for $n = 1$, which achieves the proof of the theorem.

We now evaluate the sum over the number of vertices, with the result

$$\begin{aligned} \rho_k^0(\alpha, \gamma) &= \sum_{n=1}^{\infty} S_n(\alpha, \gamma) = (8\pi^3 c)^{-1}\varphi^0(\alpha)\varphi^0(\gamma) \sum_{n=1}^{\infty} (-8\pi^3 e^2 c\beta V_k)^n \\ &= \frac{\varphi^0(\alpha)\varphi^0(\gamma)}{8\pi^3 c} \sum_{n=1}^{\infty} \left(-\frac{\kappa^2}{k^2}\right)^n = -\frac{\varphi^0(\alpha)\varphi^0(\gamma)}{8\pi^3 c}\,\frac{\kappa^2}{k^2+\kappa^2} \\ &= -\frac{e^2\beta}{2\pi^2}\varphi^0(\alpha)\varphi^0(\gamma)\frac{1}{k^2+\kappa^2} \end{aligned} \tag{52.4}$$

This is identical with the result of the previous section, eq. (51.10).

There is no difficulty in extending this summation procedure to an s-component plasma. The result is then:

$$G^{0(\sigma,\sigma')}(\alpha,\gamma) = -c^2\varphi^{0(\sigma)}(\alpha)\varphi^{0(\sigma')}(\gamma)\beta e^2 z_\sigma z_{\sigma'}\, \mathrm{e}^{-\kappa r}/r;\quad \sigma,\ \sigma' = 1,\ldots, s \qquad (52.5)$$

The inverse Debye length κ is now defined as:

$$\kappa^2 = 4\pi e^2\beta \sum_{\sigma=1}^{s} c_\sigma z_\sigma^2 \qquad (52.6)$$

§ 53. Thermodynamics of a Plasma in Equilibrium

Our purpose in this paragraph is the calculation of the thermodynamic functions of a plasma in equilibrium. Of course, interest lies only in the deviations of the properties of a plasma from those of a perfect gas. Such properties derive from the contributions of order $e^2(e^2c)^m(e^2t)^n$ to the distribution function. We have evaluated the binary correlation function up to this order, but not the velocity distribution $\varphi(\mathbf{v})$. In a calculation of the internal energy, say, we would need *both* functions calculated up to this order. However, in the special case of the *classical equilibrium distribution,* the function $\varphi^0(\mathbf{v})$ is independent of e^2 and equals the maxwellian to all orders in the charge. Hence the effect of the imperfection of the gas is completely contained in the correlation function. This special property is no longer true in quantum mechanics (see Chapter 18).

A knowledge of the two-body correlation function is sufficient for the calculation of all important thermodynamic properties of a plasma in equilibrium. However, for the evaluation of these quantities the model of an electron gas in the presence of a charged background is not very convenient. Indeed, in order to avoid trivial divergences (such as the compensation of the total electrostatic energy of the gas by the self-energy of the continuous background), one needs elaborate and rather artificial arguments. It is much better to start this calculation directly for a realistic s-component plasma.

We calculate first the *internal energy density* by using formula (1.16′). Let us collect the quantities which enter this formula:

$$\begin{aligned} V^{(\sigma\sigma')} &= z_\sigma z_{\sigma'}/|\mathbf{x}_1-\mathbf{x}_2| \\ f_1^{0(\sigma)}(1) &= c_\sigma \varphi^{0(\sigma)}(1) \\ f_2^{0(\sigma\sigma')}(1, 2) &= c_\sigma c_{\sigma'} \varphi^{0(\sigma)}(1)\varphi^{0(\sigma')}(2)\{1-\beta e^2 z_\sigma z_{\sigma'} \mathrm{e}^{-\kappa r}/r\} \\ \kappa &= 4\pi e^2 \beta \sum_{\sigma=1}^{s} c_\sigma z_\sigma^2 \end{aligned} \tag{53.1}$$

The average kinetic energy yields the well known internal energy of a perfect gas. In the second term we take $\mathbf{x}_1$ and $\mathbf{r} = \mathbf{x}_1-\mathbf{x}_2$ as integration variables and go over to polar coordinates for the $\mathbf{r}$-integration.

$$E = \tfrac{3}{2}kT + \tfrac{1}{2}c^{-1}e^2 4\pi \sum_{\sigma\sigma'} \int_0^\infty dr\, r^2 (z_\sigma z_{\sigma'}/r) c_\sigma c_{\sigma'} \{1-\beta e^2 z_\sigma z_{\sigma'} \mathrm{e}^{-\kappa r}/r\} \tag{53.2}$$

We now note that the first term in the integrand gives rise to a linear divergence ($\int dr\, r$). This term is just the total electrostatic energy of a set of N_σ ions σ and $N_{\sigma'}$ ions σ'. (This is the place where difficulties occur for a single-component plasma.) However, it must be kept in mind that the plasma is electrically neutral on the whole. This implies the following condition:

$$\sum_{\sigma=1}^{s} c_\sigma z_\sigma = 0 \tag{53.3}$$

If in (53.2) the summations over σ and σ' are performed *before* the integration, it is immediately seen that the integrand vanishes identically as a consequence of eq. (53.3). This result is physically quite sound: the total electrostatic energy of the electrically neutral plasma is of course zero.

In the remaining term we also perform the summations first (although this term does not diverge, because of the occurrence of the exponential factor). The result is:

$$E = \tfrac{3}{2}kT - (kT/8\pi c)\kappa^3 \tag{53.4}$$

This is our final result. We now calculate the free energy density, by using the well known thermodynamic formula *

* See any textbook on thermodynamics for this and the following formulae, for instance: I. Prigogine and R. Defay, *Thermodynamique Chimique*, ed. Desoer, Liege, 1950; P. T. Landsberg, *Thermodynamics*, Interscience, New York, 1961; L. D. Landau and E. M. Lifshitz, *Statistical Physics*, Pergamon Press, London, 1958.

$$\frac{E}{T^2} = -\frac{\partial}{\partial T}\left(\frac{F}{T}\right) \tag{53.5}$$

We integrate this formula by using (53.4) and (53.1):

$$F = -T\int dT\,T^{-2}\{\tfrac{3}{2}kT - (kT/8\pi c)(4\pi e^2 \textstyle\sum c_\sigma z_\sigma^2/kT)^{\frac{3}{2}}\} + \text{const.}$$

The constant must be chosen in such a way that in the limit $e^2 \to 0$ the expression obtained equals the free energy of a perfect gas, F_p (which is well known and will not be written out explicitly):

$$F = F_p - (kT/12\pi c)\kappa^3 \tag{53.6}$$

Once we know the free energy as a function of temperature and volume, we may calculate all the other thermodynamic functions by simple algebraic and differential operations. Some of these functions are given below:

Entropy density

$$s = (E-F)/T = s_p - (k/24\pi c)\kappa^3 \tag{53.7}$$

Specific heat at constant volume

$$c_\Omega = (\partial E/\partial T)_\Omega = \tfrac{3}{2}k + (k/16\pi c)\kappa^3 \tag{53.8}$$

Equation of state

$$p/ckT = -(1/ckT)(\partial F/\partial \Omega) = 1 - (1/24\pi c)\kappa^3 \tag{53.9}$$

The changes in the thermodynamic quantities introduced by the presence of the interactions as compared to the perfect gas can easily be understood qualitatively. We first note a lowering of the energy (53.4) and of the free energy (53.6). This reflects the stabilization of the system by the interactions; we may think of an extra cohesion brought about by the collective effects. We also note that the entropy of the system (53.7) is lowered: of course, the collective effects introduce an ordered structure. Around each ion there is a polarization cloud instead of the random distribution of particles existing in a perfect gas. If we supply energy to the system, this energy is no longer used only to increase the kinetic energy of the particles, but also partly to destroy the polarization structure: this is why the specific heat (53.8) is increased.

We note that all the changes in the thermodynamic functions are proportional to the non-dimensional quantity

$$N_D^{-1} = \kappa^3 / \tfrac{4}{3}\pi c \tag{53.10}$$

N_D has the simple physical interpretation of the *number of particles in a sphere whose radius equals the Debye length.* The gas is closer to perfection the more particles there are in a Debye sphere. Note that N_D is closely related to our parameter Γ:

$$N_D^{-1} \sim \Gamma^{\frac{3}{2}}$$

The outstanding feature of all these formulae is the following. Omitting component indices, we see that $N_D^{-1} \sim c^{\frac{1}{2}}$. This means that *the thermodynamic functions of a plasma are not analytical functions of the density*; in other words, they cannot be expanded in a series of integral powers of c. Historically this was the first and main difficulty which appeared in the theory of plasmas and of electrolyte solutions. The first complete solution of this problem was given by Debye and Hückel * in a celebrated work.

Their brilliant approach was largely intuitive, combining statistical mechanics and phenomenological electrostatics. Their work is so well known that it is not necessary to review it here. The rigorous justification of the Debye–Hückel theory had however to wait 30 more years until the fundamental paper of Mayer ** clarified the situation. The readers who are familiar with the equilibrium statistical mechanics of imperfect gases will find below a sketch of Mayer's theory.

§ 54. Mayer's Theory

Mayer's theory of plasmas (or of electrolyte solutions) will be briefly sketched in this paragraph. This outline will show in particular how this derivation, based on the equilibrium theory of imperfect gases, compares with our treatment, which derives equilibrium results as asymptotic properties of a general non-equilibrium theory. We assume that the reader is familiar with the cluster expansion, and we shall therefore not prove the

* P. Debye and E. Hückel, *Physik. Z.*, **24**, 185 (1923).

** J. Mayer, *J. Chem. Phys.*, **18**, 1426 (1950).

starting formulae.* As the virial expansion for the equation of state is more generally known, we shall study this expansion here and shall derive eq. (53.9), but the same theory can easily be extended to the calculation of the correlation function $g^0(r)$ or of the potential of average force.**

In order to simplify the notations, we consider a gas consisting of N_e electrons and of N_i monovalent positive ions, $N_i+N_e = N$. The electroneutrality condition then requires $c_i = c_e$. It can be shown that the pressure of a two-component gas is given by:

$$\frac{p}{kT} = c_i+c_e+S-c_i\frac{\partial S}{\partial c_i}-c_e\frac{\partial S}{\partial c_e} \tag{54.1}$$

where

$$S = \sum_{n\geq 2} \sum_{\substack{n_i \\ (n_i+n_e=n)}} \sum_{n_e} \beta_{n_i n_e} c_i^{n_i} c_e^{n_e} \tag{54.2}$$

and the irreducible cluster integrals $\beta_{n_i n_e}$ are defined as

$$\beta_{n_i n_e} = [n_i!\, n_e!\, \Omega]^{-1} \int d\mathbf{x}_1 \ldots d\mathbf{x}_n P \tag{54.3}$$

P is a sum of products of Mayer's f_{ij} functions

$$f_{ij}(r_{ij}) = \exp\left[-e^2V_{ij}(r_{ij})/kT\right]-1 \tag{54.4}$$

Each product entering P is represented by a diagram: the particles are represented by vertices and each factor f_{ij} by a heavy line joining vertices i and j;† P is defined as the sum of all possible graphs involving n_i ion vertices and n_e electron vertices which are at least doubly connected. This means that any two vertices are connected by at least two distinct paths following the lines. (The only exceptions are β_{11}, β_{20} and β_{02}, i.e. the three possible simply connected two-vertex diagrams.) A typical diagram contributing to β_{32} is shown in Fig. 54.1a.

* See, e.g., T. L. Hill, *Statistical Mechanics*, McGraw-Hill, New York, 1956.

** E. Meeron, *J. Chem. Phys.*, **28**, 630 (1958); E. E. Salpeter, *Ann. Phys.* **5**, 183 (1958).

† The Mayer diagrams are in a sense dual to our dynamical diagrams, where the lines correspond to particles and the vertices to interactions.

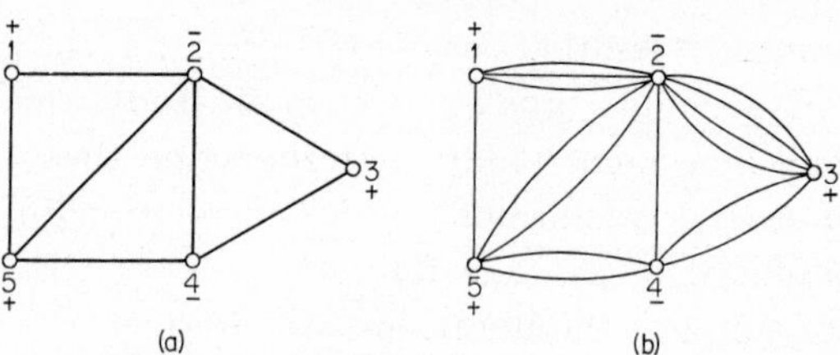

Fig. 54.1. (a) Original Mayer diagrams, $n_i = 3$, $n_e = 2$. The contribution to P is $f_{12}f_{23}f_{34}f_{45}f_{51}f_{52}f_{24}$. (b) Corresponding expanded Mayer diagram, $l = 14$. The contribution to P is proportional to $(V_{12})^3(V_{23})^4(V_{34})^2(V_{45})^2 V_{51}(V_{52})^2V_{24}$.

Let us now apply this theory to plasmas. As in Chapter 9, we shall use the following potential in the calculations

$$V(r_{st}) = z_s z_t \exp(-\alpha r_{st})/r_{st} \tag{54.5}$$

where α is a small convergence parameter. At the end of the calculation we let $\alpha \to 0$. We disregard the existence of a hard core at short distances. We now use the following "weak coupling" argument. The ratio e^2V/kT is assumed small for most distances. We may therefore expand the f_{st} factors as follows

$$\begin{aligned} f_{st} &= \sum_{n\geqq 1}^{\infty} (n!)^{-1}[-(4\pi e^2/kT)z_s z_t V(r_{st})]^n \\ &= \sum_{n\geqq 1}^{\infty} (n!)^{-1}[-\lambda z_s z_t V_\alpha(r_{st})]^n \end{aligned} \tag{54.6}$$

where

$$\begin{aligned} z_i &= +1, \; z_e = -1 \\ \lambda &= 4\pi e^2/kT \\ V_\alpha &= (z_s z_t/4\pi r_{st}) \exp(-\alpha r_{st}) \end{aligned} \tag{54.7}$$

We now introduce new "expanded" graphs in which we represent a factor $[-\lambda z_s z_t V_\alpha(r_{st})]$ by a faint line. In this way each original diagram generates a multiple infinity of new diagrams, obtained by replacing each heavy line by an arbitrary number of faint lines (see Fig. 54.1b).

Let us now discuss the orders of magnitude of the various contributions. We consider in this first discussion only diagrams involving a single type of particle; the discussion is identical for the other case. Let a given diagram have n vertices and l lines.

Such a diagram contributes to the pressure a term of order $c^n\beta'_n$, where

$$\beta'_n = (n!\Omega)^{-1}(-\lambda)^l \int \underbrace{d\mathbf{x}_1 \ldots d\mathbf{x}_n}_{n \text{ integrations}} \underbrace{V_\alpha(12) \ldots V_\alpha(rs)}_{l \text{ factors}}$$

The integrand is practically zero for distances $r > \alpha^{-1}$. Hence all integrations (except one, which cancels the factor Ω^{-1}) extend practically up to α^{-1}. A rough estimate is obtained by replacing each $V_\alpha(r)$ by a typical average value V. The result is then

$$c^n\beta'_n \approx c^n\alpha^{-3(n-1)}(-4\pi e^2 V/kT)^l = c(e^2c\alpha^{-3}V/kT)^{n-1}(e^2V/kT)^{l-n+1} \tag{54.8}$$

The ratio e^2V/kT can be assumed small at sufficiently high temperatures (the interactions are weak), but the ratio c/α^3 is a very large number. Indeed α is a parameter which tends to zero at the end of the calculation. The largeness of c/α^3 expresses the long range of the forces (physically, the range of the forces is much longer than the average interparticle distance). Therefore, although e^2V/kT is a small parameter (weak coupling), the quantity $e^2c\alpha^{-3}V/kT$ is large (long range) and must be retained to all orders. Note that we find here, with a different argument, the same rule which guided us throughout the non-equilibrium theory: quantities proportional to e^2 are considered as small, whereas quantities proportional to e^2c must be retained at all orders. Hence, in eq. (54.8), for each value of $l-n+1$ we sum over all values of n.

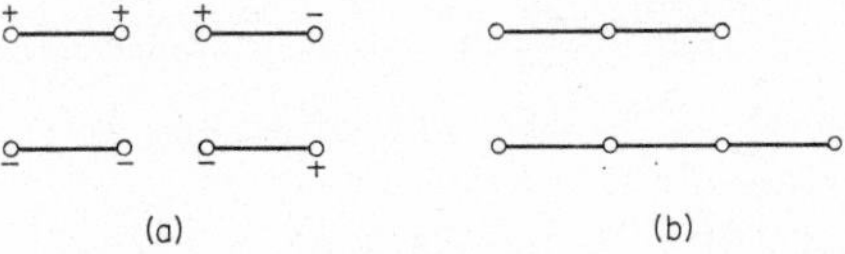

Fig. 54.2. Graphs for which $l = n-1$. Only the four graphs (a) are irreducible and contribute to P.

The dominant term in the expansion is the term with the least exponent $l-n+1$. Topologically it is easily seen that the number of lines in a graph is at least equal to the number of vertices minus one ($l = n-1$). The corresponding diagrams are shown in Fig. 54.2a. It is, however, easily seen that these four diagrams cancel pairwise because of the electroneutrality condition (the proof will not be given explicitly).

The next (and thus leading) order corresponds to $l = n$. The only irreducible diagrams of this type are the *rings* (Fig. 54.3).

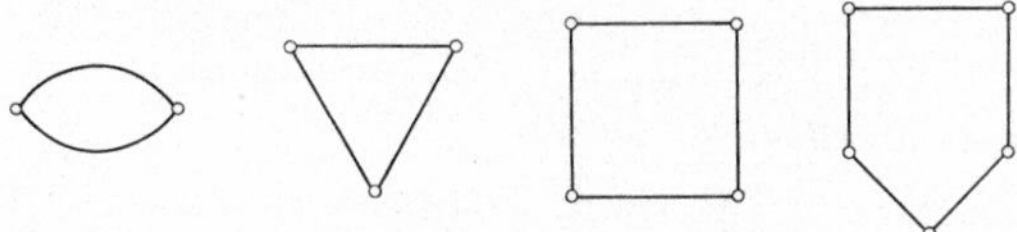

Fig. 54.3. Mayer ring diagrams.

We shall now sum the contributions of all rings. There are several rings of n particles, differing by the nature and the ordering of the vertices. The contribution to S [eq. (54.2)] of all ν-particle rings is:

$$S_\nu = \sum_{\substack{n_i \ n_e \\ (n_i+n_e=\nu)}} \frac{(\nu-1)!}{2} \frac{(-\lambda z_i^2 c_i)^{n_i}}{n_i!} \frac{(-\lambda z_e^2 c_e)^{n_e}}{n_e!} J_\nu$$

where

$$J_\nu = \Omega^{-1} \int d\mathbf{x}_1 \ldots d\mathbf{x}_\nu V_\alpha(r_{12}) V_\alpha(r_{23}) \ldots V(r_{\nu 1}) \tag{54.9}$$

The factor $(\nu-1)!$ gives the number of permutations of ν particles, holding one of them fixed (e.g. the number of distinct permutations of ν persons around a table). The factor $\frac{1}{2}$ eliminates equivalent permutations, such as 123 and 321.

S can be transformed as follows by using Newton's binomial formula

$$S_\nu = \frac{1}{2\nu} \sum_{n_i} \frac{\nu!}{n_i!(\nu-n_i)!} (-\lambda z_i^2 c_i)^{n_i} (-\lambda z_e^2 c_e)^{\nu-n_i} J_\nu$$

$$= \frac{1}{2\nu} [-\lambda(z_i^2 c_i + z_e^2 c_e)]^\nu J_\nu = \frac{1}{2\nu} (-\kappa^2)^\nu J_\nu$$

κ is the inverse Debye length defined in eq. (53.1). Hence the complete S is given by

$$S = \sum_{\nu \geqq 2} \frac{1}{2\nu} (-\kappa^2)^\nu J_\nu \tag{54.10}$$

Let us now evaluate the integral J_ν. Equation (54.9) shows that it is a convolution product: it is therefore natural to Fourier transform the potential:

$$J_\nu = (8\pi^3)^{-\nu}\Omega^{-1}\int d\mathbf{x}_1 \ldots d\mathbf{x}_\nu \int d\mathbf{l}_1 \ldots d\mathbf{l}_\nu V_{l_1} \ldots V_{l_\nu}$$
$$e^{i\mathbf{l}\cdot(\mathbf{x}_1-\mathbf{x}_2)}\, e^{i\mathbf{l}\cdot(\mathbf{x}_2-\mathbf{x}_3)} \ldots e^{i\mathbf{l}_\nu\cdot(\mathbf{x}_\nu-\mathbf{x}_1)}$$
$$= \Omega^{-1}\int d\mathbf{l}_1 \ldots d\mathbf{l}_\nu V_{l_1} \ldots V_{l_\nu}\delta(\mathbf{l}_1-\mathbf{l}_2)\delta(\mathbf{l}_2-\mathbf{l}_3) \ldots \delta(\mathbf{l}_\nu-\mathbf{l}_1)$$

A small difficulty arises here. When all integrations are performed, we are left with a factor $\delta(0)$ which is, strictly speaking, meaningless. However, we know that such a factor is to be interpreted according to the asymptotic formula:

$$\delta(\mathbf{l}) = \lim_{\Omega\to\infty} (\Omega/8\pi^3)\delta_{l_x,0}\,\delta_{l_y,0}\,\delta_{l_z,0}$$

and hence equals $\Omega/8\pi^3$. Thus the result is:

$$J_\nu = (8\pi^3)^{-1}\int d\mathbf{l}\, |V_l|^\nu$$

and

$$S = \sum_{\nu\geq 2}\int d\mathbf{l}\frac{1}{16\pi^3}\frac{(-\kappa^2 V_l)^\nu}{\nu}$$

Each term of the sum is still divergent in the limit $\alpha \to 0$. However, we apply the same argument as in § 52 and show that if the summation is performed *before* the integration the result is convergent. In order to evaluate this sum we first differentiate S with respect to κ^2; we then find an easily summable geometric progression

$$\frac{\partial S}{\partial\kappa^2} = -\int d\mathbf{l}\sum_{\nu\geq 1}\frac{1}{16\pi^3}(-\kappa^2)^\nu V_l^{\nu+1} = \frac{1}{16\pi^3}\int d\mathbf{l}\frac{\kappa^2 V_l^2}{1+\kappa^2 V_l}$$
$$= \frac{\kappa^2}{16\pi^3}\int d\mathbf{l}\frac{1}{(l^2+\alpha^2)(l^2+\alpha^2+\kappa^2)}$$

We can *now* take the limit $\alpha \to 0$ and find the convergent result

$$\frac{\partial S}{\partial\kappa^2} = \frac{4\pi\kappa^2}{16\pi^3}\int_0^\infty dl\,\frac{l^2}{l^2(l^2+\kappa^2)} = \frac{\kappa}{8\pi}$$

and hence

$$S = \int d\kappa^2 \ (\kappa/8\pi) = \kappa^3/12\pi$$

Substituting this into eq. (54.1) we find

$$p/kT = c_i + c_e - \kappa^3/24\pi$$

which agrees with eq. (53.9).

The reader will appreciate the analogy of the ideas and expressions of this derivation to those of § 52. We shall, however, not discuss here the relative advantages of the one or the other method, because this would take us outside the scope of this book.*

* I. Prigogine, *Non-Equilibrium Statistical Mechanics*, Interscience, New York, 1963.

CHAPTER 13

Non-Equilibrium Stationary States and the Theory of Transport Coefficients

§ 55. Free Relaxation and Forced Relaxation

In most problems discussed previously, we have considered systems which evolve under the influence of the free motion of their particles and of the mutual Coulomb interactions among the particles. In other words, all the driving mechanisms are of internal origin. Under such conditions, the result of experiment and of our investigation is as follows. Starting with an arbitrary distribution in phase space, the system evolves in such a way as to reach a unique final steady state called thermodynamic equilibrium. One may wonder at first sight why such stress is laid on this problem of "*free evolution*" in the theoretical literature on non-equilibrium statistical mechanics. Indeed, the usual experimental method of investigation of irreversible processes consists of imposing an external constraint on the system, such as an electric field, a temperature gradient, etc., and of studying the *forced evolution* of the system. The answer to our question comes from the observation that the process of forced evolution can be described as a superposition of two effects:

(*a*) the direct action of the constraint on the individual particles of the gas (e.g. the acceleration of the ions and electrons under the action of an electric field), an effect which drives the system systematically away from equilibrium;

(*b*) the reaction of the system owing to the existence of interactions among the particles; this effect tends to oppose the first and to bring the system back to equilibrium.

The evolution of the system under these combined effects can be roughly characterized as follows. If the external constraint is weak, the interactions can generally compensate it and the system finally reaches a steady state, which is, however, not the equilibrium state. In this state there exist steady flows (of matter, heat, electricity, . . .) which are due to the existence of the external constraint. If, on the other hand, the external constraint is very strong, the interactions may be insufficient to oppose it and the result is a complete disorganization of the system ("runaway effects").

In many problems, it is not difficult to formulate the action of the first effect (as will be shown later). The main difficulty of the problem lies in the second point. However, the mechanism producing the reaction of the system is essentially the same as the one producing its free evolution toward equilibrium. Hence a thorough knowledge of this latter process enables one in principle to solve easily the problem of forced evolution, too. The purpose of the present chapter is to show how this connection can indeed be made quantitatively.

The problem of transport coefficients can be treated from three distinct (but connected) approaches, corresponding to three fields of investigation.

A. Irreversible thermodynamics

In this approach the general laws of evolution of the macroscopic quantities are assumed to be known. The purpose of the investigation is to establish general relationships between the coefficients entering the expressions of these laws (i.e. between the transport coefficients). This type of investigation lies outside the framework of this book. Extensive treatments of this type have been given elsewhere.*

B. Kinetic theory of gases

This approach represents what can be called the "traditional"

* I. Prigogine, *Introduction to the Thermodynamics of Irreversible Processes*, 2d ed., Interscience, New York, 1961; S. R. de Groot and P. Mazur, *Non-Equilibrium Thermodynamics*, North Holland Publ. Co., Amsterdam, 1962; R. Balescu, *Thermodynamics—Irreversible Processes*, Interscience Encyclopedia of Chemical Technology, Interscience, New York, 1960. The problem of transport coefficients in a plasma in the presence of a magnetic field has been studied by T. Kihara, *J. Phys. Soc. Japan*, **14**, 128 (1959).

microscopic theory of transport coefficients. In this picture, it is assumed that a kinetic equation describing the long-time behavior of the system is given. This means essentially that the irreversibility is built into the equations from the start. We shall call such a kinetic equation a *"Boltzmann-like"* equation to stress this irreversible feature. Typical examples of such equations are the proper Boltzmann equation, the Landau equation, and the ring equation. It is then shown that the phenomenological laws (Ohm's law, Fourier's law, etc.) can be derived as consequences of these equations, and the values of the transport coefficients can be calculated. This type of theory will be reviewed in § 56.

C. Statistical mechanics

In this approach one starts from first principles (i.e. the laws of mechanics) and one *derives* the kinetic equations. Alternatively, for the problem of forced evolution, one can derive directly microscopic expressions for the macroscopic quantities of interest (such as electric current, heat flow, etc.) and study directly the evolution of these quantities. This type of approach corresponds exactly to the spirit of this book and will be developed in some detail in the rest of this chapter.

§ 56. Traditional Theories of Transport Coefficients

It is not our purpose to give here detailed calculations of transport coefficients of a plasma. This subject has been treated extensively elsewhere * and there is no point in taking over these

* Basic expositions of these theories can be found in: S. Chapman and T. Cowling, *The Mathematical Theory of Non Uniform Gases*, Cambridge University Press, 1939; J. O. Hirschfelder, C. F. Curtiss and R. B. Bird, *Molecular Theory of Gases and Liquids*, John Wiley, New York, 1951; H. Grad, *Handbuch der Physik*, **12**, 205, Springer Verlag, Berlin, 1958; L. Waldmann, *ibid.*, p. 295.

Applications of these methods to plasmas have been given by: S. Chapman and T. Cowling, *loc. cit.*; R. S. Cohen, L. Spitzer, Jr., and P. McRoutly, *Phys. Rev.*, **80**, 230 (1950); L. Spitzer, Jr., and R. Härm, *Phys. Rev.*, **89**, 977 (1953); R. K. M. Landshoff, *Phys. Rev.*, **82**, 442 (1951); J. L. Delcroix, *Introduction to the Theory of Ionized Gases*, Interscience, New York, 1960; A. N. Kaufman in *La théorie des gaz neutres et ionisés*, Hermann, Paris, 1960.

results here. Our purpose here is mainly to review the general ideas of the method and to point out its difficulties. The method will be illustrated by a simple example in § 60.

The external constraints used in the study of transport coefficients are of two essentially distinct types. The first category includes the so-called *"dynamical constraints"*: these are produced by the action of an external field, such as an electric or a magnetic field, which may be constant in time or can be arbitrarily varied. This type of problem is mathematically quite clear: the external field can be simply described by adding a term into the hamiltonian of the system. Let us, for instance, consider the case of a uniform electric field in the subsequent discussion.

The second group of constraints are sometimes called rather improperly *"thermal constraints"*. They cannot be described by a modification of the hamiltonian. They are produced by imposing externally a certain type of inhomogeneity, for instance a constant temperature gradient, a constant velocity gradient, a density gradient, etc. Mathematically, this type of constraint is expressed *in this approach* by imposing a certain form on the distribution function. As a typical constraint of this type we shall discuss the case of a constant temperature gradient.

The next step consists in writing the equations of evolution. It is assumed that the equations describing the evolution of the system *in the absence of a constraint* are known. For a plasma, according to the degree of approximation desired, these are either the Landau equation (30.11) or the ring equation (41.19). These equations can be written in the general form:

$$\partial_t f = \mathscr{C}(f) \tag{56.1}$$

where $\mathscr{C}$ is a collision operator, appearing as a non-linear functional of f. The only essential property of $\mathscr{C}$ which is needed for the argument is that $\mathscr{C}$ describes an irreversible process, i.e. *an H-theorem can be proven* for eq. (56.1). Moreover, for (56.1) to be called a generalized Boltzmann equation, it should describe *instantaneous* collisions (see § 33).

Consider now the action of the constraint *in the absence of collisions*. This action can be described by the appropriate Liouville equation for one particle (because the constraint acts in the same way on all particles):

$$\partial_t f + \mathbf{v} \cdot \nabla f + m^{-1} \mathbf{F} \cdot \partial f = 0 \tag{56.2}$$

where **F** is the external force (if any). In order to describe the complete evolution of the system under the combined action of the collisions and the constraints *it is assumed* that one can simply superpose the two effects as follows:

$$\partial_t f + \mathbf{v} \cdot \nabla f + m^{-1} \mathbf{F} \cdot \partial f = \mathscr{C}(f) \tag{56.3}$$

It is moreover *assumed* that if the system is inhomogeneous ($\nabla f \neq 0$), the collision operator has the local property, i.e. the various functions f occurring in it are all evaluated at the same space point.

The calculation of transport coefficients now proceeds as follows. It is *assumed* that the effect of the external constraint is sufficiently weak for a linear theory to be a good description of the phenomena. This assumption also ensures the existence of a stationary non-equilibrium state. Let us follow the argument separately for the case of an electric field and a temperature gradient.

We consider first the case of an external electric field **E** which we assume for simplicity to be constant in space and in time. In this problem the model of a one-component electron gas neutralized by a positive background is not adequate. The reason is that in this model there is conservation of the average velocity (§ 46); hence nothing opposes the flow of an electron current, and the conductivity is infinite. In order to obtain a finite resistivity we need at least two dynamical components: in that case a friction of the electrons relative to the ions appears and hence a transfer of momentum between the two components occurs which limits the electric current. We therefore assume that the plasma consists of *electrons*, characterized by a velocity distribution $\varphi^e(\mathbf{v}; t) \equiv \varphi(\mathbf{v}; t)$, a mass m and a valence $z_e = -1$, and of *positive ions* whose velocity distribution is $\varphi^i(\mathbf{v}; t) \equiv \Phi(\mathbf{v}; t)$, their mass being denoted by M and their valence by $z_i = +1$. Because of the overall electrical neutrality we have the relation $c_i = c_e = \frac{1}{2}c$. The "Boltzmann equation" for the electrons is:

$$\partial_t \varphi - (e/m)\mathbf{E} \cdot \partial \varphi = \mathscr{C}(\varphi, \Phi) \tag{56.4}$$

A similar equation holds for the ions.

In the absence of **E**, the stationary solution is given by

$$\mathscr{C}(\varphi, \Phi) = 0$$

and we know (*H*-theorem) that the unique solution of this equation is the equilibrium distribution for both components: $\varphi^{(\sigma)} = \varphi^{(\sigma)0}$, $\sigma = i, e$. In the presence of a small electric field the steady distribution is close to φ^0, the deviation being assumed linear in **E**. Moreover, as the only vectors appearing in the problem are **E** and **v**, the only possible scalar linear in **E** is $\mathbf{E} \cdot \mathbf{v}$ times a function of v. It can thus be assumed that the steady-state distributions are of the form:

$$\varphi_s^{(\sigma)}(\mathbf{v}) = \varphi^{(\sigma)0}(v)[1 + \mathbf{E} \cdot \mathbf{v}\psi^{(\sigma)}(v)] \tag{56.5}$$

Substituting this result into (56.4) and linearizing the equation in **E** we obtain the following equation for the steady state $(\partial_t \varphi_s^{(\sigma)} = 0)$:

$$-(e/m)\mathbf{E} \cdot \partial \varphi^0 = \mathbf{E} \cdot \mathscr{C}'(\varphi^0, \Phi^0, \psi^e, \psi^i) \tag{56.6}$$

where the vector $\mathscr{C}'$ is linear in the $\psi^{(\sigma)}$. Equation (56.6) can be solved in principle by standard methods of linear mathematics. Once the solution is known, we can calculate the relevant flow occurring in the steady state, in this case the electric current defined as:

$$\mathbf{J}_s = \sum_{\sigma = i, e} z_\sigma c_\sigma e \int d\mathbf{v}\, \mathbf{v} \varphi_s^{(\sigma)}(\mathbf{v}) = \tfrac{1}{2} ce \sum_\sigma z_\sigma \int d\mathbf{v}\, \mathbf{v}\mathbf{E} \cdot \mathbf{v} \psi^{(\sigma)}(v) \varphi^{(\sigma)0}(v) \tag{56.7}$$

Comparing this expression with the phenomenological Ohm's law:

$$\mathbf{J}_s = \boldsymbol{\sigma} \cdot \mathbf{E} \tag{56.8}$$

we identify the *electrical conductivity tensor* $\boldsymbol{\sigma}$ as:

$$\boldsymbol{\sigma} = \tfrac{1}{2} ce \sum_\sigma z_\sigma \int d\mathbf{v}\, \mathbf{v}\mathbf{v} \psi^{(\sigma)}(v) \varphi^{(\sigma)0}(v) \tag{56.9}$$

A simple example of an explicit calculation of this type is given in § 60.

The case of a thermal gradient is treated in a similar way*, but the argument runs somewhat differently. The starting equation is:

$$\partial_t f + \mathbf{v} \cdot \nabla f = \mathscr{C}(f) \tag{56.10}$$

If the gradient is weak, the Chapman–Enskog expansion, which we have already introduced in § 32, can be applied. Writing the stationary solution in the form:

$$f_s = f^0 + f^{(1)} \tag{56.11}$$

the stationary solution in zeroth approximation is again given by

$$\mathscr{C}(f^0) = 0$$

However, as discussed in § 34, the inhomogeneous character of the system allows for a richer variety of acceptable solutions: f^0 is the *local equilibrium* distribution, i.e. a Maxwell distribution with possibly space-dependent density, velocity and temperature. It is at this point that the external constraint can be formulated. In the present case, one demands that f^0 is expressed in terms of constant density and velocity, but with a temperature which depends linearly on the position in a given direction ($\nabla T = \text{const.}$). Assuming now that the deviation $f^{(1)}$ from local equilibrium is linear in the gradient, it must necessarily be of the form

$$f^{(1)} = \nabla T \cdot \mathbf{v} \phi(v) \tag{56.12}$$

One now substitutes (56.11) and (56.12) into (56.10), linearizes with respect to ∇T and looks for a stationary solution $f^{(1)}$:

$$(\partial f^0 / \partial T)\mathbf{v} \cdot \nabla T = \nabla T \cdot \mathscr{C}''(f^0, \phi) \tag{56.13}$$

where $\mathscr{C}''(f^0, \phi)$ is again linear in ϕ. With the solution of this equation, one can calculate the heat flow in the steady state:

$$\mathbf{Q}_s = \tfrac{1}{2} c^{-1} \int d\mathbf{v}\, \mathbf{v} v^2 f_s = \tfrac{1}{2} c^{-1} \int d\mathbf{v}\, \mathbf{v} v^2 \mathbf{v} \cdot \nabla T \phi(v) \tag{56.14}$$

Comparing it with Fourier's law:

$$\mathbf{Q}_s = \boldsymbol{\lambda} \cdot \nabla T \tag{56.15}$$

the *thermal conductivity tensor* $\boldsymbol{\lambda}$ is obtained in the form:

$$\boldsymbol{\lambda} = \tfrac{1}{2} c^{-1} \int d\mathbf{v}\, \mathbf{v}\mathbf{v} v^2 \phi(v) \tag{56.16}$$

* In this problem the one-component model is sufficient.

Now that we have reviewed the ideas underlying the calculation of transport coefficients from a "Boltzmann-like" equation, we shall consider the difficulties which arise in this method. We want to discuss mainly the difficulties of principle, leaving aside the technical difficulties connected with the resolution of the equations. The main problems can be classified into four groups.

A. The rigorous molecular definition of a thermal constraint

We shall just mention this point here, but shall discuss it in the next section, after an elaboration of some additional mathematical results.

B. The correlational part of transport coefficients

Besides the contribution of the velocity distribution, there is usually also a contribution of the correlation function to the transport coefficient. In plasmas this contribution is important outside the ring approximation. For instance, in the problem of the electrical conductivity, the field felt by the particles is not really the external field. It is rather an effective field including polarization effects which are due to intermolecular correlations. Also, the general heat flow is defined more completely as the transport of kinetic energy (56.14) but also of potential energy. The definition of the latter involves the correlation function $g(r)$.

Of course, the traditional kinetic-equation approach cannot handle this problem, because it only gives the velocity distribution $\varphi_s(\mathbf{v})$ in the stationary state. The extension of the kinetic equations necessary for the treatment of the correlational part of the transport equations will be discussed in Appendix 11.

C. The problem of strong fields

Let us assume that the *free* evolution of a homogeneous system is correctly described to the desired approximation by an equation of the type (56.1). Under what conditions can the *forced* evolution be described by a simple superposition of collisions and constraint, as in eq. (56.3)? It is felt intuitively that such a superposition is valid whenever the external field is sufficiently small and when we restrict our interest to phenomena which are linear in the field. This intuition can indeed be justified by a rigorous calculation. However, when we are interested in the effect of strong fields, and wish to investigate the behavior of a plasma in regions where the linear Ohm's law (say) is no longer valid, an equation of type (56.4) may be incorrect. Indeed, the latter equation implies that the collision process is calculated *as if the field were absent.* This is clearly acceptable only if the field does not significantly affect the motion of the particles over a Debye length.

More precisely, the field intensity as compared with the average thermal energy can be measured by introducing the characteristic frequency ω_E defined as follows:

$$\omega_E = eE/mv_{\text{th}} = eE\sqrt{\beta/m} \tag{56.17}$$

The criterion for the applicability of eq. (56.4), is obtained by assuming that the field does not appreciably disturb the motion during a period of the plasma oscillation (i.e. during an effective "duration of a collision"):

$$\omega_E/\omega_p = \tfrac{1}{2}E\sqrt{\beta/\pi c} \ll 1$$

This criterion introduces a critical electric field, defined as

$$E_c = 2\sqrt{\pi c/\beta} \tag{56.18}$$

For a typical thermonuclear plasma this critical field is enormous; for instance, for $T = 10^6$ °K and $c = 10^{18}$, $E_c \sim 1.2\times 10^7$ volts/cm, but for colder and more dilute plasmas this critical field decreases rapidly; for instance, at $T = 10^4$ °K and $c = 10^{12}$ the plasma has the same value of Γ but its critical field is 1200 volts/cm.

In the region of very strong fields, one actually expects an equation of the form

$$\partial_t\varphi-(e/m)\mathbf{E}\cdot\partial\varphi = \mathscr{D}(\varphi, \Phi; \mathbf{E}) \tag{56.19}$$

where the new collision operator $\mathscr{D}$ *depends on the electric field.* The normal collision operator then appears as the zeroth-order term of a Taylor expansion of $\mathscr{D}(\varphi, \Phi; \mathbf{E})$ in powers of the field.

The correct procedure for deriving such an equation appears

quite naturally within the framework of the theory developed here. Two alternatives are possible.

(*a*) *If the electric field is sufficiently simple* (in its dependence on space and time), the equation of motion of free particles in the field can be solved exactly. The Green's function of eq. (56.2), which now includes the effect of the external field exactly, can then be found. One then starts afresh the perturbation theory of Chapter 1, replacing everywhere the free motion Green's function $G^0(xvt|x'v't')$ by $G_E^0(xvt|x'v't')$.

(*b*) *If the field has a complicated space and time dependence* a double perturbation expansion considering both the interactions and the field as perturbations is used. One then has new diagrams in which some vertices represent the action of the field.

Recent progress has been achieved in this field.* In particular, G. Severne has shown that eq. (56.19) is very important even in the *linear* region of small fields. An account of this work will be presented in Appendix 11.

A related problem is the well known "*runaway effect*". This is a typical phenomenon which illustrates the fact that in strong electric fields the interactions are not sufficiently effective in organizing the system, and the electrons are eventually accelerated away by the field. The runaway effect has been extensively studied in the literature, in the first place by Dreicer.** In all these studies (which we shall not reproduce here), the starting point is essentially eq. (56.4). Dreicer's main result is that runaway sets in [i.e. there exists no stationary solution of eq. (56.4)] when the electric field is larger than a critical runaway field E_R:

$$E_R = 2\pi e^3 c\beta/B$$

where B is defined by (35.5). Comparing this field with the critical field E_c we see that

$$E_R/E_c = (\pi^{1/2}/B)\Gamma^{2/3}$$

* V. P. Silin, *J. Exptl. Theoret. Phys. U.S.S.R.*, **38**, 1771 (1960), transl. *Sov. Phys.* (*JETP*), **11**, 1277 (1960); G. Severne, *Physica*, **29** (1963), in press.

** H. Dreicer, *Phys. Rev.*, **115**, 238 (1959).

Hence, for a plasma with small Γ, this ratio is very small and the effect occurs in a region where the linear equation (56.4) is still valid. Dreicer's calculations are thus well justified. This shows that care must be taken in analyzing non-linear effects. Equation (56.4) is linear in **E**, but its *solution* is not; moreover, the non-linear solution of this equation can be significant even for fields for which the non-linear equation (56.19) need not be fully retained.

D. The problem of large Γ

We now consider a different problem. We assume that the field is sufficiently weak for a "superposition procedure" to be justified once the homogeneous kinetic equation is valid. We concentrate our attention on the latter equation. It has been shown in the previous chapter that both in the Landau approximation and in the ring approximation one can derive an equation of the form (56.1) in which the collision operator $\mathscr{C}(\varphi)$ has all the properties needed for the validity of the theory. One may wonder what happens in the case of plasmas for which Γ is no longer small enough to allow the retention of cycles or rings alone. A result of all our previous discussions is that the "Boltzmann" property of the ring equation results from the fact that the operator $\mathscr{C}(\varphi)$ is a sum of diagonal fragments in which only the leading universal part has been retained. A natural generalization of the ring equation along these lines would be to add to the collision term a set of other diagonal fragments, corresponding to higher orders in Γ, *all* these diagrams being evaluated in the leading universal approximation. For instance, the next approximation after the rings would involve the diagonal fragments shown in Fig. 50.1. It can be shown that the most general equation constructed in this way has all the properties needed for the application of the above theory.* It is therefore natural to define the *general Boltzmann-like equation* as an equation of the type

* P. Résibois, F. Henin and F. C. Andrews, *J. Math. Phys.*, **2**, 68 (1961); R. Balescu, in *Lectures in Theoretical Physics*, Summer Inst. for Theor. Phys, Univ. of Colorado, Interscience, New York, 1961.

(56.1) in which $\mathscr{C}(\varphi)$ *is defined as the sum of all diagonal fragments evaluated in the leading universal approximation.* If such an equation were valid, the transport coefficients could be calculated to all possible orders of approximation by using the method outlined above.

However, we know already that such an equation does *not* describe the evolution of the system adequately. We have shown in § 50 that as soon as one goes beyond the ring approximation, one must include non-dominant universal and non-universal contributions. This admixture is necessary because the "duration of a collision" and the relaxation time can no longer be considered as widely separated quantities (one is precisely interested in effects of the order t_p/t_r). The added terms take more and more closely into account the essentially non-markoffian character of the "fine-grained" description of the evolution * (see also Appendix 8).

Thus the most general kinetic equation is definitely not a Boltzmann-like equation. This immediately raises the question of the validity of the extension of the traditional methods of calculating transport coefficients to any higher order beyond the ring approximation. Our main result of the next section will be the proof that, notwithstanding the difficulties mentioned here, the kinetic part of the transport coefficients for a stationary field can be calculated *rigorously* by the traditional method applied to the general Boltzmann-like equation defined above. This result will be extended in Appendix 11 to cover the correlational part too.

§ 57. A General Formula for the Electric Current

The previous discussion shows the necessity of a deeper investigation of the problem of forced evolution and of the transport coefficients. Following a preliminary investigation of M. Green,** the first complete formal theory of this type was put forward

* I. Prigogine, *Non-Equilibrium Statistical Mechanics,* Interscience, New York, 1963.

** M. S. Green, *J. Chem. Phys.*, **19**, 1036 (1951).

by Kubo,* and has been followed by many others.** The derivation of the general expression for the current given here is inspired by Kubo's work. Edwards and Sanderson† independently gave a similar treatment.

We start afresh from the Liouville equation, but now consider a two-component plasma consisting of N_e electrons and N_i ions ($N_i = N_e = \frac{1}{2}N$), placed in an external electric field. The Liouville equation is then written as

$$\Lambda f_N(xv; t) = 0 \tag{57.1}$$

where the Liouville operator is now defined as:

$$\Lambda = \mathscr{L} + e\mathscr{L}^e \tag{57.2}$$

$$\mathscr{L} = \mathscr{L}^0 + e^2\mathscr{L}' = \partial_t + \sum_{j=1}^{N} \mathbf{v}_j \cdot \nabla_j - e^2 \sum\sum_{j<n} (\nabla_j V_{jn}) \cdot (m_j^{-1}\partial_j - m_n^{-1}\partial_n) \tag{57.3}$$

$$\mathscr{L}^e = \mathbf{E} \cdot \sum_{j=1}^{N} z_j m_j^{-1} \partial_j \tag{57.4}$$

$\mathscr{L}$ is the Liouville operator of the system in the absence of an external field (also called the *internal* Liouville operator), i.e. the operator studied in Chapter 1, whereas $\mathscr{L}^e$ represents the action of an external field. It is assumed for simplicity that $\mathbf{E}$ is uniform in space but may vary in time.

We now want to describe the following situation. The system is left to itself in the absence of an electric field from $t = -\infty$ up to time 0. At that time the system has reached thermal equilibrium. At time 0 the electric field is switched on. In order to study the reaction of the plasma it therefore appears natural to

* R. Kubo, *J. Phys. Soc. Japan,* **12**, 570 (1957); see also: R. Kubo, in *Lectures in Theoretical Physics,* Summer Inst. for Theor. Phys., Univ. of Colorado, vol. 1, Interscience, New York, 1959; L. van Hove, in *La Théorie des Gaz Neutres et Ionisés,* Ecole d'Eté de Phys. Théor., Les Houches, Hermann ed., Paris, 1960.

** See, for instance, H. Mori, *J. Phys. Soc. Japan,* **11**, 1029 (1956); W. Kohn and J. M. Luttinger, *Phys. Rev.,* **108**, 590 (1957); E. W. Montroll and J. C. Ward, *Physica,* **25**, 423 (1959); J. A. McLennan, Jr., *Phys. of Fluids,* **3**, 493 (1960); S. Fujita and R. Abe, *J. Math. Phys.,* **3**, 350, 359 (1962); and many others.

† S. F. Edwards and J. J. Sanderson, *Phil. Mag.,* **6**, 71 (1961).

solve the Liouville equation (57.1) with the initial condition

$$f_N(x, v; 0) = f_N^0(x, v) = \alpha e^{-\beta H_I} \tag{57.5}$$

where H_I is the complete hamiltonian of the system in the absence of an external field; f_N^0 is the equilibrium distribution. Let us introduce the retarded Green's function of the operator Λ as the solution of the equation

$$\Lambda \Gamma(xvt|x'v't') = \delta(x-x')\delta(v-v')\delta(t-t') \tag{57.6}$$

obeying the proper boundary conditions and the causal condition

$$\Gamma(xvt|x'v't') = 0 \quad \text{for} \quad t < t' \tag{57.7}$$

If the electric field is constant in time, $\Gamma(xvt|x'v't')$ depends on its time variables only through $t-t'$, but in the general case it may depend arbitrarily on t and t'. The solution of our initial-value problem is then obtained by applying eq. (4.16):

$$f_N(x, v; t) = \int dx'dv'\, \Gamma(xvt|x'v'\,0) f_N^0(x'v') \tag{57.8}$$

Let us now calculate the current. The microscopic definition of this quantity is obviously:

$$\mathbf{J}(t) = e \int dx\, dv \sum_m z_m \mathbf{v}_m f_N(x, v; t) \tag{57.9}$$

Substituting eq. (57.8) this becomes:

$$\mathbf{J}(t) = e \sum_m z_m \int dx\, dv \int dx'\, dv'\, \mathbf{v}_m \Gamma(xvt|x'v'0) f_N^0(x'v') \tag{57.10}$$

We now apply a perturbation to Γ, exactly as in § 5. However, we now consider the system of *interacting* electrons in the absence of the field as the unperturbed quantity, and the effect of the field as the perturbation. To the decomposition (57.2) corresponds a splitting of the Green's function:

$$\Gamma = \mathcal{G} + \Gamma^e \tag{57.11}$$

where $\mathcal{G}$ is the Green's function of the interacting plasma in the absence of the field, i.e. the central quantity studied in the previous chapters, and defined by eqs. (4.8–9). We start the perturbation theory as was done in § 5, using the integral equation:

$$\Gamma(xvt|x''v''t'') = \mathscr{G}(xvt|x''v''t'') - \int dx'\,dv' \int_0^t dt'\,\mathscr{G}(xvt|x'v't')e\mathscr{L}^{e'}\Gamma(x'v't'|x''v''t'') \tag{57.12}$$

where $\mathscr{L}^{e'}$ is the perturbation operator $\mathscr{L}^e$ acting on the primed variables. Hence, using (57.10), we obtain a general formula for the current:

$$\begin{aligned}\mathbf{J}(t) = e\sum_m z_m \sum_{\nu=1}^{\infty} (-e)^{(\nu-1)} \int dx_0 dv_0 \ldots dx_\nu dv_\nu \int_0^t dt_\nu \ldots \int_0^{t_3} dt_2 \int_0^{t_2} dt_1 \\ \mathbf{v}_m \mathscr{G}(xvt|x_\nu v_\nu t_\nu)\mathscr{L}^{e(\nu)}\mathscr{G}(x_\nu v_\nu t_\nu|x_{\nu-1} v_{\nu-1} t_{\nu-1})\mathscr{L}^{e(\nu-1)} \ldots \\ \ldots \mathscr{L}^{e(1)}\mathscr{G}(x_1 v_1 t_1|x_0 v_0 0)f_N^0(x_0 v_0)\end{aligned} \tag{57.13}$$

This is a quite general formula expressing the current in terms of the internal Green's function of the system and the constraint $\mathscr{L}^e$. If the Green's function is itself expanded in powers of the interaction, we obtain a double perturbation series suitable for the study of strong field effects [see § 56, remark $C(b)$]. We shall, however, be interested here only in linear transport theory, and shall therefore retain only the first order contribution to $\mathbf{J}(t)$

$$\begin{aligned}\mathbf{J}(t) = e\sum_m z_m \int dx\,dv \int dx''\,dv''\,\mathbf{v}_m \Big\{ \mathscr{G}(xvt|x''v''0)f_N^0(x''v'') \\ -e\int dx'\,dv' \int_0^t dt'\,\mathscr{G}(xvt|x'v't')\mathscr{L}^{e'}\mathscr{G}(x'v't'|x''v''0)f_N^0(x''v'')\Big\}\end{aligned} \tag{57.14}$$

We now use the fact that $f_N^0(xv)$ is the (internal) equilibrium distribution, i.e. a *stationary solution* of the internal Liouville equation. Through formula (4.16), this implies

$$\int dx''\,dv''\,\mathscr{G}(xvt|x''v''0)f_N^0(x''v'') = f_N^0(xv) \tag{57.15}$$

Noting also that f_N^0 is an even function of the velocities, the first term in (57.14) vanishes. We note moreover that

$$e\mathscr{L}^e f_N^0(xv) = -e\beta\mathbf{E}\cdot\sum_n z_n \mathbf{v}_n f_N^0(xv)$$

Formula (57.14) then reduces to

$$\mathbf{J}(t) = e^2\beta \sum_m \sum_n z_m z_n \int dx\,dv \int dx'\,dv' \int_0^t dt'\,\mathbf{v}_m \mathscr{G}(xvt|x'v't')\mathbf{E}(t') \cdot \mathbf{v}'_n f_N^0(x'v') \tag{57.16}$$

This formula has been derived independently by Edwards and Sanderson* and by the author.** We can now go a step further, noting that the time integral is of the form:

$$\int_0^t dt' f(t-t')g(t')$$

(Indeed, $\mathscr{G}(t|t')$ is a function of $t-t'$ alone). Hence we can apply the convolution theorem of Laplace transforms. We introduce the Laplace transforms of the current and of the field:

$$\mathbf{J}(t) = (2\pi)^{-1}\int_C dz\, \mathrm{e}^{-izt}\mathbf{j}(z) \tag{57.17}$$

$$\mathbf{E}(t) = (2\pi)^{-1}\int_C dz\, \mathrm{e}^{-izt}\mathbf{e}(z) \tag{57.18}$$

We can now introduce the resolvent operator defined by eq. (4.20) to write the Laplace transform of eq. (57.16) as†

$$\mathbf{j}(z) = e^2\beta \sum_m \sum_n z_m z_n \int dx\, dv \int dx'\, \mathbf{v}_m \langle x|\mathscr{R}(z)|x'\rangle \mathbf{e}(z) \cdot \mathbf{v}_n f_N^0(x', v)$$

(Remember $\langle x|\mathscr{R}(z)|x'\rangle$ is an *operator* acting on v, see § 5.) Going over now to the Fourier representation (and noting that the functions to the left of $\langle x|\mathscr{R}(z)|x'\rangle$ are x-independent):

$$\mathbf{j}(z) = e^2\beta \sum_m \sum_n z_m z_n \sum_k \int dv\, \mathbf{v}_m \langle 0|\mathscr{R}(z)|k\rangle \mathbf{e}(z) \cdot \mathbf{v}_n \tilde{\rho}_k^0(v) \tag{57.19}$$

* *loc. cit.*

** R. Balescu, *Physica,* **27**, 693 (1961).

† Note that the general formula (57.13) can, in general, not be Laplace-transformed in this way. If the field is time-dependent, the integral is of the form:

$$\int dt_\nu \int dt_{\nu-1} \ldots F(t-t_\nu)g(t_\nu)F(t_\nu-t_{\nu-1})g(t_{\nu-1}) \ldots$$

and this is not a convolution. Only for the case of *static* fields, when $g(t_\nu)$ is time-independent, does this expression reduce to a convolution. Only in this case can one use a resolvent method. This is, however, not a serious drawback, because the theory can be just as easily worked out in terms of Green's functions. The fact that the *linear* theory can be done with a resolvent method *even for time-dependent fields* is easily seen to be a consequence of the choice of f_N^0 as initial condition [through property (57.15)].

This formula may serve as a starting point for the expansion of the electrical conductivity in a series of concentration and charge, by a straightforward application of the diagram technique. Our purpose here, however, is to go further in the study of the general behavior of plasmas in the presence of constraints. Before proceeding, we shall make two remarks.

1. Separate meanings can be given to the terms with $k = 0$ and $k \neq 0$. The former do not involve any correlation function and will clearly give rise to the kinetic part of the electrical conductivity. The other terms give rise to the potential part of the transport coefficients, determined by two-body, three-body, . . ., correlations.

2. We are now able to come back to the point raised in § 56-A. It is immediately obvious that the method described here in detail for the electrical conductivity problem can be extended immediately to any other dynamical transport coefficient, by an appropriate choice of the perturbation $\mathscr{L}^e$. One may wonder if equations of the type (57.16) can be derived for the study of *thermal* transport coefficients. Several authors have attempted such an extension.* The essential idea in these papers is to define an effective hamiltonian which describes the thermodynamic flows. Such a concept is, however, in our opinion, quite artificial and does not describe the real meaning of the process. The problem is really an *initial-value problem* for an inhomogeneous system without any external perturbation. One should, for instance, express an initial situation in which there are two large regions in which the distribution function has values compatible with homogeneous temperatures T_1 and T_2; these regions are separated by a small region with arbitrary temperature distribution. The latter is the region of interest. In the course of time one would expect the system to tend rapidly ($t \sim t_r$) toward a state involving a quasi-steady heat flow through the intermediate region. For times much longer than t_r the thermostats will eventually equalize their temperatures and the whole system will reach equilibrium. The mathematical formulation of this problem

* M. Green, R. Kubo, H. Mori, J. McLennan, Jr. (see references p. 269) and others.

is, however, difficult, and has not yet been done. As long as such a problem remains to be solved, one cannot say that the theory of thermal transport coefficients is as well founded as the theory of dynamical coefficients.

§ 58. Existence and Stability of a Stationary State

From now on we consider an electric field $\mathbf{E}(t)$ which is switched on at $t = 0$ and is constant for $t > 0$. Its Laplace transform is

$$\mathbf{e}(z) = \mathbf{E}/(-iz) \tag{58.1}$$

Moreover, in the present paragraph, only the term corresponding to $k = 0$ in eq. (57.19) will be discussed. The argument can be extended very easily to the other terms, but this will not be done explicitly here (see Appendix 11). Equation (57.19) becomes, with these restrictions:

$$\mathbf{j}_K(z) = \mathbf{E} e^2 \beta \sum_m \sum_n z_m z_n \int dv\, \mathbf{v}_m \left\{ \frac{1}{-iz} \langle 0|\mathscr{R}(z)|0\rangle \right\} \mathbf{v}_n \tilde{\rho}_0^0(v) \tag{58.2}$$

According to the general philosophy of the resolvent method (and, more generally, of the Laplace transform method), information can be gained about the time dependence of the current by studying the nature and the location of the singularities of $\mathbf{j}_K(z)$. This means that the essential part of our investigation concerns the bracketed expression of eq. (58.2).

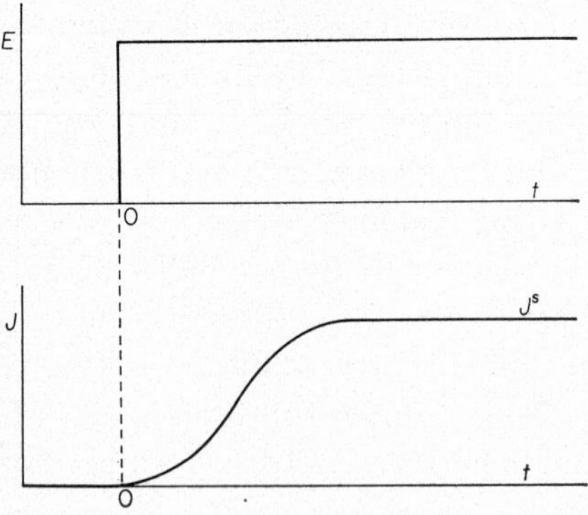

Fig. 58.1. The general relation between field and current.

Physically, in the situation under consideration, it is expected that the general type of time dependence of the current will be of the kind shown in Fig. 58.1. After a transient period, the current reaches a stationary value $\mathbf{J}^s$. Only in that stationary state can the situation be described uniquely by an electrical conductivity coefficient which expresses proportionality between $\mathbf{J}^s$ and $\mathbf{E}$. In the transient region, the behavior is much more complex and can no longer be described by a single constant.

The first property we have to prove is the *existence of a stable stationary state for a system of interacting particles.* In order to prove this statement, it is necessary to prove the following simultaneous properties of the quantity $\mathscr{A}(z)$ defined as:

$$\mathscr{A}(z) = \frac{1}{-iz} \langle 0|\mathscr{R}(z)|0\rangle \tag{58.3}$$

A) *$\mathscr{A}(z)$ has a simple pole at $z = 0$.*

B) *$\mathscr{A}(z)$ has no singularities in the upper half-plane and no poles on the real axis except at $z = 0$.*

A straightforward application of the main properties of Laplace transforms reviewed in Appendix 1 shows that $\mathbf{J}(t)$ is then the sum of a constant term and of several decaying terms. The decay of the latter is exponential if the only singularities of $\mathscr{A}(z)$ are poles in the lower half-plane, and has a more complicated form for other singularities (such as branch points, see § 28).

It has been stated previously that the most general diagram contributing to $\langle 0|\mathscr{R}(z)|0\rangle$ is a succession of an arbitrary number of diagonal fragments. Let us introduce the notation $\Psi^{(+)}(z)$ to denote the sum of all possible diagonal fragments, connected or disconnected. The 0–0 matrix element of the internal resolvent is then obtained from eq. (5.26)

$$\langle 0|\mathscr{R}(z)|0\rangle = \sum_{n=0}^{\infty} \frac{1}{-iz} \left[\Psi^{(+)}(z) \frac{1}{-iz}\right]^n \tag{58.4}$$

$\Psi^{(+)}(z)$, being defined as a Cauchy integral, is a regular function of z in the upper half-plane, and is assumed to possess an analytical continuation into the lower half-plane. It is seen from (28.4) that property A is certainly *not* true for the individual terms of the series. The existence in $\mathscr{A}(z)$ of terms in z^{-2}, z^{-3}, z^{-4}, . . . implies the existence in $\mathbf{J}_K(t)$ of terms growing as t, t^2, t^3, ... In particular, for a system of *free* electrons, $\Psi^{(+)}(z) = 0$ and $\mathscr{A}(z) = z^{-2}$. Therefore $\mathbf{J}_K(t) \sim t$, as would be expected. There is no stationary state for a system of free electrons, because they are uniformly accelerated by the field.

However, the r.h.s. of eq. (58.4) can be summed formally, being simply a geometrical progression. The fact that $\Psi^{(+)}(z)$ is really an operator is irrelevant here; the expressions can always be given a meaning if one introduces eigenfunctions of that operator. Performing the summation leads to

$$\mathscr{A}(z) = \frac{1}{-z} \frac{1}{z - i\Psi^{(+)}(z)} \tag{58.5}$$

This formula proves the property A. The proof of the second property is much more difficult in the general case. It rests upon the study of the roots of the "dispersion equation"

$$z - i\Psi^{(+)}(z) = 0 \tag{58.6}$$

We give a proof based on perturbation theory. It must be realized that, if expanded in powers of e^2, the operator $\Psi^{(+)}$ begins with a term in e^4 (because the smallest possible diagonal fragment has two vertices). A straightforward perturbational solution of (58.6) therefore leads to the root $\bar{z}$:

$$\bar{z} = i\Psi^{(+)}(0) + i^2[d\Psi^{(+)}(z)/dz]_{z=0}\Psi^{(+)}(0) + \ldots \tag{58.7}$$

The first term is at most of order e^4, the second one at most of order e^8, etc. But $\Psi^{(+)}(0)$ is precisely the "Boltzmann-like" evolution operator discussed in § 56 [see e.g. the dominant term of eq. (27.3)]. If the system obeys an H-theorem, $\Psi^{(+)}(0)$ is a self-adjoint operator which brings the system irreversibly to equilibrium. It has one zero eigenvalue, all the others being negative.*

* The proof of the stability of the non-equilibrium stationary state is thus reduced to the proof of the H-theorem. The latter itself depends crucially on the analytical properties of the resolvent. Whereas in "normal" situations a general H-theorem has been proven by Prigogine in his monograph quoted above, certain special situations can exist in which the requirements of the H-theorem are not fulfilled. A notable example is the phenomenon of superconductivity.

The eigenfunction corresponding to the eigenvalue zero is of course the equilibrium distribution ρ_0^0. As in formula (58.2) it acts on $\mathbf{v}_m \rho_0^0$, which is *not* its zero eigenfunction, $\Psi^{(+)}(0)$ can be considered as a real negative number. If the perturbation series (58.7) converges, $\bar{z}$ will be in the neighborhood of $i\Psi^{(+)}(0)$, and the corresponding pole of $\mathscr{A}(z)$ lies in the lower half-plane.

The dispersion equation (58.6) has another set of roots besides $\bar{z}$. In effect, it has been stated above that $\Psi^{(+)}(z)$ has poles in the lower half-plane. These poles z_j are located in general at a finite distance from the real axis, their imaginary part being of order τ_c^{-1}, where τ_c is the duration of a collision (see §§ 27–28). Assuming that these poles are simple, we expand $\Psi^{(+)}(z)$ as a sum of residues at these poles plus a regular term $\psi^{(+)}(z)$:

$$\Psi^{(+)}(z) = \sum_j \frac{1}{z - z_j} (\text{Res}\ \Psi^{(+)})_{z=z_j} + \psi^{(+)}(z) \tag{58.8}$$

Substituting into eq. (58.6) and solving again by perturbation theory, it is seen that each pole of $\Psi^{(+)}$ generates a root $\bar{z}_j$ of the dispersion equation:

$$\bar{z}_j = z_j + i(\text{Res}\ \Psi^{(+)})_{z=z_j} + \dots \tag{58.9}$$

The first term is independent of e^2, the remaining ones being at least proportional to e^4. Again, if the series converges, $\bar{z}_j$ will be in the neighborhood of z_j, and therefore in the lower half-plane.

This achieves the proof of the existence of a stationary state and of the tendency of the system to approach it effectively. The second point is really a proof of stability. It should be mentioned here that whereas the system has been found stable under rather general conditions in this situation, this stability is not expected in general in a non-linear theory. It has already been stated in § 55 that for strong enough fields there appears a "run-away" phenomenon, because of the inefficiency of the interactions in preventing acceleration.

§ 59. Stationary State vs. Kinetics of Approach

Having found a stationary current, it is an easy matter to calculate the conductivity. The stationary current is expressed by formula (58.2) in terms of the residue of $\mathscr{A}(z)$ at $z = 0$, which is easily evaluated from (58.5). The result is

$$\mathbf{J}_K^s = \mathbf{E} \cdot e^2\beta \sum_m \sum_n z_m z_n \int dv\, \mathbf{v}_m \frac{1}{-\Psi^{(+)}(0)} \mathbf{v}_n \tilde{\rho}_0^0(v) \tag{59.1}$$

The electrical conductivity tensor is simply the coefficient of **E** on the r.h.s. of eq. (59.1) [see eq. (56.8)]. Before discussing this formula in more detail, let us write down the form of the complete time-dependent current. Combining formulae (57.17), (58.2), (58.7) and (58.9) we obtain, by taking residues:

$$\mathbf{J}_K(t) = \mathbf{J}_K^s + \mathbf{E} \cdot e^2\beta \sum_m \sum_n z_m z_n \int dv\, \mathbf{v}_m B(t) \mathbf{v}_n \tilde{\rho}_0^0(v) \tag{59.2}$$

$$\begin{aligned} B(t) &= \mathrm{e}^{-i\bar{z}t} \frac{1}{-i\bar{z}} \left[\mathrm{Res} \frac{1}{z - i\Psi^{(+)}(z)} \right]_{z=\bar{z}} \\ &\quad + \sum_j \mathrm{e}^{-i\bar{z}_j t} \frac{1}{-i\bar{z}_j} \left[\mathrm{Res} \frac{1}{z - i\Psi^{(+)}(z)} \right]_{z=\bar{z}_j} \end{aligned} \tag{59.3}$$

The separation in (59.2) obviously corresponds to the stationary current plus the transient part of the current. The latter is in general a complicated decaying function (because $\bar{z}$ and $\bar{z}_j$ are complicated operators).

We can now clearly make the point with our discussion of § 56-D. It was stated there that the approach to equilibrium is described by three types of operators:

the leading universal terms, giving rise to the general Boltzmann-like operator $\Psi^{(+)}(0)$: it describes the action of instantaneous collisions;

the non-dominant universal terms, describing non-instantaneous collisions;

the non-universal terms, describing a non-markoffian dynamical evolution.

The same three types of evolution can be identified in eqs. (59.2–3). The second type of term of eq. (59.3) stems from the poles of $\Psi^{(+)}(z)$, and is thus responsible for the non-markoffian behavior. The first term of that equation stems from the markoffian (universal) contributions and describes the effect of "instantaneous" and "non-instantaneous" collisions. We have seen that in general systems the time scales t_r and t_p are not widely separated; as a result the various exponentials of eq. (59.3) decay at about the same rate.

It is, however, most remarkable that, even if these time scales are mixed, *the kinetic current in the stationary state* $\mathbf{J}_K^s$, *eq.* (59.1), *is determined by the Boltzmann-like operator* $\Psi^{(+)}(0)$ *alone.* This is an exact and general result.

As a consequence, we see that the study of the stationary state (and therefore the theory of transport coefficients) and the study of the kinetics of approach to the stationary state are essentially different. If the latter problem is attacked by means of kinetic equations, the full non-markoffian equations must be used. Only in a few simple situations (Landau approximation, ring approximation, low concentration) is the kinetics properly described by a Boltzmann-like equation. On the contrary, the computation of the kinetic part of the transport coefficients *to any order of approximation* involves only the general Boltzmann-like equation, derived on the (physically incorrect) assumption of instantaneous collisions. This fact simplifies the problem by many orders of magnitude. This result will be extended to the correlational part in Appendix 11.

We now show the complete equivalence of eq. (59.1) and the result of the traditional method of § 56, using the general Boltzmann-like equation. The relevant equation in this discussion is a "master equation", i.e. a kinetic equation for φ_N. The general feature of such an equation is its *linearity*. It is obviously:

$$\mathscr{B}\rho_0 \equiv \partial_t \rho_0 + e\mathscr{L}^e \rho_0 - \Psi^{(+)}(0)\rho_0 = 0 \tag{59.4}$$

Upon integration over all variables but one, this equation becomes non-linear through the factorization of ρ_0. As $\Psi^{(+)}(0)$ is defined as the sum of all diagonal fragments, the result of the reduction is precisely eq. (56.4). In the traditional method one first looks for a stationary solution ρ_0^s of this equation, then one calculates the current with that solution. Let G be the Green's function of the corresponding internal operator. The stationarity condition can be expressed by a formula similar to (57.15) (written in operator form), to first order in the field

$$G(t)\rho_0^s \equiv G^i(t)\rho_0^s - \int_0^t dt' G^i(t-t') e \mathscr{L}^e(t') G^i(t') \rho_0^s = \rho_0^s \tag{59.5}$$

Let us write

$$\rho_0^s = \rho_0^0 + \bar{\rho}_0$$

where ρ_0^0 is the equilibrium distribution in the absence of the field (i.e. a stationary solution of $\mathscr{B}^i \rho_0 = 0$), and $\bar{\rho}_0$ is first order in $\mathbf{E}$. Then (59.5) leads to an equation for $\bar{\rho}_0$:

$$[1-G^i(t)]\bar{\rho}_0 = \int_0^t dt' G^i(t-t') e \mathscr{L}^e(t') \rho_0^0 \tag{59.6}$$

Some algebra, which is quite analogous to the calculation of § 57, leads to the equation:

$$\mathbf{J}^s = e \sum_m z_m \int dv\, \mathbf{v}_m \bar{\rho}_0 = \mathbf{E} \cdot e^2 \beta \sum_m \sum_n z_m z_n \int dv\, \mathbf{v}_m B \mathbf{v}_n \rho_0^0 \tag{59.7}$$

with

$$B = [1-G^i(t)]^{-1} \int_0^t dt' G^i(t-t') \tag{59.8}$$

It is, however, well known that the Green's operator G^i of the equation

$$\partial_t \rho_0 = \Psi^{(+)}(0) \rho_0 \tag{59.9}$$

can be written symbolically in the form

$$G^i(t) = \exp [\Psi^{(+)}(0) t] \tag{59.10}$$

Therefore, eq. (59.8) becomes:

$$\begin{aligned} B &= \{1-\exp [\Psi^{(+)}(0)t]\}^{-1} \int_0^t dt' \exp [\Psi^{(+)}(0)(t-t')] \\ &= \frac{\exp [\Psi^{(+)}(0)t]\{\exp [-\Psi^{(+)}(0)t]-1\}}{-\Psi^{(+)}(0)\{1-\exp[\Psi^{(+)}(0)t]\}} = \frac{1}{-\Psi^{(+)}(0)} \end{aligned} \tag{59.11}$$

which, combined with (59.7), yields precisely the result (59.1).

§ 60. The Perfect Lorentz Gas

In order to illustrate the general theory, we shall calculate the electrical conductivity in a very simple model.* Consider a plasma consisting of a mixture of electrons and ions. Let the electron velocity distribution be $\varphi(\mathbf{v})$ and the ion distribution be $\Phi(\mathbf{v})$. The Landau equations for this system are:

$$\begin{aligned}\partial_t \varphi(\mathbf{v}_\alpha) = e^4 c \int d\mathbf{v}_j \int d\mathbf{l}(l^2+\kappa^2)^{-2} m^{-1}\mathbf{l}\cdot\partial_\alpha \delta(\mathbf{l}\cdot\mathbf{g}_{\alpha j}) \\ \cdot \{m^{-1}\mathbf{l}\cdot\partial_{\alpha j}\varphi(\alpha)\varphi(j)+(m^{-1}\mathbf{l}\cdot\partial_\alpha - M^{-1}\mathbf{l}\cdot\partial_j)\varphi(\alpha)\Phi(j)\}\end{aligned} \tag{60.1}$$

and a similar equation for the ions. One now simplifies these equations by using the following arguments:

(*a*) The ratio m/M is always a very small number. One therefore drops the term proportional to M^{-1} in (60.1).

(*b*) Because of their large mass, the average velocity of the ions is much smaller than the average velocity of the electrons. $\Phi(\mathbf{v}_j)$ is therefore a function which is very sharply peaked near zero. It is replaced by a function $\delta(\mathbf{v}_j)$ (the ions are fixed).

(*c*) One assumes that the electron–electron interactions can be neglected as compared to the electron–ion interactions.

The resulting model is called a *perfect Lorentz gas* and has been used by many authors for the calculation of the electrical conductivity of a plasma.** The equation resulting from the simplifications is:

$$\partial_t \varphi(\mathbf{v}) = e^4 c m^{-2} \int d\mathbf{l}(l^2+\kappa^2)^{-2}\mathbf{l}\cdot\partial\, \delta(\mathbf{l}\cdot\mathbf{v})\, \mathbf{l}\cdot\partial\varphi(\mathbf{v}) \tag{60.2}$$

In this way eq. (60.1) has been linearized. Equation (60.2) does not give a complete description of the approach to equilibrium. Indeed, it is immediately seen that *any* function of v^2 is a stationary solution of (60.2). However, this equation describes the disappearance of initial anisotropies of $\varphi(\mathbf{v})$, and this process is sufficient for the calculation of the electrical conductivity.

* The presentation of this section is based on some unpublished calculations of Ph. de Gottal.

** See, e.g., L. Spitzer, Jr., *Physics of Fully Ionized Gases*, Interscience, New York, 1956; J. L. Delcroix, *Introduction to the Theory of Ionized Gases*, Interscience, New York, 1960.

The integration over $\mathbf{l}$ is performed as in § 35, with the result

$$\partial_t \varphi(\mathbf{v}) = A \partial_r [v^{-3}(v^2 \delta_{rs} - v_r v_s)] \partial_s \varphi(\mathbf{v}) \tag{60.3}$$

where

$$A = \pi e^4 c B / m^2 \tag{60.4}$$

and B is defined by eq. (35.5) or (35.13). We now note a striking analogy with the quantum mechanical orbital angular momentum operator L^2 (see any textbook on quantum mechanics) written in velocity space:

$$-L^2 = (\mathbf{v} \times \partial) \cdot (\mathbf{v} \times \partial) = \partial_r (v^2 \delta_{rs} - v_r v_s) \partial_s \tag{60.5}$$

Using also the property

$$v_r (v^2 \delta_{rs} - v_r v_s) \equiv 0 \tag{60.6}$$

eq. (60.3) becomes:

$$\partial_t \varphi(\mathbf{v}) = -A v^{-3} L^2 \varphi(\mathbf{v}) \tag{60.7}$$

The great advantage of the Lorentz gas model is that the eigenvalues and eigenfunctions of the collision operator are well known from quantum mechanics. If we denote by v, θ, φ the polar coordinates of $\mathbf{v}$ in an arbitrary reference system, the eigenvalues and eigenfunctions are given by:

$$L^2 Y_l^m(\theta, \varphi) = l(l+1) Y_l^m(\theta, \varphi) \tag{60.8}$$

where $Y_l^m(\theta, \varphi)$ are the spherical harmonics. Hence the general solution of (60.7) can be written as an expansion in spherical harmonics:

$$\varphi(\mathbf{v}; t) = \sum_{l=0}^{\infty} \sum_{m=-l}^{l} c_{lm}(v) v^l Y_l^m(\theta, \varphi) e^{-\lambda_l t} \tag{60.9}$$

The eigenvalues λ_l are the inverse "relaxation times" of the system:

$$\begin{aligned} \lambda_l^{-1} \equiv \tau_l &= v^3 / A l(l+1) \\ &= m^2 v^3 / l(l+1) 2\pi e^4 c B \end{aligned} \tag{60.10}$$

It should be noted that if one takes $v = (\beta m)^{-\frac{1}{2}}$, all these relaxation times are of the form appearing in (37.2) with different coefficients. It is seen that high anisotropies (large l) disappear faster than low ones. We now calculate the electrical conductivity both by the traditional method and by the method of § 59.

A. Traditional method

We look for a stationary solution of the equation

$$\partial_t \varphi(\mathbf{v}) - (e/m)\mathbf{E}\cdot\partial\varphi(\mathbf{v}) = -Av^{-3}L^2\varphi(\mathbf{v}) \qquad (60.11)$$

Making the ansatz (56.5), eq. (56.6) becomes in the present case:

$$-(e/m)\mathbf{E}\cdot\partial\varphi^0(v) = -Av^{-3}\varphi^0(v)L^2\mathbf{E}\cdot\mathbf{v}\psi(v) \qquad (60.12)$$

Taking the polar axis along **E**, it is seen that $\mathbf{E}\cdot\mathbf{v}$ is proportional to $Y_1^0(\theta,\varphi)$, and hence is an eigenfunction of L^2. Performing also explicitly the differentiation on the l.h.s., eq. (60.12) can be written as

$$e\beta\mathbf{E}\cdot\mathbf{v}\varphi^0(v) = -Av^{-3}\varphi^0(v)1\,.\,2\,.\,\mathbf{E}\cdot\mathbf{v}\psi(v)$$

and hence

$$\psi(v) = -e\beta v^3/2A$$

From eq. (56.7) and assumption (b) above, the stationary current is given by

$$\begin{aligned}\mathbf{J}^s &= (e^2\beta c/2A)\mathbf{E}\cdot\int d\mathbf{v}\;\mathbf{v}\mathbf{v}v^3\varphi^0(v)\\ &= (2\pi e^2\beta c/A)\mathbf{E}\cdot\int_0^\infty dv\,v^7\varphi^0(v)\mathbf{I}\end{aligned} \qquad (60.13)$$

The integral is easily evaluated, with the result

$$\sigma = \frac{(2kT)^{\frac{3}{2}}}{\pi^{\frac{3}{2}}m^{\frac{1}{2}}e^2\ln(l_M/\kappa)} \qquad (60.14)$$

B. Equation (59.1)

In the present case the operator $\Psi^{(+)}(0)$ is

$$\Psi^{(+)}(0) = -\sum_j Av^{-3}L_j^2$$

We first note that, owing to the symmetry of the operator, only the terms $m = n$ contribute in the particle summation. In the reduction from N particles to one particle use is made of $\tilde{\rho}_0^0(\mathbf{v}) = \Omega^{-1}\varphi^0(\mathbf{v})$. Formula (59.1) can thus be written as:

$$\mathbf{J}^s = \mathbf{E}\cdot e^2\beta c\int d\mathbf{v}\,\mathbf{v}\frac{1}{Av^{-3}L^2}\mathbf{v}\varphi^0(v)$$

Using (60.8), this becomes

$$\mathbf{J}^s = \mathbf{E} \cdot e^2 \beta c \int d\mathbf{v}\,\mathbf{v}\,(v^3/2A)\,\mathbf{v}\,\varphi^0(v)$$

which is identical with (60.13).

Part II

Quantum-Statistical Systems

CHAPTER 14

The General Diagram Method for Quantum Gases

§ 61. The Density Matrix and the Wigner Distribution Function

In this part of the book we shall study assemblies of charged particles in conditions such that the laws of classical mechanics are no longer an adequate approximation and quantum effects become important or even dominant. The long-range Coulomb interactions retain, of course, their main properties, which have been investigated in detail in the first part of the book. However, the manifestation of these properties will in general be different because the Coulomb effects are combined with and corrected by quantum-mechanical effects. These are of two kinds.

a) The laws of dynamics are profoundly changed by the existence of *Heisenberg's uncertainty principle.* In particular, one cannot specify a path of the particles or of the system in phase space.

b) The specification of the system must take into account the *indiscernibility of identical particles.* This leads to the two possible *quantum statistics, the Bose–Einstein and the Fermi–Dirac statistics.*

Our problem is to extend the classical formalism, which was successful in handling the problem of long-range interactions, in order to incorporate it into the framework of quantum statistics. The most convenient method for doing this is the *method of second quantization,* because it takes into account automatically both effects mentioned above. We must assume here that the reader is familiar with this method.*

We consider again the model of a gas of charged particles, say an electron gas, in the presence of a continuous neutralizing

* See, for instance, V. A. Fock, *Z. Physik,* **75**, 622 (1932); P. A. M. Dirac, *The Principles of Quantum Mechanics,* 3rd ed., Oxford University Press, 1947; D. J. Blochintzev, *Grundlagen der Quantenmechanik,* Deutsch. Verl. der Wiss., Berlin, 1953; L. I. Schiff, *Quantum Mechanics,* McGraw-Hill, New York, 1949; A. Messiah, *Mécanique Quantique,* Dunod, Paris, 1959.

background. The hamiltonian of this system [i.e. the analog of the classical formula (1.3)] is:

$$\begin{aligned}\mathscr{H} &= \sum_{\mathbf{k}} \varepsilon_{\mathbf{k}} a^{+}(\hbar\mathbf{k})a(\hbar\mathbf{k}) \\ &\quad +\tfrac{1}{2}e^2 \sum_{\mathbf{k},\mathbf{l},\mathbf{p},\mathbf{q}} \langle\mathbf{kl}|V|\mathbf{pq}\rangle\delta_{\mathbf{k}+\mathbf{l}-\mathbf{p}-\mathbf{q}}\, a^{+}(\hbar\mathbf{k})a^{+}(\hbar\mathbf{l})a(\hbar\mathbf{p})a(\hbar\mathbf{q}) \\ &\equiv \mathscr{H}_0 + e^2\mathscr{V}\end{aligned} \tag{61.1}$$

We are working here in the *momentum representation* in which the free-particle hamiltonian $\mathscr{H}_0$ is diagonal:

$$\mathscr{H}_0|n\rangle \equiv \sum_{\mathbf{k}} \varepsilon_{\mathbf{k}} a^{+}(\hbar\mathbf{k})a(\hbar\mathbf{k})|n\rangle = \Big(\sum_{\mathbf{k}} \varepsilon_{\mathbf{k}} n(\hbar\mathbf{k})\Big)|n\rangle \tag{61.2}$$

Therefore the state $|n\rangle \equiv |n(\hbar\mathbf{k}_1), n(\hbar\mathbf{k}_2), \ldots\rangle$ is specified by an infinite set of occupation numbers of momentum states $\mathbf{k}_1$, $\mathbf{k}_2, \ldots$ $a^{+}(\hbar\mathbf{k})$ and $a(\hbar\mathbf{k})$ are respectively *creation and destruction operators of particles with momentum* $\hbar\mathbf{k}$. They obey commutation or anticommutation relations, according to whether the particles obey *Bose–Einstein* or *Fermi–Dirac statistics*:

$$\begin{aligned}&[a(\hbar\mathbf{k}), a^{+}(\hbar\mathbf{k}')]_\theta \equiv a(\hbar\mathbf{k})\, a^{+}(\hbar\mathbf{k}') - \theta\, a^{+}(\hbar\mathbf{k}')a(\hbar\mathbf{k}) = \delta_{\mathbf{k}-\mathbf{k}'} \\ &[a^{+}(\hbar\mathbf{k}), a^{+}(\hbar\mathbf{k}')]_\theta = [a(\hbar\mathbf{k}), a(\hbar\mathbf{k}')]_\theta = 0\end{aligned} \tag{61.3}$$

The statistical factor θ is a useful device for writing general formulae valid for both Bose–Einstein and Fermi–Dirac statistics:

$$\theta = \begin{cases} +1 \text{ for Bose–Einstein statistics} \\ -1 \text{ for Fermi–Dirac statistics} \end{cases} \tag{61.4}$$

In the final formulae, the classical (Boltzmann) statistics limit can be achieved by setting $\theta = 0$, although some care must be exercised in this limiting procedure (see later).

The operator

$$n(\hbar\mathbf{k}) = a^{+}(\hbar\mathbf{k})a(\hbar\mathbf{k}) \tag{61.5}$$

is interpreted as usual as the *occupation number of the state* $\mathbf{k}$ (i.e. the number of particles with momentum $\hbar\mathbf{k}$). *The total number of particles N is assumed constant throughout*

$$N = \sum_{\mathbf{k}} a^{+}(\hbar\mathbf{k})a(\hbar\mathbf{k}) = \text{const.} \tag{61.6}$$

The normalized one-particle wave function is

$$|\mathbf{k}\rangle\langle\mathbf{k}|\mathbf{x}\rangle = \Omega^{-\frac{1}{2}} e^{i\mathbf{k}\cdot\mathbf{x}}|\mathbf{k}\rangle \tag{61.7}$$

Therefore the matrix elements appearing in eq. (61.1) are

$$\langle \mathbf{k}|H_0|\mathbf{k}\rangle \equiv \varepsilon_{\mathbf{k}} = \hbar^2 k^2/2m \tag{61.8}$$

The matrix element of the two-body interaction must be taken between two-particle states. Here one must properly take account of the statistics; therefore one must use *symmetrized or anti-symmetrized* states, properly normalized [factor $(\sqrt{2}\Omega)^{-1}$]:

$$\begin{aligned}\langle \mathbf{kl}|V|\mathbf{pq}\rangle\delta_{\mathbf{k+l-p-q}} &= \int d\mathbf{x}_1 d\mathbf{x}_2 (\sqrt{2}\Omega)^{-1}[\mathrm{e}^{-i\mathbf{k}\cdot\mathbf{x}_1 - i\mathbf{l}\cdot\mathbf{x}_2} + \theta\,\mathrm{e}^{-i\mathbf{k}\cdot\mathbf{x}_2 - i\mathbf{l}\cdot\mathbf{x}_1}] \\ &\quad \cdot V(|\mathbf{x}_1 - \mathbf{x}_2|)(\sqrt{2}\Omega)^{-1}[\mathrm{e}^{i\mathbf{p}\cdot\mathbf{x}_1 + i\mathbf{q}\cdot\mathbf{x}_2} + \theta\,\mathrm{e}^{i\mathbf{p}\cdot\mathbf{x}_2 + i\mathbf{q}\cdot\mathbf{x}_1}] \\ &= (8\pi^3/\Omega)[V_{|\mathbf{k-q}|} + \theta V_{|\mathbf{k-p}|}]\delta_{\mathbf{k+l-p-q}}\end{aligned} \tag{61.9}$$

where $V_{|\mathbf{k}|}$ is the Fourier transform of the potential, eq. (9.8). The fact that the matrix element now has *two* terms is a direct consequence of the statistics. The two terms are called the *direct* term and the *exchange* term respectively. For simplicity we consider throughout this book spinless particles. The inclusion of spin is not a major difficulty.

Besides the momentum representation which was used above, we shall sometimes use the *position representation*. We then introduce creation and destruction operators in the following form:

$$\begin{aligned}\psi^+(\mathbf{x}) &= \Omega^{-\frac{1}{2}} \sum_{\mathbf{k}} a^+(\hbar\mathbf{k})\,\mathrm{e}^{-i\mathbf{k}\cdot\mathbf{x}} \\ \psi(\mathbf{x}) &= \Omega^{-\frac{1}{2}} \sum_{\mathbf{k}} a(\hbar\mathbf{k})\,\mathrm{e}^{i\mathbf{k}\cdot\mathbf{x}}\end{aligned} \tag{61.10}$$

The commutation relations (61.3) imply that in the limit of a large system:

$$\begin{aligned}[\psi(\mathbf{x}), \psi^+(\mathbf{x}')]_\theta &= \Omega^{-1}\sum_{\mathbf{k}}\sum_{\mathbf{k}'}[a(\hbar\mathbf{k}), a^+(\hbar\mathbf{k}')]_\theta \mathrm{e}^{i\mathbf{k}\cdot\mathbf{x} - i\mathbf{k}'\cdot\mathbf{x}'} \\ &\xrightarrow[(\Omega\to\infty)]{} (8\pi^3)^{-1}\int d\mathbf{k}\,\mathrm{e}^{i\mathbf{k}\cdot(\mathbf{x}-\mathbf{x}')} = \delta(\mathbf{x}-\mathbf{x}') \\ [\psi(\mathbf{x}), \psi(\mathbf{x}')]_\theta &= [\psi^+(\mathbf{x}), \psi^+(\mathbf{x}')]_\theta = 0\end{aligned} \tag{61.11}$$

In terms of these operators the hamiltonian takes the following form whose identity with (61.1) is readily verified:

$$\mathscr{H} = \int d\mathbf{x}\, \psi^+(\mathbf{x}) \left(-\frac{\hbar^2}{2m}\nabla^2\right) \psi(\mathbf{x}) \\ + \tfrac{1}{4}e^2 \iint d\mathbf{x}\, d\mathbf{x}'\, \psi^+(\mathbf{x})\psi^+(\mathbf{x}')V(|\mathbf{x}-\mathbf{x}'|)\psi(\mathbf{x}')\psi(\mathbf{x}) \tag{61.12}$$

Let us recall at this point the general prescriptions for the construction of the quantum-mechanical operator corresponding to a classical observable. This construction is performed in two stages. First the operator in ordinary quantum mechanics (i.e. without consideration of statistics, and thus without second quantization) is defined. A unique prescription is given by *Weyl's rule,** which proceeds via the Fourier transform of the dynamical function. Let $A(\mathbf{x}_1 \dots \mathbf{x}_N, \mathbf{p}_1 \dots \mathbf{p}_N)$ be a dynamical function [i.e. a function of the type appearing on the right-hand side of eq. (1.12)], and let it be Fourier transformed as follows:

$$A(\mathbf{x}_1 \dots \mathbf{x}_N, \mathbf{p}_1 \dots \mathbf{p}_N) \\ = (8\pi^3)^{-2N} \int (d\mathbf{k})^N (d\mathbf{s})^N \alpha(\{\mathbf{k}\}, \{\mathbf{s}\})\, e^{-i\Sigma(\mathbf{k}_r\cdot\mathbf{x}_r+\mathbf{s}_r\cdot\mathbf{p}_r)} \tag{61.13}$$

According to Weyl's rule, the corresponding operator in ordinary quantum mechanics is:

$$A_{op} = (8\pi^3)^{-2N} \int (d\mathbf{k})^N (d\mathbf{s})^N \alpha(\{\mathbf{k}\}, \{\mathbf{s}\})\, e^{-i\Sigma(\mathbf{k}_r\cdot\mathbf{x}_r-i\hbar\mathbf{s}_r\cdot\nabla_r)} \tag{61.14}$$

Hence the quantum operator is obtained simply by substituting $-i\hbar\boldsymbol{\nabla}_r$ for $\mathbf{p}_r$ in the argument of the exponential. The latter becomes, therefore, a finite displacement operator.

The next step consists in defining the quantum-statistical operator. Knowing the ordinary operator A_{op}, the rules of second quantization require the following form for the operator in the occupation number representation:

$$\mathscr{A} = (N!)^{-1} \int (d\mathbf{x})^N \psi^+(\mathbf{x}_1) \dots \psi^+(\mathbf{x}_N) A_{op} \psi(\mathbf{x}_N) \dots \psi(\mathbf{x}_1) \tag{61.15}$$

Note the *order* of the creation and destruction operators, which is important in fermion systems. The factor $(N!)^{-1}$ comes from the

* H. Weyl, *The Theory of Groups and Quantum Mechanics,* 2nd. ed. (reprinted), Dover Publ. Co., New York, 1931.

normalization of the N-particle states. In order to abridge the expressions we introduce the following *normal product* notation:

$$:\prod_{j=1}^{N} \psi^{+}(\mathbf{x}_j) A \psi(\mathbf{x}'_j): \equiv \psi^{+}(\mathbf{x}_1) \ldots \psi^{+}(\mathbf{x}_N) A \psi(\mathbf{x}'_N) \ldots \psi(\mathbf{x}'_1) \quad (61.16)$$

Using (61.14), eq. (61.15) can thus be rewritten:

$$\begin{aligned}\mathscr{A} = (8\pi^3)^{-2N} (N!)^{-1} \int (d\mathbf{x})^N (d\mathbf{k})^N (d\mathbf{s})^N \alpha(k, s) \\ :\prod_j \psi^{+}(\mathbf{x}_j)\, e^{-i\Sigma(\mathbf{k}_r \cdot \mathbf{x}_r - i\hbar \mathbf{s}_r \cdot \nabla_r)} \psi(\mathbf{x}_j):\end{aligned} \quad (61.17)$$

We now need a concept which characterizes the state of the system in the same way as did the distribution function f_N in classical mechanics. This generalization is provided by the well known *von Neumann density matrix.** We assume that the reader is familiar with this concept, which is treated in all textbooks on quantum statistics.** The density matrix $\boldsymbol{\rho}$ characterizes in general an ensemble of systems. It obeys the *von Neumann equation,* which is very similar to the Liouville equation (1.6):

$$i\hbar\, \partial_t \boldsymbol{\rho} + [\boldsymbol{\rho}, \mathscr{H}]_- = 0 \quad (61.18)$$

Here, the brackets denote the *commutator* of the operators $\boldsymbol{\rho}$ and $\mathscr{H}$:

$$[\boldsymbol{\rho}, \mathscr{H}]_- = \boldsymbol{\rho}\mathscr{H} - \mathscr{H}\boldsymbol{\rho} \quad (61.19)$$

Consider now an observable, described by the operator $\mathscr{A}$. By the definition of the density matrix, the *observable value of the quantity A in the state defined by $\boldsymbol{\rho}$* is given by

$$A(t)_{\text{macro}} = \text{Trace}\, \boldsymbol{\rho}(t) \mathscr{A} = \text{Trace}\, \mathscr{A} \boldsymbol{\rho}(t) \quad (61.20)$$

This prescription replaces formula (1.12) for quantum-mechanical systems.

It is often useful to have a formalism which is still closer to the classical formalism. We should like to avoid the trace operation appearing in (61.20), which gives to the formulae a mathematical structure very different from their classical analogs. Moreover,

* J. von Neumann, *Mathematische Grundlagen der Quantenmechanik*, Springer Verlag, Berlin, 1932.

** A very complete exposition is given in R. C. Tolman, *The Principles of Statistical Mechanics*, Oxford University Press, 1955.

trace computations are very tedious in practice. Wigner * has shown that it is possible to write the expression for $A(t)_{\text{macro}}$ exactly in the classical form (1.12), in terms of an integral, i.e. as an average calculated with a *phase-space distribution function.*

The existence of such a function might appear surprising because the concept of a phase space is in principle meaningless in quantum mechanics. The uncertainty principle excludes the simultaneous specification of positions and momenta. Hence it is to be expected that the quantum-phase-space distribution is a formal concept with a less clear physical interpretation than the classical distribution function. However, the concept is fully operational and leads to very simple expressions for the calculation of macroscopic quantities.

The Wigner function is constructed as follows. Consider the classical dynamical function $A(x, p)$ defined by eq. (61.13). Its average value in the corresponding quantum-mechanical picture is obtained by substituting eq. (61.17) into (61.20):

$$A(t)_{\text{macro}} = (8\pi^3)^{-2N} \int (d\mathbf{k})^N \int (d\mathbf{s})^N \alpha(k, s)$$
$$\cdot \operatorname{Tr} (N!)^{-1} \int (d\mathbf{x})^N [\boldsymbol{\rho}(t) : \prod_j \boldsymbol{\psi}^+(\mathbf{x}_j) \, e^{-i\Sigma(\mathbf{k}_r \cdot \mathbf{x}_r - i\hbar \mathbf{s}_r \cdot \nabla_r)} \boldsymbol{\psi}(\mathbf{x}_j) :] \tag{61.21}$$

The trace appearing on the right-hand side is a function of $\{\mathbf{k}\}$ and $\{\mathbf{s}\}$. Let us Fourier transform it:

$$\operatorname{Tr}(N!)^{-1} \int (d\mathbf{x})^N [\ldots] = \int (d\mathbf{x}')^N (d\mathbf{p}')^N e^{-i\Sigma(\mathbf{k}_r \cdot \mathbf{x}'_r + \mathbf{s}_r \cdot \mathbf{p}'_r)} f_N^W(\{\mathbf{x}'\}, \{\mathbf{p}'\}; t) \tag{61.22}$$

Substituting this into (61.21) we obtain

$$A(t)_{\text{macro}} = (8\pi^3)^{-2N} \int (d\mathbf{x}')^N (d\mathbf{p}')^N$$
$$\cdot \int (d\mathbf{k})^N (d\mathbf{s})^N \alpha(k, s) \, e^{-i\Sigma(\mathbf{k}_r \cdot \mathbf{x}'_r + \mathbf{s}_r \cdot \mathbf{p}'_r)} f_N^W(x', p'; t)$$

or, taking account of (61.13):

$$A(t)_{\text{macro}} = \int (d\mathbf{x})^N (d\mathbf{p})^N A(x, p) f_N^W(x, p; t) \tag{61.23}$$

* E. Wigner, *Phys. Rev.*, **40**, 749 (1932).

In this way we have obtained a formula which is identical with eq. (1.10). It gives the macroscopic value of A in terms of the *classical* dynamical function $A(x, p)$ [and not of the operator $\mathscr{A}$, as in (61.20)]. The function $f_N^W(x, p; t)$ is called the *Wigner distribution function (of N particles)*.* As stated above, the Wigner function cannot be expected to have the character of a true distribution function like the classical f_N. In particular, f_N^W is *not* in general a positive definite function. In spite of this difficulty, f_N^W is a tool which is quite as valuable as f_N and all the computational rules applying to f_N apply also to f_N^W. It will be seen, moreover, that *the classical limit of the quantum-mechanical expressions is particularly simply calculated when the latter are expressed in terms of* f_N^W.

Note that we have used the momentum variables rather than the velocities. As stated in § 1, the momenta are natural variables at the molecular level, whereas velocities are more natural macroscopic variables. This is even more true in quantum mechanics: the velocity has no simple meaning, whereas the momentum is a variable which can be readily quantized. We shall therefore perform most calculations in this chapter using the momenta. Once the final equations for the Fourier components of the Wigner function are obtained we shall switch over to the velocities, because at that point we have equations with the same structure as their classical counterparts.

Let us now express the relation between f_N^W and $\boldsymbol{\rho}$ in a more convenient form. Taking the inverse Fourier transform of eq. (61.22) we obtain

$$f_N^W(x, p; t) = (8\pi^3)^{-2N}(N!)^{-1}\int (d\mathbf{k})^N (d\mathbf{s})^N e^{ikx+isp} \cdot \operatorname{Tr}\int (d\mathbf{x}')^N \boldsymbol{\rho}(t) :\prod_j \psi^+(\mathbf{x}_j') e^{-ikx'-\hbar s\nabla'} \psi(\mathbf{x}_j'): \tag{61.24}$$

(We use here the condensed notations $k = \{\mathbf{k}_1 \ldots \mathbf{k}_N\}$, $kx = \sum \mathbf{k}_r \cdot \mathbf{x}_r$, etc. . . .) The exponential of $-\hbar s\nabla'$ is a finite

* Recent detailed expositions of the Wigner function can be found in: D. Massignon, *Mécanique Statistique des Fluides,* Dunod, Paris, 1957; H. Mori, I. Oppenheim and J. Ross, in *Studies in Statistical Mechanics,* vol. 1, edited by J. de Boer and G. E. Uhlenbeck, Interscience, New York, 1962.

displacement operator, as is well known:

$$e^{a(\partial/\partial x)} f(x) \equiv f(x+a) \tag{61.25}$$

In order to perform the operations, it is first necessary to express the exponential of a sum of *operators* $(-ikx'-\hbar s\nabla')$ as a product of exponentials. This is done by using a well known operator identity:

$$e^{A+B} = e^{B/2} e^{A} e^{B/2} \tag{61.26}$$

where A and B are arbitrary (generally non-commuting) operators with the restriction that the commutator of A and B is a c-number. Applied to the exponential occurring under the trace in (61.24) this yields:

$$e^{-ikx'-\hbar s\nabla'} = e^{-ikx'/2} e^{-\hbar s\nabla'} e^{-ikx'/2} = e^{-ik(x'-\hbar s/2)} e^{-\hbar s\nabla'} \tag{61.27}$$

Performing the displacement operations explicitly we obtain:

$$f_N^W(x, p; t) = (8\pi^3)^{-2N}(N!)^{-1} \int (d\mathbf{k})^N (d\mathbf{x}')^N (d\mathbf{s})^N e^{ik(x-x'+\hbar s/2)} e^{isp}$$

$$\cdot \operatorname{Tr}\{\boldsymbol{\rho}(t) : \prod_j \boldsymbol{\psi}^+(\mathbf{x}_j')\boldsymbol{\psi}(\mathbf{x}_j'-\hbar\mathbf{s}_j):\} \tag{61.28}$$

$$= (8\pi^3)^{-N}(N!)^{-1} \int (d\mathbf{s})^N e^{isp} \operatorname{Tr}\{\boldsymbol{\rho}(t) : \prod_j \boldsymbol{\psi}^+(\mathbf{x}_j+\tfrac{1}{2}\hbar\mathbf{s}_j)\boldsymbol{\psi}(\mathbf{x}_j-\tfrac{1}{2}\hbar\mathbf{s}_j):\}$$

Starting from f_N^W we can define reduced *s-particle distribution functions* by the same formulae as in the classical case, i.e. eqs. (1.13), (1.18–20). These functions are manipulated exactly as in classical mechanics. We must mention at this point an important property (without proving it): *The reduced Wigner functions of momenta alone,* $\varphi_s(\mathbf{p}_1, \ldots, \mathbf{p}_s; t)$, *and the reduced functions of positions alone,* $n_s(\mathbf{x}_1, \ldots, \mathbf{x}_s; t)$, *are true distribution functions and hence are definite positive.* These properties are in agreement with the uncertainty principle. Indeed, one can specify exactly the positions of the particles if one wants no information about their momenta, and conversely.

§ 62. The Fourier Transform of the Wigner Function

Our main purpose in this section is to set up a formalism which is as close as possible to the classical formalism developed in the first part of the book. In this way, most of the quantum-mechanical results will be obtained as a straightforward generalization of the classical ones.

One of the most important tools in Chapter 1 was the introduction and classification of the Fourier components of the distribution function. An analogous Fourier decomposition can be performed on the Wigner function. As in § 2, we must take some care with the volume dependence. Therefore, we shall perform all the calculations for a finite, but large, system consisting of N particles enclosed in a cubic box of volume Ω, in which we assume periodic boundary conditions (2.4). We can then go over to the momentum representation and introduce the creation and destruction operators $a^+(\hbar\boldsymbol{\kappa})$, $a(\hbar\boldsymbol{\kappa})$ through eq. (61.10), in which the wave-vector $\boldsymbol{\kappa}$ takes the values:

$$\boldsymbol{\kappa} = (2\pi/\Omega^{\frac{1}{3}})\mathbf{n}$$

where $\mathbf{n}$ is a vector whose components are integers. An important difference from classical systems now arises: in a finite system not only the wave-vector $\boldsymbol{\kappa}$ but also the momentum $\mathbf{p}$ is quantized, its permitted values being $\hbar\boldsymbol{\kappa}$. This fact must be taken into account in formula (61.28), in which the limits of integration for each component of s are $(-L/2\hbar)$ and $(L/2\hbar)$. Also, in a finite system the momentum integrations must be replaced by discrete sums, using the following prescription [see eq. (2.11)]

$$(8\pi^3\hbar^3/\Omega)\sum_{\mathbf{p}} \equiv (h^3/\Omega)\sum_{\mathbf{p}} \xrightarrow[\Omega\to\infty]{} \int d\mathbf{p} \tag{62.1}$$

We now substitute (61.10) into (61.28) with the result:

$$\begin{aligned} f_N^W(x, p; t) = [(8\pi^3\Omega)^{-N}/N!]\sum_{\kappa}\sum_{\kappa'}\int_{-L/2\hbar}^{L/2\hbar} ds\, e^{isp}\, e^{i(\kappa'-\kappa)x}\, e^{-i(\kappa+\kappa')\hbar s/2} \\ \cdot \operatorname{Tr}\{\rho(t) : \prod_j a^+(\hbar\boldsymbol{\kappa}_j)a(\hbar\boldsymbol{\kappa}_j'):\} \end{aligned} \tag{62.2}$$

where we used the "normal product" notation:

$$:\prod_j a^+(\mathbf{p}_j)a(\mathbf{p}_j') : \equiv a^+(\mathbf{p}_1)a^+(\mathbf{p}_2)\ldots a^+(\mathbf{p}_N)a(\mathbf{p}_N')\ldots a(\mathbf{p}_2')a(\mathbf{p}_1') \tag{62.3}$$

Perform now the change of variables:

$$\begin{cases} \varkappa_j' - \varkappa_j = \mathbf{k}_j \\ \varkappa_j' + \varkappa_j = 2\mathbf{K}_j \end{cases}; \qquad \begin{cases} \varkappa_j = \mathbf{K}_j - \frac{1}{2}\mathbf{k}_j \\ \varkappa_j' = \mathbf{K}_j + \frac{1}{2}\mathbf{k}_j \end{cases} \tag{62.4}$$

and integrate over s; the result is:

$$f_N^W(x, p; t) = h^{-3N}(N!)^{-1} \sum_{\{\mathbf{k}\}} e^{i\Sigma \mathbf{k}_j \cdot \mathbf{x}_j} \cdot \mathrm{Tr}\{\boldsymbol{\rho}(t) : \prod_n a^+(\mathbf{p}_n - \tfrac{1}{2}\hbar \mathbf{k}_n) a(\mathbf{p}_n + \tfrac{1}{2}\hbar \mathbf{k}_n) :\} \tag{62.5}$$

In this way we have obtained a first form of the Fourier decomposition of the Wigner function, corresponding to the classical "compact Fourier expansion" (2.5), and we may identify the coefficients as follows:

$$\tilde{\rho}_{\mathbf{k}_1 \ldots \mathbf{k}_N}(\mathbf{p}_1 \ldots \mathbf{p}_N; t) = h^{-3N}(N!)^{-1} \mathrm{Tr}\{\boldsymbol{\rho}(t) : \prod_n a^+(\mathbf{p}_n - \tfrac{1}{2}\hbar \mathbf{k}_n) a(\mathbf{p}_n + \tfrac{1}{2}\hbar \mathbf{k}_n) :\} \tag{62.6}$$

In order to understand easily the subsequent treatment, it is important to note the peculiar structure of these Fourier components. A component for which the wave-vectors $\mathbf{k}_1 \ldots \mathbf{k}_s$ are non-zero and all the others vanish is characterized by the following property:

If j is a particle in the set $(1, \ldots, s)$, the difference between the argument of the a^+ operator describing particle j and the argument of the corresponding a operator equals $(-\hbar \mathbf{k}_j)$.

If j is not contained in the set $(1, \ldots, s)$, the pair of corresponding a^+ and a operators have the same argument.

The important formula (62.5) permits one to verify very easily the normalization condition on f_N^W:

$$(h^3/\Omega)^N \sum_{\{\mathbf{p}\}} \int (d\mathbf{x})^N f_N^W(x, p; t) = 1 \tag{62.7}$$

For, substituting (62.5), we obtain

$$h^{3N}\Omega^{-N}(N!)^{-1} h^{-3N} \sum_{\{\mathbf{p}\}} \sum_{\{\mathbf{k}\}} \int (d\mathbf{x})^N e^{ikx}\, \mathrm{Tr}\,\{\boldsymbol{\rho} : a^+(p - \tfrac{1}{2}\hbar k) a(p + \tfrac{1}{2}\hbar k) :\}$$
$$= (N!)^{-1}\, \mathrm{Tr}\,\{\sum_{\{p\}} \boldsymbol{\rho} : a^+(p) a(p) :\} = \mathrm{Tr}\, \boldsymbol{\rho} = 1$$

We have used here a theorem on the summation of products of creation and destruction operators which is proven in Appendix 5.

From the compact expansion we can immediately go over to the explicit expansion (2.8). Indeed, that form of the Fourier expansion is an expression of the fundamental postulate of *finiteness of local macroscopic quantities*, which is of course equally valid whether the evolution of the system obeys classical or quantum mechanics. The fact that this postulate is expressed in the same form as in classical mechanics comes from our use of the Wigner function to describe quantum situations.

Moreover, the requirement of the mutual independence of widely separated particles is also a macroscopic condition which is valid in both quantum and classical large systems: therefore, the whole discussion of § 2, and in particular the factorization properties (2.23)–(2.25), can be taken over directly for quantum-statistical systems.*

Having postulated the form (2.8) of the explicit Fourier expansion, we shall write:

$$\begin{aligned} f_N^W(x, p; t) = \Omega^{-N}\{\rho_0(|p; t) &+ (8\pi^3/\Omega) \sum_j \sum_{\mathbf{k}}{}' \rho_{\mathbf{k}}(\mathbf{p}_j| \ldots; t)\, e^{i\mathbf{k}\cdot\mathbf{x}_j} \\ &+ (8\pi^3/\Omega)^2 \sum_j \sum_n \sum_{\substack{\mathbf{k} \\ (\mathbf{k}+\mathbf{k}'\neq 0)}}{}' \sum_{\mathbf{k}'}{}' \rho_{\mathbf{k}\mathbf{k}'}(\mathbf{p}_j, \mathbf{p}_n| \ldots; t)\, e^{i\mathbf{k}\cdot\mathbf{x}_j + i\mathbf{k}'\cdot\mathbf{x}_n} \\ &+ (8\pi^3/\Omega) \sum_j \sum_n \sum_{\mathbf{k}}{}' \rho_{\mathbf{k},-\mathbf{k}}(\mathbf{p}_j, \mathbf{p}_n| \ldots; t)\, e^{i\mathbf{k}\cdot(\mathbf{x}_j - \mathbf{x}_n)} + \ldots\} \end{aligned} \tag{62.8}$$

This is justified because the *Fourier components* $\rho_{\{k\}}$ *have the same physical meaning as the* $\rho_{\{k\}}$ *in classical mechanics*. All formulae of § 2 remain valid.

However, *the Fourier components* ρ *with more than one non-vanishing wave-vector have peculiar properties which are due to the quantum statistics*. This will easily be understood by discussing the case of $\rho_{\mathbf{k}_1\mathbf{k}_2}$ in detail. Using formula (62.6) we have:

* There is no contradiction with the exclusion principle in this statement. It is well known that pure quantum-statistical symmetry requirements lead to a *distance-dependent* correlation function, which dies out for distances larger than a characteristic length. These statistics effects are thus described in our formalism by a coefficient $\rho_{\mathbf{k},-\mathbf{k}}$, which is, of course, not factorized.

$$\rho_{\mathbf{k}_1\mathbf{k}_2}(\mathbf{p}_1\mathbf{p}_2|\ldots) = [\Omega^N(\Omega/8\pi^3)^2/N!\,h^{3N}]$$
$$\text{Tr}\,\{\boldsymbol{\rho}\,a^+(\mathbf{p}_N)\ldots a^+(\mathbf{p}_3)a^+(\mathbf{p}_2-\tfrac{1}{2}\hbar\mathbf{k}_2)a^+(\mathbf{p}_1-\tfrac{1}{2}\hbar\mathbf{k}_1)$$
$$\cdot\, a(\mathbf{p}_1+\tfrac{1}{2}\hbar\mathbf{k}_1)a(\mathbf{p}_2+\tfrac{1}{2}\hbar\mathbf{k}_2)a(\mathbf{p}_3)\ldots a(\mathbf{p}_N)\} \qquad (62.9)$$

As noted above, the arguments of the a^+a pair pertaining to particle 1 differ by $(-\hbar\mathbf{k}_1)$ and those of the pair relative to particle 2 differ by $(-\hbar\mathbf{k}_2)$, all other pairs having equal arguments.

Take now $\mathbf{p}_1 = \mathbf{p}_2-\frac{1}{2}\hbar(\mathbf{k}_1+\mathbf{k}_2)$; the r.h.s. becomes

$$\Omega^N(\Omega/8\pi^3)^2(N!)^{-1}h^{-3N}\,\text{Tr}\{\boldsymbol{\rho}\ldots a^+(\mathbf{p}_2-\tfrac{1}{2}\hbar\mathbf{k}_2)a^+(\mathbf{p}_2-\hbar\mathbf{k}_1-\tfrac{1}{2}\hbar\mathbf{k}_2)$$
$$a(\mathbf{p}_2-\tfrac{1}{2}\hbar\mathbf{k}_2)a(\mathbf{p}_2+\tfrac{1}{2}\hbar\mathbf{k}_2)\ldots\}$$
$$= \theta\,\Omega^N(\Omega/8\pi^3)^2(N!)^{-1}h^{-3N} \qquad (62.10)$$
$$\text{Tr}\{\boldsymbol{\rho}\ldots a^+(\mathbf{p}_2-\tfrac{1}{2}\hbar\mathbf{k}_2)a^+[\mathbf{p}_2-\tfrac{1}{2}\hbar\mathbf{k}_1-\tfrac{1}{2}\hbar(\mathbf{k}_1+\mathbf{k}_2)]$$
$$a[\mathbf{p}_2-\tfrac{1}{2}\hbar\mathbf{k}_1+\tfrac{1}{2}\hbar(\mathbf{k}_1+\mathbf{k}_2)]a(\mathbf{p}_2-\tfrac{1}{2}\hbar\mathbf{k}_2)\ldots\}$$

The last step has been obtained by inverting the order of the two destruction operators and taking account of (61.3). One notes that on the r.h.s. of this formula the central pair a^+a has arguments differing by $-\hbar(\mathbf{k}_1+\mathbf{k}_2)$, and the two other operators have the same argument.

Hence, this is a component with only *one* non-vanishing wave-vector. Taking into account the difference in volume factors of a component $\rho_{\mathbf{k}\mathbf{k}'}$ and a component $\rho_{\mathbf{k}}$ [see (62.8) or Table 2.1], formula (62.10) represents explicitly:

$$\theta(\Omega/8\pi^3)\rho_{\mathbf{k}_1+\mathbf{k}_2}(\mathbf{p}_2-\tfrac{1}{2}\hbar\mathbf{k}_1|\mathbf{p}_2-\tfrac{1}{2}\hbar\mathbf{k}_2,\ldots)$$

(Remember our notation: momentum arguments written to the left of the bar are relative to the particles with non-zero wave-vectors; arguments to the right of the bar pertain to particles with zero wave-vectors. In this case the particle described by the wave-vector $\mathbf{k}_1+\mathbf{k}_2$ has momentum $\mathbf{p}_2-\frac{1}{2}\hbar\mathbf{k}_1$; the "unexcited" particles have momenta $\mathbf{p}_2-\frac{1}{2}\hbar\mathbf{k}_2$, $\mathbf{p}_3, \ldots, \mathbf{p}_N$.)

This fact shows that the momenta $\mathbf{p}$ and the wave-vectors $\mathbf{k}$ are not independent; a Fourier component of one type becomes one of another type for certain values of the momenta; the latter values involve combinations of two momenta and the corresponding wave-vectors. This is a direct consequence of both the quantum-mechanical relationship between wave-vector and momentum and the indiscernibility of the particles (quantum statistics). Moreover, the ratio $(\rho_{\mathbf{k}_1+\mathbf{k}_2}/\rho_{\mathbf{k}_1,\mathbf{k}_2})$ is proportional to $\Omega/8\pi^3$ (see Table 2.1); therefore *we have here a real δ-singularity in momentum space in the limit of a large volume.* A similar behavior is obtained by taking

$$\mathbf{p}_1 = \mathbf{p}_2 + \tfrac{1}{2}\hbar(\mathbf{k}_1 + \mathbf{k}_2)$$

It will turn out to be convenient to *separate off these singular parts from* $\rho_{\mathbf{k}_1,\mathbf{k}_2}$ *by introducing coefficients* $\hat{\rho}_{\mathbf{k}}$, $\hat{\rho}_{\mathbf{k}_1,\mathbf{k}_2}$ *which have no singularities* in the following way:

$$\begin{aligned}
\rho_0(|\mathbf{p}_1, \ldots, \mathbf{p}_N) &= \hat{\rho}_0(|\mathbf{p}_1, \ldots, \mathbf{p}_N) \\
\rho_{\mathbf{k}}(\mathbf{p}_1|\mathbf{p}_2, \ldots, \mathbf{p}_N) &= \hat{\rho}_{\mathbf{k}}(\mathbf{p}_1|\mathbf{p}_2, \ldots, \mathbf{p}_N) \\
\rho_{\mathbf{k}_1\mathbf{k}_2}(\mathbf{p}_1, \mathbf{p}_2| \ldots) &= \hat{\rho}_{\mathbf{k}_1\mathbf{k}_2}(\mathbf{p}_1, \mathbf{p}_2| \ldots) \\
&\quad + \theta(\Omega/8\pi^3)\, \delta_{\mathbf{p}_1-\mathbf{p}_2+\hbar(\mathbf{k}_1+\mathbf{k}_2)/2}\, \hat{\rho}_{\mathbf{k}_1+\mathbf{k}_2}(\mathbf{p}_2 - \tfrac{1}{2}\hbar\mathbf{k}_1|\mathbf{p}_2 - \tfrac{1}{2}\hbar\mathbf{k}_2, \ldots) \\
&\quad + \theta(\Omega/8\pi^3)\, \delta_{\mathbf{p}_2-\mathbf{p}_1+\hbar(\mathbf{k}_1+\mathbf{k}_2)/2}\, \hat{\rho}_{\mathbf{k}_1+\mathbf{k}_2}(\mathbf{p}_1 - \tfrac{1}{2}\hbar\mathbf{k}_2|\mathbf{p}_1 - \tfrac{1}{2}\hbar\mathbf{k}_1, \ldots)
\end{aligned} \tag{62.11}$$

Before discussing the advantages of performing this splitting, we first explain the generalization of these rules.

1. The operation which consists in constructing from a Fourier component $\rho_{\mathbf{k}_1,\mathbf{k}_2,\mathbf{k}_3,\ldots,\mathbf{k}_s}$ a non-singular Fourier component with one less wave-vector $\hat{\rho}_{\mathbf{k}_1+\mathbf{k}_2,\,0,\mathbf{k}_3,\ldots,\mathbf{k}_s}$ (or with two less wave-vectors, if $\mathbf{k}_1+\mathbf{k}_2 = 0$) is called a *contraction on the pair* 1–2. (There is, of course, no loss in generality in assuming that the contracted pair is the pair 1–2.) As the particles, 3, . . ., *s* are unaffected, eq. (62.11) gives the complete definition of the contraction of any Fourier component.

A contraction involves a product of $\theta(\Omega/8\pi^3)$, a δ-symbol and a non-singular Fourier component. The arguments of the δ-symbol and of the Fourier component can be read off directly from (62.11).

Assume first that $\mathbf{k}_1+\mathbf{k}_2 \neq 0$. One then sees from (62.11) that with the pair 1–2 one can perform *two distinct contractions*, related to each other by a permutation of the indices 1 and 2.

If we consider now the case $\mathbf{k}_1+\mathbf{k}_2 = 0$, in particular the homogeneous coefficient $\rho_{\mathbf{k}\,-\mathbf{k}}(\mathbf{p}_1, \mathbf{p}_2| \ldots)$, these two contractions become identical $(\delta_{\mathbf{p}_1-\mathbf{p}_2})$. By virtue of the purpose attained by our decomposition process, *this contraction must be counted only once*. Therefore, the decomposition of $\rho_{\mathbf{k},-\mathbf{k}}$ is

$$\begin{aligned}\rho_{\mathbf{k},-\mathbf{k}}(\mathbf{p}_1, \mathbf{p}_2| \ldots) &= \hat{\rho}_{\mathbf{k},-\mathbf{k}}(\mathbf{p}_1, \mathbf{p}_2| \ldots) \\ &+\theta(\Omega/8\pi^3)\delta_{\mathbf{p}_1-\mathbf{p}_2}\hat{\rho}_0(|\mathbf{p}_1+\tfrac{1}{2}\hbar\mathbf{k}, \mathbf{p}_1-\tfrac{1}{2}\hbar\mathbf{k}, \ldots)\end{aligned} \tag{62.12}$$

2. We do not wish to write down the general decomposition formula for a Fourier component with an arbitrary number, n, of wave-vectors, because such a formula would be very complicated. It is, however, easily seen how such a component could be completely contracted. Starting with the initial component, one begins by contracting successively all possible pairs of wave-vectors: one thus obtains in general $2 \cdot \frac{1}{2}n(n-1)$ terms, each containing one factor θ, one delta symbol, one volume factor and a Fourier component with $n-1$ wave-vectors. Each of the terms obtained is in turn submitted to a second set of contractions. After this operation each of the previous terms gives rise (in general) to $2 \cdot 1\frac{1}{2}(n-1)(n-2)$ terms each containing a factor θ^2, two delta symbols, two volume factors and a Fourier component with $n-2$ wave-vectors. This process is continued until one arrives at Fourier components with a single non-vanishing wave-vector (or with all wave-vectors vanishing). Among all the components obtained in this way, each distinct contraction is retained only once.

As an example, we quote the main types of terms obtained by contraction from $\rho_{\mathbf{k}_1\mathbf{k}_2\mathbf{k}_3}$:

$$\begin{aligned}\rho_{\mathbf{k}_1\mathbf{k}_2\mathbf{k}_3}(\mathbf{p}_1, \mathbf{p}_2, \mathbf{p}_3| \ldots) &= \hat{\rho}_{\mathbf{k}_1\mathbf{k}_2\mathbf{k}_3}(\mathbf{p}_1, \mathbf{p}_2, \mathbf{p}_3| \ldots) \\ &+\theta(\Omega/8\pi^3)\delta_{\mathbf{p}_1-\mathbf{p}_2+\hbar(\mathbf{k}_1+\mathbf{k}_2)/2}\hat{\rho}_{\mathbf{k}_1+\mathbf{k}_2,\mathbf{k}_3}(\mathbf{p}_2-\tfrac{1}{2}\hbar\mathbf{k}_1, \mathbf{p}_3|\mathbf{p}_2-\tfrac{1}{2}\hbar\mathbf{k}_2, \ldots) \\ &+\theta(\Omega/8\pi^3)\delta_{\mathbf{p}_2-\mathbf{p}_1+\hbar(\mathbf{k}_1+\mathbf{k}_2)/2}\hat{\rho}_{\mathbf{k}_1+\mathbf{k}_2,\mathbf{k}_3}(\mathbf{p}_1-\tfrac{1}{2}\hbar\mathbf{k}_2, \mathbf{p}_3|\mathbf{p}_1-\tfrac{1}{2}\hbar\mathbf{k}_1, \ldots) \\ &+[13]+[23] \\ &+\theta^2(\Omega/8\pi^3)^2\delta_{\mathbf{p}_1-\mathbf{p}_2+\hbar(\mathbf{k}_1+\mathbf{k}_2)/2}\delta_{\mathbf{p}_2-\frac{1}{2}\hbar\mathbf{k}_1-\mathbf{p}_3+\hbar(\mathbf{k}_1+\mathbf{k}_2+\mathbf{k}_3)/2} \\ &\qquad \hat{\rho}_{\mathbf{k}_1+\mathbf{k}_2+\mathbf{k}_3}(\mathbf{p}_3-\tfrac{1}{2}\hbar[\mathbf{k}_1+\mathbf{k}_2]|\mathbf{p}_3-\tfrac{1}{2}\hbar\mathbf{k}_3, \mathbf{p}_2-\tfrac{1}{2}\hbar\mathbf{k}_2, \ldots) \\ &+\theta^2(\Omega/8\pi^3)^2\delta_{\mathbf{p}_1-\mathbf{p}_2+\hbar(\mathbf{k}_1+\mathbf{k}_2)/2}\delta_{\mathbf{p}_3-\mathbf{p}_2+\frac{1}{2}\hbar\mathbf{k}_1+\hbar(\mathbf{k}_1+\mathbf{k}_2+\mathbf{k}_3)/2} \\ &\hat{\rho}_{\mathbf{k}_1+\mathbf{k}_2+\mathbf{k}_3}(\mathbf{p}_2-\tfrac{1}{2}\hbar\mathbf{k}_1-\tfrac{1}{2}\hbar\mathbf{k}_3|\mathbf{p}_2-\tfrac{1}{2}\hbar\mathbf{k}_1-\tfrac{1}{2}\hbar[\mathbf{k}_1+\mathbf{k}_2], \mathbf{p}_2-\tfrac{1}{2}\hbar\mathbf{k}_2, \ldots) \\ &+\theta^2 \ldots\end{aligned} \tag{62.13}$$

3. It should be noted that a contraction is not always associated with a factor $\Omega/8\pi^3$. In certain cases a contraction leads to a Fourier com-

ponent with the same volume dependence. A typical example is a component of the type $\rho_{\mathbf{k}_1,\mathbf{k}_2,-\mathbf{k}_2}(\mathbf{p}_1, \mathbf{p}_2, \mathbf{p}_3|\ldots)$: the contraction of the two wave-vectors $(\mathbf{k}_1, \mathbf{k}_2)$ or $(\mathbf{k}_1, -\mathbf{k}_2)$ leads to a component of the type $\rho_{\mathbf{k}_1+\mathbf{k}_2,-\mathbf{k}_2}$ (or $\rho_{\mathbf{k}_1-\mathbf{k}_2,\mathbf{k}_2}$), which, according to Table 2.1, has the same volume dependence as the component with three wave-vectors two of which are equal and opposite. However, the important feature of $\rho_{\mathbf{k}_1,\mathbf{k}_2,-\mathbf{k}_2}$ of having a part which is determined by a lower order Fourier component remains true. Moreover, it will be seen that this component still gives a finite contribution, in reduced equations, although it has apparently too small a size (this is due to the special relationship between the wave-vectors).

We are now ready to understand why the separation of the Fourier components ρ_k into sums of components $\hat{\rho}_k$ is important. In fact, we could do the whole theory with the Fourier components ρ_k, which have the same physical content as the classical ρ_k's. However, in trying to extend the considerations of § 3 concerning the orders of magnitude in e^2 of the various Fourier components, we should immediately arrive at a serious difficulty. Indeed, in a quantum-mechanical system, *the molecular correlations are not due to interactions alone as in classical systems, but also to statistics.* Among the well known properties which result from this fact let us quote, for instance, the fact that the equation of state of a *perfect* boson or fermion gas is not identical to the classical one but contains virial coefficients of all orders, which are due only to quantum-statistical correlation effects. The explicit form of the equilibrium density correlation function of an ideal quantum gas has been known for a long time and is given by the *London–Placzek* formula:*

$$g^0(r) = \theta c^2 \left| \int d\mathbf{k}\, e^{i\mathbf{k}\cdot\mathbf{r}} \varphi^0(\hbar\mathbf{k}) \right|^2 \quad \text{(free particles)} \qquad (62.14)$$

where φ^0 is the equilibrium Bose–Einstein or Fermi–Dirac momentum distribution. This shows that $\rho_{\mathbf{k},-\mathbf{k}}$ differs from zero even for free particles, in contradistinction to the classical Fourier component.

Let us now introduce the decomposition (62.12), and calculate $g(r)$ from it:

* F. London, *J. Chem. Phys.*, **11**, 203 (1943); G. Placzek, *2nd. Berkeley Symp. Math. Statist. and Prob.*, Berkeley, 1950, p. 581; S. Fujita and R. Hirota, *Phys. Rev.*, **118**, 6 (1960).

$$
\begin{aligned}
g(r) &= \frac{N(N-1)}{\Omega^2}\,\frac{8\pi^3}{\Omega}\,\frac{h^{3N}}{\Omega^N}\sum_{\mathbf{p}_1}\cdots\sum_{\mathbf{p}_N}\sum_{\mathbf{k}} e^{i\mathbf{k}\cdot\mathbf{r}}\rho_{\mathbf{k},-\mathbf{k}}(\mathbf{p}_1,\mathbf{p}_2|\ldots) \\
&\xrightarrow[N,\,\Omega\to\infty]{} c^2\int d\mathbf{k}\,e^{i\mathbf{k}\cdot\mathbf{r}}\int d\mathbf{p}_1 d\mathbf{p}_2\hat{\rho}_{\mathbf{k},-\mathbf{k}}(\mathbf{p}_1,\mathbf{p}_2) \qquad (62.15)\\
&+c^2\theta\hbar^3\int d\mathbf{k}d\mathbf{p}\,e^{i\mathbf{k}\cdot\mathbf{r}}\varphi(\mathbf{p}+\tfrac{1}{2}\hbar\mathbf{k})\varphi(\mathbf{p}-\tfrac{1}{2}\hbar\mathbf{k}) \\
&= c^2\left\{\int d\mathbf{k}\,e^{i\mathbf{k}\cdot\mathbf{r}}\int d\mathbf{p}_1 d\mathbf{p}_2\hat{\rho}_{\mathbf{k},-\mathbf{k}}(\mathbf{p}_1,\mathbf{p}_2)+\theta\left|\int d\mathbf{q}\,e^{i\mathbf{q}\cdot\mathbf{r}}\varphi(\hbar\mathbf{q})\right|^2\right\}
\end{aligned}
$$

Comparing this with (62.14) we see that the equilibrium correlation for free particles is obtained by setting $\hat{\rho}_{\mathbf{k},-\mathbf{k}} = 0$. This example clearly shows that *our contraction procedure has explicitly separated out the part of the correlation function which is present only when the particles interact from the part which is present even for free particles.* It should be realized that this does not mean that interactions and statistics effects are separated. In an interacting system $\hat{\rho}_{\mathbf{k},-\mathbf{k}}$ is a function of θ (and thus contains statistics effects) and the second term depends on e^2. But the important fact is that $\hat{\rho}_{\mathbf{k},-\mathbf{k}}$ *is only present in interacting systems, and it is therefore proportional to the same power of* e^2 *as the classical coefficient* $\rho_{\mathbf{k},-\mathbf{k}}$, Table 2.1. Other, even more important properties of the contractions will appear later.

The previous discussion is easily extended to all higher order Fourier components, and leads to the conclusion that all Fourier components can be attributed the same order of magnitude in e^2 as in Table 2.1.

To sum up, our description of a quantum-statistical system is the following.

An n-particle correlation pattern of a definite type is described by a Fourier component $\rho_{\mathbf{k}_1\ldots\mathbf{k}_n}(\mathbf{p}_1, \ldots, \mathbf{p}_n| \ldots)$ of the Wigner function. This component is defined by formula (62.6), including the appropriate volume factors taken from Table 2.1. All macroscopic quantities are defined as averages by the same equations as in classical mechanics, provided one replaces everywhere the classical components $\rho_{\{k\}}$ by the corresponding Wigner components.

Each component with more than one wave-vector gives rise to

a sum of non-singular terms $\hat{\rho}_{\{k\}}$ by the contraction process. A given $\rho_{\{k\}}$ appears as a sum of the corresponding non-singular $\hat{\rho}_{\{k\}}$ plus a set of terms involving only components $\hat{\rho}_{\{k'\}}$ with fewer non-vanishing wave-vectors. In the limit of Boltzmann statistics ($\theta = 0$) the latter terms vanish and $\hat{\rho}_{\{k\}} = \rho_{\{k\}}$. We have already encountered one important advantage of studying directly $\hat{\rho}_{\{k\}}$ rather than $\rho_{\{k\}}$: pure quantum-statistics effects (i.e. correlations existing in a degenerate gas of *free* particles) have been separated out and hence the size of the components $\hat{\rho}_{\{k\}}$ is determined by the interactions. Other advantages will be seen later.

§ 63. The Equation of Evolution of the Fourier Components of the Wigner Function

We have not yet discussed the equation of evolution of f_N^W corresponding to the classical Liouville equation, because the most explicit and suggestive form of that equation had to await the exposition of the Fourier expansion and the contraction process.

The fundamental concept in quantum-statistical mechanics is the von Neumann density matrix $\boldsymbol{\rho}$ introduced in § 61: the Wigner function can be considered as a particular representation of this operator. The equation of evolution of f_N^W or of its Fourier components will be derived from the fundamental von Neumann equation, (61.18).

In order to condense the notations, let us introduce the following operators:

$$\Pi_N \equiv : \prod_{m=1}^{N} a^+(\mathbf{p}_m - \tfrac{1}{2}\hbar\mathbf{k}_m)a(\mathbf{p}_m + \tfrac{1}{2}\hbar\mathbf{k}_m) : \tag{63.1}$$

$$P_N(j\pm) \equiv a^+(\mathbf{p}_j - \tfrac{1}{2}\hbar\mathbf{k}_j \pm \hbar\mathbf{l})\Pi_{N-1}^{(j)} a(\mathbf{p}_j + \tfrac{1}{2}\hbar\mathbf{k}_j) \tag{63.2}$$

$$\begin{aligned} P_N(j+, n-) \equiv a^+(\mathbf{p}_j - \tfrac{1}{2}\hbar\mathbf{k}_j + \hbar\mathbf{l})a^+(\mathbf{p}_n - \tfrac{1}{2}\hbar\mathbf{k}_n - \hbar\mathbf{l})\Pi_{N-2}^{(jn)} \\ a(\mathbf{p}_n + \tfrac{1}{2}\hbar\mathbf{k}_n)a(\mathbf{p}_j + \tfrac{1}{2}\hbar\mathbf{k}_j) \end{aligned} \tag{63.3}$$

$$Q_N(j\pm) \equiv a^+(\mathbf{p}_j - \tfrac{1}{2}\hbar\mathbf{k}_j)\Pi_{N-1}^{(j)} a(\mathbf{p}_j + \tfrac{1}{2}\hbar\mathbf{k}_j \pm \hbar\mathbf{l}) \tag{63.4}$$

$$\begin{aligned} Q_N(j+, n-) \equiv a^+(\mathbf{p}_j - \tfrac{1}{2}\hbar\mathbf{k}_j)a^+(\mathbf{p}_n - \tfrac{1}{2}\hbar\mathbf{k}_n)\Pi_{N-2}^{(jn)} \\ a(\mathbf{p}_n + \tfrac{1}{2}\hbar\mathbf{k}_n - \hbar\mathbf{l})a(\mathbf{p}_j + \tfrac{1}{2}\hbar\mathbf{k}_j + \hbar\mathbf{l}) \end{aligned} \tag{63.5}$$

The operator $\Pi_{N-1}^{(j)}$ is obtained from Π_N by suppressing the creation and the destruction operator labeled with the suffix j; an obviously analogous definition holds for $\Pi_{N-2}^{(jn)}$.

We start from eq. (62.6) and take the time derivative of both sides. Noting that only $\boldsymbol{\rho}(t)$ depends on time on the r.h.s. of that equation, we can use eq. (61.18) and then the property of the cyclic invariance of the trace

$$\begin{aligned}\partial_t \tilde{\rho}_{\{k\}}(p;t) &= (N!\,h^{3N})^{-1}\,\mathrm{Tr}\,\{\Pi_N\,\partial_t\boldsymbol{\rho}\}\\ &= (i/\hbar)(N!\,h^{3N})^{-1}\,\mathrm{Tr}\,\{\Pi_N\,[\boldsymbol{\rho},\mathscr{H}]_-\}\\ &= (i/\hbar)(N!\,h^{3N})^{-1}\,\mathrm{Tr}\,\{[\mathscr{H},\Pi_N]_-\,\boldsymbol{\rho}\}\end{aligned} \tag{63.6}$$

In order to derive the equation of evolution, we have therefore to evaluate the commutator in the last term of eq. (63.6), using the hamiltonian (61.1). The contribution of the unperturbed hamiltonian $\mathscr{H}_0$ is very simple and is expressed in the following theorem, which is proved in Appendix 5:

$$\left[\sum_{\mathbf{l}} (\hbar^2 l^2/2m)a^+(\hbar\mathbf{l})a(\hbar\mathbf{l}),\,\Pi_N\right]_- = -\left(\sum_{j=1}^{N}(\hbar/m)\mathbf{k}_j\cdot\mathbf{p}_j\right)\Pi_N \tag{63.7}$$

Consider now the contribution of the interaction hamiltonian. We shall study separately each term of the commutator, beginning with the first. Using a theorem proved in Appendix 5 we can show that

$$\begin{aligned}\mathscr{V}\Pi_N &= \tfrac{1}{2}e^2\sum_{\mathbf{l},\mathbf{r},\mathbf{s}}\langle\mathbf{s}-\mathbf{l},\mathbf{r}+\mathbf{l}|V|\mathbf{r},\mathbf{s}\rangle a^+(\hbar\mathbf{s}-\hbar\mathbf{l})a^+(\hbar\mathbf{r}+\hbar\mathbf{l})a(\hbar\mathbf{r})a(\hbar\mathbf{s})\Pi_N\\ &= \tfrac{1}{2}e^2\sum_{\mathbf{l},\mathbf{r},\mathbf{s}}\langle\mathbf{s}-\mathbf{l},\,\mathbf{r}+\mathbf{l}|V|\mathbf{r},\,\mathbf{s}\rangle\{a^+(\hbar\mathbf{s}-\hbar\mathbf{l})a^+(\hbar\mathbf{r}+\hbar\mathbf{l})\Pi_N a(\hbar\mathbf{r})a(\hbar\mathbf{s})\\ &\quad+\sum_{j=1}^{N}a^+(\hbar\mathbf{s}-\hbar\mathbf{l})P_N(j+)a(\hbar\mathbf{s})\delta_{\hbar\mathbf{r}-\mathbf{p}_j+\hbar\mathbf{k}_j/2}\\ &\quad+\sum_{j=1}^{N}a^+(\hbar\mathbf{r}+\hbar\mathbf{l})P_N(j-)a(\hbar\mathbf{r})\delta_{\hbar\mathbf{s}-\mathbf{p}_j+\hbar\mathbf{k}_j/2}\\ &\quad+\sum_{j<n}P_N(j-,\,n+)\delta_{\hbar\mathbf{s}-\mathbf{p}_j+\hbar\mathbf{k}_j/2}\delta_{\hbar\mathbf{r}-\mathbf{p}_n-\hbar\mathbf{k}_n/2}\\ &\quad+\sum_{j<n}P_N(j+,\,n-)\delta_{\hbar\mathbf{s}-\mathbf{p}_n+\hbar\mathbf{k}_n/2}\delta_{\hbar\mathbf{r}-\mathbf{p}_j-\hbar\mathbf{k}_j/2}\}\end{aligned} \tag{63.8}$$

We now note that in formula (63.6) the operator $\mathscr{V}\Pi_N$ is multiplied by $\boldsymbol{\rho}$, and then the trace of the resulting operator is taken. But we have imposed the restriction (61.6) on the *constancy of the number of particles*; therefore the trace is taken only over states in which the total number of particles is N. An operator of the type $a^+a^+\Pi_N aa$ contains $N+2$ destruction operators (and $N+2$ creation operators): acting on an N-particle state its result is zero. The same is true of the operators $a^+P_N a$, which contain $N+1$ destruction operators. Therefore, only the two last terms of the r.h.s. of (63.8) contribute to the trace of (63.6). Performing the summations over $\mathbf{r}$ and $\mathbf{s}$ and using the obvious symmetry property (see 61.9):

$$\langle \mathbf{rs}|V|\mathbf{tu}\rangle = \langle \mathbf{sr}|V|\mathbf{ut}\rangle \tag{63.9}$$

the two latter terms are seen to be equal, and we get the result:

$$\mathscr{V}\Pi_N = e^2 \sum_{j<n} \sum \sum_{\mathbf{l}} \tag{63.10}$$
$$\langle \mathbf{p}_j - \tfrac{1}{2}\hbar\mathbf{k}_j + \hbar\mathbf{l},\, \mathbf{p}_n - \tfrac{1}{2}\hbar\mathbf{k}_n - \hbar\mathbf{l}|V|\mathbf{p}_n - \tfrac{1}{2}\hbar\mathbf{k}_n,\, \mathbf{p}_j - \tfrac{1}{2}\hbar\mathbf{k}_j\rangle P_N(j+,\, n-)$$

A similar calculation leads to the following result for the second term of the commutator:

$$\Pi_N\mathscr{V} = e^2 \sum_{j<n} \sum \sum_{\mathbf{l}} \tag{63.11}$$
$$\langle \mathbf{p}_j + \tfrac{1}{2}\hbar\mathbf{k}_j,\, \mathbf{p}_n + \tfrac{1}{2}\hbar\mathbf{k}_n|V|\mathbf{p}_n + \tfrac{1}{2}\hbar\mathbf{k}_n + \hbar\mathbf{l},\, \mathbf{p}_j + \tfrac{1}{2}\hbar\mathbf{k}_j - \hbar\mathbf{l}\rangle Q_N(j-,\, n+)$$

It is now easy to substitute the results obtained above into eq. (63.6), and to derive the equation for $\tilde{\rho}_{\{k\}}$. We must note that:

$$\begin{aligned}
&(N!)^{-1}h^{-3N}\,\mathrm{Tr}\{P_N(j+,\, n-)\rho\} \\
&= (N!\,h^{3N})^{-1}\,\mathrm{Tr}\,\{a^+(\mathbf{p}_j + \tfrac{1}{2}\hbar\mathbf{l} - \tfrac{1}{2}\hbar\mathbf{k}_j + \tfrac{1}{2}\hbar\mathbf{l})a^+(\mathbf{p}_n - \tfrac{1}{2}\hbar\mathbf{l} - \tfrac{1}{2}\hbar\mathbf{k}_n - \tfrac{1}{2}\hbar\mathbf{l}) \\
&\qquad \Pi_N^{(jn)}\, a(\mathbf{p}_n - \tfrac{1}{2}\hbar\mathbf{l} + \tfrac{1}{2}\hbar\mathbf{k}_n + \tfrac{1}{2}\hbar\mathbf{l})a(\mathbf{p}_j + \tfrac{1}{2}\hbar\mathbf{l} + \tfrac{1}{2}\hbar\mathbf{k}_j - \tfrac{1}{2}\hbar\mathbf{l})\rho\} \qquad (63.12) \\
&= \tilde{\rho}_{\mathbf{k}_1,\ldots,\mathbf{k}_j-\mathbf{l},\ldots,\mathbf{k}_n+\mathbf{l},\ldots,\mathbf{k}_N}(\mathbf{p}_1,\, \ldots,\, \mathbf{p}_j + \tfrac{1}{2}\hbar\mathbf{l},\, \ldots,\, \mathbf{p}_n - \tfrac{1}{2}\hbar\mathbf{l},\, \ldots,\, \mathbf{p}_N) \\
&= \exp\,[\tfrac{1}{2}\hbar\mathbf{l}\cdot\mathfrak{D}_{jn}]\tilde{\rho}_{\mathbf{k}_1,\ldots,\mathbf{k}_j-\mathbf{l},\ldots,\mathbf{k}_n+\mathbf{l},\ldots,\mathbf{k}_N}(\mathbf{p}_1,\, \ldots,\, \mathbf{p}_N)
\end{aligned}$$

where $\mathfrak{D}_{jn} = \partial/\partial\mathbf{p}_j - \partial/\partial\mathbf{p}_n$.

The trace involving Q_N is calculated in a similar way, and finally all the results are collected into the following equation (including the correct volume factors):

$$
\begin{aligned}
&\partial_t \rho_{\mathbf{k}_1\ldots\mathbf{k}_r}(p) + i \sum_n \mathbf{k}_n \cdot \mathbf{v}_n\, \rho_{\mathbf{k}_1\ldots\mathbf{k}_r}(p) \\
&= (ie^2/\hbar) \sum_{j<n} \sum \sum_{\mathbf{l}} \{\langle \mathbf{p}_j - \tfrac{1}{2}\hbar\mathbf{k}_j + \hbar\mathbf{l}, \\
&\qquad \mathbf{p}_n - \tfrac{1}{2}\hbar\mathbf{k}_n - \hbar\mathbf{l} | V | \mathbf{p}_n - \tfrac{1}{2}\hbar\mathbf{k}_n, \mathbf{p}_j - \tfrac{1}{2}\hbar\mathbf{k}_j \rangle \exp[\tfrac{1}{2}\hbar\mathbf{l}\cdot\mathfrak{D}_{jn}] \\
&- \langle \mathbf{p}_j + \tfrac{1}{2}\hbar\mathbf{k}_j, \mathbf{p}_n + \tfrac{1}{2}\hbar\mathbf{k}_n | V | \mathbf{p}_n + \tfrac{1}{2}\hbar\mathbf{k}_n + \hbar\mathbf{l}, \mathbf{p}_j + \tfrac{1}{2}\hbar\mathbf{k}_j - \hbar\mathbf{l} \rangle \qquad (63.13) \\
&\qquad \exp[-\tfrac{1}{2}\hbar\mathbf{l}\cdot\mathfrak{D}_{jn}]\} \\
&\qquad (8\pi^3/\Omega)^{\nu'-\nu} \rho_{\mathbf{k}_1,\ldots,\mathbf{k}_j-\mathbf{l},\ldots,\mathbf{k}_n+\mathbf{l},\ldots,\mathbf{k}_r}(p)
\end{aligned}
$$

The exponent ν has been defined in eq. (5.25).

Equation (63.13) is already very interesting in itself. One has to compare it with eq. (5.25), which is its exact classical counterpart. Two remarks are obvious.

A. The left-hand sides of both equations are identical: in the Wigner formalism the free motion of the particles is described in exactly the same way as in classical mechanics.

B. The right-hand sides of the equations are formally very similar: they are of the form of a sum of matrix elements connecting different Fourier components. In both cases, these matrix elements are *operators acting on the momenta*. However, whereas in the classical case the matrix elements are *differential operators*, in the quantum-mechanical case they are *finite displacement operators*. We shall come back to this essential feature in § 64.

However, eq. (63.13) cannot be considered as the final stage of our investigation, because of the peculiar structure of the coefficients ρ_k which was explained in detail in § 62. It has already been shown that a perturbation method using (63.13) would be very complex because all coefficients $\rho_{\{k\}}$ are of the same order in e^2. More important, if we want to derive from (63.13) *equations for reduced Fourier components* (by integration over momenta) we shall run into difficulties. Indeed, we have seen that the coefficients $\rho_{\{k\}}$ have quantum-statistical *δ-singularities in the momentum variables*. Consequently, the reduction process is much more complex than in the classical case. For all these reasons it is much more convenient to study the evolution of the non-singular components $\hat{\rho}_{\{k\}}$. The tedious calculations involved in the reduction process are then done once and for all. The resulting expressions can then be taken as they stand for the calculation of any physical quantity. Having determined these coefficients we can reconstitute immediately the complete correlation functions $\rho_{\{k\}}$ by using the contraction formulae. In conclusion, our next task will be the *derivation of equations of evolution for the coefficients* $\hat{\rho}_{\{k\}}$. The procedure to be followed here is very simple, although the calculations are very long and tedious. We shall thus only give the rules to be used in this calculation and tabulate the results obtained. The procedure is illustrated by a simple example which is carried out in some detail in Appendix 6.

Consider successively the equations for $\rho_0, \rho_{\mathbf{k}}, \rho_{\mathbf{k},\mathbf{k}'}, \ldots$

1. On the left-hand side, the Fourier component ρ is decomposed completely, and the contracted parts are separated off.

2. The Fourier components occurring on the right-hand side are also decomposed by successive contractions.

3. The terms of the r.h.s. contributing to the *contracted* parts of the l.h.s. (which have been separated off) are discarded. These terms are easily recognizable.

Example: the equation for $\partial_t \rho_{\mathbf{k},-\mathbf{k}}(\mathbf{p}_1, \mathbf{p}_2| \ldots)$ is separated into a sum of two equations by applying (62.12):

$$\partial_t \hat{\rho}_{\mathbf{k},-\mathbf{k}}(\mathbf{p}_1, \mathbf{p}_2| \ldots) + \theta(\Omega/8\pi^3)\delta_{\mathbf{p}_2-\mathbf{p}_1} \partial_t \hat{\rho}_0(|\mathbf{p}_1 - \tfrac{1}{2}\hbar\mathbf{k}_2, \mathbf{p}_1 - \tfrac{1}{2}\hbar\mathbf{k}_1, \ldots)$$

The two contributions are easily recognized by the occurrence of the δ-function $\delta_{\mathbf{p}_2-\mathbf{p}_1}$ in the second term. Therefore, in order to obtain an equation for $\partial_t \hat{\rho}_{\mathbf{k},-\mathbf{k}}$ alone, all terms of the r.h.s. in which the factor $\delta_{\mathbf{p}_2-\mathbf{p}_1}$ appears through the contraction process are discarded.

4. One must be careful to count each distinct contraction only once.

5. At this point it is convenient to change variables from the momenta to the velocities. This change is trivial: it involves only a few factors of m in the right places.

6. After all contractions have been performed, all the terms containing the same Fourier components on the r.h.s. are collected, and the results are expressed in the form:

$$\partial_t \hat{\rho}_{\{k\}}(v;t) + i \sum_j \mathbf{k}_j \cdot \mathbf{v}_j \hat{\rho}_{\{k\}}(v;t) = \sum_{\{\mathbf{k}'\}} e^2 \langle \{k\} | \mathscr{L}' | \{k'\} \rangle (8\pi^3/\Omega)^{\nu'-\nu} \hat{\rho}_{\{k'\}}(v;t) \tag{63.14}$$

The simplest way of representing the results is to introduce diagrams again exactly as in the classical case; in particular, a vertex is associated with each distinct matrix element. The results can be summarized as follows.

I. The number of essentially distinct matrix elements is finite, as it is in the classical case. [We call "essentially distinct" those matrix elements characterized by distinct vertices; matrix elements differing only by lines which do not go through the vertex are not considered as essentially distinct (see also p. 44).] However, instead of 6 basic vertices, there are now 9 *basic vertices for quantum-statistical systems*. The corresponding matrix elements are listed in Table 63.1 at the end of the book.

II. As has been noted already, the various matrix elements are *finite displacement operators* acting on every function of the momenta written to their right.

III. The structure of eq. (63.14) is identical to the structure of the classical Liouville equation (5.25). This is why we use the same notations for the operator $\mathscr{L}'$ (which is defined by its matrix elements listed in Table 63.1) as for the classical Liouville operator; eq. (63.14) will be called the "*quantum-statistical Liouville equation*".

IV. It is interesting to discuss the meaning and origin of the three purely quantum-statistical vertices. The vertex G is due to the existence of the exchange term in the matrix element of the potential [see (61.9)]: the direct term does not contribute. The existence of this vertex is thus a direct consequence of the indiscernibility of the particles. The important feature of this vertex is its *diagonal* character. It is closely analogous to the well known process of "*forward collision*".

Vertex H has a different origin. The process which it describes is very similar to the one of vertex F, i.e. a transfer of correlation from particle n to particle 1. However, the existence of another line influences the expression for the corresponding matrix element: this is, of course, again a consequence of the indiscernibility of the particles. We prefer to separate the change in the vertex F owing to the presence of an extra line and define it as a separate vertex; for this purpose we introduce the wavy line connecting the vertex $1-n$ with the line 2.

Vertex J has the same origin as H: it is the modification of vertex B owing to the presence of an extra line. This vertex has the peculiarity that *three* wave-vectors change their values during the transition, although there is no three-body interaction. This apparent three-body interaction is another characteristic quantum-statistical effect.

Having separated off the vertices H and J, all nine vertices are independent of extra lines. This fact is very important and is the very reason for the practicability of the method: if each vertex were modified by the presence of each extra line, the number of basic vertices would be infinite, making explicit calculations extremely difficult.

§ 64. The Limit of Boltzmann Statistics and the Classical Limit

The formalism developed in the previous sections is just one of a number of possible methods of treating quantum-statistical systems. It is closely related to the formalism developed by Résibois,* which is based on a study of the von Neumann density matrix in the occupation number representation. The formalism of Van Hove,** and the numerous other diagram techniques developed for the quantum-mechanical many-body problem † differ more from the present one in their motivation, purpose or emphasis.

The formalism based on the Wigner function, however, has the unique advantage over the others of a complete formal identity with the classical theory. This enables us to treat all systems,

* P. Résibois, *Physica,* **27**, 541 (1961).

** L. Van Hove, *Physica,* **21**, 517 (1955); **23**, 441 (1957).

† I. Prigogine and P. Résibois, *Physica,* **24**, 705 (1958); E. W. Montroll and J. C. Ward, *Physica,* **25**, 423 (1959); S. Fujita, *Physica,* **27**, 940 (1961). A number of important reprints in this field are collected in D. Pines, *The Many Body Problem,* Benjamin, New York, 1962.

from extreme classical situations to extreme quantum situations, within the same framework. Another important advantage of this formalism is the fact that *the Wigner function is directly connected to the macroscopic quantities*, which are calculated by taking a simple average weighted with f_s^W. Any explicit calculation starting with $\boldsymbol{\rho}$ leads to long and tedious trace computations, which must be repeated in each particular example. In our formalism these long calculations are done once and for all.

Because of the close analogy with the classical theory, the classical limit of any quantum-statistical expression can be calculated very easily. This is done in two steps.

A. The first step consists in evaluating expressions which are valid for *quantum-mechanical systems which satisfy Boltzmann (classical) statistics.*

In order to obtain the corresponding expressions, it is sufficient to set $\theta = 0$ in all matrix elements of Table 63.1. It has already been stated that one has to be careful in using this limiting procedure; as long as the second quantization formalism is being manipulated one can switch without ambiguity from $\theta = +1$ to $\theta = -1$, but taking $\theta = 0$ results in ambiguous or even meaningless expressions. On the other hand, the coefficients $\rho_{\{k\}}$ have an intrinsic physical meaning, as in classical statistics. In all equations involving these coefficients, the quantum-statistical effects are very simply eliminated by setting $\theta = 0$. The resulting matrix elements are listed in Table 64.1 at the end of the book.

The quantum-mechanical theory with Boltzmann statistics was first elaborated in the present form by Prigogine and Ono.* We note that the purely quantum-statistical vertices G, H, J have disappeared in this limit, as would be expected (they are proportional to θ). By comparison with Table 6.1, it is seen that the quantum expressions are obtained from the corresponding classical ones by making the replacement

$$\underset{\text{classical}}{m^{-1}\mathbf{1}\cdot\partial_{jn}} \rightarrow \underset{\text{quantum-mechanical}}{\hbar^{-1}[\exp(\tfrac{1}{2}\hbar m^{-1}\mathbf{1}\cdot\partial_{jn}) - \exp(-\tfrac{1}{2}\hbar m^{-1}\mathbf{1}\cdot\partial_{jn})]} \quad (64.1)$$

It is also interesting to note that the Fourier components $\rho_{\{k\}}$

* I. Prigogine and S. Ono, *Physica*, **25**, 171 (1959).

of the Wigner function are in this case very closely related to the von Neumann density matrix. In order to see this connection we start with the following form of the Wigner function, valid for Boltzmann statistics * and analogous to (61.24):

$$f_N^W = \int (d\mathbf{k})^N (d\mathbf{s})^N \exp\left[i \sum_r (\mathbf{k}_r \cdot \mathbf{x}_r + \mathbf{s}_r \cdot \mathbf{p}_r)\right] \\ \mathrm{Tr}\,\{\boldsymbol{\rho}(t) \exp\,[-i \sum_r (\mathbf{k}_r \cdot \mathbf{x}'_r - i\hbar \mathbf{s}_r \cdot \boldsymbol{\nabla}'_r)]\} \tag{64.2}$$

or, after some simple transformations:

$$f_N^W = \int (d\mathbf{k})^N e^{i\Sigma \mathbf{k}_r \cdot \mathbf{x}_r} \tilde{\rho}_{\{k\}}(\{p\}) \tag{64.3}$$

with

$$\tilde{\rho}_{\{k\}}(\{p\}) = \int (d\mathbf{s})^N e^{isp}\, \mathrm{Tr}\, \boldsymbol{\rho}(t) B \tag{64.4}$$

$$B = e^{-ikx'} e^{i\hbar ks/2} e^{-\hbar s\nabla'} \tag{64.5}$$

Using the momentum representation, we have

$$\langle\kappa|B|\kappa'\rangle = e^{-i\hbar s\kappa' + i\hbar ks/2} \delta_{k-\kappa'-\kappa}$$

and therefore

$$\tilde{\rho}_k(p) = \sum_{\kappa,\kappa'} \langle\kappa'|\boldsymbol{\rho}|\kappa\rangle \delta_{k-\kappa'+\kappa} \delta_{p-\hbar(\kappa+\kappa')/2} \tag{64.6}$$

Thus $\tilde{\rho}_k(p)$ is nothing other than an off-diagonal matrix element of the density matrix, in the momentum representation, such that $\kappa'-\kappa = k$ and $\hbar(\kappa+\kappa')/2 = p$. This correspondence principle has been suggested by Heisenberg ** and was formulated explicitly in this case by Prigogine and Ono. †

B. *The classical limit* is obtained from the previous formulae by letting $\hbar \to 0$. In view of the correspondence principle (64.1) this limit is very simply obtained. It is sufficient to expand the exponentials in powers of $\hbar$: the first non-vanishing term is

* See, e.g., D. Massignon, *Mécanique Statistique des Fluides,* Dunod, Paris, 1957.

** W. Heisenberg, *The Physical Principles of the Quantum Theory,* Univ. of Chicago, 1930; reprinted by Dover Publ. Co., New York.

† I. Prigogine and S. Ono, *loc. cit.*

precisely the classical operator of the left-hand side. This expansion procedure, pushed further, is also a method of investigating weak quantum effects.

The fact that a differential operator $\mathbf{l} \cdot \partial_{jn}$ is replaced in quantum mechanics by a finite displacement operator is a consequence of the uncertainty principle. The elementary dynamical act in classical mechanics is an infinitesimal displacement of momenta. In quantum mechanics, however, even a very weak cause can produce a finite change of momentum.

§ 65. The Quantum-Statistical Diagram Technique

At the present point of the elaboration of the theory, the analogy with the classical formalism is so strong that most of the results of Chapter 1 can be taken over without any change. In particular, the results of §§ 4 and 5, the important concepts of the Green's function and of the resolvent and the whole scheme of perturbation theory and of the diagram technique, which forms the basis of the theory, are exactly the same in classical and in quantum statistics. Moreover, the unperturbed Green's functions (and the unperturbed resolvents) are identical for classical and quantum systems and, in particular, the fundamental formulae (5.19) and (5.21) are valid throughout. Also, formula (5.26), which is the starting point of the perturbation theory, remains valid as it stands, replacing of course the matrix elements of Table 6.1 by those of Table 63.1. The diagrams are also constructed by the same rules as in § 6.

An important difference arises, however, in the evaluation of the orders of magnitude of the diagrams. Whereas the e^2-dependence of both types of diagrams is the same, the N- and Ω-dependence (or better, the c-dependence) is different, and more complicated, in the quantum-statistical case. It will be shown now that *statistics effects introduce into the contributions to the reduced* $\hat{\rho}_k$ *arbitrary powers of the concentration in the combination* h^3c. The significance of this combination will be discussed in a later section. In order to see how this feature appears, consider in detail a simple example. The first term of the matrix element F, substituted into (5.26), gives the following contribution to the reduced inhomogeneity factor $\rho_{\mathbf{k}}(\mathbf{v}_\alpha; t)$:

$$\rho_{\mathbf{k}}(\alpha; t)|_F = \frac{1}{2\pi}\left(\frac{h^3}{m^3\Omega}\right)^{N-1} \sum_{\mathbf{v}_1}\cdots{\sum_{(\alpha)}}' \cdots \sum_{\mathbf{v}_N} \int dz\, \mathrm{e}^{-izt} \frac{1}{i(\mathbf{k}\cdot\mathbf{v}_\alpha - z)} \sum_n \sum_{\mathbf{k}'_n}$$

$$\frac{8\pi^3}{\Omega}(-e^2)\delta_{\mathbf{k}'_n-\mathbf{k}}\, i\hbar^{-1}[V_k + \theta V_{|m(\mathbf{v}_\alpha-\mathbf{v}_n)\hbar+\mathbf{k}/2|}](1+\theta\sum_s \delta_{\mathbf{v}_s-\mathbf{v}_\alpha+\hbar\mathbf{k}/2m})$$

$$\exp\left[\tfrac{1}{2}\hbar m^{-1}\mathbf{k}\cdot\partial_{\alpha n}\right]\frac{1}{i(\mathbf{k}\cdot\mathbf{v}_n - z)}\rho_{\mathbf{k}}(\mathbf{v}_n|\mathbf{v}_\alpha\ldots)$$

$$= \frac{1}{2\pi}\int dz\, \mathrm{e}^{-izt}\frac{1}{i(\mathbf{k}\cdot\mathbf{v}_\alpha - z)} 8\pi^3\frac{N}{\Omega}(-e^2)\frac{h^3}{m^3\Omega}\sum_{\mathbf{v}_n}$$

$$\frac{i}{\hbar}[V_k + \theta V_{|m(\mathbf{v}_\alpha-\mathbf{v}_n)/\hbar+\mathbf{k}/2|}]\frac{1}{i[\mathbf{k}\cdot(\mathbf{v}_n - \hbar\mathbf{k}/2m) - z]}. \tag{65.1}$$

$$[\rho_{\mathbf{k}}(\mathbf{v}_n - \tfrac{1}{2}\hbar m^{-1}\mathbf{k})\varphi(\mathbf{v}_\alpha + \tfrac{1}{2}\hbar m^{-1}\mathbf{k})$$

$$+\theta h^3 m^{-3}(N/\Omega)\sum_{\mathbf{v}_s}\delta_{\mathbf{v}_s-\mathbf{v}_\alpha+\hbar\mathbf{k}/2m}\,\rho_{\mathbf{k}}(\mathbf{v}_n - \tfrac{1}{2}\hbar m^{-1}\mathbf{k})\varphi(\mathbf{v}_\alpha+\tfrac{1}{2}\hbar m^{-1}\mathbf{k})\varphi(\mathbf{v}_s)]$$

$$\xrightarrow[\substack{N\to\infty\\ \Omega\to\infty}]{} \frac{1}{2\pi}\int dz\, \mathrm{e}^{-izt}\frac{1}{i(\mathbf{k}\cdot\mathbf{v}_\alpha - z)}(-e^2)8\pi^3 c$$

$$\int d\mathbf{v}_n\, i\hbar^{-1}[V_k + \theta V_{|m(\mathbf{v}_\alpha-\mathbf{v}_n)/\hbar+\mathbf{k}/2|}]$$

$$\frac{1}{i[\mathbf{k}\cdot(\mathbf{v}_n - \hbar\mathbf{k}/2m) - z]}$$

$$[1+\theta h^3 c m^{-3}\varphi(\mathbf{v}_\alpha - \tfrac{1}{2}\hbar m^{-1}\mathbf{k})]\rho_{\mathbf{k}}(\mathbf{v}_n - \tfrac{1}{2}\hbar m^{-1}\mathbf{k})\varphi(\mathbf{v}_\alpha + \tfrac{1}{2}\hbar m^{-1}\mathbf{k})$$

This example shows that, in the reduced functions, each delta symbol arising from a contraction (such as $\delta_{\mathbf{v}_s-\mathbf{v}_\alpha+\hbar\mathbf{k}/2m}$ above) introduces a factor h^3cm^{-3}. The form of the statistical corrections thus obtained is very familiar $(1+\theta h^3 c m^{-3}\varphi)$: they are a direct consequence of the exclusion principle for Fermi–Dirac statistics, and of the enhancement of the weight of multiply occupied states in Bose–Einstein statistics.

As a consequence of this property, we can state the following theorem, which generalizes theorem IV of § 8:

A quantum-statistical diagram consists of a sum of terms, all proportional to a power of c given by the classical topological index, and moreover to some power of h^3c.

An important property which is maintained for the quantum-statistical diagrams concerns the rules of § 8 about reduced distribution functions. The basis of these properties in the quantum diagrams is provided by the following theorems, which generalize the ordinary rules of integration by parts.

Theorem 1.

$$\int d\mathbf{v} \exp (\mathbf{l} \cdot \partial) f(\mathbf{v}) = \int d\mathbf{v} f(\mathbf{v}) \tag{65.2}$$

Theorem 2.

$$\int d\mathbf{v} g(\mathbf{v}) \exp (\mathbf{l} \cdot \partial) f(\mathbf{v}) = \int d\mathbf{v} f(\mathbf{v}) \exp (-\mathbf{l} \cdot \partial) g(\mathbf{v}) \tag{65.3}$$

These theorems are very easily proved by a substitution of variables:

$$\int d\mathbf{v} \exp (\mathbf{l} \cdot \partial) f(\mathbf{v}) = \int d\mathbf{v} f(\mathbf{v}+\mathbf{l}) = \int d\mathbf{v}' f(\mathbf{v}')$$

$$\int d\mathbf{v} g(\mathbf{v}) \exp (\mathbf{l} \cdot \partial) f(\mathbf{v}) = \int d\mathbf{v} g(\mathbf{v}) f(\mathbf{v}+\mathbf{l}) = \int d\mathbf{v}' g(\mathbf{v}'-\mathbf{l}) f(\mathbf{v}')$$

$$= \int d\mathbf{v}' f(\mathbf{v}') \exp (-\mathbf{l} \cdot \partial') g(\mathbf{v}')$$

Let us now extend Theorem I of p. 49 to quantum-mechanical systems. It is necessary to show that a diagram containing vertex C, in which both momenta $\mathbf{v}_j$ and $\mathbf{v}_n$ are integrated over, vanishes. Using Table 63.1 and (65.2), such a diagram gives a contribution of the form:

$$\begin{aligned} \int d\mathbf{v}_j d\mathbf{v}_n \{ & \exp [\tfrac{1}{2}\hbar m^{-1} \mathbf{l} \cdot \partial_{nj}] [V_l + \theta V_{m|\mathbf{v}_j - \mathbf{v}_n|/\hbar}] \\ & - \exp [-\tfrac{1}{2}\hbar m^{-1} \mathbf{l} \cdot \partial_{nj}] [V_l + \theta V_{m|\mathbf{v}_j - \mathbf{v}_n|/\hbar}] \} F(\mathbf{v}_j \mathbf{v}_n) \\ = \int d\mathbf{v}_j d\mathbf{v}_n \{ & [V_l + \theta V_{m|\mathbf{v}_j - \mathbf{v}_n|/\hbar}] - [V_l + \theta V_{m|\mathbf{v}_j - \mathbf{v}_n|/\hbar}] \} F(\mathbf{v}_j \mathbf{v}_n) = 0 \end{aligned} \tag{65.4}$$

These properties are sufficient to prove, as in § 8, that the diagrams contributing to a reduced distribution function f consist of at most s disconnected parts, each of which is a connected or semiconnected diagram (see Theorem II, p. 50). This then implies that the reduced distribution functions, if initially finite, remain finite for all times in the limit $N \to \infty$, $\Omega \to \infty$, $N/\Omega = c$ (basic theorem, p. 52).

We have now elaborated the complete quantum-statistical formalism, and are ready to study its applications to charged particles.

CHAPTER 15

Short-Time Behavior of Quantum Plasmas and the Quantum Vlassov Equation

§ 66. The Choice of Diagrams

The rules of choice for quantum diagrams are based in the first place on a correspondence principle which is easily formulated in the formalism developed until now. The basic fact here is the correspondence between classical and quantum vertices.

If a problem has been solved in classical mechanics by summing a certain group of diagrams, it is necessary in solving the corresponding quantum-statistical problem to sum all the same diagrams (in which the vertices represent, of course, the quantum matrix elements, Table 63.1). Indeed, if a single diagram of the group is left out, the classical limit cannot be correctly recovered. From this point of view the choice is easy. *We must, however, not forget that there are three quantum-statistical vertices which have no classical counterpart.* Their contribution has to be discussed separately.

We shall now apply this general principle to the problem of the electron gas. *From here on we shall thus only consider a gas of charged fermions.* We shall, however, keep the parameter θ (whose value is -1) explicitly in the expressions in order to make apparent the transition to the Boltzmann limit $\theta = 0$. The first question to be raised is the following: *Is the basic assumption*

$$\Gamma = e^2 c^{\frac{1}{3}} \beta \ll 1 \tag{66.1}$$

compatible with the requirements of quantum statistics?

The problem is made more precise if we note that the introduction of quantum mechanics implies the existence of an additional parameter for the characterization of a charged gas: this parameter is the Planck constant h. The existence of this extra quantity radically changes the dimensional analysis performed in § 2.1.* In particular, Γ is no longer the only dimension-

* A dimensional analysis similar to the one below has been performed by E. W. Montroll and J. C. Ward, *Phys. of Fluids*, **1**, 55 (1958).

less quantity which can be constructed with the parameters at our disposal. It is immediately verified that the following quantity is also dimensionless, and is independent of Γ:

$$\Sigma = \hbar^2 c^{\frac{2}{3}} \beta m^{-1} \tag{66.2}$$

The parameter Σ is independent of the interactions (e^2) and measures the importance of quantum-statistical effects, or the "*degree of degeneration*" of the system. It is seen to be very small in the classical region of high temperatures and small densities, as well as for heavy particles. The existence of an additional parameter changes the principle of corresponding states which was derived in classical mechanics. The new principle states:

Any reduced (i.e. dimensionless) property of the system can depend on e^2, c, β, m and $\hbar$ only through the following combination:

$$\gamma = \Gamma f(\Sigma) \tag{66.3}$$

The function $f(\Sigma)$ cannot be exactly specified, except for its limiting properties. For very small values of Σ, i.e. in the classical

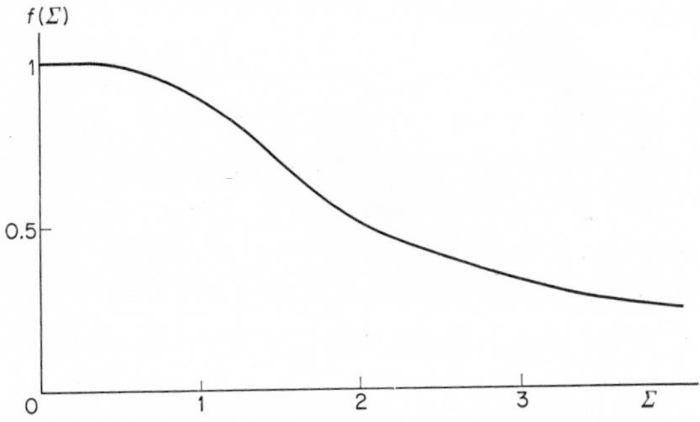

Fig. 66.1. General shape of the function $f(\Sigma)$.

limit, we know that the reduced properties can only depend on Γ, and hence:

$$f(\Sigma) \to 1 \quad \text{for} \quad \Sigma \to 0 \tag{66.4}$$

In the limit of very low temperatures, $\beta \to \infty$ (and thus $\Sigma \to \infty$), the properties of the system must tend toward a definite finite value. This value must be independent of temperature. We therefore require that $f(\Sigma)$ compensates the temperature dependence of Γ, hence:

$$f(\Sigma) \to \Sigma^{-1} \tag{66.5}$$

The function $f(\Sigma)$ has the general form shown in Fig. 66.1. This is all that can be said from purely dimensional arguments.

Let us now examine the meaning of the classical criteria for the choice of diagrams in the quantum region. The basic argument here was the separation of contributions proportional to e^2 and to e^2c. The basis of this distinction lies in the following two arguments:

(*a*) The plasma frequency is $\omega_p = (4\pi e^2 c/m)^{\frac{1}{2}}$; if we wish to obtain a theory valid for times of the order of or longer than ω_p^{-1} this quantity must be retained to all orders, and hence all diagrams proportional to powers of e^2c must be summed over.

(*b*) The parameter $\Gamma = e^2c^{\frac{1}{3}}\beta$ contains uncompensated factors of e^2; thus every diagram proportional to e^2 leads to physical quantities proportional to some power of Γ, and such quantities are assumed to be small.

In the quantum-statistical region, the essential features of the collective behavior are not expected to be very different from the classical ones. As ω_p is independent of the temperature, it may be expected to play the same central role in the quantum region as in the classical region. This expectation will be confirmed *a posteriori* by the theory. But argument (*b*) does certainly not apply in the quantum region. In effect, comparing the forms of the parameters Γ and Σ:

$$\Gamma = e^2c^{\frac{1}{3}}\beta$$
$$\Sigma = \hbar^2 c^{\frac{2}{3}}\beta m^{-1}$$

it is immediately seen that if the temperature is lowered or the density is raised (in order to obtain appreciable quantum effects) the value of Γ is necessarily increased. The answer to our initial question is thus negative: *A highly degenerate electron gas necessarily has a large Γ.*

However, our previous dimensional argument shows that Γ

is no longer the characteristic parameter of the system. The discussion must rather be based on the number γ, eq. (66.3), which measures simultaneously the effects of the interactions and of quantum statistics. In particular, in the extreme quantum region, where (66.5) is applicable, the appropriate parameter is:

$$\gamma_0 = \Gamma/\Sigma = e^2 m/\hbar^2 c^{\frac{1}{3}} \tag{66.6}$$

The natural assumption for a perturbation theory is to take γ_0 as a smallness parameter for a degenerate electron gas:

$$\gamma_0 \ll 1 \tag{66.7}$$

It is remarkable to note that assumption (66.7) is a *high density approximation,* whereas the classical assumption (66.1) gives a low density theory.

We now note that γ_0, like Γ, is also proportional to *uncompensated powers of* e^2. This fact shows that if in our quantum-statistical perturbation expansion we adopt the rule of treating e^2 as "small" but e^2c as a "finite" quantity, this means precisely that γ_0 is treated as a small quantity in accordance with (66.7).

This rather lengthy discussion was essentially meant to show why *the unique rule of choice of diagrams, based on the distinction between* e^2 *and* e^2c, *leads to correct results in the classical region as well as in the quantum region, although the basic assumption* $\gamma \ll 1$ *covers essentially different types of systems in these two regions.*

We now specify our problem more precisely. We shall be interested in the evolution of $\rho_{\mathbf{k}}(\mathbf{v}_\alpha; t)$ with the initial condition

$$\rho_{\mathbf{k}}(\mathbf{v}_\alpha; t=0) = q_{\mathbf{k}}(\mathbf{v}_\alpha) \tag{66.8}$$

We also assume that the deviations from homogeneity are sufficiently small to justify a linearized treatment. We thus consider only diagrams with one outgoing line at left and one at right.

If, moreover, we are interested in *short times,* of order t_p, this is the exact quantum-statistical counterpart of the linearized Vlassov problem studied in Chapter 2. The correspondence rule discussed above states immediately that in this approximation one must sum all the quantum-statistical diagrams which are of the same form as the classical diagrams, i.e. those of Fig. 66.2a.

It is also immediately seen, as in the classical case, that vertex C cannot contribute in the present approximation, and thus $\varphi(\mathbf{v}_\alpha)$ is a constant of the motion in this approximation.

Fig. 66.2. Diagrams in the quantum Vlassov approximation.

We need now a separate discussion of the purely quantum-statistical vertices of Table 63.1. It is immediately seen, for topological reasons, that the only relevant diagram in the present discussion is the "forward scattering" diagram G. It is also seen that this vertex is of order e^2c and should be retained within the Vlassov approximation (see Fig. 66.2b).

Having thus chosen the relevant diagrams, we now proceed to simplify the problem. The following two approximations will be made here and in all subsequent chapters:

1) *In all the vertices, the exchange part of the interaction matrix element will be neglected.*

2) *The forward scattering diagrams will be neglected.*

These two approximations are actually related, because we have seen that only the exchange part of the potential contributes to diagram G. These approximations are made in all the present treatments of the electron gas although, in our opinion, there exists no very clear justification for their use, at least in the short-time behavior. (It will be seen in later chapters that the long-time study introduces an additional divergence argument which gives a clearer reason for their neglect.) The only convincing answer would come from a treatment which includes, at least approximately, these vertices and estimates their contribution to physical quantities. This problem is very simple formally for vertex G, which is *diagonal*; its inclusion amounts to a modification of the unperturbed resolvent $R^0(z)$. The inclusion of the exchange part of vertex F is more difficult but feasible. We shall, however, not go into these questions here.

§ 67. Derivation and Solution of the Linear Quantum Vlassov Equation

The treatment here is completely analogous to the classical treatment of §§ 10 and 13. It is first seen that $\varphi(\alpha)$ remains constant in time for short times, because there is no diagram of the desired order which would contribute to $\rho_0(t)$. We therefore examine more closely the interesting part of $f_1(\mathbf{x}_\alpha, \mathbf{v}_\alpha; t)$, i.e. $\rho_\mathbf{k}(\mathbf{v}_\alpha; t)$. We assume that initially the Wigner distribution function is:

$$f(\alpha; 0) = c\left\{\varphi(\mathbf{v}_\alpha) + \int d\mathbf{k}\, q_\mathbf{k}(\mathbf{v}_\alpha)\, e^{i\mathbf{k}\cdot\mathbf{x}_\alpha}\right\} \tag{67.1}$$

We now write a formula for the reduced inhomogeneity factor corresponding to the diagram Fig. 66.2a by using formula (5.26) with the matrix element F of Table 63.1 [in which the exchange contributions $\theta V(|m(\mathbf{v}_1 - \mathbf{v}_n)/\hbar \pm \frac{1}{2}\mathbf{k}_1)$ are neglected].

The analog of (13.2) is then:

$$\rho_\mathbf{k}(\mathbf{v}_\alpha; t) = \frac{1}{2\pi}\int_C dz\, e^{-izt} \frac{1}{i(\mathbf{k}\cdot\mathbf{v}_\alpha - z)} \left(\frac{h^3}{m^3\Omega}\right)^N \sum_{\mathbf{v}_1}\cdots\sum_{\mathbf{v}_N} \Bigg\{ q_\mathbf{k}(\mathbf{v}_\alpha|\ldots)$$
$$+ \frac{ie^2}{\hbar}\frac{8\pi^3}{\Omega}\sum_n V_k[(1+\theta\sum_s \delta_{\mathbf{v}_s - \mathbf{v}_\alpha + \hbar\mathbf{k}/2m}) \exp(\tfrac{1}{2}\hbar m^{-1}\mathbf{k}\cdot\partial_{\alpha n})$$
$$-(1+\theta\sum_s \delta_{\mathbf{v}_s - \mathbf{v}_\alpha - \hbar\mathbf{k}/2m}) \exp(-\tfrac{1}{2}\hbar m^{-1}\mathbf{k}\cdot\partial_{\alpha n})] \frac{1}{i(\mathbf{k}\cdot\mathbf{v}_n - z)} q_\mathbf{k}(\mathbf{v}_n|\mathbf{v}_\alpha, \mathbf{v}_s\ldots)$$
$$+ \frac{i^2 e^4}{\hbar^2}\left(\frac{8\pi^3}{\Omega}\right)^2 \sum_n\sum_m V_k^2[(1+\theta\sum_s \delta_{\mathbf{v}_s - \mathbf{v}_\alpha + \hbar\mathbf{k}/2m}) \exp(\tfrac{1}{2}\hbar m^{-1}\mathbf{k}\cdot\partial_{\alpha n}) \tag{67.2}$$
$$-(1+\theta\sum_s \delta_{\mathbf{v}_s - \mathbf{v}_\alpha - \hbar\mathbf{k}/2m}) \exp(-\tfrac{1}{2}\hbar m^{-1}\mathbf{k}\cdot\partial_{\alpha n})]$$
$$\frac{1}{i(\mathbf{k}\cdot\mathbf{v}_n - z)} [(1+\theta\sum_r \delta_{\mathbf{v}_r - \mathbf{v}_n + \hbar\mathbf{k}/2m}) \exp(\tfrac{1}{2}\hbar m^{-1}\mathbf{k}\cdot\partial_{nm})$$
$$-(1+\theta\sum_r \delta_{\mathbf{v}_r - \mathbf{v}_n - \hbar\mathbf{k}/2m}) \exp(-\tfrac{1}{2}\hbar m^{-1}\mathbf{k}\cdot\partial_{nm})] \frac{1}{i(\mathbf{k}\cdot\mathbf{v}_m - z)}$$
$$q_\mathbf{k}(\mathbf{v}_m|\mathbf{v}_\alpha, \mathbf{v}_n, \mathbf{v}_r, \mathbf{v}_s, \ldots) + \ldots \Bigg\}$$

Performing the velocity summations and then going over to the limit $\Omega \to \infty$, $N \to \infty$, $N/\Omega = c$, we obtain [using (62.1) and (65.2), see also (65.1)]

$$\rho_{\mathbf{k}}(\mathbf{v}_\alpha;\, t) = \frac{1}{2\pi}\int_C dz\, \mathrm{e}^{-izt}\frac{1}{i(\mathbf{k}\cdot\mathbf{v}_\alpha - z)}\Big\{q_{\mathbf{k}}(\mathbf{v}_\alpha)$$
$$+ i\hbar^{-1}8\pi^3 e^2 c V_k\{[1+\theta h^3 cm^{-3}\varphi(\mathbf{v}_\alpha - \tfrac{1}{2}\hbar m^{-1}\mathbf{k})]\exp(\tfrac{1}{2}\hbar m^{-1}\mathbf{k}\cdot\partial_\alpha)$$
$$-[1+\theta h^3 cm^{-3}\varphi(\mathbf{v}_\alpha + \tfrac{1}{2}\hbar m^{-1}\mathbf{k})]\exp(-\tfrac{1}{2}\hbar m^{-1}\mathbf{k}\cdot\partial_\alpha)\}\varphi(\mathbf{v}_\alpha)$$
$$\int d\mathbf{v}_1 \frac{1}{i(\mathbf{k}\cdot\mathbf{v}_1 - z)}\, q_{\mathbf{k}}(\mathbf{v}_1) + i\hbar^{-1}8\pi^3 e^2 c V_k\{[1+\theta h^3 cm^{-3}\varphi(\mathbf{v}_\alpha - \tfrac{1}{2}\hbar m^{-1}\mathbf{k})]$$
$$\exp(\tfrac{1}{2}\hbar m^{-1}\mathbf{k}\cdot\partial_\alpha) - [1+\theta h^3 cm^{-3}\varphi(\mathbf{v}_\alpha + \tfrac{1}{2}\hbar m^{-1}\mathbf{k})] \tag{67.3}$$
$$\exp(-\tfrac{1}{2}\hbar m^{-1}\mathbf{k}\cdot\partial_\alpha)\}\varphi(\mathbf{v}_\alpha) + i\hbar^{-1}8\pi^3 e^2 c V_k \int d\mathbf{v}_1 \frac{1}{i(\mathbf{k}\cdot\mathbf{v}_1 - z)}$$
$$\{[1+\theta h^3 cm^{-3}\varphi(\mathbf{v}_1 - \tfrac{1}{2}\hbar m^{-1}\mathbf{k})]\exp(\tfrac{1}{2}\hbar m^{-1}\mathbf{k}\cdot\partial_1) - [1+\theta h^3 cm^{-3}$$
$$\varphi(\mathbf{v}_1 + \tfrac{1}{2}\hbar m^{-1}\mathbf{k})]\exp(-\tfrac{1}{2}\hbar m^{-1}\mathbf{k}\cdot\partial_1)\varphi(\mathbf{v}_1)\int d\mathbf{v}_2 \frac{1}{i(\mathbf{k}\cdot\mathbf{v}_2 - z)}\, q_{\mathbf{k}}(\mathbf{v}_2) + \ldots\Big\}$$

We note, by performing the displacement operations explicitly, that *the terms containing θ cancel pairwise within each curly bracket* of eq. (67.3).

At this point we can write down the equation obeyed by this function. Taking the time derivative of both sides, and applying the same argument as in § 11, we easily obtain

$$\partial_t \rho_{\mathbf{k}}(\mathbf{v}_\alpha;\, t) + i\mathbf{k}\cdot\mathbf{v}_\alpha\, \rho_{\mathbf{k}}(\mathbf{v}_\alpha;\, t) \tag{67.4}$$
$$= 8\pi^3 e^2 c i\hbar^{-1} V_k\{\varphi(\mathbf{v}_\alpha + \tfrac{1}{2}\hbar m^{-1}\mathbf{k}) - \varphi(\mathbf{v}_\alpha - \tfrac{1}{2}\hbar m^{-1}\mathbf{k})\}\int d\mathbf{v}_j\, \rho_{\mathbf{k}}(\mathbf{v}_j;\, t)$$

This is the analog of the Fourier-transformed linearized Vlassov equation, i.e. eq. (11.5) in which the second term of the r.h.s. is discarded.

We now derive the solution of eq. (67.4), which can be treated exactly as in the classical case. The series (67.3) is easily seen to exhibit a repeating unit, analogous to that of (13.5):

$$J_{\mathbf{k}}(z) = \frac{8\pi^3 e^2 c}{\hbar} V_k \int d\mathbf{v}_1 \frac{1}{\mathbf{k}\cdot\mathbf{v}_1 - z}[\varphi(\mathbf{v}_1 + \tfrac{1}{2}\hbar m^{-1}\mathbf{k}) - \varphi(\mathbf{v}_1 - \tfrac{1}{2}\hbar m^{-1}\mathbf{k})] \tag{67.5}$$

Therefore the series (67.3) is a geometrical series which can be summed as in the classical case (using $8\pi^3 V_k = 4\pi/k^2$)

$$\rho_{\mathbf{k}}(\mathbf{v}_\alpha;t) = \frac{1}{2\pi i}\int_C dz\, e^{-izt}\left\{\frac{q_{\mathbf{k}}(\mathbf{v}_\alpha)}{\mathbf{k}\cdot\mathbf{v}_\alpha - z}\right. \tag{67.6}$$

$$\left. + \frac{4\pi e^2 c}{\hbar k^2}\frac{[\varphi(\mathbf{v}_\alpha+\tfrac{1}{2}\hbar m^{-1}\mathbf{k})-\varphi(\mathbf{v}_\alpha-\tfrac{1}{2}\hbar m^{-1}\mathbf{k})]}{\mathbf{k}\cdot\mathbf{v}_\alpha - z}\frac{1}{\varepsilon_+(k;z)}\int d\mathbf{v}_1 \frac{q_{\mathbf{k}}(\mathbf{v}_1)}{\mathbf{k}\cdot\mathbf{v}_1 - z}\right\}$$

with

$$\varepsilon_+(k;z) = 1 - \frac{4\pi e^2 c}{\hbar k^2}\int d\mathbf{v}\frac{1}{\mathbf{k}\cdot\mathbf{v}-z}[\varphi(\mathbf{v}+\tfrac{1}{2}\hbar m^{-1}\mathbf{k})-\varphi(\mathbf{v}-\tfrac{1}{2}\hbar m^{-1}\mathbf{k})] \tag{67.7}$$

We note that the dielectric constant could also be written as follows:

$$\varepsilon_+(k;z) = 1 - \frac{4\pi e^2 c}{\hbar k^2}\int d\mathbf{v}(\mathbf{k}\cdot\mathbf{v}-z)^{-1}[\varphi(\mathbf{v}+\tfrac{1}{2}\hbar m^{-1}\mathbf{k})\psi(\mathbf{v}-\tfrac{1}{2}\hbar m^{-1}\mathbf{k})$$
$$-\varphi(\mathbf{v}-\tfrac{1}{2}\hbar m^{-1}\mathbf{k})\psi(\mathbf{v}+\tfrac{1}{2}\hbar m^{-1}\mathbf{k})] \tag{67.7a}$$

where

$$\psi(\mathbf{v}) \equiv 1 - h^3 c m^{-3}\varphi(\mathbf{v}) \tag{67.8}$$

Although the ψ factors are actually redundant, it is convenient, in some transformations, to write them explicitly.

Integrating (67.6) over the velocities $\mathbf{v}_\alpha$, we obtain an expression for the Fourier transform of the local density excess [see also eq. (14.1)]

$$h_{\mathbf{k}}(t) = \frac{1}{2\pi i}\int_C dz\, e^{-izt}\frac{1}{\varepsilon_+(k;z)}\int d\mathbf{v}\frac{q_{\mathbf{k}}(\mathbf{v})}{\mathbf{k}\cdot\mathbf{v}-z} \tag{67.9}$$

§ 68. General Features of the Quantum Vlassov Description

The analogy between eqs. (67.6–67.7) and the corresponding classical results (13.8–13.9) is so close that most of the general properties derived in Chapters 3 and 4 remain valid without any change for quantum-statistical systems. We summarize these results here and refer the reader to the classical discussion for details of the derivation.

A) The Laplace transform of the local density excess

$$h_{\mathbf{k}}(z) = \frac{1}{\varepsilon_+(k;z)}\int d\mathbf{v}\frac{q_{\mathbf{k}}(\mathbf{v})}{i(\mathbf{k}\cdot\mathbf{v}-z)} \tag{68.1}$$

is a product of two functions. The first one is a Cauchy integral and hence is regular in the upper half-plane but has singularities in the lower half-plane. The position of these singularities depends on the shape of the initial perturbation $q_{\mathbf{k}}(\mathbf{v})$. The second factor, $1/\varepsilon_+(k; z)$, may have singularities located anywhere in the complex plane; their location depends only on the shape of the velocity distribution, and thus on the permanent properties of the electron gas. More precisely, the position of the poles is determined by the solutions of the dispersion equation:

$$\varepsilon_+(k; z) = 0 \tag{68.2}$$

B) The time dependence of the density excess $h_{\mathbf{k}}(t)$ is determined through eq. (67.9) by the singularities of $h_{\mathbf{k}}(z)$. Assuming for simplicity that all these singularities are poles, $h_{\mathbf{k}}(t)$ is the sum of the residues of the integrand of (67.9) at these poles.

(*a*) The Cauchy integral leads always to damped oscillating terms (if the initial perturbation is reasonably regular). These terms represent transient motions set up by the initial perturbations and bodily carried away by the individual particles. The damping is due to destructive interference produced by the distribution of velocities.

(*b*) The roots of the dispersion equation give rise to steady, damped or growing oscillations. If the dispersion equation has no roots in the upper half-plane, the plasma is stable. The appearance of an instability coincides actually with a breakdown of the theory. We will come back to this point in § 69. The steady or slightly "Landau damped" modes, which normally occur for long wave-lengths, are manifestations of an organized collective behavior. Short wave-length oscillations on the other hand are strongly damped.

C) The evolution of the plasma in the present approximation can be entirely described in terms of an effective potential whose Fourier–Laplace transform is (see § 18)

$$U_k(z) = \frac{1}{\varepsilon_+(k; z)} V_k \tag{68.3}$$

where V_k is the true Coulomb potential. The inverse Fourier–Laplace transform of this function represents a screened and retarded interaction. The function $\varepsilon_+(k; z)$ is therefore interpreted as a frequency-dependent *dielectric constant.*

In the Vlassov description, it contains complete information about the reaction of the plasma to external disturbances.

D) It is worth stressing that the quantum Vlassov equation (67.4) has the same form for systems obeying Boltzmann statistics or quantum statistics: the Vlassov operator does not involve the statistical parameter θ. This does not mean, however, that statistics effects are absent from the equation. Indeed, the velocity distribution $\varphi(\mathbf{v})$ may depend on θ. If it describes, for instance, an equilibrium state it could be taken either as the Maxwell–Boltzmann or as the Fermi–Dirac equilibrium. Thus, *in the Vlassov approximation quantum-statistical effects do not enter through the laws of evolution but only through the intrinsic properties of the medium.*

§ 69. The Stability Condition for Quantum-Statistical Systems

We shall discuss here the extension of the results derived in § 19, and formulate simple criteria of stability based on the shape of the velocity distribution.

The first part of § 19 is extended straightforwardly. We can introduce the concept of a hodograph of the function $\varepsilon_+(z)$ for a given value of k: it is defined again as the contour described by the complex number $\varepsilon_+(z)$ when z travels along the real axis from $-\infty$ to $+\infty$. It is then shown that if the hodograph does not encircle the origin, the system is stable.

Let us write the dielectric constant (67.7) in the form:

$$\varepsilon_+(k; z) = 1 - \frac{4\pi e^2 c}{\hbar k^2} \int dv \frac{\bar{\varphi}(v+\hbar k/2m) - \bar{\varphi}(v-\hbar k/2m)}{v - z/k} \tag{69.1}$$

The barred quantities are defined as usual by eq. (14.6). We now prove that Nyquist's theorem is valid for such systems:

If the distribution function $\bar{\varphi}(v)$ is continuous and has a single maximum, the system is stable.

For real values $z = ky$ of the frequency, the dielectric constant is written as

$$\varepsilon_+(ky) = 1 - \frac{4\pi e^2 c}{\hbar k^2} \mathscr{P} \int d\nu \frac{\bar{\varphi}(\nu + \hbar k/2m) - \bar{\varphi}(\nu - \hbar k/2m)}{\nu - y}$$
$$- \frac{4\pi^2 e^2 c}{\hbar k^2} i[\bar{\varphi}(y + \hbar k/2m) - \bar{\varphi}(y - \hbar k/2m)] \tag{69.2}$$

(we write $\varepsilon_+(ky)$ for $\varepsilon_+(k; ky)$; k is fixed).

It is first seen from this formula that when $y \to \pm\infty$ the imaginary part of ε_+ vanishes, and its real part is equal to $+1$. Hence the hodograph is a closed curve; it cuts the real axis at the point $+1$. We now investigate the location of the other intersection points. It is seen from (69.2) that these points are defined as the solutions ν_0 of the equation:

$$\bar{\varphi}(\nu_0 + \hbar k/2m) = \bar{\varphi}(\nu_0 - \hbar k/2m) \tag{69.3}$$

If the distribution function $\bar{\varphi}(\nu)$ has a single maximum at ν_M, has no inflection point, is continuous and vanishes for $\nu = \pm\infty$, it takes all positive values between 0 and $\bar{\varphi}(\nu_M)$ exactly twice: once to the left of ν_M and once to the right. Hence (69.3) has one and only one solution for finite ν_M. We now show that this point is such that

$$\varepsilon_+(k\nu_0) > 1 \tag{69.4}$$

This property achieves the proof of stability: as the hodograph crosses the real axis only at the points $+1$ and $\varepsilon_+(k\nu_0) > 1$, it cannot encircle the origin.

In order to prove our statement, we write $\varepsilon_+(k\nu_0)$ (which is real) in the form:

$$\varepsilon_+(k\nu_0) = 1 - \lim_{\eta \to 0} \frac{4\pi e^2 c}{\hbar k^2} \left\{ \int_{-\infty}^{\nu_0 - \eta} d\nu \frac{\bar{\varphi}(\nu + \hbar k/2m) - \bar{\varphi}(\nu - \hbar k/2m)}{\nu - \nu_0} \right.$$
$$\left. + \int_{\nu_0 + \eta}^{\infty} d\nu \frac{\bar{\varphi}(\nu + \hbar k/2m) - \bar{\varphi}(\nu - \hbar k/2m)}{\nu - \nu_0} \right\} \tag{69.5}$$

In the first integral, $\nu < \nu_0$ throughout, whereas in the second, $\nu > \nu_0$. We now show that

$$\begin{aligned} &\bar{\varphi}(\nu + \hbar k/2m) > \bar{\varphi}(\nu - \hbar k/2m) \quad \text{for} \quad \nu < \nu_0 \qquad \text{(a)} \\ &\bar{\varphi}(\nu + \hbar k/2m) < \bar{\varphi}(\nu - \hbar k/2m) \quad \text{for} \quad \nu > \nu_0 \qquad \text{(b)} \end{aligned} \tag{69.6}$$

In order to avoid heavy analytical discussions, we shall verify these properties graphically. The point ν_0 is constructed as follows (Fig. 69.1).

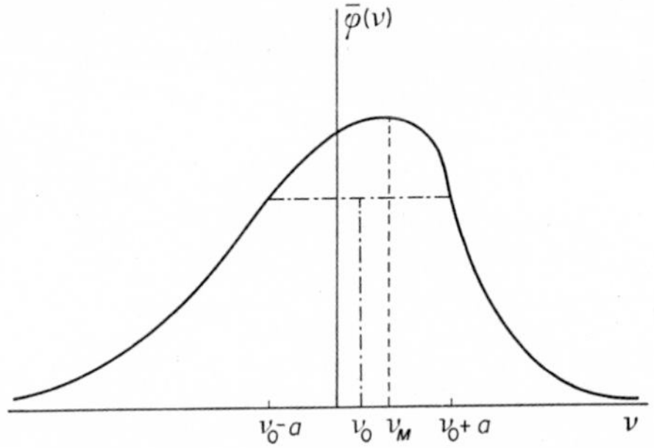

Fig. 69.1. Construction of the point ν_0 where the hodograph of $\varepsilon_+(k\nu_0)$ crosses the real axis. $a \equiv \hbar k/2m$.

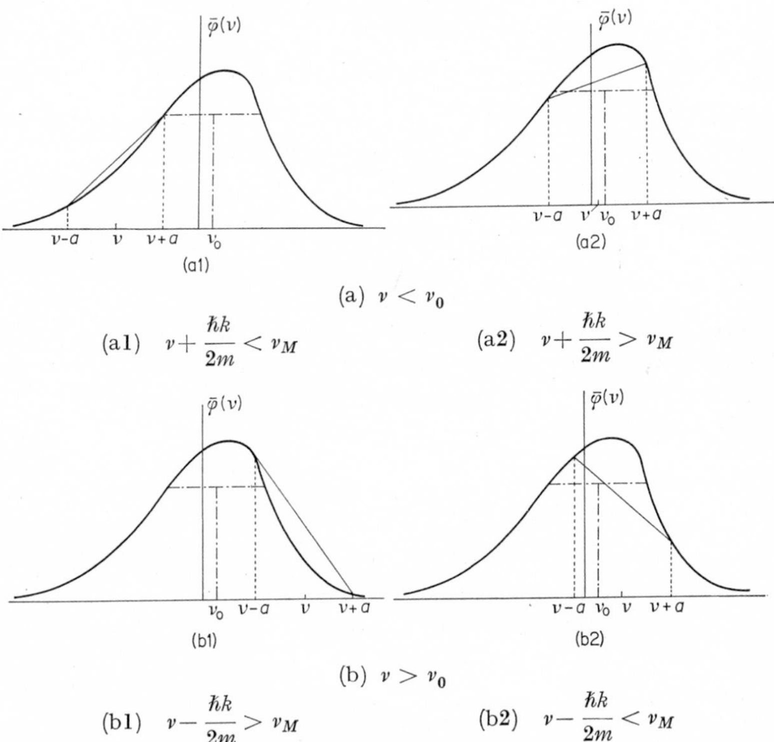

Fig. 69.2. Graphical verification of the inequalities (69.6). In each graph the directed segment goes from $\bar{\varphi}(\nu - \hbar k/2m)$ to $\bar{\varphi}(\nu + \hbar k/2m)$.

Consider a segment of length $\hbar k/m$, and attach one of its ends to the curve representing the distribution $\bar{\varphi}(\nu)$. This end-point is moved along the curve, keeping the segment parallel to the real axis, until the second end-point also touches the curve $\bar{\varphi}(\nu)$. The abscissa of the middle of the segment in this position is the point ν_0. Let us first assume that $\nu_0 < \nu_M$ (the discussion of the case $\nu_0 \geqq \nu_M$ is identical). It is then easily verified graphically in Fig. 69.2 that the properties (69.6) are satisfied for every $\bar{\varphi}(\nu)$ which is continuous and has a single maximum.

It then follows that the expression enclosed in curly brackets in eq. (69.5) is definite-negative, and hence $\varepsilon_+(\nu_0) > 1$. Q.E.D.

§ 70. The Quantum Dielectric Constant for an Electron Gas at Zero Temperature

We now investigate the dielectric constant of a dense degenerate electron gas in equilibrium at zero temperature. The distribution function is then the well known Fermi–Dirac distribution:

$$\varphi_F^0(\mathbf{v}) = (m^3/h^3c)\,\theta(v_F - v) \tag{70.1}$$

where, as usual:

$$\theta(v_F - v) = \begin{cases} 1, & v < v_F \\ 0, & v > v_F \end{cases}$$

and the Fermi velocity is

$$v_F = (3h^3c/8\pi m^3)^{\frac{1}{3}} \tag{70.2}$$

The occurrence of the factor (m^3/h^3c) in eq. (70.1) may be unfamiliar. In most textbooks the Fermi distribution appears as the mean occupation number $\bar{n}(\varepsilon_i)$ of a given quantum state ε_i:

$$\bar{n}(\varepsilon_i) = \theta(\varepsilon_F - \varepsilon_i)$$

$\bar{n}(\varepsilon_i)$ is a dimensionless number. The normalization condition is

$$\sum_i \bar{n}(\varepsilon_i) = N$$

We now want to change variables and consider the velocities rather than the energies as independent variables. Let $\bar{n}(\mathbf{v}_i)$ be the mean occupation number of the velocity state $\mathbf{v}_i$:

$$\bar{n}(\mathbf{v}_i) = \theta(v_F - v_i)$$

The normalization condition now becomes

$$N^{-1} \sum_i \bar{n}(\mathbf{v}_i) = 1$$

Suppose now that $N \to \infty$; then the sum over velocity states becomes an integral over the velocities, according to eq. (62.1). The previous equation becomes:

$$(m^3\Omega/h^3N)\int d\mathbf{v}\,\bar{n}(\mathbf{v}) = 1$$

The mean occupation number of the velocity state $\mathbf{v}$ is by definition proportional to the velocity distribution function. In order to determine the proportionality constant we compare the last equation with the normalization condition of the distribution function:

$$\int d\mathbf{v}\,\varphi(\mathbf{v}) = 1$$

Whence we obtain the following relation between $\bar{n}(\mathbf{v})$ and $\varphi(\mathbf{v})$, which accounts for the factor (m^3/h^3c) in eq. (70.1):

$$\bar{n}(\mathbf{v}) = (h^3c/m^3)\,\varphi(\mathbf{v}) \tag{70.3}$$

In formula (67.7) for ε_+, change $\mathbf{v}$ into $-\mathbf{v}$ in the first term of the difference occurring in the integral. As $\varphi_F^0(\mathbf{v})$ is an even function of $\mathbf{v}$, the dielectric constant becomes:

$$\varepsilon_+(k;z) = 1+\frac{4\pi e^2c}{\hbar k^2}\int d\mathbf{v}\,\varphi_F^0(\mathbf{v}-\hbar\mathbf{k}/2m)\left\{\frac{1}{\mathbf{k}\cdot\mathbf{v}+z}+\frac{1}{\mathbf{k}\cdot\mathbf{v}-z}\right\} \tag{70.4}$$

Taking now $\mathbf{v}-\hbar\mathbf{k}/2m$ as a new integration variable, introducing polar coordinates and using (70.1) we obtain

$$\begin{aligned}\varepsilon_+(k;z) &= 1+\frac{4\pi e^2c}{\hbar k^2}\,\frac{2\pi}{h^3c}\\ &\int_0^{v_F}dv\,v^2\int_{-1}^{1}d\xi\left\{\frac{1}{kv\xi+(\hbar k^2/2m)-z}+\frac{1}{kv\xi+(\hbar k^2/2m)+z}\right\}\end{aligned} \tag{70.5}$$

where, as usual, z is assumed to lie in the upper half-plane S_+. The integrations are very easily performed. The result is most simply expressed in a set of reduced variables. The obvious reduction standard for velocities is v_F, and for wave-vectors $k_F = mv_F/\hbar$. We thus introduce:

$$\chi = \frac{\hbar}{mv_F}k, \quad y = \frac{1}{v_F}\,\frac{z}{k}, \quad a^2 = \frac{2e^2}{\pi\hbar v_F} \tag{70.6}$$

The result of the integration of (70.5) is then:

$$\varepsilon_+(\chi;\chi y) = 1+\frac{a^2}{\chi^2}\,m(\chi;y) \tag{70.7}$$

with

$$\begin{aligned} m(\chi; y) = 1 + \frac{1}{2\chi} [1-(y-\tfrac{1}{2}\chi)^2] \ln \frac{\frac{1}{2}\chi+1-y}{\frac{1}{2}\chi-1-y} \\ + \frac{1}{2\chi} [1-(y+\tfrac{1}{2}\chi)^2] \ln \frac{\frac{1}{2}\chi+1+y}{\frac{1}{2}\chi-1+y}; \quad y \in S_+ \end{aligned} \tag{70.8}$$

This expression for the dielectric constant of a Fermi gas was first obtained by Lindhard * and later by many authors in various forms.**

The limit of *real* y can be obtained directly from (70.8) by discussing carefully the behavior of the logarithms, or else from (70.4) by interpreting $\lim (\mathbf{k}\cdot\mathbf{v}-z)^{-1}$ as $\delta_-(\mathbf{k}\cdot\mathbf{v}-\text{Re}\, z)$. The result is:

$$m(\chi; y) = m_1(\chi; y) + i m_2(\chi; y) \tag{70.9}$$

with

$$\begin{aligned} m_1(\chi; y) = 1 + \frac{1}{2\chi} [1-(y-\tfrac{1}{2}\chi)^2] \ln \left| \frac{\frac{1}{2}\chi+1-y}{\frac{1}{2}\chi-1-y} \right| \\ + \frac{1}{2\chi} [1-(y+\tfrac{1}{2}\chi)^2] \ln \left| \frac{\frac{1}{2}\chi+1+y}{\frac{1}{2}\chi-1+y} \right| \end{aligned} \tag{70.10}$$

The function $m_2(\chi; y)$ is defined in Table 70.1. A plot of the dielectric constant $\varepsilon_+(y)$ is given in Fig. 70.1.

The remarkable feature of this result is that for frequencies larger than $(\frac{1}{2}\chi+1)$ the imaginary part of the dielectric constant vanishes identically. If the real part $\varepsilon_1(\chi; \chi y)$ has a zero in that region, the corresponding value of y is the frequency of an *undamped* mode of oscillation. It will be shown below that such a zero does indeed exist for small χ, i.e. large wave-lengths.

Indeed, suppose that

$$\begin{aligned} \chi &< 2 \\ y &\gg 1 \end{aligned} \tag{70.11}$$

* J. Lindhard, *Kgl. Danske Math-fys. Medd.*, **28** no. 8 (1954).

** See, e.g., D. Bohm and D. Pines, *Phys. Rev.*, **92**, 609 (1953); K. Sawada, K. A. Brueckner, N. Fukuda and R. Brout, *Phys. Rev.*, **108**, 507 (1957); Chen Chun-Sian and Choh-Shih-Hsun, *J. Exptl. Theoret. Phys. U.S.S.R.*, **34**, 1566 (1958); transl. *Sov. Phys. JETP*, **34** (7), 1080 (1959); J. Hubbard, *Proc. Phys. Soc. London*, **68A**, 441 (1955); R. H. Ritchie, *Phys. Rev.* **114**, 644 (1959); P. Nozières and D. Pines, *Nuovo Cimento*, **9**, 470 (1958).

Table 70.1. The imaginary part of the polarizability, $m_2(\chi;\, y)$

χ	y	$m_2(\chi;\, y)$
	$y \geqq \frac{1}{2}\chi+1$	0
$\chi > 2$	$\frac{1}{2}\chi-1 \leqq y \leqq \frac{1}{2}\chi+1$	$\frac{\pi}{2\chi}[1-(\frac{1}{2}\chi-y)^2]$
	$0 \leqq y \leqq \frac{1}{2}\chi-1$	0
	$y \geqq \frac{1}{2}\chi+1$	0
$0 < \chi < 2$	$1-\frac{1}{2}\chi \leqq y \leqq 1+\frac{1}{2}\chi$	$\frac{\pi}{2\chi}[1-(y-\frac{1}{2}\chi)^2]$
	$0 \leqq y \leqq 1-\frac{1}{2}\chi$	πy
all χ	$y \leqq 0$	$m_2(\chi;\, y) = -m_2(\chi;\, -y)$

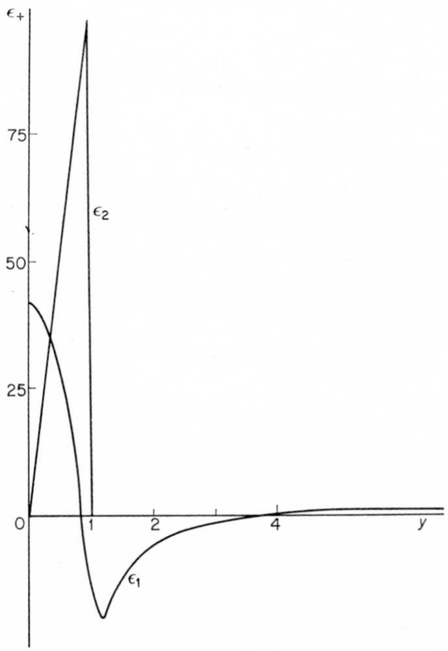

Fig. 70.1. Real and imaginary parts of the dielectric constant $\varepsilon_+ = \varepsilon_1+i\varepsilon_2$ of an electron gas at zero temperature.

Expanding the logarithms of (70.10) in inverse powers of y, and substituting the result in (70.7) we obtain

$$\varepsilon_+(\chi; \chi y) \approx 1 - \tfrac{2}{3}a^2\chi^{-2}y^{-2} - a^2(\tfrac{2}{5}\chi^{-2} + \tfrac{1}{6})y^{-4} + \ldots + ia^2\chi^{-2}m_2 \quad (70.12)$$

Keeping only the first two terms of the expansion, we see that the approximate condition for the zero y_0 of ε_1 to lie in the domain $y \gg 1$ where $m_2 = 0$ and also where the expansion is valid is:

$$\chi^2 \ll \tfrac{2}{3}a^2 \quad (70.13)$$

Comparing the definition (70.6) of a^2 with (70.2) and (66.6), we see that

$$a^2 \sim \gamma_0 \ll 1 \quad (70.14)$$

Hence, in the limit of very small wave-vectors (compared to $\sqrt{\tfrac{2}{3}}a\hbar^{-1}mv_F$) there exists an *undamped* mode of oscillation of frequency

$$\chi^2 y^2 = \tfrac{2}{3}a^2 \quad (70.15)$$

A Fermi gas in its ground state is thus able to sustain a permanent collective oscillation. This is in contrast to a gas at non-zero temperature in which, even for long wave-lengths, there always exists a Landau damping which eventually destroys the collective oscillations. The permanent character of the zero-point collective oscillations justifies their interpretation as independent entities which can be quantized separately. The resulting view of an electron gas as being described by a set of "collective variables" plus a set of "individual variables" forms the basic starting point of the Bohm and Pines theory.* We shall, however, not give a detailed account of this theory here.

It is extremely suggestive to transform eq. (70.12) back into ordinary variables k and $z = \omega$, using (70.6):

$$\varepsilon_+(k; \omega) = 1 - \frac{\omega_p^2}{\omega^2}\left\{1 + \frac{3}{5}\frac{k^2 v_F^2}{\omega^2} + \frac{1}{4}\frac{\hbar^2 k^4}{m^2\omega^2} + \ldots\right\} \quad (70.16)$$

Condition (70.13) becomes:

$$k \ll \pi\omega_p/v_F \quad (70.17)$$

* D. Bohm and D. Pines, *Phys. Rev.*, **92**, 609 (1953).

The most remarkable feature of eq. (70.16) is the fact that in the limit $k \to 0$, the frequency of the collective oscillation is:

$$\omega = \omega_p \tag{70.18}$$

This clearly shows the basic role played by the plasma oscillation frequency throughout all the possible physical situations, from zero temperature up to the highest temperatures. *For all long-range phenomena a system of charged particles always reacts as a continuous medium, forgetting completely the laws of evolution of its component particles.*

For shorter wave-lengths, the dispersion equation derived from (70.16) is, up to $O(k^4)$:

$$\omega^2 = \omega_p^2 + \frac{3}{5} \frac{k^2 v_F^2}{\omega_p^2} + \frac{1}{4} \frac{\hbar^2 k^4}{m^2 \omega_p^2} \tag{70.19}$$

It should be stressed that there exists a complete range of values of k for which the plasma oscillations are *rigorously undamped.* The collective behavior is therefore enhanced in a zero-temperature Fermi gas as compared with a classical gas. This may seem surprising if one thinks of the existence of a large zero-point energy. It is even more surprising that the condition of validity of the "Vlassov" approximation (70.14) can be written as follows:

$$a \sim \hbar\omega_p/E_F \ll 1 \tag{70.20}$$

As the energy of the plasma oscillation is much smaller than the Fermi energy, there exist many particles within the Fermi sea which have enough energy to perform plasma oscillations. This means that *in the ground state of the interacting system, part of the zero-point energy comes from organized modes of motion in which a large number of particles oscillate in phase.*

It is also interesting to consider at this point a remark already made previously. It was stated that in real metals $\hbar\omega_p/E_F > 1$. Hence the dielectric constant approximation does not apply to these systems. However, using the language of this approximation, one might state that the electrons in the ground state do not have enough energy to perform plasma oscillations and consequently there exist no "plasmons" at zero temperature. The only effect of the long-range Coulomb interactions is to provide the screening. The description of an electron gas in a metal is that of a set of particles interacting through short-range effective forces. This is the argument put forward by Pines * in order to justify the surprising success of independent-particle models in explaining the properties of the metals.

* D. Pines, *Solid State Physics,* **1**, 368 (1955).

CHAPTER 16

Long-Time Behavior of Quantum Plasmas—The Cycle Approximation

§ 71. Choice of Diagrams

We now want to study the long-time behavior of the one-particle reduced Wigner function along the same lines as in Chapters 6–8. For simplicity, we shall only consider homogeneous systems, so that the discussion is restricted to the velocity distribution $\varphi(\mathbf{v}_\alpha)$.

The cycle approximation in the quantum case is defined by the same correspondence principle as in § 66. We want to select diagrams of order:

$$[e^4 ct(h^3c)^n]^m \tag{71.1}$$

where the corrections h^3c take account of all quantum-statistical effects. In this case the choice is especially simple.

We have first to discuss the origin of the factor t^m. It was shown in § 27 that in the classical case, terms of this type arise from the existence of a pole of order $m+1$ at $z = 0$ in the resolvent. But in the quantum-statistical formalism we have developed, the unperturbed resolvent is identical to the classical one, and therefore a factor t^m will exist in all diagrams which have the structure of a succession of m diagonal fragments, exactly as in the classical case. This shows again that this type of term is *universal*, because it depends only on the structure of the diagram and not on the nature of the interactions. However, a difference will arise as compared to the classical case in the discussion of the *time scales*, i.e. of the time after which the non-universal terms decay. This question will be examined in the next section.

We now know that a contribution of order (71.1) is given by a succession of m diagonal fragments. Each of these must have two vertices (factor e^4). By the correspondence principle explained in § 66, we know that we must retain the quantum-statistical cycle which, in the limit $\hbar \to 0$, gives the correct classical Landau equation. Moreover, none of the purely quantum-statistical vertices of Table 63.1 can be used to build a two-vertex diagonal fragment, and thus the cycles are the only diagrams to be retained in this approximation.

In order not to complicate the notations and to remain consistent with the other calculations performed in the quantum-mechanical part of this book, we shall moreover neglect the exchange matrix element of the potential. For the theory of the cycle approximation, however, it is a straightforward matter to include this contribution, and this will be indicated in due place.

§ 72. The Quantum-Mechanical Cycle Diagram

We shall evaluate and study in some detail the contribution to $\varphi(\alpha; t)$ of a single cycle. We proceed as in eq. (27.1), using eq. (5.26), the matrix elements A and C of Table 63.1 and formula (2.23):

$$\varphi(\mathbf{v}_\alpha; t) = \frac{8\pi^3 e^4 c}{2\pi}\left(\frac{h^3}{m^3\Omega}\right)^N \left(\frac{i}{\hbar}\right)^2 \sum_j \sum_{\mathbf{v}_1} \cdots \sum_{\mathbf{v}_N} \int_C dz\, \mathrm{e}^{-izt} \frac{8\pi^3}{\Omega} \sum_{\mathbf{l}} \frac{1}{-iz}$$

$$\cdot \left[\exp\left(-\tfrac{1}{2}\hbar m^{-1}\mathbf{l}\cdot\partial_{\alpha j}\right) - \exp\left(\tfrac{1}{2}\hbar m^{-1}\mathbf{l}\cdot\partial_{\alpha j}\right)\right] \frac{V_l^2}{i(\mathbf{l}\cdot\mathbf{g}_{\alpha j}-z)} \qquad (72.1)$$

$$\cdot \Big[\Big(1+\theta \sum_s \delta_{\mathbf{v}_s-\mathbf{v}_\alpha+\hbar\mathbf{l}/2m}\Big)\Big(1+\theta\sum_r \delta_{\mathbf{v}_r-\mathbf{v}_j-\hbar\mathbf{l}/2m}\Big) \exp\left(\tfrac{1}{2}\hbar m^{-1}\mathbf{l}\cdot\partial_{\alpha j}\right)$$

$$-\Big(1+\theta\sum_s \delta_{\mathbf{v}_s-\mathbf{v}_\alpha-\hbar\mathbf{l}/2m}\Big)\Big(1+\theta\sum_r \delta_{\mathbf{v}_r-\mathbf{v}_j+\hbar\mathbf{l}/2m}\Big)\exp\left(-\tfrac{1}{2}\hbar m^{-1}\mathbf{l}\cdot\partial_{\alpha j}\right)\Big]$$

$$\cdot \frac{1}{-iz}\rho_0(|\mathbf{v}_\alpha, \mathbf{v}_j, \mathbf{v}_r, \mathbf{v}_s, \ldots; 0)$$

In the limit $N\to\infty$, $\Omega\to\infty$, $N/\Omega = c$, this expression becomes:

$$\varphi(\mathbf{v}_\alpha; t) = \int d\mathbf{v}_n \frac{1}{2\pi}\int_C dz\, \mathrm{e}^{-izt}\frac{1}{-z^2}\Psi_+(\mathbf{v}_\alpha \mathbf{v}_n; z)\varphi(\mathbf{v}_\alpha; 0)\varphi(\mathbf{v}_n; 0) \qquad (72.2)$$

where the operator $\Psi_+(\mathbf{v}_\alpha, \mathbf{v}_n; z)$ is defined as:

$$\Psi_+ = \frac{8\pi^3 e^4 c i^2}{\hbar^2} \int d\mathbf{l}\, [\exp(-\tfrac{1}{2}\hbar m^{-1}\mathbf{l}\cdot\partial_{\alpha n}) - \exp(\tfrac{1}{2}\hbar m^{-1}\mathbf{l}\cdot\partial_{\alpha n})]$$

$$\cdot \frac{V_l^2}{i(\mathbf{l}\cdot\mathbf{g}_{\alpha n} - z)} ([\psi(\mathbf{v}_\alpha - \tfrac{1}{2}\hbar m^{-1}\mathbf{l})\psi(\mathbf{v}_n + \tfrac{1}{2}\hbar m^{-1}\mathbf{l})\exp(\tfrac{1}{2}\hbar m^{-1}\mathbf{l}\cdot\partial_{\alpha n}) \\ - \psi(\mathbf{v}_\alpha + \tfrac{1}{2}\hbar m^{-1}\mathbf{l})\psi(\mathbf{v}_n - \tfrac{1}{2}\hbar m^{-1}\mathbf{l})\exp(-\tfrac{1}{2}\hbar m^{-1}\mathbf{l}\cdot\partial_{\alpha n})] \tag{72.3}$$

We shall often use henceforth the abbreviation:

$$\psi(\mathbf{v}) \equiv 1 + \theta h^3 c m^{-3} \varphi(\mathbf{v}) \tag{72.4}$$

In the limit of Boltzmann statistics ($\theta \to 0$),

$$\psi(\mathbf{v}) \to 1 \quad \text{(Boltzmann statistics)} \tag{72.5}$$

Here appears an essential difference between classical and quantum-mechanical systems. A specific feature of classical systems appears in formula (27.1) [which is the analog of (72.2)]: the tensor $\mathbf{\Phi}$ is an integral over $\mathbf{l}$ which involves only the interaction potential. It has been shown in detail in § 28 that the analytical behavior of this function of z determines entirely the time scale of decay of the non-universal terms. This is no longer true for quantum systems. Although the form of eq. (72.3) is very similar to (27.2) the similarity is only formal, because Ψ_+ contains displacement operators which replace $\mathbf{v}$ by $\mathbf{v} \pm \tfrac{1}{2}\hbar m^{-1}\mathbf{l}$ in the distribution functions $\varphi(\mathbf{v})$. In order to perform the $\mathbf{l}$-integration, it is thus necessary *first* to perform the displacement operations. The result is then:

$$\Psi_+ \varphi(\alpha; 0)\varphi(n; 0) = \frac{8\pi^3 e^4 c i^2}{\hbar^2} \int d\mathbf{l}\, V_l^2$$

$$\times \left\{ \frac{1}{i[\mathbf{l}\cdot\mathbf{g}_{\alpha n} - (\hbar l^2/m) - z]} [\varphi(\mathbf{v}_\alpha)\varphi(\mathbf{v}_n)\psi(\mathbf{v}_\alpha - \hbar\mathbf{l}/m)\psi(\mathbf{v}_n + \hbar\mathbf{l}/m) \\ - \varphi(\mathbf{v}_\alpha - \hbar\mathbf{l}/m)\varphi(\mathbf{v}_n + \hbar\mathbf{l}/m)\psi(\mathbf{v}_\alpha)\psi(\mathbf{v}_n)] \\ - \frac{1}{i[\mathbf{l}\cdot\mathbf{g}_{\alpha n} + (\hbar l^2/m) - z]} [\varphi(\mathbf{v}_\alpha + \hbar\mathbf{l}/m)\varphi(\mathbf{v}_n - \hbar\mathbf{l}/m)\psi(\mathbf{v}_\alpha)\psi(\mathbf{v}_n) \\ - \varphi(\mathbf{v}_\alpha)\varphi(\mathbf{v}_n)\psi(\mathbf{v}_\alpha + \hbar\mathbf{l}/m)\psi(\mathbf{v}_n - \hbar\mathbf{l}/m)] \right\} \tag{72.6}$$

This integral is a *functional of* $\varphi(\mathbf{v})$: this fact is one of the main differences between classical and quantum non-equilibrium statistical mechanics. The physical reason for this fact is rather clear. The laws of evolution of quantum systems are determined not only by the interactions but also by *statistics* (or indiscernibility) effects. Moreover, even for low density systems for which statistics effects are negligible, we have to take account of *diffraction* effects. It is clear that both types of effects depend on the distribution of the particles. The first conclusion of this discussion is therefore that *the way in which non-universal contributions decay in time depends both on the interactions and on the shape of the initial distribution function.*

The detailed study of the analytical properties of an expression such as (72.6) is a difficult and still open problem. We shall limit ourselves here to a few qualitative remarks.

(*a*) $\Psi_+\varphi(\alpha)\varphi(n)$ *regarded as an integral over* **1** is not a Cauchy integral of the same type as in the classical case. Indeed, the denominators in (72.6) contain a term proportional to the square of **1**.* But an important point to note is that *when* $\Psi_+\varphi(\alpha)\varphi(n)$ *is integrated over* $\mathbf{v}_n$ in order to complete the calculation of (72.2), *the resulting expression, regarded as an integral over* $\mathbf{v}_n$, *is a Cauchy integral over the whole real axis* [the denominators are $\{\mathbf{1}\cdot\mathbf{v}_n-(\mathbf{1}\cdot\mathbf{v}_\alpha+hl^2/m+z)\}$]. As we are always interested in reduced distributions, we are sure that the properties of Cauchy integrals needed for the general theory are always satisfied in the quantum-mechanical expressions.

(*b*) *A characteristic type of singularity of the function* $\Psi_+\varphi(\alpha)\varphi(n)$ *is the algebraic branch-point of second order.* The occurrence of this type of singularity is specifically due to the fact that the quantum propagator is *quadratic in l*; this fact is due in turn to the displacement operator and hence is a specific quantum effect. If the l-integration is to be performed by the method of residues, the residues at the *two* poles (in l) due to this propagator have to be evaluated. Factorization of a typical denominator leads to:

* C. George (private communication) has shown that at least part of the integrals can be transformed into Cauchy integrals taken between a finite point and infinity. Such Cauchy integrals have peculiar properties rather different from the ones of the Cauchy integral over the whole real axis (see Appendix 2).

$$\frac{1}{i[-(\hbar l^2/m)+lg-z]} = \frac{1}{-(i\hbar/m)(l-l_1)(l-l_2)}$$

where the roots l_1, l_2 are:

$$l_{1,2} = -\tfrac{1}{2}g \pm \sqrt{[\tfrac{1}{4}g^2 - \hbar z/m]} \tag{72.7}$$

It is now clear that by taking the residues at the poles l_1, l_2, one obtains a function of z containing expressions of the type $\sqrt{(a-z)}$, and hence algebraic branch-points of second order. The location of these branch-points actually depends on the distribution function. [Remember that eq. (72.2) involves an integration over $\mathbf{v}_n$.] The existence of a spread of velocities (due to thermal motion or to the quantum zero-point energy) will usually shift these branch-points into the complex plane.

(*c*) *The time scales of decay of non-universal quantum contributions are related both to the range of the interactions and the de Broglie wave-length.* The interactions enter the problem as in classical mechanics through the factor V_l^2. The de Broglie wave-length appears in two ways: through the distribution functions of type $\varphi(\mathbf{v}-\hbar\mathbf{l}/m)$, and through the branch-points discussed above. According to the type of system considered, there are two relevant de Broglie wave-lengths:

The thermal de Broglie wave-length, associated with the average thermal momentum:

$$\lambda_T = \hbar(\beta/2m)^{\frac{1}{2}} \tag{72.8}$$

The wave-length associated with the Fermi momentum, which characterizes the quantum zero-point energy:

$$\lambda_F = \hbar/mv_F \tag{72.9}$$

The latter is dominant for dense systems at very low temperatures, in which statistics effects are important. One can construct three basic times in a quantum system of charged particles:

$$t_P = (m/4\pi e^2 c)^{\frac{1}{2}} \tag{72.10}$$

$$t_T = \hbar\beta \tag{72.11}$$

$$t_F = 2\hbar/mv_F^2 \tag{72.12}$$

The non-universal contributions to the cycle decay with time scales which are complicated superpositions of these three basic times. Note also that the decay is not always exponential (because of the occurrence of singularities other than poles). Actually, in the quasi-classical region (see Fig. 66.1) the time scale t_F does not appear, whereas in the extreme quantum region (degenerate gases) the time t_T can be disregarded.

In these two extreme cases one can then state:

One is justified in retaining only the "leading universal contributions" to the cycle [*i.e.* terms of order (71.1)] *if one considers times much longer than the largest of t_P on the one hand, and t_T or t_F respectively, on the other hand.*

§ 73. The Quantum "Cycle" Equation

The summation of the cycle diagrams is just as easy in our formalism as in the classical case studied in § 30. We shall first indicate a small formal point which might be misleading. In the previous section, we have evaluated the expression for the operator $\Psi_+(z)$ and have already noted the important point that in quantum statistics it is a functional of the distribution function. We would expect (from the classical theory) that the right side of the kinetic equation would consist (essentially) of the operator $\Psi_+(0)$ acting on a product of distributions $\varphi(t)$ at time t (see Appendix 4). However, eq. (72.3) exhibits the distribution functions at time 0, whereas we would expect to see these functions evaluated at time t. In order to show that our first guess is indeed correct, it is convenient to write the expression (72.2) in a somewhat different form. We write:

$$\Psi_+(\mathbf{v}_\alpha, \mathbf{v}_n; z) = \int d\mathbf{v}_r d\mathbf{v}_s \Delta_+(\mathbf{v}_\alpha, \mathbf{v}_n, \mathbf{v}_r, \mathbf{v}_s; z)\varphi(\mathbf{v}_r; 0)\varphi(\mathbf{v}_s; 0) \quad (73.1)$$

with

$$\begin{aligned} \Delta_+ = {} & \frac{8\pi^3 e^4 c i}{\hbar^2}\int d\mathbf{l}[\exp(-\tfrac{1}{2}\hbar m^{-1}\mathbf{l}\cdot\partial_{\alpha n}) - \exp(\tfrac{1}{2}\hbar m^{-1}\mathbf{l}\cdot\partial_{\alpha n})]\frac{V_l^2}{i(\mathbf{l}\cdot\mathbf{g}_{\alpha n} - z)} \\ & \cdot\{[1+\theta h^3 c m^{-3}\delta(\mathbf{v}_r - \mathbf{v}_\alpha + \tfrac{1}{2}\hbar m^{-1}\mathbf{l})][1+\theta h^3 c m^{-3}\delta(\mathbf{v}_s - \mathbf{v}_n - \tfrac{1}{2}\hbar m^{-1}\mathbf{l})] \\ & \exp(\tfrac{1}{2}\hbar m^{-1}\mathbf{l}\cdot\partial_{\alpha n}) - [1+\theta h^3 c m^{-3}\delta(\mathbf{v}_r - \mathbf{v}_\alpha - \tfrac{1}{2}\hbar m^{-1}\mathbf{l})] \\ & [1+\theta h^3 c m^{-3}\delta(\mathbf{v}_s - \mathbf{v}_n + \tfrac{1}{2}\hbar m^{-1}\mathbf{l})]\exp(-\tfrac{1}{2}\hbar m^{-1}\mathbf{l}\cdot\partial_{\alpha n})\} \end{aligned} \quad (73.2)$$

We have introduced in this way an operator $\Delta_+(z)$ which is no longer a functional of the distribution function. On the other hand, it is an operator which involves *four* particles instead of *two*. This is, of course, another manifestation of the typical quantum-statistical effect according to which the initial states and final states of the particles involved in a collision are not independent, because of the exclusion principle. We shall come back to this point later.

Using the operator Δ_+, we shall have no difficulty in repeating the argument of § 30. The only difference is that the operator Δ_+ is analogous to a diagonal fragment involving four particles (see also Appendix 4). The result, similar to (30.8), is

$$\partial_t \varphi(\alpha; t) = \int d\mathbf{v}_n d\mathbf{v}_r d\mathbf{v}_s \Delta_+(0) \varphi(\alpha; t) \varphi(n; t) \varphi(r; t) \varphi(s; t) \tag{73.3}$$

It is now an easy matter to write down the explicit kinetic equation. Using (73.2), the integrations over $\mathbf{v}_r$ and $\mathbf{v}_s$ are easily carried out. We use the limiting value of the propagator occurring in (73.2) [see eq. (A2.2.8) of Appendix 2]

$$\lim_{\substack{z \to 0 \\ (z \in S_+)}} \frac{1}{i(\mathbf{l} \cdot \mathbf{g}_{\alpha n} - z)} = \pi \delta_-(\mathbf{l} \cdot \mathbf{g}_{\alpha n})$$

We then obtain:

$$\begin{aligned} &\partial_t \varphi(\mathbf{v}_\alpha; t) \\ &= \frac{8\pi^4 e^4 c}{\hbar^2} \int d\mathbf{v}_n \int d\mathbf{l} [\exp(-\tfrac{1}{2}\hbar m^{-1} \mathbf{l} \cdot \partial_{\alpha n}) - \exp(\tfrac{1}{2}\hbar m^{-1} \mathbf{l} \cdot \partial_{\alpha n})] \delta_-(\mathbf{l} \cdot \mathbf{g}_{\alpha n}) \\ &\cdot V_l^2 \{\psi(\mathbf{v}_\alpha + \tfrac{1}{2}\hbar m^{-1}\mathbf{l}) \psi(\mathbf{v}_n - \tfrac{1}{2}\hbar m^{-1}\mathbf{l}) \exp(-\tfrac{1}{2}\hbar m^{-1}\mathbf{l} \cdot \partial_{\alpha n}) \\ &- \psi(\mathbf{v}_\alpha - \tfrac{1}{2}\hbar m^{-1}\mathbf{l}) \psi(\mathbf{v}_n + \tfrac{1}{2}\hbar m^{-1}\mathbf{l}) \exp(\tfrac{1}{2}\hbar m^{-1}\mathbf{l} \cdot \partial_{\alpha n})\} \varphi(\mathbf{v}_\alpha) \varphi(\mathbf{v}_n) \end{aligned} \tag{73.4}$$

In this form, it is immediately verified that, setting $\theta = 0$ (i.e. $\psi = 1$) and then letting $\hbar \to 0$, we obtain the correct classical result (30.11).*

* It is easily verified that if the exchange part of the matrix element of V is taken into account, one should merely replace V_l by $V_l + \theta V_{m|\mathbf{v}_\alpha - \mathbf{v}_n|/\hbar}$ in this equation (it is then important to keep the latter expression in the position where it stands in (73.4): the exchange matrix element does not commute with the displacement operators!).

We now write this equation in a still more explicit form, by performing the displacement operations. We then obtain:

$$\partial_t \varphi(\mathbf{v}_\alpha; t) = \frac{8\pi^4 e^4 c}{\hbar^2} \int d\mathbf{v}_n \int d\mathbf{l}$$

$$\cdot \{\delta_-[\mathbf{l} \cdot (\mathbf{g}_{\alpha n} - \hbar\mathbf{l}/m)] V_l^2 [-\varphi(\mathbf{v}_\alpha)\varphi(\mathbf{v}_n)\psi(\mathbf{v}_\alpha - \hbar\mathbf{l}/m)\psi(\mathbf{v}_n + \hbar\mathbf{l}/m)$$
$$+\varphi(\mathbf{v}_\alpha - \hbar\mathbf{l}/m)\varphi(\mathbf{v}_n + \hbar\mathbf{l}/m)\psi(\mathbf{v}_\alpha)\psi(\mathbf{v}_n)]$$
$$-\delta_-[\mathbf{l} \cdot (\mathbf{g}_{\alpha n} + \hbar\mathbf{l}/m)] V_l^2 [-\varphi(\mathbf{v}_\alpha + \hbar\mathbf{l}/m)\varphi(\mathbf{v}_n - \hbar\mathbf{l}/m)\psi(\mathbf{v}_\alpha)\psi(\mathbf{v}_n)$$
$$+\varphi(\mathbf{v}_\alpha)\varphi(\mathbf{v}_n)\psi(\mathbf{v}_\alpha + \hbar\mathbf{l}/m)\psi(\mathbf{v}_n - \hbar\mathbf{l}/m)]\}$$

In the first group of terms we change $\mathbf{l}$ into $-\mathbf{l}$ and use the obvious properties [see eqs. (A2.2–4) of Appendix 2]

$$\delta_-(-x) = \delta_+(x)$$
$$\delta_-(x) + \delta_+(x) = 2\delta(x) \tag{73.5}$$

We then obtain the familiar form of the quantum cycle equation:

$$\partial_t \varphi(\mathbf{v}_\alpha; t) = \frac{16\pi^4 e^4 c}{\hbar^2} \int d\mathbf{v}_n \int d\mathbf{l}\; \delta(\mathbf{l} \cdot \mathbf{g}_{\alpha n} + \hbar l^2/m) V_l^2 \tag{73.6}$$
$$\{-\varphi(\mathbf{v}_\alpha)\varphi(\mathbf{v}_n)\psi(\mathbf{v}_\alpha + \hbar\mathbf{l}/m)\psi(\mathbf{v}_n - \hbar\mathbf{l}/m)$$
$$+\varphi(\mathbf{v}_\alpha + \hbar\mathbf{l}/m)\varphi(\mathbf{v}_n - \hbar\mathbf{l}/m)\psi(\mathbf{v}_\alpha)\psi(\mathbf{v}_n)\}$$

§ 74. General Properties of the Cycle Equation

Equation (73.4) [or (73.6)] is the quantum-mechanical analog of the classical Landau equation (30.11): it is a weak coupling equation. It will be shown below that it has the required properties of an irreversible (kinetic) equation. However, it exhibits two marked differences from its classical analog, and these differences will be discussed first.

A) *In quantum-mechanical systems, the approximation of weak coupling is not equivalent to the approximation of small momentum transfers (or small deflections).*

This is immediately obvious from the fact that the kinetic equation (73.4) is a finite difference equation, quite analogous from this point of view to the classical Boltzmann equation (33.1). We have already stressed repeatedly that this is a consequence of Heisenberg's uncertainty principle: in quantum mechanics even a weak interaction can produce a very strong momentum transfer. The quantum-mechanical analog of the Boltzmann equation (33.1) is known as the *Uehling–Uhlenbeck equation.** As a consequence of the present remark, eq. (73.4) cannot be derived from the Uehling–Uhlenbeck equation by Landau's method (§ 33). It is rather obtained by expanding the collision cross-section appearing in the latter equation in powers of e^2 and retaining only the first non-vanishing term (i.e. the first Born approximation). Another important consequence of this remark is the fact that the **l** *integral converges at the upper limit.* We have seen in § 28 that the short-distance divergence (i.e. large l divergence) of the classical cycle diagram is due to an improper description of the strong deflections. In quantum mechanics, however, the weak-coupling approximation includes strong deflections and as a result the **l**-integral is automatically cut off at large values of l. This will be seen explicitly in more detailed calculations such as those of § 75.

B) *In Fermi statistics the probability of a collision depends not only on the initial momenta but also on the final ones.* This can be seen quite clearly from eq. (73.6), which can be interpreted as follows (see also the discussion of the Boltzmann equation, § 33).

The collision term expresses the fact that the time rate of change of $\varphi(\alpha)$ is due to two types of processes: a loss due to direct collisions of particles of velocity $\mathbf{v}_\alpha$ with particles "$\mathbf{v}_n$", resulting in final velocities $\mathbf{v}_\alpha+\hbar\mathbf{l}/m$, $\mathbf{v}_n-\hbar\mathbf{l}/m$; and a gain due to the restituting collision (in which the final state is $\mathbf{v}_\alpha$, $\mathbf{v}_n$). The cross-section of these processes is $V_l^2\delta[\mathbf{l}\cdot(\mathbf{v}_\alpha-\mathbf{v}_n)+\hbar l^2/m]$. Note that the argument of the δ-function is just the difference in energies of the initial and final states; therefore the cross-section vanishes if the energy is not conserved in the collision. The number of direct collisions is proportional to the probability of the occurrence of the velocities $\mathbf{v}_\alpha$ and $\mathbf{v}_n$ [i.e. $\varphi(\mathbf{v}_\alpha)\varphi(\mathbf{v}_n)$] and *also* to the probability that the final states $\mathbf{v}_\alpha+\hbar\mathbf{l}/m$, $\mathbf{v}_n-\hbar\mathbf{l}/m$ are *empty* (i.e. $[1-h^3cm^{-3}\varphi(\mathbf{v}_\alpha+\hbar\mathbf{l}/m)][1-h^3cm^{-3}\varphi(\mathbf{v}_n-\hbar\mathbf{l}/m)]$). The latter

* E. A. Uehling and G. E. Uhlenbeck, *Phys. Rev.*, **43**, 552 (1933).

effect is clearly an expression of the Pauli exclusion principle, which dominates the Fermi–Dirac statistics. Thus, the probability of collisions is in general smaller in Fermi systems than in the corresponding classical systems.

We now investigate the irreversible properties of the kinetic equation (73.6). We first show that the equilibrium Fermi distribution $\varphi_F(\mathbf{v})$ is a stationary solution of eq. (73.6):

$$\varphi_F(\mathbf{v}) = \frac{m^3}{h^3 c} \frac{1}{\exp\left[-\zeta + \frac{1}{2}\beta m v^2\right] + 1} \tag{74.1}$$

ζ is the Fermi energy divided by kT. We now note the following identity

$$\varphi_F(\mathbf{v})\psi_F(\mathbf{v}+\hbar\mathbf{l}/m) = \exp\left[\hbar\beta(\mathbf{l}\cdot\mathbf{v}+\tfrac{1}{2}\hbar l^2/m)\right]\varphi_F(\mathbf{v}+\hbar\mathbf{l}/m)\psi_F(\mathbf{v}) \tag{74.2}$$

This identity is easily proved by using (74.1). We now substitute $\varphi_F(\mathbf{v})$ for $\varphi(\mathbf{v})$ in eq. (73.6) and apply identity (74.2) twice. The result is:

$$\begin{aligned} \partial_t \varphi_F(\mathbf{v}_\alpha) &= \frac{16\pi^4 e^4 c}{\hbar^2} \int d\mathbf{v}_n \int d\mathbf{l}\, V_l^2 \delta(\mathbf{l}\cdot\mathbf{g}_{\alpha n} + \hbar l^2/m) \\ &\cdot \{1 - \exp[\hbar\beta(\mathbf{l}\cdot\mathbf{g}_{\alpha n} + \hbar l^2/m)]\}\varphi_F(\mathbf{v}_\alpha + \hbar\mathbf{l}/m)\varphi_F(\mathbf{v}_n - \hbar\mathbf{l}/m) \\ &\qquad \psi_F(\mathbf{v}_\alpha)\psi_F(\mathbf{v}_n) = 0 \end{aligned} \tag{74.3}$$

Indeed, due to the δ-function, the argument of the exponential must be zero, and the expression in the curly brackets vanishes.

We now prove an H-theorem, i.e. we construct a functional of $\varphi(\alpha)$ which is a monotonous increasing function of time and remains stationary only at the Fermi–Dirac equilibrium. Hence $\varphi_F(\alpha)$ is the only stationary solution of eq. (73.6). The entropy density is defined for a Fermi system as follows:

$$s(t) = -k \int d\mathbf{v}\{\varphi(\mathbf{v}) \ln h^3 c m^{-3}\, \varphi(\mathbf{v}) + [\psi(\mathbf{v}) m^3/h^3 c] \ln \psi(\mathbf{v})\} \tag{74.4}$$

The familiar definition of the entropy density, found in the textbooks, is in terms of mean occupation numbers:

$$s(t) = -(k/N) \sum_{\{\mathbf{p}\}} \{\bar{n}(\mathbf{p}) \ln \bar{n}(\mathbf{p}) + [1-\bar{n}(\mathbf{p})] \ln [1-\bar{n}(\mathbf{p})]\}$$

This definition is of the Boltzmann type, but contains an additional term related to the probability that the state $\mathbf{p}$ is empty. We do not want to go into a detailed discussion of the origin of this feature here.*

We now take the limit $N \to \infty$, using eq. (62.1), and introduce the velocity distribution through eq. (70.3)

$$s(t) = -(km^3/h^3c)\int d\mathbf{v}\{h^3cm^{-3}\,\varphi(\mathbf{v})\ln h^3cm^{-3}\varphi(\mathbf{v}) + [1-h^3cm^{-3}\varphi(\mathbf{v})]\ln[1-h^3cm^{-3}\varphi(\mathbf{v})]\}$$

This formula is identical to (74.4).

We now calculate the time rate of change of the entropy:

$$\begin{aligned}\partial_t s(t) &= -k\int d\mathbf{v}_\alpha\{(\ln h^3cm^{-3}\varphi+1)\partial_t\varphi+(m^3/h^3c)\,(\ln\psi+1)\partial_t\psi\}\\ &= -k\int d\mathbf{v}_\alpha\{\ln h^3cm^{-3}\varphi(\mathbf{v}_\alpha)-\ln\psi(\mathbf{v}_\alpha)\}\partial_t\varphi(\mathbf{v}_\alpha)\end{aligned}$$

Using explicitly eq. (73.6) and performing a symmetrization similar to the one used in eq. (34.4), we obtain

$$\begin{aligned}\partial_t s &= -Ak\int d\mathbf{v}_\alpha d\mathbf{v}_n d\mathbf{l}\, V_l^2\delta(\mathbf{l}\cdot\mathbf{g}_{\alpha n}+\hbar l^2/m)[\ln\varphi(\mathbf{v}_\alpha)-\ln\psi(\mathbf{v}_\alpha)]\\ &\quad\cdot[-\varphi(\mathbf{v}_\alpha)\varphi(\mathbf{v}_n)\psi(\mathbf{v}_\alpha+\hbar\mathbf{l}/m)\psi(\mathbf{v}_n-\hbar\mathbf{l}/m)\\ &\qquad+\varphi(\mathbf{v}_\alpha+\hbar\mathbf{l}/m)\varphi(\mathbf{v}_n-\hbar\mathbf{l}/m)\psi(\mathbf{v}_\alpha)\psi(\mathbf{v}_n)]\\ &= -Ak\int d\mathbf{v}_\alpha d\mathbf{v}_n d\mathbf{l}\, V_l^2\delta(\mathbf{l}\cdot\mathbf{g}_{\alpha n}+\hbar l^2/m) \qquad (74.5)\\ &\quad\left\{\ln\frac{\varphi(\mathbf{v}_\alpha)\varphi(\mathbf{v}_n)\psi(\mathbf{v}_\alpha+\hbar\mathbf{l}/m)\psi(\mathbf{v}_n-\hbar\mathbf{l}/m)}{\psi(\mathbf{v}_\alpha)\psi(\mathbf{v}_n)\varphi(\mathbf{v}_\alpha+\hbar\mathbf{l}/m)\varphi(\mathbf{v}_n-\hbar\mathbf{l}/m)}\right\}\\ &\quad\cdot[-\varphi(\mathbf{v}_\alpha)\varphi(\mathbf{v}_n)\psi(\mathbf{v}_\alpha+\hbar\mathbf{l}/m)\psi(\mathbf{v}_n-\hbar\mathbf{l}/m)\\ &\qquad+\varphi(\mathbf{v}_\alpha+\hbar\mathbf{l}/m)\varphi(\mathbf{v}_n-\hbar\mathbf{l}/m)\psi(\mathbf{v}_\alpha)\psi(\mathbf{v}_n)] \geqq 0\end{aligned}$$

This quantity is semi-definite positive, because of the identity:

$$(y-x)\ln(x/y) \leqq 0 \qquad (74.6)$$

The equality sign holds only when $x = y$, i.e. at the Fermi–Dirac equilibrium [see (74.3)].

§ 75. Motion of a Test Particle in an Electron Gas at Zero Temperature

As an application of the "cycle" equation we shall discuss

* See, for instance, A. Sommerfeld, *Vorlesungen über theoretische Physik*, vol. 5 (Thermodynamik und Statistik), Dietrich Verlag, Wiesbaden, 1952.

now the problem of a test particle injected into an electron gas which is in equilibrium. This is the problem corresponding to the classical brownian motion problem treated in §§ 37 and 38. However, we prefer not to call it a "brownian motion" problem, because we have seen that in quantum-mechanical systems, even if the interactions are weak, the deflections are finite so that the mechanism of approach to equilibrium is rather different from a brownian motion process. In particular, it is obvious from eq. (73.6) that the problem cannot be reduced to a Fokker–Planck equation (i.e. to a partial differential equation of the diffusion type). This means that the motion can no longer be completely specified by means of only two functions (friction constant and diffusion tensor) as in the classical case; an infinite number of "transition moments" are now necessary. Nevertheless, the first few moments retain their physical importance, and the friction coefficient especially is a quantity which determines approximately the electrical conductivity and which can also be measured directly in some cases.

We consider the motion of a test electron through an electron gas under conditions in which the quantum-statistical effects are essential. An experimental situation which approaches the conditions of this problem is realized by sending a beam of electrons through a metal foil and measuring the energy loss of the electrons as a function of their initial energy. Of course, a metal is not an ideal Fermi gas, nor is the density of electrons in a metal high enough for a straightforward application of the cycle (or ring) approximation. However, if quantitative agreement with experiment cannot be expected, at least one can hope that several interesting qualitative features of the phenomenon will be put forward by the present model.

Our starting point is eq. (73.6). Several important simplifications arise in the present situation.

(*a*) The particles of the medium are assumed to be in equilibrium at zero temperature; thus their velocity distribution is

$$\varphi_F^0(\mathbf{v}_n) = \frac{m^3}{h^3 c}\,\theta(v_F - v_n) \tag{75.1}$$

with the Fermi velocity defined as:

$$v_F = (3h^3c/8\pi m^3)^{\frac{1}{3}} \tag{75.2}$$

(*b*) The test particle can, by definition, be distinguished from the medium. It follows that it obeys Boltzmann statistics, and hence:

$$\psi(\mathbf{v}_\alpha) \equiv 1 \tag{75.3}$$

(*c*) We must also discuss the choice of an effective potential V_l because we know that the true Coulomb potential leads to a logarithmic divergence in the cycle approximation. It will be shown in the next chapter, from an analysis of the ring approximation, that, to a first approximation, a Debye-like potential can be used which automatically ensures a correct behavior at small wave-vectors:

$$V_l = (2\pi^2)^{-1}(l^2+\kappa_q^2)^{-1} \tag{75.4}$$

and it turns out that the value of κ_q is given by

$$\kappa_q = \frac{4e^2m^2v_F}{\pi\hbar^3} = \left(\frac{192}{\pi}\right)^{\frac{1}{3}}\frac{e^2c^{\frac{1}{3}}m}{\hbar^2} \tag{75.5}$$

Before proceeding further, it is convenient to simplify the notations by using a set of reduced variables, defined as usual with respect to the characteristics of the medium. In the present case these variables are:

$$\begin{aligned} \mathbf{w} &= \mathbf{v}_\alpha/v_F \\ \mathbf{u} &= \mathbf{v}_n/v_F \\ \boldsymbol{\lambda} &= \mathbf{l}/(mv_F\hbar^{-1}) \\ \kappa &= \kappa_q/(mv_F\hbar^{-1}) \\ \tau &= (4e^4c/m^2v_F^3)t \\ &\equiv (\tfrac{4}{3}e^4m/\pi^2\hbar^3)t \end{aligned} \tag{75.6}$$

It is interesting to point out a peculiar property of the reduction coefficient for the time variable, which is actually a measure of the relaxation time of the system:

$$\tau_r = \pi^2\hbar^3/\tfrac{4}{3}e^4m \tag{75.7}$$

The interesting point is that the *relaxation time is independent of the temperature and of the density.* This must be so if the relaxation time is to keep a finite limit as $\beta \to \infty$ and $c \to \infty$. Moreover, under these conditions the thermal equilibrium coincides with the dynamical equilibrium, viz. the ground state of the Fermi gas. Hence the relaxation time (which is determined by the parameters of the *medium*) is fixed by the purely mechanical parameters e^3 and m and excludes any statistical parameters such as β or c.

Using the results of our previous discussion, the kinetic equation (73.6) reduces to the following expression:

$$\begin{aligned}\partial_\tau \varphi(\mathbf{w}) = \int d\mathbf{u} \int d\boldsymbol{\lambda}\, (\lambda^2+\kappa^2)^{-2} \delta(\boldsymbol{\lambda}\cdot\mathbf{w} - \boldsymbol{\lambda}\cdot\mathbf{u} + \lambda^2) \\ \{-\varphi(\mathbf{w})\varphi_F^0(\mathbf{u})\psi_F^0(\mathbf{u}-\boldsymbol{\lambda}) + \varphi(\mathbf{w}+\boldsymbol{\lambda})\varphi_F^0(\mathbf{u}-\boldsymbol{\lambda})\psi_F^0(\mathbf{u})\}\end{aligned} \tag{75.8}$$

with

$$\varphi_F^0(\mathbf{u}) = \theta(1-u), \quad \psi_F^0(\mathbf{u}) = \theta(u-1) \tag{75.9}$$

We now multiply both sides of (75.8) by $\mathbf{w}$ and integrate over $\mathbf{w}$. We moreover take as new integration variables $\mathbf{u}-\boldsymbol{\lambda} \to \mathbf{u}'$ in the first term of the r.h.s. and $\mathbf{w}+\boldsymbol{\lambda} \to \mathbf{w}'$ in the second term; dropping the dashes after the substitution we obtain:

$$\begin{aligned}\langle \partial_\tau \mathbf{w} \rangle = \int d\mathbf{w}\, d\mathbf{u}\, d\boldsymbol{\lambda}\, (\lambda^2+\kappa^2)^{-2} \delta(\boldsymbol{\lambda}\cdot\mathbf{w} - \boldsymbol{\lambda}\cdot\mathbf{u}) \\ \{-\mathbf{w}\varphi_F^0(\mathbf{u}+\boldsymbol{\lambda})\psi_F^0(\mathbf{u}) + (\mathbf{w}-\boldsymbol{\lambda})\varphi_F^0(\mathbf{u}-\boldsymbol{\lambda})\psi_F^0(\mathbf{u})\}\varphi(\mathbf{w})\end{aligned} \tag{75.10}$$

After some elementary algebra, this expression can be written in the form:

$$\langle \partial_\tau \mathbf{w} \rangle = -\int d\mathbf{w}\, \varphi(\mathbf{w})\mathbf{w}\eta(w) \tag{75.11}$$

In this form the friction coefficient appears explicitly as:

$$\begin{aligned}\eta(w) = w^{-2} \int d\boldsymbol{\lambda}\, (\lambda^2+\kappa^2)^{-2} \boldsymbol{\lambda}\cdot\mathbf{w} \int d\mathbf{u}\, \delta(\boldsymbol{\lambda}\cdot\mathbf{u} - \boldsymbol{\lambda}\cdot\mathbf{w} + \lambda^2) \\ \varphi_F^0(\mathbf{u})\psi_F^0(\mathbf{u}+\boldsymbol{\lambda})\end{aligned} \tag{75.12}$$

It will be shown now that this expression can be related to the imaginary part of the dielectric constant discussed in § 70. Indeed, the imaginary part of the dielectric constant $\varepsilon_+(\chi; \chi y)$

or of the related function $m_2(\chi; y)$, eqs. (70.7) and (70.9), for the real frequency χy can be written, from (70.4) as:

$$m_2(\lambda; y) = \int d\mathbf{u}\, \varphi_F^0(\mathbf{u})\psi_F^0(\mathbf{u}+\boldsymbol{\lambda}) \\ \{\delta(\boldsymbol{\lambda}\cdot\mathbf{u}+\tfrac{1}{2}\lambda^2-\lambda y)-\delta(\boldsymbol{\lambda}\cdot\mathbf{u}+\tfrac{1}{2}\lambda^2+\lambda y)\} \tag{75.13}$$

We have included here the factor $\psi_F^0(\mathbf{u}+\boldsymbol{\lambda})$ in the integral, although it is actually redundant [see (67.7a)]. We now note that at zero temperature the expression $\boldsymbol{\lambda}\cdot\mathbf{u}+\frac{1}{2}\lambda^2$ must be positive for $\varphi_F^0(\mathbf{u})\psi_F^0(\mathbf{u}+\boldsymbol{\lambda})$ to be non-zero. It then follows that for positive y only the first term on the r.h.s. contributes to $m_2(\lambda; y)$ whereas for negative y this term vanishes. But the **u**-integral in (75.12) is identical with this first term if one takes $\lambda y = \boldsymbol{\lambda}\cdot\mathbf{w}-\frac{1}{2}\lambda^2$. Hence:

$$\int d\mathbf{u}\, \varphi_F^0(\mathbf{u})\psi_F^0(\mathbf{u}+\boldsymbol{\lambda})\delta(\boldsymbol{\lambda}\cdot\mathbf{u}-\boldsymbol{\lambda}\cdot\mathbf{w}+\lambda^2) = \begin{cases} m_2[\lambda;\, (\boldsymbol{\lambda}\cdot\mathbf{w}/\lambda)-\frac{1}{2}\lambda] \\ \qquad \text{for } \boldsymbol{\lambda}\cdot\mathbf{w}-\frac{1}{2}\lambda^2 > 0 \\ 0 \qquad \text{for } \boldsymbol{\lambda}\cdot\mathbf{w}-\frac{1}{2}\lambda^2 < 0 \end{cases} \tag{75.14}$$

where $m_2(\lambda; y)$ is the function defined in Table 70.1. The friction coefficient can thus be written as:

$$\eta(w) = w^{-2}\int d\boldsymbol{\lambda}\, (\lambda^2+\kappa^2)^{-2}(\boldsymbol{\lambda}\cdot\mathbf{w})\, m_2[\lambda;\, (\boldsymbol{\lambda}\cdot\mathbf{w}/\lambda)-\tfrac{1}{2}\lambda]\theta[\boldsymbol{\lambda}\cdot\mathbf{w}-\tfrac{1}{2}\lambda^2] \tag{75.15}$$

We now go over to polar coordinates in a reference system whose z-axis is directed along **w**. In order to satisfy the θ-function, the following condition must be satisfied

$$\lambda w\,\xi-\tfrac{1}{2}\lambda^2 \geqq 0 \tag{75.16}$$

where $\xi = \cos\theta$. Hence

$$(\lambda/2w) \leqq \xi \leqq 1$$

This condition in turn can only be satisfied if $\lambda \leqq 2w$. Hence

$$\eta(w) = 2\pi w^{-1}\int_0^{2w} d\lambda\, \lambda^3(\lambda^2+\kappa^2)^{-2}\int_{\lambda/2w}^{1} d\xi\, \xi m_2(\lambda;\, w\xi-\tfrac{1}{2}\lambda) \tag{75.17}$$

Substitution of the explicit form of $m_2(\lambda; y)$ taken from Table 70.1 leads to a simple but tedious discussion of the limits of integration in this formula. Introducing the functions:

$$\frac{\lambda^3}{(\lambda^2+\kappa^2)^2}(2\pi^2\xi/w)(w\xi-\tfrac{1}{2}\lambda) = R_1$$

$$\frac{\lambda^2}{(\lambda^2+\kappa^2)^2}(\pi^2\xi/w)(1-\lambda^2+2\lambda w\xi-w^2\xi^2) = R_2$$

the result is:

For $w \leqq 1$

$$\eta(w) = \int_0^{2w} d\lambda \int_{\lambda/2w}^{1} d\xi\, R_1 \tag{75.18}$$

For $1 \leqq w \leqq 3$

$$\begin{aligned}\eta(w) = &\int_0^{2} d\lambda \int_{\lambda/2w}^{1} d\xi\, R_1 + \int_0^{w-1} d\lambda \int_{1/w}^{(\lambda+1)/w} d\xi\, R_2 \\ &+ \int_{w-1}^{2} d\lambda \int_{1/w}^{1} d\xi\, R_2 + \int_2^{w+1} d\lambda \int_{(\lambda-1)/w}^{1} d\xi\, R_2\end{aligned} \tag{75.19}$$

For $w \geqq 3$

$$\begin{aligned}\eta(w) = &\int_0^{2} d\lambda \int_{\lambda/2w}^{1/w} d\xi\, R_1 + \int_0^{2} d\lambda \int_{1/w}^{(1+\lambda)/w} d\xi\, R_2 \\ &+ \int_2^{w-1} d\lambda \int_{(\lambda-1)/w}^{(\lambda+1)/w} d\xi\, R_2 + \int_{w-1}^{w+1} d\lambda \int_{(\lambda-1)/w}^{1} d\xi\, R_2\end{aligned} \tag{75.20}$$

There is absolutely no difficulty in the evaluation of these integrals, although the calculations are rather lengthy. It turns out that the contributions from the regions ($1 \leqq w \leqq 3$) and ($w \geqq 3$) lead to the same functional dependence.

The final result is:

For $w \leqq 1$

$$\pi^{-2}\eta(w) = -\tfrac{11}{9}-\tfrac{5}{24}\frac{\kappa^2}{w^2}+\tfrac{1}{3}\ln\frac{4w^2+\kappa^2}{\kappa^2}+\left(\frac{3}{4}\frac{\kappa}{w}+\frac{5}{48}\frac{\kappa^3}{w^3}\right)\tan^{-1}\frac{2w}{\kappa} \tag{75.21}$$

For $w \geqq 1$

$$\pi^{-2}\eta(w) = -\frac{7}{18w^3} - \frac{\kappa^2}{8w^3} - \frac{5}{6w} + \frac{w+1}{24w^3}$$

$$\cdot\,[3(w^2-1)^2+2\kappa^2(w^2-4w+1)-\kappa^4]\,\frac{1}{\kappa^2+(w+1)^2}$$

$$+\frac{w-1}{24w^3}\,[-3(w^2-1)^2-2\kappa^2(w^2+4w+1)+\kappa^4]\,\frac{1}{\kappa^2+(w-1)^2}$$

$$+\frac{1}{8w^3}\ln\frac{[\kappa^2+(w-1)^2][\kappa^2+(w+1)^2]}{\kappa^2(\kappa^2+4)} + \tfrac{1}{3}\ln\frac{\kappa^2+(w+1)^2}{\kappa^2+(w-1)^2} \tag{75.22}$$

$$-\frac{\kappa}{w^3}\left(\frac{5\kappa^2}{48}+\frac{3}{4}\right)\tan^{-1}\frac{2}{\kappa}$$

$$+\frac{1}{24\kappa w^3}[3(w^2-1)^2-18\kappa^2(w^2+1)-5\kappa^4]\left[\tan^{-1}\frac{w-1}{\kappa}-\tan^{-1}\frac{w+1}{\kappa}\right]$$

One checks immediately that $\eta(w)$ is continuous at $w = 1$. A very illuminating insight into the nature of the phenomenon is obtained by investigating the behavior of $\eta(w)$ for very small and for very large velocities. Straightforward expansions yield the results:

$$\pi^{-2}\eta(w) \approx \frac{8}{35}\left(\frac{w}{\kappa}\right)^4 - \frac{128}{189}\left(\frac{w}{\kappa}\right)^6 + \ldots, \qquad w \ll \kappa \tag{75.23}$$

$$\pi^{-2}\eta(w) \approx \tfrac{4}{3}w^{-3}\ln w$$

$$+\left\{\frac{5}{9}+\frac{\kappa^2}{24}-\tfrac{1}{3}\ln[\kappa^2(\kappa^2+4)]-\left(\frac{3\kappa}{4}+\frac{5\kappa^3}{48}\right)\tan^{-1}\frac{2}{\kappa}\right\}w^{-3}$$

$$+\left\{-\tfrac{2}{5}+\kappa^2+\frac{\kappa^4}{6}\right\}w^{-5}+\ldots, \qquad w \gg \kappa \tag{75.24}$$

A plot of the friction coefficient $\eta(w)$ is given in Fig. 75.1.

Comparing eq. (75.24) with the corresponding classical result, it is seen that both functions behave in a similar way, viz. the friction coefficient tends to zero for large velocities. This implies in particular that the runaway effect is by no means a purely classical effect (see

Chapter 13). The trend to zero is, however, somewhat slower in the quantum case ($w^{-3} \ln w$) than in the classical case (w^{-3}).

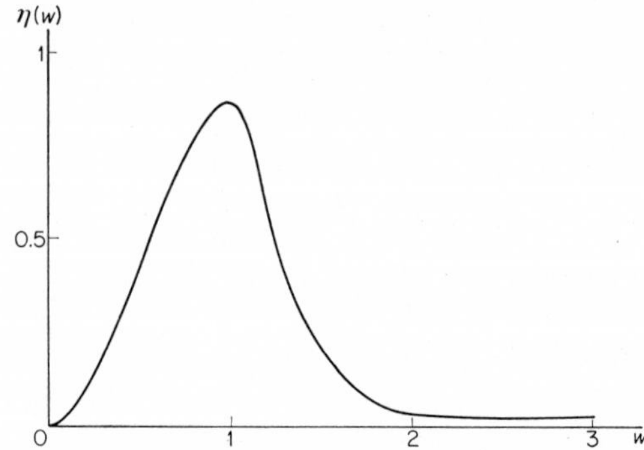

Fig. 75.1. The friction coefficient for a test particle moving through an electron gas at zero temperature.

But the most interesting feature appears in the behavior for small velocities. Whereas the classical friction coefficient tends toward a finite constant as $w \to 0$, the quantum coefficient vanishes strongly, as w^4. *This is a typical Fermi statistical effect* which can be understood rather easily. For an interaction between the test particle and a particle of the medium to be possible, a particle of the medium must be excited into an unoccupied state, i.e., its energy after the collision must be higher than the Fermi energy. If the energy of the test particle is very small, only particles very close to the Fermi level can be involved in such a collision. The number of particles which can interact with the test particle is thus a decreasing function of the velocity of this particle. This is why a very slow test particle which travels through the medium almost ignores the presence of the medium.

CHAPTER 17

The Quantum-Statistical Ring Approximation

§ 76. The Quantum-Statistical Ring Diagrams

We now study the quantum extension of the results of Chapter 9. We shall, however, restrict ourselves to a study of homogeneous systems, and shall derive a kinetic equation for $\varphi(\mathbf{v}_\alpha)$. The order of approximation will be defined by retaining all powers of the combination:

$$e^2(e^2c)^n(h^3c)^mt \tag{76.1}$$

We therefore choose a set of diagrams by using the correspondence principle explained in § 66. By virtue of this principle, we have first to retain all diagrams which have the same topological structure as their classical analogs, i.e. the set of all *rings*. We have still to discuss the possible contribution of the purely quantum-statistical vertices, the only ones which enter the discussion being vertices G and H of Table 63.1.

In order to clarify the argument, we go back first to the classical case (Chapter 9). We have seen that the classical cycle diagram diverges logarithmically at $\mathbf{l} \to 0$ in the case of a Coulomb interaction. Considering diagrams of higher order in the charge, we note that some of them converge and others diverge in various ways. Of course, a perturbation calculation cannot be valid if the individual coefficients of the powers of e^2 are large, or even infinite. More exactly, it shows that the initial choice of the expansion parameter is not appropriate. It can happen, however, that a regrouping of the terms obtained by first performing a partial summation of a subseries leads to a convergent result. Therefore, we shall adopt the following rule: Among all diagrams of a given order in the *charge* e^2, we select those which are the most strongly divergent. The infinite series obtained by picking these diagrams for all orders in e^2 is then summed. The terms left out from this partial summation are small (or can be made small after a regrouping of the same type) compared with those considered first. This type of argument was first used by Mayer * in his equilibrium theory of electrolyte solutions. The same argument was later used by Macke, Gell-Mann and Brueckner ** for the calculation of the correlation energy of an electron gas.

In the classical case, it is easily seen that this procedure leads to the selection of the rings. Indeed, the divergence comes from the piling up of factors l^{-2} (i.e. V_l). The rings have a structure such that the wave-vectors are the same ($\mathbf{l}$ and $-\mathbf{l}$) in all intermediate states. This leads to an integral of the type

$$\int d\mathbf{l}(V_l)^n \ldots \sim \int d\mathbf{l} \frac{1}{l^{2n}} \ldots$$

[see, for instance, eq. (39.1)]. Any diagram other than a ring involves a change in the wave-vector $\mathbf{l}$ in some intermediate state. For instance, a four-vertex diagram (Fig. 76.1b) contains an integral of the type

$$\int d\mathbf{l} \int d\mathbf{l}' \frac{1}{(l^2)^2} \frac{1}{|\mathbf{l}-\mathbf{l}'|^2} \frac{1}{l'^2} \ldots$$

which is less divergent than the four-vertex ring (a):

$$\int d\mathbf{l} \frac{1}{(l^2)^4}$$

Thus, in the classical case, the argument based on the charge and concentration dependence $[(e^2t)^n(e^2c)^m]$ and the argument based on the degree of divergence are completely equivalent: they both lead independently to the choice of the rings. In the

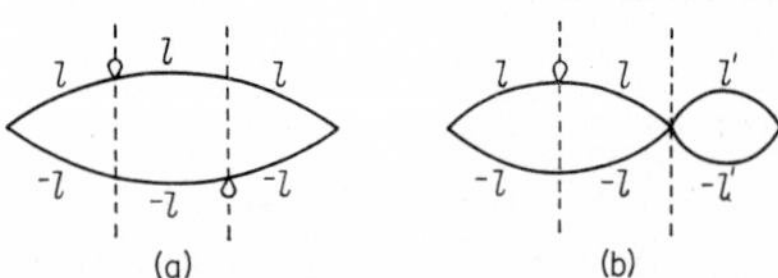

Fig. 76.1. Wave-vectors in the intermediate states of a four-vertex classical ring (a) and in another four-vertex classical diagram (b).

* J. Mayer, *J. Chem. Phys.*, **18**, 1426 (1950).

** W. Macke, *Z. Naturforsch.*, **5a**, 192 (1950); M. Gell-Mann and K. A. Brueckner, *Phys. Rev.*, **106**, 364 (1957).

quantum case this is no longer so; the second argument is more restrictive than the first, as will now be shown.

The general correspondence principle used in the study of quantum systems requires the choice of the ring diagrams, as well as that of any diagonal fragments involving vertices G and H of Table 63.1 (see Fig. 76.2). All these fragments are of the order (76.1). We now require that *among all contributions selected by this criterion of a given order in* e^2, *we retain only those which are the most divergent.*

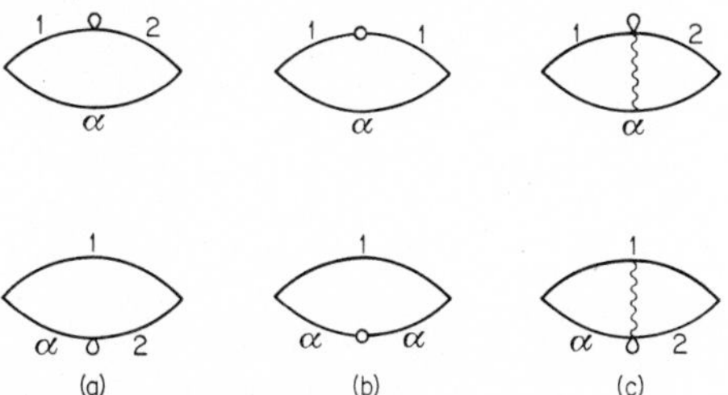

Fig. 76.2. The set of all the three-vertex quantum-statistical diagrams of order $e^2(e^2c)^2(h^3c)^n t$.

Consider first the rings (Fig. 76.2a). We have seen that each quantum-statistical vertex involves a direct and an exchange term. In other words, each quantum diagram with n vertices can be considered as a sum of 2^n terms. Among these, the single term which involves the direct matrix element at each vertex is the most divergent. Indeed, each exchange vertex lowers the degree of divergence of the contribution, as can be easily seen from Table 63.1. For instance, for a ring of Fig. 76.2a, we have the following types of integral:

$$\int d\mathbf{l} \frac{1}{(l^2)^3} \ldots \text{ only direct vertices}$$

$$\int d\mathbf{l} \frac{1}{(l^2)^2} \frac{1}{|\frac{1}{2}\mathbf{l}+m(\mathbf{v}_1-\mathbf{v}_n)/\hbar|^2} \ldots \text{ one exchange vertex.}$$

Consider now diagrams of set (b), including a vertex G. A glance at Table 63.1 shows that the inclusion of such a vertex has the same effect on the $\mathbf{l}$-dependence as the inclusion of an exchange contribution of vertex F. The same structure occurs in diagrams of type (76.2c) involving a vertex H. Thus, if the criterion (76.1) is supplemented by the "maximum divergence" criterion, we are led to retain only the rings (with exchange matrix elements discarded).

To sum up this discussion, the quantum-statistical ring approximation will be defined as *the sum of all ring diagrams in which the exchange matrix element of the interaction is discarded at each vertex.*

If this sum acting on a product of distribution functions $\varphi(\mathbf{v}_j; t)$ is called $\mathscr{R}[\varphi(\mathbf{v}_j; t)]$, then one can show, as in Appendix 4, that the kinetic equation is written formally as

$$\partial_t \varphi(\mathbf{v}_\alpha; t) = \mathscr{R}[\varphi(\mathbf{v}_j; t)] \tag{76.2}$$

We now go over to the explicit summation of the ring diagrams.

§ 77. Derivation of the Kinetic Equation

We first establish a series of rules enabling us to write down at once the contribution of a ring. Consider again the example of Fig. 39.2. Using the matrix elements A, C and F of Table 63.1, in which $[V_l + \theta V \ldots]$ is replaced by V_l, we obtain:

$$\begin{aligned}
\mathscr{R}_1^1 = \left(\frac{8\pi^3}{\Omega}\right)^3 \sum_1 \sum_2 \sum_3 \left(\frac{8\pi^3}{\Omega}\right) \sum_1 \left(\frac{h^3}{m^3\Omega}\right)^N \sum_{\mathbf{v}_1} \cdots \sum_{\mathbf{v}_N} (-e^2 i/\hbar) \\
V_l[\exp(\tfrac{1}{2}\hbar m^{-1}\mathbf{l}\cdot\partial_{\alpha 1}) - \exp(-\tfrac{1}{2}\hbar m^{-1}\mathbf{l}\cdot\partial_{\alpha 1})] \\
\cdot \pi\delta_-(\mathbf{l}\cdot\mathbf{g}_{\alpha 1})(-e^2 i/\hbar) V_l[(1+\theta\sum_n \delta_{\mathbf{v}_n - \mathbf{v}_\alpha - \hbar\mathbf{l}/2m}) \exp(-\tfrac{1}{2}\hbar m^{-1}\mathbf{l}\cdot\partial_{\alpha 2}) \\
-(1+\theta\sum_n \delta_{\mathbf{v}_n - \mathbf{v}_\alpha + \hbar\mathbf{l}/2m}) \exp(\tfrac{1}{2}\hbar m^{-1}\mathbf{l}\cdot\partial_{\alpha 2})] \\
\cdot \pi\delta_-(\mathbf{l}\cdot\mathbf{g}_{21})(-e^2 i/\hbar) V_l[(1+\theta\sum_m \delta_{\mathbf{v}_m - \mathbf{v}_1 + \hbar\mathbf{l}/2m}) \exp(\tfrac{1}{2}\hbar m^{-1}\mathbf{l}\cdot\partial_{13}) \\
-(1+\theta\sum_m \delta_{\mathbf{v}_m - \mathbf{v}_1 - \hbar\mathbf{l}/2m}) \exp(-\tfrac{1}{2}\hbar m^{-1}\mathbf{l}\cdot\partial_{13})] \\
\cdot \pi\delta_-(\mathbf{l}\cdot\mathbf{g}_{23})(-e^2 i/\hbar) V_l[(1+\theta\sum_r \delta_{\mathbf{v}_r - \mathbf{v}_2 - \hbar\mathbf{l}/2m}) \\
(1+\theta\sum_s \delta_{\mathbf{v}_s - \mathbf{v}_3 + \hbar\mathbf{l}/2m}) \exp(\tfrac{1}{2}\hbar m^{-1}\mathbf{l}\cdot\partial_{32}) \\
-(1+\theta\sum_r \delta_{\mathbf{v}_r - \mathbf{v}_2 + \hbar\mathbf{l}/2m})(1+\theta\sum_s \delta_{\mathbf{v}_s - \mathbf{v}_3 - \hbar\mathbf{l}/2m}) \exp(-\tfrac{1}{2}\hbar m^{-1}\mathbf{l}\cdot\partial_{32})] \\
\rho_0(|\mathbf{v}_\alpha, \mathbf{v}_1, \ldots \mathbf{v}_N)
\end{aligned}$$

In the limit $N \to \infty$, $\Omega \to \infty$, $N/\Omega = c$, this can be rewritten [using (65.2)]:

$$
\begin{aligned}
\mathscr{R}_1^1 = \int d\mathbf{v}_1 d\mathbf{v}_2 d\mathbf{v}_3 \int d\mathbf{l}\{&-8\pi^3 e^2 c i\hbar^{-1} V_l[\exp(\tfrac{1}{2}\hbar m^{-1}\mathbf{l}\cdot\partial_\alpha) \\
&- \exp(-\tfrac{1}{2}\hbar m^{-1}\mathbf{l}\cdot\partial_\alpha)]\}\pi\delta_-(\mathbf{l}\cdot\mathbf{g}_{\alpha 1}) \\
\cdot\{8\pi^3 e^2 c i\hbar^{-1} V_l[&\psi(\mathbf{v}_\alpha - \tfrac{1}{2}\hbar m^{-1}\mathbf{l})\exp(\tfrac{1}{2}\hbar m^{-1}\mathbf{l}\cdot\partial_\alpha) \\
&-\psi(\mathbf{v}_\alpha + \tfrac{1}{2}\hbar m^{-1}\mathbf{l})\exp(-\tfrac{1}{2}\hbar m^{-1}\mathbf{l}\cdot\partial_\alpha)]\}\pi\delta_-(\mathbf{l}\cdot\mathbf{g}_{21}) \\
\cdot\{-8\pi^3 e^2 c i\hbar^{-1} V_l[&\psi(\mathbf{v}_1 - \tfrac{1}{2}\hbar m^{-1}\mathbf{l})\exp(\tfrac{1}{2}\hbar m^{-1}\mathbf{l}\cdot\partial_1) \\
&-\psi(\mathbf{v}_1 + \tfrac{1}{2}\hbar m^{-1}\mathbf{l})\exp(-\tfrac{1}{2}\hbar m^{-1}\mathbf{l}\cdot\partial_1)]\}\pi\delta_-(\mathbf{l}\cdot\mathbf{g}_{23}) \\
\cdot\{-e^2 i\hbar^{-1} V_l[&\psi(\mathbf{v}_2 + \tfrac{1}{2}\hbar m^{-1}\mathbf{l})\psi(\mathbf{v}_3 - \tfrac{1}{2}\hbar m^{-1}\mathbf{l})\exp(\tfrac{1}{2}\hbar m^{-1}\mathbf{l}\cdot\partial_{32}) \\
&-\psi(\mathbf{v}_2 - \tfrac{1}{2}\hbar m^{-1}\mathbf{l})\psi(\mathbf{v}_3 + \tfrac{1}{2}\hbar m^{-1}\mathbf{l})\exp(-\tfrac{1}{2}\hbar m^{-1}\mathbf{l}\cdot\partial_{32})]\} \\
\cdot\,\varphi(\mathbf{v}_\alpha)&\varphi(\mathbf{v}_1)\varphi(\mathbf{v}_2)\varphi(\mathbf{v}_3)
\end{aligned}
\tag{77.1}
$$

$\psi(\mathbf{v})$ is defined by eq. (72.4). We can now introduce the abbreviations

$$
\begin{aligned}
D_\alpha &= -8\pi^3 e^2 c i\hbar^{-1} V_l[\exp(\tfrac{1}{2}\hbar m^{-1}\mathbf{l}\cdot\partial_\alpha) - \exp(-\tfrac{1}{2}\hbar m^{-1}\mathbf{l}\cdot\partial_\alpha)] \\
d_j &= -8\pi^3 e^2 c i\hbar^{-1} V_l[\psi(\mathbf{v}_j - \tfrac{1}{2}\hbar m^{-1}\mathbf{l})\exp(\tfrac{1}{2}\hbar m^{-1}\mathbf{l}\cdot\partial_j) \\
&\qquad -\psi(\mathbf{v}_j + \tfrac{1}{2}\hbar m^{-1}\mathbf{l})\exp(-\tfrac{1}{2}\hbar m^{-1}\mathbf{l}\cdot\partial_j)] \\
d_{jm} &= -e^2 i\hbar^{-1} V_l[\psi(\mathbf{v}_j + \tfrac{1}{2}\hbar m^{-1}\mathbf{l})\psi(\mathbf{v}_m - \tfrac{1}{2}\hbar m^{-1}\mathbf{l})\exp(\tfrac{1}{2}\hbar m^{-1}\mathbf{l}\cdot\partial_{mj}) \\
&\qquad -\psi(\mathbf{v}_j - \tfrac{1}{2}\hbar m^{-1}\mathbf{l})\psi(\mathbf{v}_m + \tfrac{1}{2}\hbar m^{-1}\mathbf{l})\exp(-\tfrac{1}{2}\hbar m^{-1}\mathbf{l}\cdot\partial_{mj})] \\
\delta_-^{jm} &= \pi\delta_-(\mathbf{l}\cdot\mathbf{g}_{jm})
\end{aligned}
\tag{77.2}
$$

We can then rewrite formula (77.1) as:

$$
\partial_t\varphi(\alpha)\Big|_1^1 = \int d\mathbf{l}\int d\mathbf{v}_1 d\mathbf{v}_2 d\mathbf{v}_3 D_\alpha \delta_-^{\alpha 1}(-d_\alpha)\delta_-^{21} d_1 \delta_-^{23} d_{32}\varphi(\alpha)\varphi(1)\varphi(2)\varphi(3) \tag{77.3}
$$

The analogy with eq. (39.3) is striking. We have thus shown that for writing the contribution of a ring we may use exactly the same rules as the classical ones given in § 39, with only the change of rule A to A′:

A′) To the extreme left vertex corresponds the expression $D_\alpha\delta_-^{\alpha 1}$.

This remarkable formal analogy between the classical and the quantum-mechanical expressions saves us repeating the calculations of § 40. Indeed, a glance at those calculations shows that they do not depend on the *explicit meaning* of the operators d_j, d_{jm} as long as these operators depend on the same variables. Therefore, the calculations can all be taken over without any change, with the result:

$$\partial_t \varphi(\alpha) = \int d\mathbf{l}\, D_\alpha F(\alpha) \tag{77.4}$$

$$\varepsilon(\alpha)F(\alpha) = q(\alpha) + i\pi d(\alpha) \int d\mathbf{v}_1 \delta_-(\mathbf{l}\cdot\mathbf{v}_\alpha - \mathbf{l}\cdot\mathbf{v}_1) F^*(1) \tag{77.5}$$

The coefficients of the integral equation (77.5) are defined in terms of d_j and d_{jn} exactly as in (40.6–8); their explicit form is easily seen to be:

$$\begin{aligned}\varepsilon(\alpha) = 1 + 8\pi^4 e^2 c\hbar^{-1} i V_l \int d\mathbf{v}_1\, \delta_-(\mathbf{l}\cdot\mathbf{g}_{\alpha 1})\\ \cdot [\varphi(\mathbf{v}_1 + \tfrac{1}{2}\hbar m^{-1}\mathbf{l})\psi(\mathbf{v}_1 - \tfrac{1}{2}\hbar m^{-1}\mathbf{l}) - \varphi(\mathbf{v}_1 - \tfrac{1}{2}\hbar m^{-1}\mathbf{l})\psi(\mathbf{v}_1 + \tfrac{1}{2}\hbar m^{-1}\mathbf{l})]\end{aligned} \tag{77.6}$$

$$\begin{aligned}q(\alpha) = \pi e^2 \hbar^{-1} i V_l \int d\mathbf{v}_1 \delta_-(\mathbf{l}\cdot\mathbf{g}_{\alpha 1})[\varphi(\mathbf{v}_\alpha + \tfrac{1}{2}\hbar m^{-1}\mathbf{l})\varphi(\mathbf{v}_1 - \tfrac{1}{2}\hbar m^{-1}\mathbf{l})\\ \psi(\mathbf{v}_\alpha - \tfrac{1}{2}\hbar m^{-1}\mathbf{l})\psi(\mathbf{v}_1 + \tfrac{1}{2}\hbar m^{-1}\mathbf{l})\\ -\varphi(\mathbf{v}_\alpha - \tfrac{1}{2}\hbar m^{-1}\mathbf{l})\varphi(\mathbf{v}_1 + \tfrac{1}{2}\hbar m^{-1}\mathbf{l})\psi(\mathbf{v}_\alpha + \tfrac{1}{2}\hbar m^{-1}\mathbf{l})\psi(\mathbf{v}_1 - \tfrac{1}{2}\hbar m^{-1}\mathbf{l})]\end{aligned} \tag{77.7}$$

$$\begin{aligned}d(\alpha) = 8\pi^3 e^2 c\hbar^{-1} V_l [\varphi(\mathbf{v}_\alpha + \tfrac{1}{2}\hbar m^{-1}\mathbf{l})\psi(\mathbf{v}_\alpha - \tfrac{1}{2}\hbar m^{-1}\mathbf{l})\\ -\varphi(\mathbf{v}_\alpha - \tfrac{1}{2}\hbar m^{-1}\mathbf{l})\psi(\mathbf{v}_\alpha + \tfrac{1}{2}\hbar m^{-1}\mathbf{l})]\end{aligned} \tag{77.8}$$

These complex numbers can be separated into real and imaginary parts, but we shall not write down the resulting expressions. The subsequent treatment of eq. (77.5) is exactly as in § 41. The important fact here is that properties (41.7) and (41.10) are preserved in the quantum-statistical case. Therefore, the method of solution is exactly identical to the classical case and we can directly quote the final result:

$$\begin{aligned}\partial_t \varphi(\mathbf{v}_\alpha; t) = \frac{8\pi^4 e^4 c}{\hbar^2} \int d\mathbf{l} \int d\mathbf{v}_1 [\exp(\tfrac{1}{2}\hbar m^{-1}\mathbf{l}\cdot\partial_\alpha)\\ -\exp(-\tfrac{1}{2}\hbar m^{-1}\mathbf{l}\cdot\partial_\alpha)]\{V_l^2/|\varepsilon(\alpha)|^2\}\delta(\mathbf{l}\cdot\mathbf{g}_{\alpha 1})\\ \cdot [\psi(\mathbf{v}_\alpha - \tfrac{1}{2}\hbar m^{-1}\mathbf{l})\psi(\mathbf{v}_1 + \tfrac{1}{2}\hbar m^{-1}\mathbf{l}) \exp(\tfrac{1}{2}\hbar m^{-1}\mathbf{l}\cdot\partial_{\alpha 1})\\ -\psi(\mathbf{v}_\alpha + \tfrac{1}{2}\hbar m^{-1}\mathbf{l})\psi(\mathbf{v}_1 - \tfrac{1}{2}\hbar m^{-1}\mathbf{l}) \exp(-\tfrac{1}{2}\hbar m^{-1}\mathbf{l}\cdot\partial_{\alpha 1})] \cdot \varphi(\alpha)\varphi(1)\end{aligned} \tag{77.9}$$

This equation was first obtained by the author * using a slightly different formalism. Particular cases of this equation have been obtained by other investigators. Klimontovich and Temko ** have derived the equation for a test particle in the ring approximation by using Bogoliubov's theory. Konstantinov and Perel' † obtained the equation to which (77.9) reduces in the limit of Boltzmann statistics by using a Green's function diagram technique. More recently, Guernsey †† derived eq. (77.9), as well as its non-markovian extension and the expression (79.2) of the correlations, by a generalization of his classical formalism.

In the form (77.9) it is immediately verifiable that the classical limit of this equation is indeed (41.19). A simple glance at this formula shows that the ring equation (77.9) has the same status compared to the cycle equation (73.4) as the classical ring equation (41.19) has compared to the classical cycle equation (30.11). Hence, all general properties discussed in Chapter 11 remain valid for eq. (77.9). The latter appears as a kinetic equation describing explicitly many-body collisions, or "collective collisions". These appear in the equation through an effective potential $U_l = V_l/\varepsilon(\alpha)$ which is a functional of the distribution function. Rather than giving here trivial extensions of the classical calculations, we shall treat in detail one very simple situation and thus put forward the main new features introduced by eq. (77.9).

§ 78. Collective Effects in the Motion of a Test Particle through an Electron Gas at Zero Temperature

We now consider again the problem of the test particle studied in the cycle approximation in § 75, and consider the new effects introduced by refining the approximation. We again use assump-

* R. Balescu, *Phys. of Fluids*, **4**, 94 (1961).

** Iu. Klimontovich and S. V. Temko, *J. Exptl. Theoret. Phys. U.S.S.R.*, **33**, 132 (1957); transl. *Sov. Phys. JETP*, **6**, 102 (1958).

† O. V. Konstantinov and V. I. Perel', *J. Exptl. Theoret. Phys. U.S.S.R.*, **39**, 861 (1960); transl. *Sov. Phys. JETP*, **12**, 597 (1961).

†† R. L. Guernsey, *Phys. Rev.*, **127**, 1446 (1962).

tions (*a*) and (*b*) of § 75, which define the test particle problem, i.e.

$$\varphi(\mathbf{v}_n) = \varphi_F^0(\mathbf{v}_n) = (m^3/h^3c)\theta(v_F - v_n), \quad n \neq \alpha \tag{78.1}$$

$$\varphi(\mathbf{v}_\alpha) = 1 \tag{78.2}$$

We use here, however, the true Coulomb potential rather than a quantum Debye potential, because the divergence for small $\mathbf{l}$ has been suppressed by using the ring approximation. We intend to calculate again the dynamical friction coefficient $\eta(w)$ defined by an equation of the form (75.11). The transformations of eq. (77.9) which lead to the definition of this friction coefficient are identical to those performed in § 75 in order to derive eq. (75.15) and will not be repeated here. Using the reduced variables defined in (75.6), the friction coefficient is defined in the ring approximation as [see also eq. (75.17)]

$$\eta(w) = 2\pi w^{-1} \int_0^{2w} d\lambda\, \lambda^3 \int_{\lambda/2w}^{1} d\xi\, \xi \frac{m_2(\lambda;\, w\xi - \frac{1}{2}\lambda)}{|\lambda^2 \varepsilon(\lambda;\, \lambda w\, \xi - \frac{1}{2}\lambda^2)|^2} \tag{78.3}$$

where $\varepsilon(\chi;\, \chi y)$ is the dielectric constant of a Fermi gas at zero temperature, defined by eq. (70.7).

We note moreover that the separation of eq. (75.15) into a sum of explicit integrals performed in eqs. (75.18–20) depends only on the fact that the function $m_2(\lambda;\, y)$ defined in Table 70.1 has different functional forms in various domains of λ and y and vanishes identically in some of these domains. As the argument, y, of the dielectric function appearing in the denominator is the same $(w\xi - \frac{1}{2}\lambda)$ as that of m_2 appearing in the numerator, the same discussion can be carried out in the present case. As a result, the friction coefficient can be expressed by the same formulae (75.18–20) provided the integrands R_1 and R_2 are replaced by S_1 and S_2:

$$\begin{aligned} S_1 &= \frac{2\pi^2 \lambda^3 (\xi/w)(w\xi - \frac{1}{2}\lambda)}{[\lambda^2 + a^2 m_1(\lambda;\, w\xi - \frac{1}{2}\lambda)]^2 + a^4 \pi^2 (w\xi - \frac{1}{2}\lambda)^2} \\ S_2 &= \frac{\pi^2 \lambda^2 (\xi/w)(1 - \lambda^2 + 2\lambda w\xi - w^2 \xi^2)}{[\lambda^2 + a^2 m_1(\lambda;\, w\xi - \frac{1}{2}\lambda)]^2 + a^4 (\pi^2/4\lambda^2)(1 - \lambda^2 + 2\lambda w\xi - w^2 \xi^2)^2} \end{aligned} \tag{78.4}$$

where $m_1(\lambda;\, y)$ has been defined in eq. (70.10). In making this identification with eqs. (75.18–20), we make, however, a proviso for a possible extra contribution to $\eta(w)$ which has not been included in these equations. Such a contribution could be due to a singular behavior of the integrand. We shall discuss this point below.

Before proceeding with these calculations, it is appropriate at this point to make some comments about the effective potential appearing in these equations.

Comparing eqs. (78.3) and (75.17) we note that the Debye-like potential

$$U_\lambda^D = 1/(\lambda^2+\kappa^2)$$

is replaced in the more refined version by the potential

$$U_\lambda(w) = 1/[\lambda^2\varepsilon(\lambda;\, \lambda w\xi-\tfrac{1}{2}\lambda^2)] \tag{78.5}$$

This is precisely what would be expected from the general features of the ring approximation. The effective potential takes into account the dynamical screening as well as the plasma oscillation effects. From our previous experience of the classical case, we would expect the velocity-dependent effective potential $U_\lambda(w)$ to approach the Debye potential for low velocities. Let us see if this is true in the quantum case. The real part of the dielectric constant, taken from eqs. (70.7) and (70.10) for $y = w\xi-\frac{1}{2}\lambda$, yields

$$\varepsilon_1 = 1+\frac{a^2}{\lambda^2}\, m_1(\lambda;\, w\xi-\tfrac{1}{2}\lambda) \tag{78.6}$$

with

$$\begin{aligned} m_1(\lambda;\, w\xi-\tfrac{1}{2}\lambda) &= 1+(2\lambda)^{-1}[1-(w\xi-\lambda)^2]\ln\left|\frac{\lambda+1-w\xi}{\lambda-1-w\xi}\right| \\ &\quad +(2\lambda)^{-1}(1-w^2\xi^2)\ln\left|\frac{1+w\xi}{1-w\xi}\right| \end{aligned} \tag{78.7}$$

which, for small velocities, becomes approximately

$$m_1(\lambda;\, -\tfrac{1}{2}\lambda) = 1-\frac{1-\lambda^2}{2\lambda}\ln\left|\frac{1+\lambda}{1-\lambda}\right| \tag{78.8}$$

The corresponding limiting value of $m_2(\lambda; -\frac{1}{2}\lambda)$ is taken from Table 70.1. This calculation definitely shows that, in contrast to the classical effective potential, $U_\lambda(w)$ does *not* seem to tend to a Debye potential in the limit of small velocities. Instead of being a real constant (as it should be if $U_\lambda(0)$ were Debye-like), the function m is still a rather complicated complex function of λ. However, a glance at eq. (75.17) shows that the integration over λ in the expression for $\eta(w)$ goes from 0 to $2w$. Hence, for small w only the behavior of the integrand at small values of the wave-vector contributes to the friction. The relevant part of the dielectric constant will, therefore, be given by the first terms of a λ-expansion. The behavior of $m(\lambda; -\frac{1}{2}\lambda)$ for small λ is given by:

$$m(\lambda; -\tfrac{1}{2}\lambda) \approx 2+\tfrac{1}{3}\lambda^2+ \dots +\tfrac{1}{2}i\pi\lambda$$

and hence the dominant term will be $m = 2$, which, substituted into (78.6) and (78.5), yields the Debye-like potential:

$$U_\lambda^D(w) \approx \frac{1}{\lambda^2+2a^2} \quad (\text{small } \lambda, \text{ small } w) \tag{78.9}$$

Hence eq. (75.21), which is valid for the range $w \leq 1$, can be regarded as a fairly good approximation. The low velocity approximation (75.23) especially, with the characteristic behavior in w^4, is an exact limiting result within the ring approximation. On the other hand, the result (75.24) is much rougher.

We now go back to the complete expressions (75.18–20) evaluated with the integrands S_1, S_2, eq. (78.4). These integrals are fairly complicated and cannot be calculated exactly. They represent refinements of the results (75.21–22) and could be calculated numerically. Their physical interpretation is rather clear: they are contributions to the friction coefficient due to "individual particle collisions" in which the dynamical screening effect is considered exactly. We shall now investigate whether one can isolate a "collective plasmon emission" effect, analogous to the one found in § 49.

We know from the discussion of § 70 that the dielectric constant of a zero-temperature electron gas possesses a zero *on the real axis* for small values of λ and high values of y. This property ensures the existence of undamped plasma oscillations of frequency ω_P. *We can therefore expect from* (78.3) *that particles whose velocity is such that* $w\xi - \frac{1}{2}\lambda = \omega_P$ *will excite plasma oscillations and hence lose energy to the gas as a whole (rather than to a specific collision partner).* This phenomenon provides a characteristic *collective contribution to the dynamical friction.* Let us investigate this process mathematically.

Comparing eqs. (78.3) and (75.18–20) (with R replaced by S), we first note that the effect under discussion can occur only in the integrals

$$\int_0^{w-1} d\lambda \int_{1/w}^{(\lambda+1)/w} d\xi\, S_2 \quad \text{for} \quad 1 \leqq w \leqq 3 \tag{78.10a}$$

and

$$\int_0^{2} d\lambda \int_{1/w}^{(1+\lambda)/w} d\xi\, S_2 + \int_2^{w-1} d\lambda \int_{(\lambda-1)/w}^{(\lambda+1)/w} d\xi\, S_2 \quad \text{for} \quad w \geqq 3 \tag{78.10b}$$

In these integrals, the ξ-integration is not carried out up to $\xi = 1$ [as prescribed in (78.3)] because $m_2(\lambda;\, w\xi - \frac{1}{2}\lambda) = 0$ for $(1+\lambda)/w \leqq \xi \leqq 1$ (see Table 70.1). Hence we could add formally to the r.h.s. of both eqs. (75.19) and (75.20) a term representing the contribution coming from the domain in which m_2 vanishes:

$$\eta_c(w) = 2\pi w^{-1} \int_0^{w-1} d\lambda\, \lambda^3 \int_{(\lambda+1)/w}^{1} d\,\xi\,\xi \frac{m_2}{(\lambda^2 + a^2 m_1)^2 + a^4 m_2^2}, \quad w \geqq 1 \tag{78.11}$$

As the numerator $m_2(\lambda;\, w\xi - \frac{1}{2}\lambda)$ equals zero in the whole domain of integration, the integral vanishes, *except* if the denominator also has a zero within the domain. In that case a closer inspection of the undetermined expression 0/0 is necessary. Let us therefore investigate whether the expression $\lambda^2 + a^2 m_1$ can vanish within the domain. We shall actually assume that the velocity is large compared to 1: $w \gg 1$. We can then use the expansion (70.12) for m_1, and obtain approximately the equation:

$$\lambda^2 - \tfrac{2}{3} a^2 (w\xi_0 - \tfrac{1}{2}\lambda)^{-2} = 0$$

Using the auxiliary condition $w\xi - \frac{1}{2}\lambda > 0$ [eq. (75.16)] we find the unique root:

$$\xi_0 = (\lambda/2w) + (b/\lambda w) \tag{78.12}$$

where $b = \sqrt{2/3}a$. For ξ_0 to be within the domain of integration, it is necessary that

$$\lambda + 1 \leqq \tfrac{1}{2}\lambda + b/\lambda \leqq w \tag{78.13}$$

Working out these inequalities, we find the following condition on λ:

$$w[1 - \sqrt{\{1 + (2b/w^2)\}}] \leqq \lambda \leqq -1 + \sqrt{(1+2b)} \tag{78.14}$$

Referring now to eqs. (70.2), (70.6), (66.6) and (13.11), it is seen that b can be expressed as follows:

$$b = \tfrac{1}{2}\pi(\hbar\omega_P/E_F) \sim \gamma_0^{\frac{1}{2}} \ll 1 \tag{78.15}$$

The coefficient b is thus proportional to the ratio of the energy of a quantum of plasma oscillation to the Fermi energy. Within the ring approximation this ratio must be considered as small.* One can then simplify condition (78.14) by expanding the square roots:

$$b/w \leqq \lambda \leqq b \tag{78.16}$$

Thus, only if λ satisfies this condition does there exist a zero of the real part of the dielectric constant (and hence a singularity of the integrand) within the range of integration in (78.11). The latter formula therefore reduces to:

$$\eta_c(w) = \frac{2\pi w^{-1}}{a^2}\int_{b/w}^{b} d\lambda\, \lambda^3 \int_{(\lambda+1)/w}^{1} d\xi\, \xi\, \frac{a^2 m_2}{(\lambda^2 + a^2 m_1)^2 + a^4 m_2^2}, \quad w \gg 1 \tag{78.17}$$

The behavior of the integrand at the singularity is obtained by using the well known asymptotic formula **

* For typical real metals, $\gamma_0 \approx 1$; hence one sees here that the ring approximation can only give a qualitative description of metals, as mentioned before.

** This formula is obtained by taking the imaginary part of both sides of eq. (A2.2.7) of Appendix 2.

$$\lim_{\sigma\to 0} \frac{\sigma}{x^2+\sigma^2} = \pi\delta(x) \tag{78.18}$$

Using the large velocity approximation for m_1, we obtain

$$\eta_c(w) = \frac{2\pi^2}{wa^2}\int_{b/w}^{b} d\lambda \int_{(\lambda+1)/w}^{1} d\xi\, \xi\, \delta[\lambda^2 - \tfrac{2}{3}a^2(w\xi - \tfrac{1}{2}\lambda)^{-2}] \tag{78.19}$$

Using also the property:

$$\delta[1-(a^2/x^2)] = \tfrac{1}{2}a[\delta(x+a)+\delta(x-a)] \tag{78.20}$$

as well as the condition (75.16), we get

$$\eta_c(w) = \frac{\pi^2 b}{a^2 w}\int_{b/w}^{b} d\lambda \int_{(\lambda+1)/w}^{1} d\xi\, \xi\, \delta\left(w\xi - \tfrac{1}{2}\lambda - \frac{b}{\lambda}\right) = \frac{\pi^2 b}{a^2 w^3}\int_{b/w}^{b} d\lambda \left\{\frac{\lambda}{2} + \frac{b}{\lambda}\right\}$$

Neglecting the first term in the integrand, which gives a contribution of relative order b compared with the second, we finally obtain:

$$\eta_c(w) \approx \tfrac{2}{3}\pi^2 w^{-3} \ln w, \quad w \gg 1 \tag{78.21}$$

This expression for the friction coefficient leads to an expression for the energy loss per unit path length (in reduced variables):

$$-\frac{dW}{dt} = \eta w \approx \tfrac{1}{3}\pi^2 w^{-2} \ln w^2, \quad w \gg 1 \tag{78.22}$$

This type of velocity dependence has been found previously by several authors using various methods.* In the present treatment it comes out of a unified theory of plasmas within the ring approximation.

It is worth noting that the asymptotic collective contribution to the friction coefficient equals exactly half the asymptotic "individual" contribution, eq. (75.22). This contrasts with the classical situation, where the collective collision is smaller by an order $\ln \Gamma$ than the individual contribution.

* D. Bohm and D. Pines, *Phys. Rev.*, **92**, 609 (1953); U. Fano, *Phys. Rev.*, **103**, 1202 (1956); R. H. Ritchie, *Phys. Rev.*, **114**, 644 (1959).

Another important difference from the classical case is in the fact that at zero temperature there exist undamped collective oscillations. Hence, in the expression for $\eta_c(w)$ there is a sharp resonance, i.e. a true δ-function representing the emission of plasmons, if λ lies between the limits (78.16). Outside these limits, the oscillations are rapidly damped and one can no longer speak of a collective behavior.

CHAPTER 18

Binary Correlations and the Equilibrium State for an Electron Gas

§ 79. The Density Correlation Function

It has been shown in Chapter 14 that the Fourier transform of the correlation functions can be separated into a sum of terms such that the pure statistics effect is separated out; in the case of the two-body correlation function in a homogeneous system, this separation is given by formula (62.12). All the problems studied previously have illustrated the striking analogy between the classical and the quantum theory when the latter is studied from the point of view of the coefficients resulting from this separation. In the case of charged particles in the ring approximation, it has been shown that, after a suitable redefinition of the symbols, all the classical equations can be taken over without any change. The same applies to the study of the quantum correlation function, so that it becomes completely superfluous to repeat the calculations of Chapter 10: absolutely everything which was done there remains unchanged, provided the symbols d_j, d_{jn}, $q(\alpha)$, $d(\alpha)$, $\varepsilon(\alpha)$ are suitably redefined as was done in Chapter 17. We can therefore quote the final result, giving the part of the two-body correlation function due to the interactions [see (62.12), (44.6) and (45.6)]:

$$\hat{\rho}_k(\mathbf{v}_\alpha, \mathbf{v}_\gamma; t) = \delta_-^{\alpha\gamma}\{d_{\gamma\alpha}\varphi(\gamma)\varphi(\alpha)+d_\gamma\varphi(\gamma)F(\alpha)+d_\alpha^*\varphi(\alpha)F^*(\gamma)\} \tag{79.1}$$

$$F(\mathbf{v}_\alpha) = \frac{q(\mathbf{v}_\alpha)}{\varepsilon(\nu_\alpha)} + \frac{i\pi}{k}\, d(\mathbf{v}_\alpha)\int d\nu_1\,\delta_-(\nu_\alpha-\nu_1)\,\frac{\bar{q}(\nu_1)}{|\varepsilon(\nu_1)|^2} \tag{79.2}$$

where the various symbols are defined by eqs. (77.2) and (77.6–8).

We shall mainly be interested in a function which contains less information than $\hat{\rho}_k(\mathbf{v}_\alpha, \mathbf{v}_\gamma)$, but which is sufficient for deriving most thermodynamic and hydrodynamic functions. This function is the density correlation function, $g(r)$; it is the integral of $G(r, \mathbf{v}_\alpha, \mathbf{v}_\gamma)$ over the velocities (see Table 2.2). One of our purposes is to show that the density correlation function calculated out of equilibrium by the sum of creation diagrams acting on ρ_0 reduces to the correct equilibrium value when the velocity distribution $\varphi(\mathbf{v})$ attains the Fermi distribution. This is a proof of the generalized quantum-mechanical *H*-theorem in the particular case of plasmas.

In order to achieve this result, we proceed essentially as in § 52. However, as mentioned already, the velocity-dependent correlation function is not simple in equilibrium; as opposed to the classical case, it does not factorize into a Fermi distribution for the momenta and a radial correlation. Only after integration over the momenta do the results simplify. We shall therefore proceed as follows.

1. We sum all creation diagrams with a given number of vertices acting on the Fermi equilibrium distribution $\varphi_F(\mathbf{v})$ [defined in eq. (74.1)] and integrate them over the velocities $\mathbf{v}_\alpha$, $\mathbf{v}_\gamma$. Let $S_n(k)$ be the result of this operation.

2. We sum: $\sum_{n=1}^{\infty} S_n(k)$.

In the first step we evaluate in detail the first few S_n's. The simplest is, of course, S_1, given by vertex A. Applying the general construction of § 44 we have

$$\begin{aligned} S_1(k) &= \int d\mathbf{v}_\alpha d\mathbf{v}_\gamma \delta_-^{\alpha\gamma} d_{\gamma\alpha} \varphi_F(\alpha)\varphi_F(\gamma) \\ &= e^2 V_k \int d\mathbf{v}_\alpha d\mathbf{v}_\gamma \pi\delta_-(\mathbf{k}\cdot\mathbf{g}_{\alpha\gamma})(i/\hbar) \qquad (79.3) \\ &\cdot \{\varphi_F(\mathbf{v}_\alpha+\tfrac{1}{2}\hbar\mathbf{k}/m)\varphi_F(\mathbf{v}_\gamma-\tfrac{1}{2}\hbar\mathbf{k}/m)\psi_F(\mathbf{v}_\alpha-\tfrac{1}{2}\hbar\mathbf{k}/m)\psi_F(\mathbf{v}_\gamma+\tfrac{1}{2}\hbar\mathbf{k}/m) \\ &-\varphi_F(\mathbf{v}_\alpha-\tfrac{1}{2}\hbar\mathbf{k}/m)\varphi_F(\mathbf{v}_\gamma+\tfrac{1}{2}\hbar\mathbf{k}/m)\psi_F(\mathbf{v}_\alpha+\tfrac{1}{2}\hbar\mathbf{k}/m)\psi_F(\mathbf{v}_\gamma-\tfrac{1}{2}\hbar\mathbf{k}/m)\} \end{aligned}$$

We now use the following important property of the Fermi distribution, which has already been introduced previously [see eq. (74.2)]

$$\varphi_F(\mathbf{v}+\tfrac{1}{2}\hbar\mathbf{k}/m)\psi_F(\mathbf{v}-\tfrac{1}{2}\hbar\mathbf{k}/m) = \mathrm{e}^{-\beta\hbar\mathbf{k}\cdot\mathbf{v}}\varphi_F(\mathbf{v}-\tfrac{1}{2}\hbar\mathbf{k}/m)\psi_F(\mathbf{v}+\tfrac{1}{2}\hbar\mathbf{k}/m) \qquad (79.4)$$

We then rewrite (79.3) as:

$$S_1(k) = e^2 V_k \int d\mathbf{v}_\alpha d\mathbf{v}_\gamma \pi\delta_-(\mathbf{k}\cdot\mathbf{g}_{\alpha\gamma})(i/\hbar)[e^{-\beta\hbar\mathbf{k}\cdot\mathbf{g}_{\alpha\gamma}}-1] \tag{79.5}$$
$$\cdot\varphi_F(\mathbf{v}_\alpha-\tfrac{1}{2}\hbar\mathbf{k}/m)\varphi_F(\mathbf{v}_\gamma+\tfrac{1}{2}\hbar\mathbf{k}/m)\psi_F(\mathbf{v}_\alpha+\tfrac{1}{2}\hbar\mathbf{k}/m)\psi_F(\mathbf{v}_\gamma-\tfrac{1}{2}\hbar\mathbf{k}/m)$$

Note, however, that the δ-function part of the δ_--function does not contribute [see also (74.3)]:

$$\delta(\mathbf{k}\cdot\mathbf{g})[e^{-\beta\hbar\mathbf{k}\cdot\mathbf{g}}-1) = 0 \tag{79.6}$$

Noting also that the principal part of $1/x$ reduces to an ordinary integral, we can write

$$S_1(k) = e^2 V_k \int d\mathbf{v}_\alpha d\mathbf{v}_\gamma \frac{e^{-\beta\hbar\mathbf{k}\cdot\mathbf{g}_{\alpha\gamma}}-1}{\hbar\mathbf{k}\cdot\mathbf{g}_{\alpha\gamma}} \tag{79.7}$$
$$\varphi_F(\mathbf{v}_\alpha-\tfrac{1}{2}\hbar\mathbf{k}/m)\varphi_F(\mathbf{v}_\gamma+\tfrac{1}{2}\hbar\mathbf{k}/m)\psi_F(\mathbf{v}_\alpha+\tfrac{1}{2}\hbar\mathbf{k}/m)\psi_F(\mathbf{v}_\gamma-\tfrac{1}{2}\hbar\mathbf{k}/m)$$

Let us now introduce the following function:

$$G_k(\tau) = \int d\mathbf{v} \exp(-\tau\hbar\mathbf{k}\cdot\mathbf{v})\varphi_F(\mathbf{v}-\tfrac{1}{2}\hbar\mathbf{k}/m)\psi_F(\mathbf{v}+\tfrac{1}{2}\hbar\mathbf{k}/m) \tag{79.8}$$

It is easily seen to be a function of the absolute value of k alone. It has the following property, which is a consequence of (79.4):

$$G_k(\tau) = G_k(\beta-\tau) \tag{79.9}$$

In terms of this function, (79.7) can be written as:

$$S_1(k) = -e^2 V_k \int_0^\beta d\tau\, G_k(\tau)G_k(\tau) = -e^2 V_k \int_0^\beta d\tau\, G_k(\beta-\tau)G_k(\tau) \tag{79.10}$$

We now evaluate the next order sum, S_2.

$$S_2 = \int d\mathbf{v}_\alpha d\mathbf{v}_\gamma d\mathbf{v}_1 \{\delta_-^{\alpha\gamma} d_\gamma \delta_-^{\alpha 1} d_{1\alpha} + \delta_-^{\alpha\gamma} d_\alpha^* \delta_-^{1\gamma} d_{\gamma 1}\}$$
$$= \int d\mathbf{v}_\alpha d\mathbf{v}_\gamma d\mathbf{v}_1 \{\delta_-^{\alpha\gamma} d_\gamma \delta_-^{\alpha 1} d_{1\alpha} + \delta_-^{\gamma\alpha} d_\gamma \delta_-^{1\alpha} d_{1\alpha}\}$$
$$= 8\pi^3 e^4 c V_k^2 \int d\mathbf{v}_\alpha d\mathbf{v}_\gamma d\mathbf{v}_1 i\hbar^{-1}[\delta_-(\mathbf{k}\cdot\mathbf{g}_{\alpha\gamma})-\delta_-(\mathbf{k}\cdot\mathbf{g}_{\gamma\alpha})]$$
$$\cdot[\varphi_F(\mathbf{v}_\gamma-\tfrac{1}{2}\hbar\mathbf{k}/m)\psi_F(\mathbf{v}_\gamma+\tfrac{1}{2}\hbar\mathbf{k}/m) - \varphi_F(\mathbf{v}_\gamma+\tfrac{1}{2}\hbar\mathbf{k}/m)\psi_F(\mathbf{v}_\gamma-\tfrac{1}{2}\hbar\mathbf{k}/m)]\frac{e^{-\beta\hbar\mathbf{k}\cdot\mathbf{g}_{\alpha 1}}-1}{\hbar\mathbf{k}\cdot\mathbf{g}_{\alpha 1}}$$
$$\cdot\varphi_F(\mathbf{v}_\alpha-\tfrac{1}{2}\hbar\mathbf{k}/m)\varphi_F(\mathbf{v}_1+\tfrac{1}{2}\hbar\mathbf{k}/m)\psi_F(\mathbf{v}_\alpha+\tfrac{1}{2}\hbar\mathbf{k}/m)\psi_F(\mathbf{v}_1-\tfrac{1}{2}\hbar\mathbf{k}/m)$$

$$S_2 = 16\pi^3 e^4 c V_k^2 \int d\mathbf{v}_\alpha d\mathbf{v}_\gamma d\mathbf{v}_1 \frac{1}{\hbar\mathbf{k}\cdot\mathbf{g}_{\alpha\gamma}} \frac{e^{-\beta\hbar\mathbf{k}\cdot\mathbf{g}_{\alpha 1}}-1}{\hbar\mathbf{k}\cdot\mathbf{g}_{\alpha 1}} \tag{79.11}$$
$$\cdot [\varphi_F(\mathbf{v}_\gamma - \tfrac{1}{2}\hbar\mathbf{k}/m)\psi_F(\mathbf{v}_\gamma + \tfrac{1}{2}\hbar\mathbf{k}/m) - \varphi_F(\mathbf{v}_\gamma + \tfrac{1}{2}\hbar\mathbf{k}/m)\psi_F(\mathbf{v}_\gamma - \tfrac{1}{2}\hbar\mathbf{k}/m)]$$
$$\cdot \varphi_F(\mathbf{v}_\alpha - \tfrac{1}{2}\hbar\mathbf{k}/m)\varphi_F(\mathbf{v}_1 + \tfrac{1}{2}\hbar\mathbf{k}/m)\psi_F(\mathbf{v}_\alpha + \tfrac{1}{2}\hbar\mathbf{k}/m)\psi_F(\mathbf{v}_1 - \tfrac{1}{2}\hbar\mathbf{k}/m)$$

In going from the first to the second line, the integration variables $\mathbf{v}_\alpha$ and $\mathbf{v}_\gamma$ have been interchanged, and the signs of d_γ^* and $d_{\alpha 1}$ have been changed. It is now easily verified, by direct integration, that (79.11) can be expressed in terms of the function $G_k(\tau)$ as follows:

$$\begin{aligned} S_2(k) &= 8\pi^3 e^4 c V_k^2 \int_0^\beta d\tau_2 \int_0^{\tau_2} d\tau_1 G_k(\beta-\tau_2) G_k(\tau_2-\tau_1) G_k(\tau_1) \\ &\quad + 8\pi^3 e^4 c V_k^2 \int_0^\beta d\tau_2 \int_{\tau_2}^\beta d\tau_1 G_k(\beta-\tau_2) G_k(\tau_1-\tau_2) G_k(\tau_1) \\ &= 8\pi^3 e^4 c V_k^2 \int_0^\beta d\tau_2 \int_0^\beta d\tau_1 G_k(|\beta-\tau_2|) G_k(|\tau_2-\tau_1|) G_k(|\tau_1|) \end{aligned} \tag{79.12}$$

The result (79.12) can easily be extended to all higher orders, the general formula being:

$$\begin{aligned} S_n(k) = &-e^2 V_k (-8\pi^3 e^2 c V_k)^{n-1} \int_0^\beta d\tau_n \ldots \int_0^\beta d\tau_1 \\ &\cdot G_k(|\beta-\tau_n|) G_k(|\tau_n-\tau_{n-1}|) \ldots G_k(|\tau_2-\tau_1|) G_k(|\tau_1|) \end{aligned} \tag{79.13}$$

We now have an expression for S_n in the form of a convolution integral. This is not yet convenient for the summation over n. In order to perform this operation, we show that $\sum S_n$ can be reduced to the summation of a geometric series. We follow a method due to Montroll and Ward * and slightly modified by de Witt.** (These authors have used it for the evaluation of the grand partition function.) We note that $G_k(|x|)$ is by definition an even function of x. This function can be considered as the kernel of an integral transformation. Eigenvalues λ_n and eigen-

* E. W. Montroll and J. C. Ward, *Phys. of Fluids*, **1**, 55 (1958).
** H. de Witt, *J. Nucl. Energy*, **C2**, 27 (1961).

functions $\xi_n(\tau)$ of this integral operator can then be defined as follows

$$\lambda_n \xi_n(\tau) = \int_0^\beta d\tau' G_k(|\tau-\tau'|)\xi_n(\tau') \tag{79.14}$$

The eigenfunctions $\xi_n(\tau)$ and the eigenvalues λ_n will be studied in more detail in the next paragraph. For the moment, we need only consider their very general properties.

The kernel $G_k(|x|)$ being symmetrical, one can apply here the well known theorem concerning Fredholm-type integral equations: *

(*a*) The eigenfunctions $\xi_n(\tau)$ form a complete orthonormal set in the domain $0 \leqq \tau \leqq \beta$. We therefore have

$$\int_0^\beta d\tau\, \xi_n^*(\tau)\xi_m(\tau) = \delta_{nm} \tag{79.15}$$

(*b*) The kernel can be expanded as follows:

$$G_k(|\tau'-\tau''|) = \sum_{n=-\infty}^{\infty} \lambda_n \xi_n^*(\tau')\xi_n(\tau'') \tag{79.16}$$

Substituting this expansion into (79.10) and using (79.15) we obtain:

$$\begin{aligned} S_1(k) &= -e^2 V_k \int_0^\beta d\tau \sum_m \sum_p \lambda_m \lambda_p \xi_m^*(\beta)\xi_m(\tau)\xi_p^*(\tau)\xi_p(0) \\ &= -e^2 V_k \sum_p \lambda_p^2 \xi_p^*(\beta)\xi_p(0) \end{aligned}$$

In the same way, we obtain the general expression:

$$S_n(k) = -e^2 V_k \sum_p \lambda_p^2 (-8\pi^3 e^2 c V_k \lambda_p)^{n-1} \xi_p^*(\beta)\xi_p(0) \tag{79.17}$$

It can easily be shown (see § 80) that

$$\xi_p(\beta) = \xi_p(0) = \beta^{-\frac{1}{2}} \tag{79.18}$$

* See, e.g., R. Courant and D. Hilbert, *Methoden der mathematischen Physik*, vol. 1, Springer Verlag, Berlin, 1931; P. M. Morse and H. Feshbach, *Methods of Theoretical Physics*, McGraw-Hill, New York, 1953; or any other book dealing with integral equations.

This achieves our reduction. The summation of the series is now elementary, with the result:

$$\hat{\rho}_k^0 = -\frac{e^2}{\beta} \sum_{p=-\infty}^{\infty} \frac{\lambda_p^2(k) V_k}{1+8\pi^3 e^2 c \lambda_p(k) V_k} \tag{79.19}$$

This formula was previously obtained by Fujita and Hirota * from equilibrium statistical mechanics.

§ 80. Eigenvalues and Eigenfunctions of the Kernel $G_k(|\tau|)$

We study here in more detail some properties of the eigenvalues λ_n and the eigenfunctions $\xi_n(\tau)$ of the kernel $G_k(|\tau|)$ introduced in eq. (79.14). It is immediately verified that the eigenfunctions are of the following form:

$$\xi_n(\tau) = \beta^{-\frac{1}{2}} \exp(-2\pi i n\tau/\beta) \tag{80.1}$$

They are periodic functions of period β. The eigenfunction expansion therefore coincides with a Fourier expansion. One can also verify immediately the property used in the previous section:

$$\xi_n(0) = \xi_n(\beta) = \beta^{-\frac{1}{2}} \tag{80.2}$$

In order to evaluate the eigenvalue λ_n, eq. (80.1) is substituted into the r.h.s. of (79.14), with the result

$$\begin{aligned} &\beta^{-\frac{1}{2}} \int_0^\beta d\tau' G_k(|\tau-\tau'|) \exp(-2\pi i n\tau'/\beta) \\ &= \beta^{-\frac{1}{2}} \int_{\tau-\beta}^{\tau} d\tau' G_k(|\tau'|) \exp(2\pi i n\tau'/\beta) \exp(-2\pi i n\tau/\beta) \\ &= \xi_n(\tau) \int_0^\beta d\tau' G_k(\tau') \exp(2\pi i n\tau'/\beta) \end{aligned}$$

Hence the eigenvalue is

$$\lambda_n = \int_0^\beta d\tau' G_k(\tau') \exp(2\pi i n\tau'/\beta) \tag{80.3}$$

We can obtain an interesting physical interpretation of the eigenvalues by calculating them more explicitly. Substitute eq. (79.8)

* S. Fujita and R. Hirota, *Phys. Rev.*, **118**, 6 (1960).

into (80.3), integrate over β and then use eq. (79.4) to obtain:

$$\begin{aligned}
\lambda_n &= \int_0^\beta d\tau \int d\mathbf{v}\, \varphi_F(\mathbf{v}-\tfrac{1}{2}\hbar\mathbf{k}/m)\psi_F(\mathbf{v}+\tfrac{1}{2}\hbar\mathbf{k}/m)\exp[2\pi in\beta^{-1}\tau-\hbar\mathbf{k}\cdot\mathbf{v}\tau] \\
&= \int d\mathbf{v}\, \varphi_F(\mathbf{v}-\tfrac{1}{2}\hbar\mathbf{k}/m)\psi_F(\mathbf{v}+\tfrac{1}{2}\hbar\mathbf{k}/m) \\
&\qquad \{\exp(-\hbar\mathbf{k}\cdot\mathbf{v}\beta)-1\}[2\pi in\beta^{-1}-\hbar\mathbf{k}\cdot\mathbf{v}]^{-1} \qquad (80.4) \\
&= \int d\mathbf{v}[2\pi in\beta^{-1}-\hbar\mathbf{k}\cdot\mathbf{v}]^{-1}[\varphi_F(\mathbf{v}+\tfrac{1}{2}\hbar\mathbf{k}/m)\psi_F(\mathbf{v}-\tfrac{1}{2}\hbar\mathbf{k}/m) \\
&\qquad -\varphi_F(\mathbf{v}-\tfrac{1}{2}\hbar\mathbf{k}/m)\psi_F(\mathbf{v}+\tfrac{1}{2}\hbar\mathbf{k}/m)]
\end{aligned}$$

Hence, comparing this with the definition (67.7a) of the dielectric constant, we obtain the following suggestive formula:

$$8\pi^3 e^2 c V_k \lambda_n(k) = \varepsilon_+[k; i(2\pi n/\beta\hbar)]-1 \qquad (80.5)$$

The eigenvalue $\lambda_n(k)$ is thus simply related to the *dielectric constant for an imaginary value of the frequency.* The expression (80.4) for the eigenvalue can also be written in the following form (changing $\mathbf{v}\to-\mathbf{v}$ in the second bracketed term):

$$\begin{aligned}
\lambda_n(k) &= -\int d\mathbf{v}\, \varphi_F(\mathbf{v}+\tfrac{1}{2}\hbar\mathbf{k}/m)\psi_F(\mathbf{v}-\tfrac{1}{2}\hbar\mathbf{k}/m) \\
&\qquad \left\{\frac{1}{\hbar\mathbf{k}\cdot\mathbf{v}+2\pi in\beta^{-1}}+\frac{1}{\hbar\mathbf{k}\cdot\mathbf{v}-2\pi in\beta^{-1}}\right\} \qquad (80.6)
\end{aligned}$$

In this form, it is clearly seen that $\lambda_n(k)$ is real, as expected. Indeed, $\lambda_n(k)$ is an eigenvalue of a symmetrical operator. From (80.5) follows then the interesting property: *the dielectric constant* $\varepsilon_+(k; z)$ *is real for imaginary frequencies,* $z = iy$.*

In the next section, we shall be interested in the behavior of an electron gas at zero temperature. In that case $\beta\to\infty$. It is seen from (80.1) and (80.3) that the spectrum of eigenvalues becomes continuous and the following prescriptions hold:

* It can be shown that this property is true for all *stable* momentum distributions: see L. D. Landau and E. M. Lifshitz, *Electrodynamics of Continuous Media*, Pergamon Press, Oxford 1960.

$$2\pi n/\beta \to s, \quad \text{a continuous variable}$$

$$(2\pi/\beta) \sum_n \to \int ds \tag{80.7}$$

$$\lambda_n(k) \to \lambda(k; s) = \int_0^\infty d\tau \, G_k(\tau) e^{i\tau s}$$

We shall now prove two very interesting *theorems which relate the dielectric constant for an imaginary argument to the dielectric constant for a real frequency*

$$\int_0^\infty d\omega \left\{ 1 - \frac{1}{\varepsilon_+(i\omega)} \right\} = \int_0^\infty d\omega \, \operatorname{Im} \frac{1}{\varepsilon_+(\omega)} \tag{80.8}$$

$$\int_0^\infty d\omega \, \{\varepsilon_+(i\omega) - 1\} = \int_0^\infty d\omega \, \operatorname{Im} \varepsilon_+(\omega) \tag{80.9}$$

The proof of eq. (80.8) goes as follows. Consider the complex integral

$$J = \int_G dz \left\{ 1 - \frac{1}{\varepsilon_+(z)} \right\} \frac{z}{z^2 + \omega^2} \tag{80.10}$$

where the contour G consists of the real axis completed by a semi-circle at infinity in the upper half-plane (see Fig. 19.1). The function $[\varepsilon_+(z)]^{-1}$ tends to 1 as $z \to \infty$; hence the integrand in eq. (80.10) tends to zero faster than z^{-1} and therefore J can be evaluated by the method of residues. *If the velocity distribution is stable,* $\varepsilon_+(z)$ has no zeros in the upper half-plane. Hence J equals $2\pi i$ times the residue of the integrand at the pole $z = +i\omega$. As the integral along the semi-circle vanishes, we obtain:

$$\int_{-\infty}^\infty dz \left\{ 1 - \frac{1}{\varepsilon_+(z)} \right\} \frac{z}{z^2 + \omega^2} = \pi i \left\{ 1 - \frac{1}{\varepsilon_+(i\omega)} \right\}$$

But the real part of $\varepsilon_+(z)$ is an even function for real z, whereas its imaginary part is odd; hence this formula reduces to

$$2 \int_0^\infty dz \, \operatorname{Im} \frac{1}{\varepsilon_+(z)} \, \frac{z}{z^2 + \omega^2} = \pi \left\{ 1 - \frac{1}{\varepsilon_+(i\omega)} \right\} \tag{80.11}$$

[we have used here the fact that $\varepsilon_+(i\omega)$ is real]. Now integrate both sides over ω from 0 to ∞ and the result (80.8) is obtained at once. Equation (80.9) is proven in a similar way; note, however, that in this case the stability condition is not necessary.

§ 81. The Ground-State Energy of an Electron Gas

As an application of the general theory discussed in the previous paragraphs, we shall now calculate the internal energy per particle of an electron gas in equilibrium at zero temperature. This function is obtained from formula (1.16). As in the classical case (§ 53), we are interested here in the deviations of the thermodynamic functions from those of a perfect gas. We must therefore know both $\varphi(\mathbf{v})$ and $\rho_k(\mathbf{v}_1, \mathbf{v}_2)$ up to the order $e^2(e^2c)^n(h^3c)^m(e^2t)^p$. But, unlike the classical case, the quantum-statistical momentum distribution depends on e^2 even at equilibrium. In order to apply formula (1.16) we should be able to calculate $\varphi(\mathbf{v})$ up to the next order beyond the ring approximation.

This difficulty can, however, be circumvented by using another expression for the internal energy, derived by Fujita and Hirota,* whose proof is given in Appendix 7:

$$E = \int d\mathbf{v}\, \tfrac{1}{2}mv^2\varphi_F(\mathbf{v}) + \tfrac{1}{2}c^{-1}\int d\mathbf{r}\, d\mathbf{v}_1 d\mathbf{v}_2 \int_0^{e^2} d\eta\, V(r)\frac{\partial}{\partial\beta}\beta f_2^{\text{eq}}(r, \mathbf{v}_1, \mathbf{v}_2; \beta, \eta) \tag{81.1}$$

where $\beta = (kT)^{-1}$ as usual, and η is the coupling parameter (in other words, in the expression for f_2^{eq}, the square of the charge e^2 is called η and becomes an integration variable). In the limit of zero temperature, it can be shown that

$$\lim_{\beta\to\infty} \beta(\partial/\partial\beta)f_2^{\text{eq}} = 0$$

and hence eq. (81.1) reduces to **

* S. Fujita and R. Hirota, *Phys. Rev.*, **118**, 6 (1960).

** This particular formula is attributed to W. Pauli. Another simple proof can be found in K. Sawada, *Phys. Rev.*, **106**, 372 (1957).

$$E^0 = \int d\mathbf{v}\, \tfrac{1}{2} m v^2 \varphi_F^0(\mathbf{v}) + \tfrac{1}{2} c^{-1} \int d\mathbf{r}\, d\mathbf{v}_1 d\mathbf{v}_2 \int_0^{e^2} d\eta\, V(r) f_2^0(r, \mathbf{v}_1, \mathbf{v}_2; \infty, \eta) \tag{81.2}$$

These formulae are remarkable, because they express the internal energy in terms of the momentum distribution of the *perfect gas* and of the perturbed correlation function. It can be shown that eq. (81.1) reduces to (1.16) in the classical limit.

From the general theory of § 62 it results that

$$f_2^0(r, \mathbf{v}_1, \mathbf{v}_2) = c^2 \left\{ \varphi_F^0(\mathbf{v}_1) \varphi_F^0(\mathbf{v}_2) + \int d\mathbf{k}\, \rho_k^0(\mathbf{v}_1, \mathbf{v}_2) e^{i\mathbf{k}\cdot\mathbf{r}} \right\} \tag{81.3}$$

Moreover, from the general separation procedure described in § 62 it follows that the Fourier component $\rho_k^0(\mathbf{v}_1, \mathbf{v}_2)$ can be separated according to eq. (62.12) [see also eq. (62.15)]:

$$\rho_k^0(\mathbf{v}_1, \mathbf{v}_2) = \hat{\rho}_k^0(\mathbf{v}_1, \mathbf{v}_2) - \hbar^3 m^{-3} \delta(\mathbf{v}_1 - \mathbf{v}_2) \varphi_F^0(\mathbf{v}_1 + \tfrac{1}{2}\hbar \mathbf{k}/m) \varphi_F^0(\mathbf{v}_1 - \tfrac{1}{2}\hbar \mathbf{k}/m) \tag{81.4}$$

Performing the Fourier transformation of $V(r)$ and of $f_2^0(r)$ in (81.2) and using (81.3) and (81.4), we find that the energy can be written as a sum of terms which will be discussed successively:

$$E = E_K + E_B + E_X + E_C \tag{81.5}$$

$$E_K = \int d\mathbf{v}\, \tfrac{1}{2} m v^2 \varphi_F^0(\mathbf{v}) \tag{81.6}$$

$$E_B = \tfrac{1}{2} e^2 c \int d\mathbf{r} \int d\mathbf{l} \int d\mathbf{v}_1 d\mathbf{v}_2 V_l e^{i\mathbf{l}\cdot\mathbf{r}} \varphi_F^0(\mathbf{v}_1) \varphi_F^0(\mathbf{v}_2) \tag{81.7}$$

$$E_X = -\tfrac{1}{2} e^2 c \hbar^3 m^{-3} \int d\mathbf{r} \int d\mathbf{l} \int d\mathbf{k} \int d\mathbf{v}_1 d\mathbf{v}_2 V_l e^{i(\mathbf{l}+\mathbf{k})\cdot\mathbf{r}} \delta(\mathbf{v}_1 - \mathbf{v}_2) \varphi_F^0(\mathbf{v}_1 - \tfrac{1}{2}\hbar\mathbf{k}/m) \varphi_F^0(\mathbf{v}_1 + \tfrac{1}{2}\hbar\mathbf{k}/m) \tag{81.8}$$

$$E_C = \tfrac{1}{2} c \int_0^{e^2} d\eta \int d\mathbf{r} \int d\mathbf{l} \int d\mathbf{k} \int d\mathbf{v}_1 d\mathbf{v}_2 V_l e^{i(\mathbf{l}+\mathbf{k})\cdot\mathbf{r}} \hat{\rho}_k^0(\mathbf{v}_1, \mathbf{v}_2) \tag{81.9}$$

Consider first the term E_K. This is clearly the average kinetic energy and exists even in the absence of interactions. In other words, it is the *ground-state energy of an ideal Fermi gas.* Its evaluation is quite elementary using the Fermi distribution:

$$\varphi_F^0(\mathbf{v}) = (m^3/h^3c)\theta(v_F - v) \qquad (81.10)$$

We obtain

$$E_K = (2\pi m^4/h^3\, c)\int_0^\infty dv\, v^4\, \theta(v_F - v) = (2\pi/5)m^4 v_F^5/h^3 c \qquad (81.11)$$

This contribution is well known and can be found in every elementary textbook on quantum statistics. We now turn to the second term. Using the normalization property of $\varphi_F^0(\mathbf{v})$:

$$\int d\mathbf{v}\, \varphi_F^0(\mathbf{v}) = 1 \qquad (81.12)$$

we find:

$$E_B = \tfrac{1}{2} 8\pi^3 e^2 c V_0 \equiv \tfrac{1}{2} e^2 c \int d\mathbf{r}\, V(r) \qquad (81.13)$$

This term is therefore $\frac{1}{2}c$ times the *unweighted average of the interaction energy over the whole space.* It obviously diverges. One must, however, keep in mind the presence of the neutralizing positive background which is implicit in the whole theory. An elementary argument shows that the electrostatic self-energy of this continuous medium equals $\frac{1}{2}8\pi^3 e^2 c V_0$, and that the average interaction energy between the positive medium and the electrons equals $-8\pi^3 e^2 c V_0$. Hence the sum of E_B and of these two contributions vanishes, and the divergence is suppressed. In more realistic systems the term E_B will also be canceled by corresponding terms involving the positive ions.

Consider now the third term (81.8), which is written as:

$$E_X = -\tfrac{1}{2} e^2 c\, h^3 m^{-3} \int d\mathbf{k}\, d\mathbf{v}\, V_k \varphi_F^0(\mathbf{v} + \tfrac{1}{2}\hbar\mathbf{k}/m)\varphi_F^0(\mathbf{v} - \tfrac{1}{2}\hbar\mathbf{k}/m) \qquad (81.14)$$

This term is usually called the *exchange energy.* It appears in traditional treatments as the expectation value of the interaction energy in the unperturbed ground state (after elimination of the direct term E_B). Its presence is due to the fact that the wave function is antisymmetric and thus E_X is a purely statistical effect. In our formalism it appears as the contribution to the ground-state energy owing to the existence of purely quantum-statistical correlations. This shows again how natural appears our separation procedure for ρ_k. The evaluation of the integral in (81.14) is simple but tedious. The result is well known:

$$E_X = -2\pi e^2 m^4 v_F^4/ch^4 \tag{81.15}$$

The last term, (81.9), is the truly non-trivial contribution to E. It is due to the existence of correlations originating from the interactions (and the statistics). In other words, it is due to the change in structure of the ground state produced by the Coulomb interactions. It is called the *correlation energy*:

$$E_C = \tfrac{1}{2}8\pi^3 c \int_0^{e^2} d\eta \int d\mathbf{k} V_k \hat{\rho}_k^0 \tag{81.16}$$

We can now use our results of § 79 for $\hat{\rho}_k^0$, in particular formula (79.19) in which e^2 is replaced by the integration variable η. It must be kept in mind that at zero temperature the summation over eigenvalues becomes an integration, according to (80.7)

$$E_C = -\tfrac{1}{2}(2\pi)^{-1} 4\pi \int_0^{e^2} d\eta \int_0^{\infty} dk k^2 \int_{-\infty}^{\infty} ds \frac{8\pi^3 \eta c \lambda^2(k;s) V_k^2}{1 + 8\pi^3 \eta c \lambda(k;s) V_k} \tag{81.17}$$

The η-integration can be performed very simply by making the change of variables $8\pi^3 \eta c \lambda(k;s) V_k \to \eta'$, and noting that $\lambda(k;s)$ is an even function of s:

$$E_C = -2(8\pi^3 c)^{-1} \int_0^{\infty} ds \int_0^{\infty} dk k^2 \{8\pi^3 e^2 c \lambda(k;s) V_k - \ln [1 + 8\pi^3 e^2 c \lambda(k;s) V_k]\} \tag{81.18}$$

This formula was first obtained by Gell-Mann and Brueckner * in a famous paper. It was rederived and further discussed later by other authors.** It is interesting to note that it can be written

* M. Gell-Mann and K. A. Brueckner, *Phys. Rev.*, **106**, 364 (1957).

** K. Sawada, *Phys. Rev.*, **106**, 372 (1957); K. Sawada, K. A. Brueckner, N. Fukuda and R. Brout, *Phys. Rev.*, **108**, 507 (1957); R. Brout, *Phys. Rev.*, **108**, 515 (1957); E. W. Montroll and J. C. Ward, *Phys. of Fluids*, **1**, 55 (1958).

entirely in terms of the dielectric constant of imaginary frequency by using the relation (80.5):

$$E_C = -2(8\pi^3 c)^{-1}\int_0^\infty dk\,k^2\int_0^\infty ds\{\varepsilon_+(k;\,is\hbar^{-1})-1-\ln\varepsilon_+(k;\,is\hbar^{-1})\} \tag{81.19}$$

We now want to show more explicitly the role of the collective effects in the correlation energy. In order to do this conveniently, we shall put the energy of the ground state into a different form, due to Hubbard.* In order to achieve this we first regroup the terms in (81.5) (dropping E_B) as follows:

$$E = E_K + E'_C - E_V \tag{81.20}$$

where

$$E_V = \tfrac{1}{2}e^2\int d\mathbf{k}\,V_k \tag{81.21}$$

$$E'_C = E_V + E_X + E_C \tag{81.22}$$

We now note that E_C, taken from eq. (81.17), can be transformed as follows

$$\begin{aligned} E_C &= -2\int_0^{e^2} d\eta\int_0^\infty dk\,k^2\int_0^\infty ds\,V_k\lambda(k;\,s)\,\frac{8\pi^3\eta c\lambda(k;\,s)V_k}{1+8\pi^3\eta c\lambda(k;\,s)V_k}\\ &= -2\int_0^{e^2} d\eta\int_0^\infty dk\,k^2\int_0^\infty ds\,V_k\lambda(k;\,s)\\ &\qquad+\int_0^{e^2} d\eta\int dk\,k^2\int ds\,\frac{V_k\lambda(k;\,s)}{1+8\pi^3\eta c\lambda(k;\,s)V_k} \end{aligned} \tag{81.23}$$

We now transform the first term of the r.h.s. by applying eqs. (80.5) and (80.9)

$$\begin{aligned} &-2\int_0^{e^2} d\eta\int_0^\infty dk\,k^2\int_0^\infty ds\,V_k\lambda(k;\,s)\\ &= -2(8\pi^3 e^2 c)^{-1}\int_0^{e^2} d\eta\int_0^\infty dk\,k^2\int_0^\infty ds\{\varepsilon_+(k;\,is\hbar^{-1})-1\} \end{aligned}$$

* J. Hubbard, *Proc. Roy. Soc. London*, **A240**, 539 (1957); **A243**, 336 (1958).

$$= -2\hbar(8\pi^3e^2c)^{-1}\int d\eta\int_0^\infty dk\,k^2\int_0^\infty d\omega\ \mathrm{Im}\ \varepsilon_+(k;\,\omega) \tag{81.24}$$

$$= -2\pi\int_0^{e^2} d\eta\int_0^\infty dk\,k^2V_k\int_0^\infty d\omega\int d\mathbf{v}\,\varphi_F^0(\mathbf{v}+\tfrac{1}{2}\hbar\mathbf{k}/m)\psi_F^0(\mathbf{v}-\tfrac{1}{2}\hbar\mathbf{k}/m)$$
$$\cdot\{\delta(\hbar\mathbf{k}\cdot\mathbf{v}-\omega)-\delta(\hbar\mathbf{k}\cdot\mathbf{v}+\omega)\}$$

$$= -2\pi e^2\int_0^\infty dk\,k^2V_k\int d\mathbf{v}\,\varphi_F^0(\mathbf{v}+\tfrac{1}{2}\hbar\mathbf{k}/m)[1-h^3cm^{-3}\varphi_F^0(\mathbf{v}-\tfrac{1}{2}\hbar\mathbf{k}/m)]$$

$$= -E_V-E_X$$

Substituting into (81.23) and then into (81.22) we are left with:

$$E'_C = 2\int_0^{e^2} d\eta\int dk\,k^2\int ds\,\frac{V_k\lambda(k;\,s)}{1+8\pi^3\eta c\lambda(k;\,s)V_k} \tag{81.25}$$

We now use (80.5) and the transformation (80.8) to get

$$E'_C = 2(8\pi^3c)^{-1}\int_0^{e^2} d\eta\,\eta^{-1}\int_0^\infty dk\,k^2\int_0^\infty ds\{1-[\varepsilon_+(k;\,is\hbar^{-1})]^{-1}\}$$
$$= (2\hbar/i8\pi^3c)\int_0^{e^2} d\eta\,\eta^{-1}\int_0^\infty dk\,k^2\int_0^\infty d\omega\ \mathrm{Im}\ [\varepsilon_+(k;\,\omega)]^{-1} \tag{81.26}$$

Writing $\varepsilon_+ = \varepsilon_1+i\varepsilon_2$ and combining (81.26) and (81.20) we finally get Hubbard's formula for the ground-state energy:

$$E = E_K+(2\hbar/8\pi^3c)\int_0^{e^2} d\eta\,\eta^{-1}\int_0^\infty dk\,k^2\int_0^\infty d\omega\,\{\varepsilon_2(k;\,\omega)/[\varepsilon_1^2+\varepsilon_2^2]\}$$
$$-2\pi e^2\int_0^\infty dk\,k^2V_k \tag{81.27}$$

The divergence at the upper limit of the last term in this formula is not dangerous. It can be shown that the second term also diverges in such a way as to compensate the former and lead to a convergent sum.

The most suggestive feature of Hubbard's formula is the occurrence of a resonant integrand, which we have already met in the discussion of the friction coefficient, eq. (78.17). We saw there that such a function can give a contribution even from the region where the numerator ε_2 vanishes, because it has a δ-

singularity at the points where the real part of the dielectric constant vanishes within that domain. Such zeros exist for very small values of k, up to a critical value, say k_c. In that region we can use the dominant part of the small k expansion (70.16) of the dielectric constant. Treating the singularity as in § 78, the collective part of the correlation energy becomes:

$$E_{\text{coll}} \simeq (2\hbar/8\pi^3 c)(4\pi)^{-1}\int_0^{e^2} d\eta\,\eta^{-1}\int_{k<k_c} d\mathbf{k}\int_0^{\infty} d\omega\,\pi\delta[1-(\omega_P^2(\eta)/\omega^2)]$$

$$= (\hbar/8\pi^3 c)\tfrac{1}{4}\int_0^{e^2} d\eta\,\eta^{-1}\omega_P(\eta)\int_{k<k_c} d\mathbf{k}\int_0^{\infty} d\omega\{\delta(\omega+\omega_P)+\delta(\omega-\omega_P)\}$$

$$= \tfrac{1}{4}(\hbar/8\pi^3 c)\int_0^{e^2} d\eta\,\eta^{-1}\omega_P(\eta)\int_{k<k_c} d\mathbf{k}$$

Noting that $\omega_P(\eta) = (4\pi c/m)^{\frac{1}{2}}\eta^{\frac{1}{2}}$, the η-integration can be performed

$$\int_0^{e^2} d\eta\,\eta^{-1}(4\pi\eta c/m)^{\frac{1}{2}} = 2(4\pi e^2 c/m)^{\frac{1}{2}} = 2\omega_P$$

Hence the final result is

$$E_{\text{coll}} = c^{-1}\int_{k<k_c}(d\mathbf{k}/8\pi^3)\,\tfrac{1}{2}\hbar\omega_P \tag{81.28}$$

Although this result is approximate, it has an extremely suggestive form. It shows that the collective contribution to the ground-state energy is precisely the zero-point energy of a collection of quantized oscillators of frequency ω_P. It confirms quantitatively the remarks made in § 78, showing how deep is the change of structure of the ground state in the presence of long-range interactions. The latter organize the motion of the particles in such a way that a significant part of the ground-state energy comes from large groups of particles oscillating in phase. This extra cohesion is perhaps the most characteristic feature of the collective behavior of charged particles.

APPENDICES

APPENDIX 1

Properties of the Laplace Transformation

Several important properties of the Laplace transformation which are extensively used in this book will be reviewed here. This appendix is meant to be an "aide-mémoire" rather than a rigorous mathematical exposition. Therefore no proofs will be given. For more details we refer the reader to one of the several existing standard textbooks on the subject.*

Let $f(t)$ be a function of the *real* variable t, having the following properties

$$f(t) = 0 \qquad \text{for } t < 0 \tag{A1.1}$$

$$\int dt\, e^{-\sigma t}|f(t)| < \infty \qquad \sigma\text{: some real number} \tag{A1.2}$$

Then the *one-sided Laplace transform* $\varphi(z)$ of $f(t)$ is defined as**

$$\varphi(z) = \int_0^\infty dt\, e^{izt} f(t) \tag{A1.3}$$

There are two main differences between this formula and the related Fourier transformation:

(*a*) *The domain of integration goes here from* 0 *to* ∞, whereas in the Fourier integral it goes from $-\infty$ to $+\infty$.

(*b*) *The variable* z *appearing here is supposed to be complex*, whereas in the Fourier integral it is assumed to be real. This point is extremely important. It gives to the Laplace transfor-

* E.g. G. Doetsch, *Die Laplace Transformation*, Dover Publ. Co., New York, 1943.

** In most textbooks the notation is different from the one chosen here. Our complex variable iz is usually denoted by $-s$. This implies some trivial changes of sign and of factors i. Also, the terms "ordinate of convergence", "upper half-plane", "above . . ." appearing here in connection with the Laplace integral correspond to the terms "abscissa of convergence", "right half-plane", "to the right of . . ." in the traditional notation.

mation an enormous flexibility due to the powerful methods of the theory of functions of complex variables.

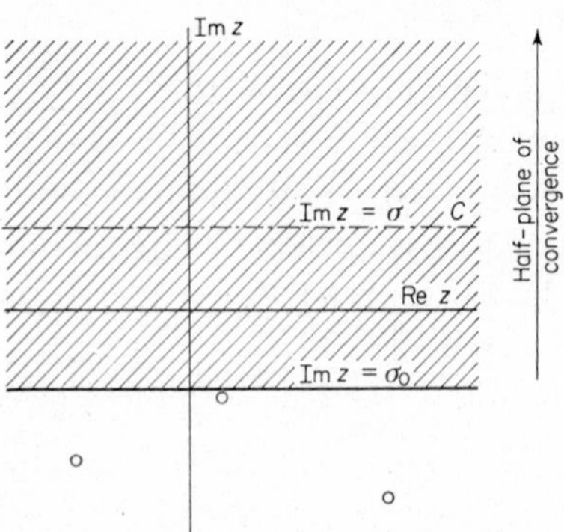

Fig. A1.1. The plane of the complex variable z appearing in eq. (A1.3). The open circles represent singularities of the analytical continuation of $\varphi(z)$.

From our assumptions the following properties can be deduced.

(α) Assumption (A1.2) implies that the integral (A1.3) converges absolutely for $\text{Im}\, z = \sigma$.

(β) Assumption (A1.1) then implies that the integral converges absolutely and uniformly for *every* z such that $\text{Im}\, z \geqq \sigma$. It is important to realize that this statement would not be true if we were allowed to take into consideration negative values of t.

(γ) So far the number σ is largely arbitrary. Let now σ_0 be the smallest number for which eq. (A1.2) is satisfied [i.e. (A1.2) is true for $\sigma_0+\varepsilon$ and it is untrue for $\sigma_0-\varepsilon$, for $\varepsilon > 0$]. σ_0 is called the *ordinate of convergence*, and the half-plane $\text{Im}\, z > \sigma_0$ is called the *half-plane of convergence*.

The location of σ_0 depends, of course, on the behavior of the function $f(t)$. If the latter has, for instance, a growing exponential behavior, $f(t) \sim e^{at}$, $a > 0$, then σ_0 must at least equal a for (A1.2) to converge; in this case the ordinate of convergence lies in the upper half-plane. If on the contrary $f(t) \sim e^{-at}$, then one may take negative values of σ ($|\sigma| < a$) without endangering the convergence of (A1.2).

The following *basic theorem* follows from (α), (β) and (γ):

The Laplace transform $\varphi(z)$ *of a function* $f(t)$ *satisfying conditions* (A1.1–2) *is a regular function of* z *throughout the half-plane* $\operatorname{Im} z > \sigma_0$.

(δ) We have so far defined the function $\varphi(z)$ in the half-plane above the line $\operatorname{Im} z = \sigma_0$. Having found that it is an analytical (and even regular) function in that region, we can extend its definition outside this initial domain by virtue of the principle of analytical continuation (we disregard the exceptional cases in which such a continuation is impossible). *We thus obtain an analytical function* $\varphi(z)$ *defined in the whole complex plane. The function* $\varphi(z)$ *is regular in the half-plane* $\operatorname{Im} z > \sigma_0$ (as was shown), *and hence necessarily has singularities in the complementary half-plane* $\operatorname{Im} z < \sigma_0$. *One of these singularities lies just below the line* $\operatorname{Im} z = \sigma_0$. The nature (pole, branch-point, essential singularity) and the location of these singularities cannot further be predicted from general principles.

Equation (A1.3) transforms a function $f(t)$ of a real variable t into an analytical function $\varphi(z)$ of the complex variable z. To this equation corresponds an *inversion formula* (A1.4) which associates with every analytical function $\varphi(z)$, regular in a certain half-plane $\operatorname{Im} z > \sigma_0$, a function $f(t)$ having the properties (A1.1–2). Moreover, this formula is reciprocal to (A1.3): i.e. the function $f(t)$ obtained by (A1.4) is the same as the one giving rise to $\varphi(z)$ through eq. (A1.3).

$$f(t) = (2\pi)^{-1} \int_C dz\, \mathrm{e}^{-izt} \varphi(z) \tag{A1.4}$$

The contour of integration C is a parallel to the real axis lying above the ordinate of convergence σ_0. Its exact location is otherwise arbitrary and $f(t)$ is independent of its location.

The integral (A1.4) can be evaluated by using any of the powerful methods of complex analysis. In particular, the original contour C can be deformed into various equivalent contours which make the evaluation easier. The most convenient type of deformation is of course dictated by the properties of the function $\varphi(z)$. We shall consider here the especially important case in which the singular points of $\varphi(z)$ are all poles of various orders (or even essential singularities).* In this case $\varphi(z)$ can be written as:

$$\varphi(z) = \sum_j \frac{a_j}{z-z_j} + \sum_k \frac{b_k}{(z-z_k)^2} + \sum_l \frac{c_l}{(z-z_l)^3} + \ldots + \psi(z) \qquad \text{(A1.5)}$$

The summations run over all the various poles z_j, z_k, z_l of various orders; $\psi(z)$ is a regular function. In this case we shall deform the contour C by pulling it downwards to $-i\infty$. According to the rules of complex integration the deformed contour "sticks" to the poles and its final shape is shown in Fig. A1.2. The integral

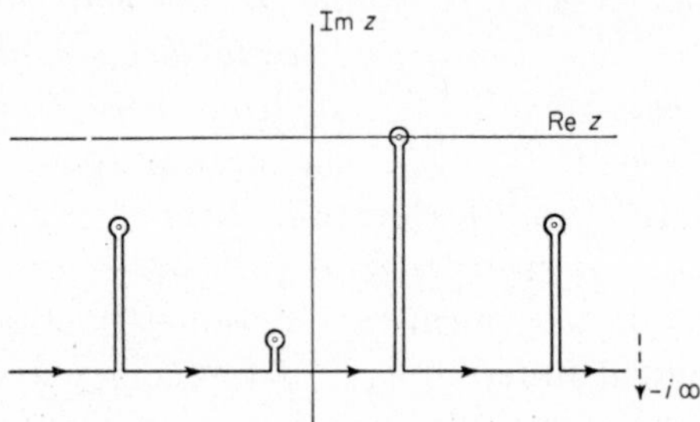

Fig. A1.2. Deformation of the contour C in eq. (A1.4) for an integrand of type (A1.5).

along the parallel to the real axis gives a vanishing contribution as the line is moved toward $-i\infty$; the contributions of the parallels to the imaginary axis cancel pairwise. The integral (A1.4) therefore equals $2\pi i$ times the sum of residues of the integrand at the various poles. These residues are calculated by expanding the exponential around each pole, with the result:

$$f(t) = i\left\{\sum_j a_j e^{iz_j t} + \sum_k b_k it e^{iz_k t} + \sum_l \tfrac{1}{2} i^2 c_l t^2 e^{iz_l t} + \ldots\right\} \qquad \text{(A1.6)}$$

The connection between eqs. (A1.5) and (A1.6) is exceedingly important for the problems studied in this book. In these problems t always represents a time variable. The two previous equations show that it is possible to obtain information about the time

* An example treating other types of singularities (logarithmic branch-points) can be found in § 28.

dependence of a given function from a study of the nature and of the location of the singularities of its Laplace transform. In particular, if all the singularities of $\varphi(z)$ are poles of various orders, we see that $f(t)$ can be represented as a sum of terms, each of which is a product of a *constant*, a *power of t* and an *exponential*.

The *order of the pole* determines the *power of t*, as follows:

Simple pole $\to 1$
Double pole $\to t$
Triple pole $\to t^2$
n-fold pole $\to t^{n-1}$

The *location of the pole* determines the *type of exponential*:

Pole at the origin $\to 1$
Pole on the real axis $\to$ steady oscillation
Pole in the upper half-plane $\to$ exponentially growing oscillation
Pole in the lower half-plane $\to$ exponentially decaying oscillation

Applications of this property appear under various forms throughout the whole book.

We finally quote another important theorem. Let $f_1(t)$ and $f_2(t)$ be two functions satisfying conditions (A1.1–2). The *convolution product* of these functions, denoted by $F(t)$ or $f_1 * f_2$ is defined as

$$F(t) \equiv f_1 * f_2 = \int_0^t d\tau \, f_1(t) f_2(\tau - t) \tag{A1.7}$$

An important property of the Laplace transformation is the fact that it maps a convolution product into an ordinary product. In other words, let $\varphi_1(z)$, $\varphi_2(z)$, $\Phi(z)$ be respectively the Laplace transforms of $f_1(t)$, $f_2(t)$ and $F(t)$. The *convolution theorem* then states:

$$\Phi(z) = \varphi_1(z)\varphi_2(z) \tag{A1.8}$$

APPENDIX 2

The Cauchy Integral

A.2.1. Analytical Properties of the Cauchy Integral

Let us consider a class of complex functions $f(x)$ defined by the following three properties.

a′) *$f(x)$ is defined for values of x lying on a certain contour Γ.* The word "contour" means a closed curve. However, in most cases of interest in connection with our problems, Γ is a curve closed at infinity. Such a generalized contour is provided, for instance, by the real axis from $-\infty$ to $+\infty$. As all the Cauchy integrals encountered in the text concern this type of contour we shall specialize assumption a′):

a) *$f(x)$ is defined for all real values of x: $-\infty < x < +\infty$.*

b) *$f(x)$ satisfies a Lipschitz condition* (also called a *Hölder condition*) *on Γ*, i.e.

$$|f(x_2)-f(x_1)| \leqq A|x_2-x_1|^{\mu}, \quad A > 0, \quad 0 < \mu \leqq 1 \tag{A2.1.1}$$

This condition is stronger than the mere requirement of continuity. Indeed, a function is continuous if $|f(x_2)-f(x_1)| \to 0$ when $|x_2 - x_1| \to 0$. For a Lipschitz function, *the exact way* in which it goes to zero is specified, i.e. as a power of $|x_2-x_1|$.

c) *$f(x)$ has the following behavior at infinity*: $f(x)$ tends toward a definite limiting value $f(\infty)$ in such a way that

$$|f(x)-f(\infty)| < A/|x|^{\mu}, \quad \mu > 0, \quad A > 0$$

When not explicitly stated otherwise, we assume here that $f(\infty) = 0$.

These conditions are sufficient for the validity of the results obtained below; they are presumably not necessary.

Consider now a complex number z, and the expression $\Phi(z)$ defined as:

$$\Phi(z) = \frac{1}{2\pi i}\int_{-\infty}^{\infty} dx\, \frac{f(x)}{x-z} \tag{A2.1.2}$$

The function $\Phi(z)$ is called a *Cauchy integral.* Equation (A2.1.2) can be regarded as expressing an operation which associates with each function $f(x)$ defined on the real axis a function $\Phi(z)$ defined in the whole complex plane. The function $\Phi(z)$ will sometimes be called the *Cauchy integral associated with the function* $f(x)$. The properties of this function are studied briefly in most textbooks on complex analysis. The most recent and detailed account is given in Gakhov's monograph.* A very extensive treatment is also given in Muskhelishvili's monograph ** and, to less extent, in Mikhlin's book.† We shall outline here the properties which are most interesting for our purpose without claiming full mathematical rigor.

It is very easily shown that $\Phi(z)$ *is a regular function in any domain D which does not include points on the real axis. It tends to zero for* $z \to \infty$.

The function $\Phi(z)$ thus possesses a derivative [because $(x-z)^{-1}$ is bounded for all non-real values of z]. More generally, the pth derivative of $\Phi(z)$ is

$$\Phi^{(p)}(z) = \frac{p!}{2\pi i}\int_{-\infty}^{\infty} dx \frac{f(x)}{(x-z)^{p+1}} \tag{A2.1.3}$$

The previous property implies in our case that $\Phi(z)$ is regular in the whole upper half-plane and in the whole lower half-plane. (From now on, these regions will be called, respectively, S_+ and S_-.) The integral (A2.1.2) has, however, no meaning as it stands for z real. We shall now show that it can be given a meaning in terms of the following fundamental theorem. All the properties discussed in this appendix are direct consequences of it.

When z *tends toward the real value* y *by taking only values in* S_+, *the function* $\Phi(z)$ *tends toward a definite limit* $\Phi^{(+)}(y)$.

When z *tends toward the real value* y *by taking only values in* S_-, *the function* $\Phi(z)$ *tends toward a definite limit* $\Phi^{(-)}(y)$.

* F. D. Gakhov, *Boundary Problems* (Kraevye Zadatchi) (in Russian), Fizmatgiz, Moscow, 1958.

** N. I. Muskhelishvili, *Singular Integral Equations,* Noordhoff, Groningen, 1953.

† S. G. Mikhlin, *Singular Integral Equations,* Amer. Math. Soc. Translations, no. 24, 1950.

The limits $\Phi^{(+)}(y)$ and $\Phi^{(-)}(y)$ are generally different.

In other words, $\Phi(z)$ is discontinuous on the real axis; however, it is continuous from above and from below.

Consider first the behavior for $z \to y$ from above. We may take the limit by slightly deforming the contour of integration in order to avoid the point y (see Fig. A2.1). The limit is correctly

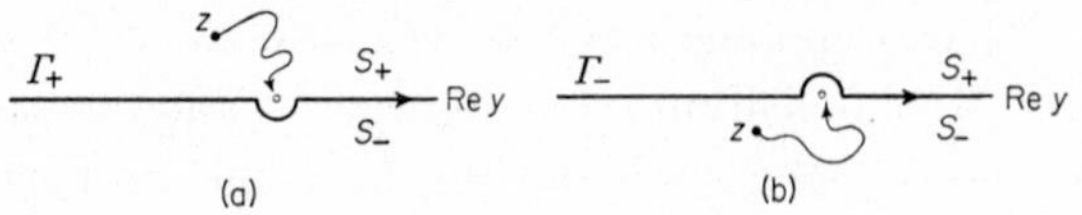

Fig. A2.1. Contours of integration: (a) for $\Phi^{(+)}(y)$; (b) for $\Phi^{(-)}(y)$.

taken if the topological relationship between the point z and the contour of integration is preserved, i.e. if in the process $z \to y$ the point z always remains *above* the contour. This implies that the latter must be indented as shown in (a). The evaluation of the integral for $z = y$ no longer presents any ambiguities with this modified contour. Introduce the concept of a *Cauchy principal part*:

$$\mathscr{P}\int_{-\infty}^{\infty} dx\, \frac{f(x)}{x-y} = \lim_{\varepsilon\to 0}\left\{\int_{-\infty}^{y-\varepsilon} dx\, \frac{f(x)}{x-y} + \int_{y+\varepsilon}^{\infty} dx\, \frac{f(x)}{x-y}\right\}, \quad \varepsilon > 0 \tag{A2.1.4}$$

The integration along the modified contour Γ_+ is then easily performed by residues. There is a first term equal to $\frac{1}{2}(2\pi i)$ times the residue of the integrand at $x = y$ (this is the contribution of the dent around y) plus a principal part integral:

$$\Phi^{(+)}(y) = \tfrac{1}{2}f(y) + \frac{1}{2\pi i}\mathscr{P}\int_{-\infty}^{\infty} dx\, \frac{f(x)}{x-y}, \quad y \text{ real} \tag{A2.1.5}$$

A similar calculation (see Fig. A2.1b) leads to the following limiting value of Φ for $z \to y$ from below:

$$\Phi^{(-)}(y) = -\tfrac{1}{2}f(y) + \frac{1}{2\pi i}\mathscr{P}\int_{-\infty}^{\infty} dx\, \frac{f(x)}{x-y}, \quad y \text{ real} \tag{A2.1.6}$$

Formulae (A2.1.5–6) are fundamental to the whole theory:

they are called the *Plemelj formulae*. Another equivalent and very useful form is:

$$\Phi^{(+)}(y)-\Phi^{(-)}(y) = f(y), \qquad y \text{ real} \tag{A2.1.7}$$

$$\Phi^{(+)}(y)+\Phi^{(-)}(y) = \frac{1}{\pi i}\mathscr{P}\int_{-\infty}^{\infty} dx\,\frac{f(x)}{x-y} \tag{A2.1.8}$$

The Plemelj theorem clearly shows that $\Phi(z)$, defined by (A2.1.2), is a discontinuous function of a particular type called a "*sectionally regular function*". This concept is defined as a function which is: (1) regular in each region not containing points of the contour Γ; (1′) continuous from above and from below on the contour Γ.

There is also another way of viewing the object $\Phi(z)$. Let us recall the general definition of an analytical function (see for details any textbook on complex analysis). One starts with a function which is regular in a circle whose radius of convergence is r_0. One then takes a point on the boundary of the circle and performs an analytical continuation into a region overlapping the circle and extending outside it. If the process is possible, it is iterated systematically in all directions: The function is thus defined in the vicinity of all points attained in this process. This function is called an *analytical function*.

We now come back to the function $\Phi(z)$ and start with $z \in S_+$. Let z go downwards until it attains the real axis at y. We then give it a negative imaginary increment. The analytical continuation of $\Phi(z)$, if it exists, is obtained by a deformation of the contour Γ_+ analogous to the one which defined $\Phi^{(+)}(y)$ (i.e. a

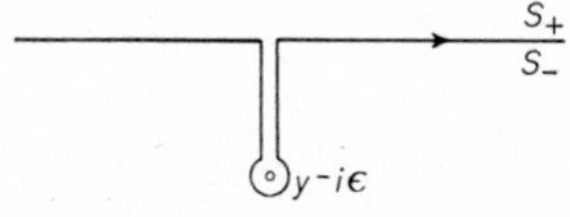

Fig. A2.2. Contour for the analytical continuation of $\Phi^{(+)}(z)$.

deformation such that z is always above the contour, see Fig. A2.2). The resulting function is then:

$$\Phi^{(+)}(y-i\varepsilon) = \frac{1}{2\pi i}\int_{-\infty}^{\infty} dx\, \frac{f(x)}{x-y+i\varepsilon} + f(y-i\varepsilon), \quad \varepsilon > 0 \qquad \text{(A2.1.9)}$$

This function defines the *analytical continuation into* S_- *of the function defined by* (A2.1.2) *for* $z \in S_+$. It is *distinct* from the Cauchy integral (A2.1.2) evaluated at $y-i\varepsilon$. A similar reasoning can be applied for z initially in S_- and going to S_+. We therefore conclude that the Cauchy integral defines *two distinct analytical functions*: a function $\Phi^{(+)}(z)$ if z in (A2.1.2) lies in S_+, and a function $\Phi^{(-)}(z)$ for $z \in S_-$. These two functions are distinct, because neither of them can be attained from the other by the process of analytical continuation described above. This is another way of saying that $\Phi^{(+)}(z)$ and $\Phi^{(-)}(z)$ are not elements of the same analytical function. Some additional comments will help clarify the situation.

I. The origin of the discontinuity of $\Phi(z)$ can be easily understood in terms of familiar concepts. In many cases, the function $f(x)$, which is defined initially on the real axis alone, can be continued analytically into the whole plane of x just by replacing the real variable x in the function by a complex x. The function thus obtained has, in general, poles in the complex plane, at points $x = x_j$. Assume moreover that the integral (A2.1.2) can be evaluated by the method of residues by completing the real axis with a semi-circle at infinity in the upper half-plane. First let $z \in S_+$; then the integral $\Phi^{(+)}(z)$ is:

$$\begin{aligned}\Phi^{(+)}(z) &= 2\pi i \sum_j \left(\text{residues of } \frac{f(x)}{x-z} \text{ at } x = x_j\right) \\ &\quad + 2\pi i \left(\text{residue of } \frac{f(x)}{x-z} \text{ at } x = z\right)\end{aligned} \qquad \text{(A2.1.10)}$$

Now let $z \in S_-$. In the integral (A2.1.2) (where the real axis is completed as above), the pole $x = z$ is no longer inside the contour, therefore

$$\Phi^{(-)}(z) = 2\pi i \sum_j \left(\text{residues of } \frac{f(x)}{x-z} \text{ at } x = x_j\right) \qquad \text{(A2.1.11)}$$

If, however, the function $\Phi^{(+)}(z)$ is continued into S_- by means of the contour Γ_+ of Fig. A2.2, the pole $x = z$ is again within the contour and the result of the integration is again (A2.1.10). Thus, the discontinuity is due to the fact that the pole $x = z$ moves from the inside to the outside of the contour as z goes from S_+ into S_-. When it crosses the real axis, $\Phi(z)$ changes discontinuously because one must no longer take the residue at $x = z$ into consideration.

II. In order to perform the analytical continuation of $\Phi^{(+)}(z)$ into S_- (or of $\Phi^{(-)}(z)$ into S_+), the conditions (a) to (c) are not sufficient. Formula (A2.1.9) shows that $\Phi^{(+)}(z)$ has an analytical continuation into S_- if and only if $f(x)$ itself has such a continuation. If this is so, the function $\Phi^{(+)}(z)$ is defined in the whole complex plane by the formulae:

$$\Phi^{(+)}(z) = \frac{1}{2\pi i}\int_{-\infty}^{\infty} dx \frac{f(x)}{x-z}, \qquad z \in S_+$$

$$\Phi^{(+)}(z) = \frac{1}{2\pi i}\mathscr{P}\int_{-\infty}^{\infty} dx \frac{f(x)}{x-z} + \tfrac{1}{2}f(z), \quad z \text{ real}$$

$$\begin{aligned}\Phi^{(+)}(z) &= \frac{1}{2\pi i}\int_{-\infty}^{\infty} dx \frac{f(x)}{x-z} + f(z), \qquad z \in S_- \\ &= \Phi^{(-)}(z) + f(z)\end{aligned} \tag{A2.1.12}$$

We have seen that $\Phi^{(+)}(z)$ is regular in the upper half-plane. We may add now that *if it possesses an analytical continuation, the latter necessarily has singularities in* S^-, otherwise $\Phi^{(+)}(z)$ would be a constant, and even, more precisely, zero (because $\Phi^{(+)}(z)$ vanishes at infinity). Moreover, from the last formula (A2.1.12) it is seen that $\Phi^{(+)}(z)$ equals $\Phi^{(-)}(z)+f(z)$ in the lower half-plane. But $\Phi^{(-)}(z)$ is regular in S_-; hence *the singularities of* $\Phi^{(+)}(z)$ *in* S_- *coincide with the singularities of* $f(z)$ *itself.*

III. It is of interest to discuss the analytical behavior of $\Phi(z)$ from the point of view of Riemann surfaces. We assume that both $\Phi^{(+)}(z)$ and $\Phi^{(-)}(z)$ can be defined in the whole complex plane by means of analytical continuation. $\Phi(z)$ is then a two-valued function; its Riemann surface has a cut along which its two sheets go through one another (Fig. A2.3). There is, however, no branch-point in the finite region of the plane, because the cut goes from $-\infty$ to ∞.

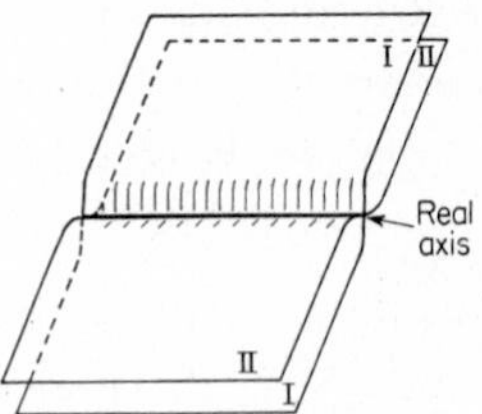

Fig. A2.3. Riemann surface for the function $\Phi^{(+)}(z)$.

It follows therefore that if one starts with a point on one sheet of the surface and moves it on the surface without leaving the sheet, one can by no means reach a point on the other sheet. This is why the two functions $\Phi^{(+)}(z)$ and $\Phi^{(-)}(z)$ are *distinct* analytical functions. However, if the limits of integration in (A2.1.2) were, say, 0 and ∞, the point 0 would be a branch-point. The nature of this singularity is close to that of a logarithmic branch-point. Indeed, let us make a cut from 0 to ∞ and consider a path which encircles the origin counterclockwise repeatedly. By virtue of the Plemelj formulae, each time the path crosses the cut, the value of the function is increased by $-f(y)$. Hence $\Phi(z)$ is a single analytical function with an infinite number of branches represented by

$$\Phi(z) = \Phi^{(0)}(z)+nf(z)$$

where n is an integer. In the case $f(z) = 1$ the well known behavior of $(2\pi i)^{-1} \ln z$ is, of course, recovered.

A2.2. Plus-Functions and Minus-Functions

The Plemelj formula (A2.1.7) can be viewed from another point of view: it provides a decomposition of an arbitrary function $F(y)$ of a *real variable* [satisfying conditions (a) to (c) of the previous section] into a difference of two functions, which we rewrite as follows:

$$F(y) = F_+(y)-F_-(y) \qquad \text{(A2.2.1)}$$

In this connection, $F_+(y)$ will be called the *plus part* of $F(y)$ and $F_-(y)$ its *minus part*. The plus part of $F(y)$ is uniquely characterized by the requirement that its analytical continuation into S_+ is regular. The same is true for the analytical continuation of $F_-(y)$ into S_-.

This decomposition is by no means trivial. A function of a real variable can often be continued into S_+ and S_-, but these continuations are in general *not* regular, either in S_+ or in S_-. If, however, it happens that the continuation of $F(y)$ into S_+ is regular then $F(y)$ is called a *plus-function*. Its minus part is then identically zero. For, if $F_-(y)$ did not vanish, its analytical continuation into S_+ would necessarily have singularities; the analytical continuation of F, written in the form (A2.2.1), would therefore not be regular, as was assumed.

The Plemelj formulae (A2.1.5–6) give a unique prescription for the construction of the plus and minus parts of a function. These formulae can be rewritten by introducing the following important symbols:

$$\delta_+(x) = \delta(x) + \frac{i}{\pi} \mathscr{P} \frac{1}{x} \tag{A2.2.2}$$

$$\delta_-(x) = \delta(x) - \frac{i}{\pi} \mathscr{P} \frac{1}{x} \tag{A2.2.3}$$

From this definition, we note immediately the following symmetry properties:

$$\delta_+(x) = \delta_-(-x) = [\delta_-(x)]^* \tag{A2.2.4}$$

These objects are highly singular "functions", or rather distributions (in the sense of Schwartz), which only have a meaning in connection with an integration. They can also be regarded symbolically as providing the decomposition of the δ-function into plus and minus parts:

$$\delta(x) = \tfrac{1}{2}\delta_+(x) - [-\tfrac{1}{2}\delta_-(x)]$$

This is, however, not a decomposition in the previous sense, because $\delta(x)$ has no analytical continuation. However, its meaning in the distribution sense is closely related to (A2.2.1). In effect, the Plemelj formulae (A2.1.5–6) can be written as:

$$F_+(y) = \frac{1}{2}\int dx\ \delta_+(y-x)F(x) \qquad \text{(A2.2.5)}$$

$$F_-(y) = -\frac{1}{2}\int dx\ \delta_-(y-x)F(x) \qquad \text{(A2.2.6)}$$

Comparing these equations with the Plemelj formulae (A2.1.5–6) we obtain the following important representations of the $\delta_\pm$ functions:

$$\lim_{\varepsilon\to 0} \frac{1}{x+i\varepsilon} = -\pi i\,\delta_+(x) \qquad \text{(A2.2.7)}$$

x real, $\varepsilon > 0$

$$\lim_{\varepsilon\to 0} \frac{1}{x-i\varepsilon} = \pi i\,\delta_-(x) \qquad \text{(A2.2.8)}$$

Another useful representation of the $\delta_\pm$ functions is the following

$$\begin{aligned} \delta_+(x) &= \pi^{-1}\int_0^\infty dk\ e^{ikx} \\ \delta_-(x) &= \pi^{-1}\int_0^\infty dk\ e^{-ikx} \end{aligned} \qquad \text{(A2.2.9)}$$

The decomposition given by eq. (A2.2.1) can also be viewed as a decomposition of the functional space of the functions F into two mutually orthogonal subspaces. (A2.2.5–6) are functional transformations by which a function F is projected onto one or the other subspace. One can symbolize this operation by introducing the operators J_+ and J_- by the equations

$$F_+ = J_+F, \qquad F_- = J_-F \qquad \text{(A2.2.10)}$$

These are really *projection operators,* because they satisfy the relations (whose proof is left to the reader):

$$\begin{aligned} &J_+J_+F = J_+F, \quad J_-J_-F = J_-F \\ &J_+J_-F = J_-J_+F = 0 \end{aligned} \qquad \text{(A2.2.11)}$$

We cannot here go into details about these questions, but refer the reader to a beautiful paper by Tchersky * who generalized the concept

* Iu. I. Tchersky, *Mat. Sbornik,* **41**, 277 (1957).

of singular operators and developed a general method for the solution of singular equations.

Before closing this section we should like to draw attention to a rather delicate point. It is well known that in calculations involving *products* of singular functions (Schwartz distributions) much care must be exercised in the interpretation of the results. We want to point out a peculiarity which arises in expressions of the type

$$\int dx \int dy\ \delta_{\pm}(u-x)\delta_{\pm}(x-y)f(x, y)$$

Nothing special occurs for the terms involving the δ-part of the $\delta_{\pm}$ functions or one δ-part and one principal part:

$$\int dx \int dy\ \delta(u-x)\delta(x-y)f(x, y) = f(u, u)$$

$$\int dx \int dy\ \delta(u-x)\frac{\mathscr{P}}{x-y}f(x, y) = \mathscr{P}\int dy\frac{f(u, y)}{u-y}$$

But the term involving two principal parts must be handled with much caution. The reason is that in such an expression *the order of integrations cannot be changed freely*. This results from a famous *theorem due to Poincaré and Bertrand*, which we give without proof: *

$$\begin{aligned}\int dx\frac{\mathscr{P}}{x-u}\int dy\frac{\mathscr{P}}{y-x}f(x, y)\\ = -\pi^2 f(u, u)+\int dy\int dx\frac{\mathscr{P}}{(x-u)}\frac{\mathscr{P}}{(y-x)}f(x, y)\end{aligned} \tag{A2.2.12}$$

An application of this theorem will be found in Appendix 3.

A2.3. Singular Integral Equations of the Cauchy Type

We now show that the formalism developed in the two previous sections is beautifully adapted to the solution of a certain type of integral equation which plays an important role in plasma physics. We shall only outline here the treatment of a particular

* See F. D. Gakhov and N. I. Muskhelishvili, refs. quoted on page 391.

type of such an equation (called a "dominant type"), referring for fuller details to the books by Gakhov, Muskhelishvili and Mikhlin quoted on page 391. The equations we are interested in are of the following general form:

$$a(x)f(x)+b(x)\frac{1}{\pi}\mathscr{P}\int_{-\infty}^{\infty}ds\,\frac{f(s)}{s-x}=c(x) \tag{A2.3.1}$$

where $a(x)$, $b(x)$, $c(x)$ are given complex functions of the real variable x and $f(x)$ is the unknown. The general solution is performed in three stages.

I. The homogeneous Hilbert problem

This preliminary problem can be formulated as follows:

*To find a sectionally regular function** $\Phi(z)$ *(discontinuous on the real axis), of finite degree at infinity, whose boundary values on the real axis satisfy the condition:*

$$\Phi^{(+)}(x)=g(x)\Phi^{(-)}(x) \tag{A2.3.2}$$

where $g(x)$ *satisfies the same conditions as* $f(x)$ *in* § A2.1, *and is* $\neq 0$ *on the real axis.*

We first introduce the *index* χ of the function $g(x)$ or of the equation (A2.3.2), defined as the variation of $\ln g(z)$ as z travels along the real axis from $-\infty$ to $+\infty$ and then back to $-\infty$ along a semi-circle at infinity:

$$\chi=\frac{1}{2\pi i}[\ln g(z)]_G \tag{A2.3.3}$$

In view of the central role played by the concept of the index, we give some more details which should make its calculation easier. We note first that, by virtue of the principle of the argument which was used in eq. (19.1), the index can be expressed as follows [if the function $g(x)$ is continuous and differentiable and has no real zeros]:

$$\chi=\frac{1}{2\pi i}\int_{-\infty}^{\infty}d\ln g(x)=\frac{1}{2\pi i}\int_{-\infty}^{\infty}dx\,\frac{g'(x)}{g(x)}$$

* *Not necessarily a Cauchy integral*! In order that the solution $\Phi(z)$ be a Cauchy integral we should further ask that $\Phi(z)$ vanishes at infinity (see § III of this section).

The properties given below follow immediately from this definition:

A. The index of a product of functions equals the sum of the indices of the factors. The index of a ratio of two functions equals the difference of the indices of the numerator and of the denominator.

B. If $g(x)$ is the limiting value of a function regular in the upper half-plane [in particular, if $g(x)$ is a plus-function], its index equals its number of zeros in S_+. If $g(x)$ is the limiting value of a function regular in the lower half-plane, its index equals *minus* its number of zeros in S_-.

C. If the analytical continuation $g(z)$ of $g(x)$ into S_+ is regular, with the exception of a finite number of poles, its index equals the difference between its number of zeros and its number of poles in the upper half-plane.

D. The index of a function differs only by its sign from the index of its complex conjugate.

We shall discuss here in detail the case where $\chi = 0$; then $\ln (g)z$ is a one-valued function. We first look for a particular solution $\Phi = X(z)$. We take logarithms of both sides of (A2.3.2):

$$\ln X^{(+)}(x) - \ln X^{(-)}(x) = \ln g(x) \tag{A2.3.4}$$

Knowing the difference of a plus function and a minus function, the Plemelj formula (A2.1.8) immediately gives their sum

$$\ln X^{(+)} + \ln X^{(-)} = \frac{1}{\pi i} \mathscr{P} \int ds \frac{\ln g(s)}{s-x} \equiv 2\Gamma(x)$$

or

$$\begin{aligned} X^{(+)}(x) &= \exp\left[\Gamma(x) + \tfrac{1}{2} \ln g(x)\right] \\ X^{(-)}(x) &= \exp\left[\Gamma(x) - \tfrac{1}{2} \ln g(x)\right] \end{aligned} \tag{A2.3.5}$$

We now clearly see from (A2.1.7–8) that

$$X(z) = e^{\Gamma(z)}, \quad \Gamma(z) = \frac{1}{2\pi i} \int ds \frac{\ln g(s)}{s-z} \tag{A2.3.6}$$

is a particular solution of the problem. Indeed, it is regular in the whole complex plane but for a discontinuity on the real axis. The boundary values from above and from below are determined by the behavior of the Cauchy integral in the exponential, and are given by (A2.3.5). They satisfy (A2.3.2).

The solution $X(z)$ *has the remarkable property of being dif-*

ferent from zero in the whole complex plane. For $z \to \infty$, $X(z) \to 1$.* Its boundary values are also $\neq 0$ on the real axis, as can be seen from (A2.3.5). This particular solution is called the *fundamental solution,* because all solutions of the Hilbert problem can be generated from it, as will now be shown.

The general solution of the homogeneous Hilbert problem (for $\chi = 0$*) of finite degree at infinity is:*

$$\Phi(z) = X(z)P(z) \tag{A2.3.7}$$

where $P(z)$ *is an arbitrary polynomial.*

Indeed, we have

$$\Phi^{(+)}(x) = g(x)\Phi^{(-)}(x), \qquad X^{(+)}(x) = g(x)X^{(-)}(x)$$

As $X^{(+)}(x)$ and $X^{(-)}(x)$ are $\neq 0$ for all x, it follows that

$$\Phi^{(+)}(x)/X^{(+)}(x) = \Phi^{(-)}(x)/X^{(-)}(x)$$

and thus the function $\Phi(z)/X(z)$ is regular in the whole plane (it has no discontinuity on the real axis). As it has finite degree at infinity, it is a polynomial.

II. The inhomogeneous Hilbert problem

To find a sectionally regular function $\Phi(z)$*, of finite degree at infinity, whose boundary values on the real axis satisfy the condition*

$$\Phi^{(+)}(x) = g(x)\Phi^{(-)}(x)+h(x) \tag{A2.3.8}$$

where $g(x)$ *and* $h(x)$ *are given non-vanishing functions of the real variable* x.

This problem is easily reduced to the previous one. Let $X(z)$ be the fundamental solution of the corresponding homogeneous problem (A2.3.2); then

$$g(x) = X^{(+)}(x)/X^{(-)}(x)$$

Substituting this expression into (A2.3.8) and dividing throughout by $X^{(+)}(x)$ (which is $\neq 0$ on the real axis) we get:

* This property shows that the fundamental solution $X(z)$ of the Hilbert problem cannot be expressed as a Cauchy integral (which would vanish at infinity). This is also the reason why the limiting values (A2.3.5) do not satisfy the Plemelj formulae, but rather the Hilbert condition (A2.3.2).

$$\frac{\Phi^{(+)}(x)}{X^{(+)}(x)}-\frac{\Phi^{(-)}(x)}{X^{(-)}(x)}=\frac{h(x)}{X^{(+)}(x)}$$

The sectionally regular function $\Phi(z)/X(z)$ is then determined by the Plemelj formulae:

$$\frac{\Phi(z)}{X(z)}=\frac{1}{2\pi i}\int ds\frac{h(s)}{X^{(+)}(s)(s-z)}+P(z)$$

The arbitrary polynomial $P(z)$ takes account of the finite degree at infinity. Therefore the general solution is:

$$\Phi(z)=\frac{X(z)}{2\pi i}\int ds\frac{h(s)}{X^{(+)}(s)(s-z)}+X(z)P(z) \qquad \text{(A2.3.9)}$$

III. Cauchy singular integral equations of dominant type

We now show that the solution of eq. (A2.3.1) can be reduced to the solution of a Hilbert problem. Consider the Cauchy integral

$$\Phi(z)=\frac{1}{2\pi i}\int_{-\infty}^{\infty} ds\frac{f(s)}{s-z} \qquad \text{(A2.3.10)}$$

Its boundary values on the real axis are related to f by the Plemelj formulae:

$$f(x)=\Phi^{(+)}(x)-\Phi^{(-)}(x)$$

$$\frac{1}{\pi}\mathscr{P}\int ds\frac{f(s)}{s-x}=i[\Phi^{(+)}(x)+\Phi^{(-)}(x)]$$

Making this substitution in (A2.3.1) we find:

$$[a(x)+ib(x)]\Phi^{(+)}(x)-[a(x)-ib(x)]\Phi^{(-)}(x)=c(x) \qquad \text{(A2.3.11)}$$

We now assume that neither of the coefficients $(a+ib)$ and $(a-ib)$ vanishes on the real axis; then (A2.3.11) defines a Hilbert problem of the general type (A2.3.7), which has already been solved. Knowing the solution $\Phi(z)$, the Plemelj formulae immediately yield $f(x)$. We shall not go through the calculations, because they are a mere repetition of the previous ones.

It must, however, be pointed out that an additional restriction must be added to the Hilbert problem in order to apply it to the present case. *As the sectionally regular function $\Phi(z)$ is to* *represent a Cauchy integral* (A2.3.10), *it must vanish at infinity.* Thus, among all solutions (A2.3.7) or (A2.3.9) [with arbitrary $P(z)$], we must choose the ones which have that property. Consider first the homogeneous case, (A2.3.7). The fundamental solution $X(z)$ was shown to be non-zero in the whole complex plane, including the point at infinity; we must therefore choose a $P(z)$ which vanishes at infinity. But the only polynomial which has that property is the trivial constant $P(z) \equiv 0$. Consider now the solution of the inhomogeneous problem, (A2.3.9). The first term is the product of the non-zero function $X(z)$ and a Cauchy integral which vanishes at infinity; applying the previous argument to the second term it leads again to the unique choice $P(z) \equiv 0$. We thus arrive at the following important conclusion:

If the index is zero, the homogeneous equation [(A2.3.1) *with* $c \equiv 0$] *has only the trivial solution* $f(x) \equiv 0$, *whereas the inhomogeneous equation* ($c \neq 0$) *has a unique non-trivial solution.*

This important theorem shows that in this respect singular integral equations with zero index behave like ordinary linear equations. This analogy can be shown to be even stronger. Noether has proved a set of theorems which generalize to singular integral equations the well known "Fredholm alternative" of the theory of ordinary linear integral equations.* These theorems show that if $\chi = 0$, Fredholm's alternative applies exactly to eq. (A2.3.1).

On the other hand, singular equations with $\chi > 0$ or $\chi < 0$ have a completely different type of behavior. The following theorem generalizes the previous one:

1*a*. *If* $\chi > 0$, *the homogeneous equation has* χ *linearly independent non-trivial solutions.*

1*b*. *If* $\chi \leqq 0$, *the homogeneous equation has no solution (apart from the trivial one).*

2*a*. *If* $\chi \geqq 0$, *the inhomogeneous equation is soluble whatever the r.h.s.,* $c(x)$; *its general solution depends linearly on* χ *arbitrary constants.*

* See, e.g., R. Courant and D. Hilbert, *Methoden der mathematischen Physik*, Springer Verlag, Berlin, 1931; P. M. Morse and H. Feshbach, *Methods of Theoretical Physics*, McGraw-Hill, New York, 1953.

2c. *If $\chi < 0$, the inhomogeneous equation is soluble if and only if the r.h.s., $c(x)$, satisfies $(-\chi)$ integral conditions:*

$$\int_{-\infty}^{\infty} ds \frac{s^k c(s)}{[a(s)+b(s)]X^{(+)}(s)} = 0; \; k = 0, 1, \ldots, -\chi-1 \quad \text{(A2.3.12)}$$

($X^{(+)}$ being the boundary value from above of the fundamental solution of the associated homogeneous Hilbert problem). If these conditions are satisfied, the solution is unique.

The proof of this fundamental theorem can be found in the books by Gakhov, Muskhelishvili and Mikhlin, quoted on page 391.

Orthogonality Properties of Vlassov Eigenfunctions

We shall prove directly the relationship (24.3):

$$\int d\nu\, \bar{\chi}_v(\nu)\tilde{\chi}_{v'}(\nu) = C_v \delta(v-v')$$

and evaluate the constant C_v. Much care must be exercised in this type of expression, because we deal here with integrals of *products* of distributions. In order to avoid misinterpretations, we multiply the integral on the left by an arbitrary well behaved function $F(v')$ and integrate over v'. [We drop the symbols $\mathscr{P}$: all integrals involving $(\nu-v)^{-1}$ and $(\nu-v')^{-1}$ are principal parts.]

$$\begin{aligned} J &\equiv \int dv'\, F(v') \int d\nu\, \bar{\chi}_v(\nu)\tilde{\chi}_{v'}(\nu) \\ &= \int dv' \int d\nu\, F(v') \Big[\pi^{-1} \frac{\varepsilon_2(\nu)}{(\nu-v)(\nu-v')} + \varepsilon_1(v) \frac{\delta(v-\nu)}{\nu-v'} \\ &\quad + \frac{\varepsilon_1(v')}{\varepsilon_2(v')} \frac{\varepsilon_2(\nu)}{\nu-v} \delta(\nu-v') + \pi^{-1} \varepsilon_1(v) \frac{\varepsilon_1(v')}{\varepsilon_2(v')} \delta(\nu-v)\delta(\nu-v') \Big] \end{aligned}$$

We now use the definitions (23.13–14); care must be taken not to exchange the order of multiple integrations in terms involving products of principal parts: such an exchange is forbidden by the Poincaré–Bertrand theorem (see Appendix 2).

$$\begin{aligned} J &= \int dv'\, d\nu\, F(v') \pi^{-1} \frac{\varepsilon_2(\nu)}{(\nu-v)(\nu-v')} + \int dv'\, d\nu\, F(v') \frac{1}{\nu-v'} \delta(\nu-v) \\ &\quad - \int dv'\, d\nu\, F(v') \frac{1}{\nu-v'} \int d\nu''\, \pi^{-1} \frac{\varepsilon_2(\nu'')}{\nu''-v} \delta(\nu-v) \\ &\qquad + \int dv'\, d\nu\, F(v') \frac{\varepsilon_2(\nu)\delta(\nu-v')}{\varepsilon_2(v')(\nu-v)} \\ &\quad - \int dv'\, F(v') \int d\nu''\, \pi^{-1} \frac{\varepsilon_2(\nu'')}{\nu''-v'} \frac{1}{\varepsilon_2(v')} \int d\nu\, \frac{\varepsilon_2(\nu)}{\nu-v} \delta(\nu-v') \\ &\quad + \int dv'\, d\nu\, \pi^{-1} F(v') \frac{\varepsilon_1(v)\,\varepsilon_1(v')}{\varepsilon_2(v')} \delta(\nu-v)\delta(\nu-v') \end{aligned}$$

We now apply the Poincaré–Bertrand theorem (A2.2.12) to the third term, in order to permute the integrations over $d\nu$ and $d\nu''$

$$-\pi^{-1}\int dv'd\nu d\nu'' F(v')\varepsilon_2(\nu'')\delta(\nu-v)\frac{1}{\nu''-\nu}\frac{1}{\nu-v'}$$

$$= -\pi^{-1}\int dv'\,d\nu''\,d\nu\,F(v')\varepsilon_2(\nu'')\delta(\nu-v)\frac{1}{\nu''-\nu}\frac{1}{\nu-v'}$$

$$+\pi\int dv'\,F(v')\varepsilon_2(v')\delta(v'-v)$$

Perform now all the integrations over the δ-functions:

$$J = \pi^{-1}\int dv'\,d\nu\,F(v')\frac{\varepsilon_2(\nu)}{(\nu-v)(\nu-v')} + \int dv'\,\frac{F(v')}{v-v'}$$

$$-\pi^{-1}\int dv'\,d\nu''\,F(v')\frac{\varepsilon_2(\nu'')}{(\nu''-v)(v-v')} + \pi F(v)\varepsilon_2(v) + \int dv'\,\frac{F(v')}{v'-v}$$

$$-\pi^{-1}\int dv'\,d\nu''\,F(v')\frac{\varepsilon_2(\nu'')}{(\nu''-v')(v'-v)} + \pi\frac{\varepsilon_1^2(v)}{\varepsilon_2(v)}\,F(v)$$

The second and fifth terms cancel; we combine the first, third and sixth terms and suppress the double dash in the latter two. We note that

$$\frac{1}{(\nu-v)(\nu-v')} - \frac{1}{(\nu-v)(v-v')} - \frac{1}{(\nu-v')(v'-v)} = 0$$

Hence we are left with:

$$J = \pi\left[\varepsilon_2(v) + \frac{\varepsilon_1^2(v)}{\varepsilon_2(v)}\right]F(v)$$

This result proves the initial formula and provides the normalization constant:

$$C_v = \pi\frac{\varepsilon_1^2(v)+\varepsilon_2^2(v)}{\varepsilon_2(v)} = \pi\frac{|\varepsilon_-(v)|^2}{\varepsilon_2(v)}$$

APPENDIX 4

Derivation of a General Type of Kinetic Equation

Suppose that in a problem one is led to choose for $\varphi(\mathbf{v}; t)$ all diagrams consisting of a succession of an arbitrary number of diagonal fragments of type 1, 2, . . , s, in any order (see Fig. A4.1).

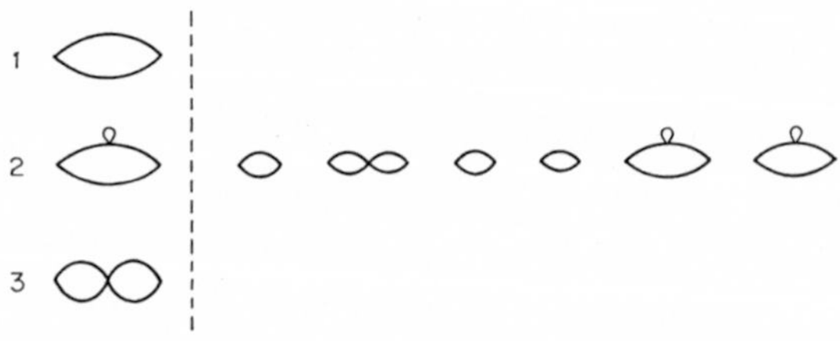

Fig. A4.1. In this example the diagrams contributing to $\rho_0(v; t)$ are supposed to be successions of an arbitrary number of the three diagonal fragments shown at left. One of these diagrams is drawn at right.

We shall give a short derivation of the equation of evolution in this approximation. In this type of general argument it is easier first to derive formally an equation for the N-body distribution function, and to reduce this equation afterwards. We then do not need to discuss semiconnections, as we did in the detailed derivation of § 30 [the contributions to $\rho_0(\mathbf{v}_1, \ldots, \mathbf{v}_N; t)$ come from disconnected diagonal fragments]. On the other hand, this type of derivation is less rigorous, because one is faced with a contradiction. The equations for $\rho_0(\mathbf{v}_1, \ldots, \mathbf{v}_N; t)$ do not have any meaning in the limit $N \to \infty$ (they diverge, as was shown in § 8); on the other hand, this limit must be taken in order that the sums over wave-vectors go over into Cauchy integrals. We shall thus take the limit formally: the contradiction is suppressed after reduction of the N-particle distribution.

Let the contribution of a single diagonal fragment of type ν to $\rho_0(\mathbf{v}_1, \ldots, \mathbf{v}_N; t)$ in the leading universal approximation be

$$\rho_0^{(\nu)}(v; t) = t\,\mathscr{D}_\nu(0)\rho_0(v; 0) \tag{A4.1}$$

$\mathscr{D}_\nu(z)$ is an operator acting on the velocities; as a function of z it is a certain Cauchy integral. The leading universal term is obtained as usual (see § 27) by taking the residue at the double pole $z = 0$ of the diagonal fragment, whence the occurrence of the factor $\mathscr{D}_\nu(0)$. We shall prove that the kinetic equation determining $\rho_0(v; t)$ in this approximation is

$$\partial_t \rho_0(v; t) = \sum_{\nu=1}^{s} \mathscr{D}_\nu(0)\rho_0(v; t) \tag{A4.2}$$

Let us classify all the diagrams into s classes, grouping in a given class (ν) all the diagrams whose leftmost diagonal fragment is of type ν. Remember that a term of order t^m is contributed by a succession of m diagonal fragments; hence the general term of order t^m is given by:

$$\begin{aligned}\rho_0(v; t)|_{t^m} = (t^m/m!)\{&\mathscr{D}_1(0)P_{m-1} \\ &+\mathscr{D}_2(0)P_{m-1} + \ldots + \mathscr{D}_s(0)P_{m-1}\}\rho_0(v; 0)\end{aligned} \tag{A4.3}$$

P_m is the sum of all possible products of m factors $\mathscr{D}_\nu(0)$ taken in

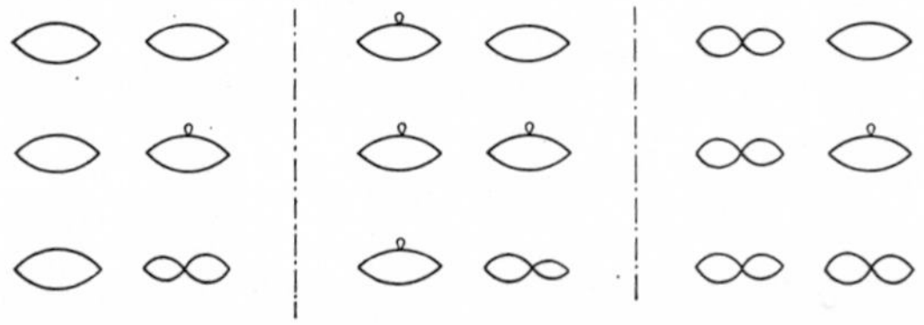

Fig. A4.2. The terms which define P_2 in the case of the example of Fig. A4.1. Note that for the general case of s fundamental diagonal fragments the sum of products P_m involves s^m terms.

any order (see an example in Fig. A4.2). Hence the complete $\rho_0(v; t)$ is

$$\rho_0(v; t) = \sum_{m=0}^{\infty} \sum_{\nu=1}^{s} (t^m/m!)\mathscr{D}_\nu(0)P_{m-1}\rho_0(v; 0) \tag{A4.4}$$

Taking the time derivative of this expression we obtain

$$\partial_t \rho_0(v; t) = \sum_{\nu=1}^{s} \mathscr{D}_\nu(0) \sum_{m=1}^{\infty} [t^{m-1}/(m-1)!]P_{m-1}\rho_0(v; 0) \tag{A4.5}$$

It is, however, easily seen from the structure of the sum of products P_m that

$$\sum_{m=0}^{\infty} (t^m/m!)P_m \rho_0(v; 0) = \rho_0(v; t) \tag{A4.6}$$

Substituting this into eq. (A4.5) we obtain eq. (A4.2) and have proven the desired property. Integrating this equation over all velocities but one, the kinetic equation for $\varphi(\mathbf{v}_\alpha; t)$ is readily derived.

APPENDIX 5

Miscellaneous Theorems Concerning Normal Products of Creation and Destruction Operators

A5.1. Sums of Normal Products

In defining reduced distribution functions, the problem arises of performing integrations (or summations) over momenta or positions of normal products of creation or destruction operators. The only rules of integration which are given *a priori* concern the normal product of one creation and one destruction operator:

$$\sum_{\mathbf{k}} a^+(\hbar\mathbf{k})a(\hbar\mathbf{k}) = N \tag{A5.1.1}$$

$$\int d\mathbf{x}\,\psi^+(\mathbf{x})\psi(\mathbf{x}) = N \tag{A5.1.2}$$

We shall now show that these rules, together with the commutation relations (61.3) and (61.11), imply the following generalized theorems:

$$\begin{aligned}&\sum_{\mathbf{k}_{s+1}} \cdot\cdot \sum_{\mathbf{k}_r} a^+(\hbar\mathbf{k}_1)\ldots a^+(\hbar\mathbf{k}_s)\\ &a^+(\hbar\mathbf{k}_{s+1})\ldots a^+(\hbar\mathbf{k}_r)a(\hbar\mathbf{k}_r)\ldots a(\hbar\mathbf{k}_{s+1})a(\hbar\mathbf{k}_s)\ldots a(\hbar\mathbf{k}_1)\\ &\quad=\frac{(N-s)!}{(N-r)!}\,a^+(\hbar\mathbf{k}_1)\ldots a^+(\hbar\mathbf{k}_s)a(\hbar\mathbf{k}_s)\ldots a(\hbar\mathbf{k}_1), \qquad s \leqq r \leqq N\end{aligned} \tag{A5.1.3}$$

$$\begin{aligned}&\int d\mathbf{x}_{s+1}\ldots d\mathbf{x}_r\,\psi^+(\mathbf{x}_1)\ldots\psi^+(\mathbf{x}_s)\\ &\quad\psi^+(\mathbf{x}_{s+1})\ldots\psi^+(\mathbf{x}_r)\psi(\mathbf{x}_r)\ldots\psi(\mathbf{x}_{s+1})\psi(\mathbf{x}_s)\ldots\psi(\mathbf{x}_1)\\ &=\frac{(N-s)!}{(N-r)!}\,\psi^+(\mathbf{x}_1)\ldots\psi^+(\mathbf{x}_s)\psi(\mathbf{x}_s)\ldots\psi(\mathbf{x}_1), \qquad s \leqq r \leqq N\end{aligned} \tag{A5.1.4}$$

We shall give the explicit proof of theorem (A5.1.3), using the abbreviations:

$$a_s^+ \equiv a^+(\hbar\mathbf{k}_s)$$

$$\delta_{s,s'} = \delta_{\mathbf{k}_s-\mathbf{k}_{s'}}$$

The theorem is proved by induction. Suppose it is valid for given

r and $s = r, r-1, \ldots s$; we prove that it holds for $s-1$. Then

$$J_{(s-1)} \equiv \sum_{k_s} \sum_{k_{s+1}} \cdots \sum_{k_r} a_1^+ \ldots a_s^+ a_{s+1}^+ \ldots a_r^+ a_r \ldots a_{s+1} a_s \ldots a_1$$
$$= \frac{(N-s)!}{(N-r)!} \sum_{k_s} a_1^+ \ldots a_s^+ a_s \ldots a_1$$

In order to perform the integration, we bring the operator $a_s^+ a_s$ to the left of the product, in order to isolate an expression of the form (A5.1.1). Using (61.3) we obtain successively:

$$J_{(s-1)} = \theta^{s-1} \frac{(N-s)!}{(N-r)!} \sum_{k_s} a_s^+ a_1^+ \ldots a_{s-1}^+ a_s a_{s-1} \ldots a_1$$
$$= \frac{(N-s)!}{(N-r)!} \theta^{s-1} \sum_{k_s} \theta^{s-1} \Big\{ a_s^+ a_s a_1^+ \ldots a_{s-1}^+ a_{s-1} \ldots a_1 - \sum_{j=1}^{s-1} \theta^{j-1} \delta_{s,j} a_s^+ a_1^+ \underset{(j)}{\ldots}{}' \, a_{s-1}^+ a_{s-1} \ldots a_1 \Big\}$$

where the dash means that the factor a_j^+ is suppressed from the normal product. The summation over k_s can now be performed:

$$J_{(s-1)} = \theta^{2(s-1)} \frac{(N-s)!}{(N-r)!} \Big\{ N a_1^+ \ldots a_{s-1}^+ a_{s-1} \ldots a_1 - \sum_{j=1}^{s-1} \theta^{j-1} a_j^+ a_1^+ \underset{(j)}{\ldots}{}' \, a_{s-1}^+ a_{s-1} \ldots a_1 \Big\}$$
$$= \frac{(N-s)!}{(N-r)!} [N-(s-1)] a_1^+ \ldots a_{s-1}^+ a_{s-1} \ldots a_1$$
$$= \frac{(N-s+1)!}{(N-r)!} : \prod_{n=1}^{s-1} a_n^+ a_n : \qquad \text{Q.E.D.}$$

One proves in the same way that if the theorem is valid for given s and r it remains valid when r is augmented by 1; the individual case $s = r = 1$ is trivial.

A5.2. A Certain Type of Commutator

We shall prove the following theorem, which is necessary for the proof of the unperturbed quantum Liouville equation [the notations are defined in eqs. (63.1–5)]:

$$C_N \equiv [\hbar^2 \sum_{\mathbf{l}} l^2 a^+(\hbar\mathbf{l})a(\hbar\mathbf{l}), \Pi_N]_- = -\sum_{j=1}^{N} (2\hbar\mathbf{k}_j \cdot \mathbf{p}_j)\Pi_N \qquad \text{(A5.2.1)}$$

Suppose the theorem is true for N particles, we show that it remains true for $N+1$ particles.

$$\begin{aligned}
C_{N+1} = &\Big\{\sum_{\mathbf{l}} \hbar^2 l^2 a^+(\hbar\mathbf{l})a(\hbar\mathbf{l})a^+(\mathbf{p}_{N+1}-\tfrac{1}{2}\hbar\mathbf{k}_{N+1})\Pi_N a(\mathbf{p}_{N+1}+\tfrac{1}{2}\hbar\mathbf{k}_{N+1}) \\
&-\sum_{\mathbf{l}} \hbar^2 l^2 a^+(\mathbf{p}_{N+1}-\tfrac{1}{2}\hbar\mathbf{k}_{N+1})\Pi_N a(\mathbf{p}_{N+1}+\tfrac{1}{2}\hbar\mathbf{k}_{N+1})a^+(\hbar\mathbf{l})a(\hbar\mathbf{l})\Big\} \\
= &\sum_{\mathbf{l}} \{\hbar^2 l^2 \delta_{\hbar\mathbf{l}-\mathbf{p}_{N+1}+\frac{1}{2}\hbar\mathbf{k}_{N+1}}\, a^+(\hbar\mathbf{l})\Pi_N a(\mathbf{p}_{N+1}+\tfrac{1}{2}\hbar\mathbf{k}_{N+1}) \\
&+\theta\hbar^2 l^2 a^+(\hbar\mathbf{l})a^+(\mathbf{p}_{N+1}-\tfrac{1}{2}\hbar\mathbf{k}_{N+1})a(\hbar\mathbf{l})\Pi_N a(\mathbf{p}_{N+1}+\tfrac{1}{2}\hbar\mathbf{k}_{N+1}) \\
&-\hbar^2 l^2 \delta_{\hbar\mathbf{l}-\mathbf{p}_{N+1}-\frac{1}{2}\hbar\mathbf{k}_{N+1}}\, a^+(\mathbf{p}_{N+1}-\tfrac{1}{2}\hbar\mathbf{k}_{N+1})\Pi_N a(\hbar\mathbf{l}) \\
&-\theta\hbar^2 l^2 a^+(\mathbf{p}_{N+1}-\tfrac{1}{2}\hbar\mathbf{k}_{N+1})\Pi_N a^+(\hbar\mathbf{l})a(\mathbf{p}_{N+1}+\tfrac{1}{2}\hbar\mathbf{k}_{N+1})a(\hbar\mathbf{l})\} \\
= &[(\mathbf{p}_{N+1}-\tfrac{1}{2}\hbar\mathbf{k}_{N+1})^2-(\mathbf{p}_{N+1}+\tfrac{1}{2}\hbar\mathbf{k}_{N+1})^2]\Pi_{N+1} \\
&-\theta^2 \sum_{j=1}^{N} (2\hbar\mathbf{k}_j \cdot \mathbf{p}_j)\Pi_{N+1} \\
= &-2\sum_{j=1}^{N+1} (\hbar\mathbf{k}_j \cdot \mathbf{p}_j)\Pi_{N+1} \qquad \text{Q.E.D.}
\end{aligned}$$

The theorem is obviously true for $N = 1$.

A5.3. Proof of Theorem (63.8)

We need the following lemma:

$$J_N \equiv a^+(q-l)a(q)\Pi_N = a^+(q-l)\Pi_N a(q) + \sum_{j=1}^{N} P_N(j-)\delta_{q-p_j+k_j/2} \qquad \text{(A5.3.1)}$$

(We have taken here $\hbar = 1$ in order to simplify the notations.) Assuming its truth for N particles, we obtain for $N+1$:

$$\begin{aligned}
J_{N+1} &= a^+(q-l)a(q)a^+(p_{N+1}-\tfrac{1}{2}k_{N+1})\Pi_N a(p_{N+1}+\tfrac{1}{2}k_{N+1}) \\
&= a^+(q-l)\Pi_N a(p_{N+1}+\tfrac{1}{2}k_{N+1})\delta_{q-p_{N+1}+\frac{1}{2}k_{N+1}} \\
&\quad +\theta a^+(q-l)a^+(p_{N+1}-\tfrac{1}{2}k_{N+1})a(q)\Pi_N a(p_{N+1}+\tfrac{1}{2}k_{N+1}) \\
&= a^+(p_{N+1}-\tfrac{1}{2}k_{N+1}-l)\Pi_N a(p_{N+1}+\tfrac{1}{2}k_{N+1})\delta_{q-p_{N+1}+\frac{1}{2}k_{N+1}} \\
&\quad +\theta^2 a^+(p_{N+1}-\tfrac{1}{2}k_{N+1})a^+(q-l)a(q)\Pi_N a(p_{N+1}+\tfrac{1}{2}k_{N+1})
\end{aligned}$$

$$= P_{N+1}(N+1-)\delta_{q-p_{N+1}+\frac{1}{2}k_{N+1}}$$
$$+a^{+}(p_{N+1}-\tfrac{1}{2}k_{N+1})a^{+}(q-l)\Pi_{N}\,a(q)a(p_{N+1}+\tfrac{1}{2}k_{N+1})$$
$$+\sum_{j=1}^{N} a^{+}(p_{N+1}-\tfrac{1}{2}k_{N+1})P_{N}(j-)a(p_{N+1}+\tfrac{1}{2}k_{N+1})\delta_{q-p_j+k_j/2}$$
$$=\sum_{j=1}^{N+1} P_{N+1}(j-)\delta_{q-p_j+k_j/2}+a^{+}(q-l)\Pi_{N+1}a(q) \qquad \text{Q.E.D.}$$

The theorem is easily proved for $N = 1$.

Theorem (63.8) is proved in exactly the same way, using the result (A5.3.1).

APPENDIX 6

Derivation of the Quantum-Statistical Liouville Equation

As an example illustrating the rules of derivation of eq. (63.14) from eq. (63.13), we consider the simpler example of the equation for $\partial_t \hat{\rho}_{\mathbf{k}_1, \mathbf{k}_2}$, which contains all the characteristic features of the calculation. In order to simplify the notation, we set $\hbar = 1$.

Equation (63.13) reduces explicitly in this case to:

$$\begin{aligned}
&\partial_t \hat{\rho}_{\mathbf{k}_1, \mathbf{k}_2}(\mathbf{p}_1, \mathbf{p}_2| \ldots) + i(\mathbf{k}_1 \cdot \mathbf{v}_1 + \mathbf{k}_2 \cdot \mathbf{v}_2) \rho_{\mathbf{k}_1, \mathbf{k}_2}(\mathbf{p}_1, \mathbf{p}_2| \ldots) \\
&= i(8\pi^3/\Omega) \sum_{\mathbf{l}} \{[V_l + \theta V_{|\mathbf{p}_1 - \mathbf{p}_2 + \mathbf{l} + (\mathbf{k}_2 - \mathbf{k}_1)/2|}] \\
&\qquad \rho_{\mathbf{k}_1 - \mathbf{l}, \mathbf{k}_2 + \mathbf{l}}(\mathbf{p}_1 + \tfrac{1}{2}\mathbf{l}, \mathbf{p}_2 - \tfrac{1}{2}\mathbf{l}| \ldots) - \ldots \\
&\quad + \sum_{n>2} [V_l + \theta V_{|\mathbf{p}_1 - \mathbf{p}_n + \mathbf{l} - \mathbf{k}_1/2|}] (8\pi^3/\Omega) \\
&\qquad \rho_{\mathbf{k}_1 - \mathbf{l}, \mathbf{k}_2, \mathbf{l}}(\mathbf{p}_1 + \tfrac{1}{2}\mathbf{l}, \mathbf{p}_2, \mathbf{p}_n - \tfrac{1}{2}\mathbf{l}| \ldots) - \ldots \\
&\quad + \sum_{n>2} [V_l + \theta V_{|\mathbf{p}_2 - \mathbf{p}_n + \mathbf{l} - \mathbf{k}_2/2|}] (8\pi^3/\Omega) \\
&\qquad \rho_{\mathbf{k}_1, \mathbf{k}_2 - \mathbf{l}, \mathbf{l}}(\mathbf{p}_1, \mathbf{p}_2 + \tfrac{1}{2}\mathbf{l}, \mathbf{p}_n - \tfrac{1}{2}\mathbf{l}| \ldots) - \ldots \\
&\quad + \sum_{j<n}\sum [V_l + \theta V_{|\mathbf{p}_j - \mathbf{p}_n + \mathbf{l}|}] (8\pi^3/\Omega) \\
&\qquad \rho_{\mathbf{k}_1, \mathbf{k}_2, -\mathbf{l}, \mathbf{l}}(\mathbf{p}_1, \mathbf{p}_2, \mathbf{p}_j + \tfrac{1}{2}\mathbf{l}, \mathbf{p}_n - \tfrac{1}{2}\mathbf{l}| \ldots) - \ldots\}
\end{aligned} \tag{A6.1}$$

We wrote explicitly only one term of the commutator in each term of the r.h.s. Contracting completely the l.h.s. we get:

$$\begin{aligned}
&[\partial_t + i(\mathbf{k}_1 \cdot \mathbf{v}_1 + \mathbf{k}_2 \cdot \mathbf{v}_2)][\hat{\rho}_{\mathbf{k}_1, \mathbf{k}_2}(\mathbf{p}_1, \mathbf{p}_2| \ldots) \\
&+ \theta(8\pi^3/\Omega)\delta_{\mathbf{p}_1 - \mathbf{p}_2 + (\mathbf{k}_1 + \mathbf{k}_2)/2}\, \hat{\rho}_{\mathbf{k}_1 + \mathbf{k}_2}(\mathbf{p}_2 - \tfrac{1}{2}\mathbf{k}_1 | \mathbf{p}_2 - \tfrac{1}{2}\mathbf{k}_2, \ldots) \\
&+ \theta(8\pi^3/\Omega)\delta_{\mathbf{p}_2 - \mathbf{p}_1 + (\mathbf{k}_1 + \mathbf{k}_2)/2}\, \hat{\rho}_{\mathbf{k}_1 + \mathbf{k}_2}(\mathbf{p}_1 - \tfrac{1}{2}\mathbf{k}_2 | \mathbf{p}_1 - \tfrac{1}{2}\mathbf{k}_1, \ldots)]
\end{aligned} \tag{A6.2}$$

We are not interested in the terms containing θ; therefore all terms containing $\delta_{\mathbf{p}_1 - \mathbf{p}_2 + (\mathbf{k}_1 + \mathbf{k}_2)/2}$ or $\delta_{\mathbf{p}_2 - \mathbf{p}_1 + (\mathbf{k}_1 + \mathbf{k}_2)/2}$ on the r.h.s. have to be discarded. We now evaluate all the contractions of the terms occurring on the r.h.s. The first term leads to

$$\begin{aligned}
&i(8\pi^3/\Omega) \sum_{\mathbf{l}} [V_l + \theta V_{|\mathbf{p}_1 - \mathbf{p}_2 + \mathbf{l} + (\mathbf{k}_2 - \mathbf{k}_1)/2|}]\, \hat{\rho}_{\mathbf{k}_1 - \mathbf{l}, \mathbf{k}_2 + \mathbf{l}}(\mathbf{p}_1 + \tfrac{1}{2}\mathbf{l}, \mathbf{p}_2 - \tfrac{1}{2}\mathbf{l}| \ldots) \\
&+ i\theta \sum_{\mathbf{l}} \delta_{\mathbf{p}_2 - \mathbf{p}_1 - \mathbf{l} + (\mathbf{k}_1 + \mathbf{k}_2)/2} [V_{|\mathbf{p}_2 - \mathbf{p}_1 + (\mathbf{k}_1 - \mathbf{k}_2)/2|} + \theta V_{k_2}] \\
&\qquad \hat{\rho}_{\mathbf{k}_1 + \mathbf{k}_2}(\mathbf{p}_1 - \tfrac{1}{2}\mathbf{k}_2 | \mathbf{p}_2 + \tfrac{1}{2}\mathbf{k}_2, \ldots) \\
&+ i\theta \sum_{\mathbf{l}} \delta_{\mathbf{p}_1 - \mathbf{p}_2 + \mathbf{l} + (\mathbf{k}_1 + \mathbf{k}_2)/2} [V_{|\mathbf{p}_2 - \mathbf{p}_1 - (\mathbf{k}_1 + \mathbf{k}_2)/2|} + \theta V_{k_1}] \\
&\qquad \hat{\rho}_{\mathbf{k}_1 + \mathbf{k}_2}(\mathbf{p}_2 - \tfrac{1}{2}\mathbf{k}_1 | \mathbf{p}_1 + \tfrac{1}{2}\mathbf{k}_1, \ldots)
\end{aligned} \tag{A.6.3}$$

In the two last terms the factor θ is introduced into the square brackets, leading e.g. to $[\theta V_{|\mathbf{p}_2-\mathbf{p}_1-(\mathbf{k}_1+\mathbf{k}_2)/2|}+\theta^2 V_{k_1}]$ and one notes that $\theta^2 = 1$ for both quantum statistics.

The second term of (A6.1) leads to the following contractions

$$
\begin{aligned}
& i(8\pi^3/\Omega)^2 \sum_{\mathbf{l}}\sum_{n} [V_l+\theta V_{|\mathbf{p}_1-\mathbf{p}_n+\mathbf{l}-\mathbf{k}_1/2|}]\hat{\rho}_{\mathbf{k}_1-\mathbf{l},\mathbf{k}_2,\mathbf{l}}(\mathbf{p}_1+\tfrac{1}{2}\mathbf{l},\ \mathbf{p}_2,\ \mathbf{p}_n-\tfrac{1}{2}\mathbf{l}|\ldots) \\
& +i\theta(8\pi^3/\Omega) \sum_{\mathbf{l}}\sum_{n} [V_{|\mathbf{p}_2-\mathbf{p}_1+(\mathbf{k}_1+\mathbf{k}_2)/2|}+\theta V_{|\mathbf{p}_2-\mathbf{p}_n+\mathbf{k}_2/2|}]\delta_{\mathbf{p}_2-\mathbf{p}_1-\mathbf{l}+(\mathbf{k}_1+\mathbf{k}_2)/2} \\
& \qquad \hat{\rho}_{\mathbf{k}_1+\mathbf{k}_2-\mathbf{l},\mathbf{l}}(\mathbf{p}_1+\tfrac{1}{2}\mathbf{l}-\tfrac{1}{2}\mathbf{k}_2,\ \mathbf{p}_n-\tfrac{1}{2}\mathbf{l}|\mathbf{p}_2+\tfrac{1}{2}\mathbf{k}_2,\ldots) \\
& +i\theta(8\pi^3/\Omega) \sum_{n} [V_{|\mathbf{p}_1-\mathbf{p}_n-\mathbf{k}_1/2|}+\theta V_0]\hat{\rho}_{\mathbf{k}_1,\mathbf{k}_2}(\mathbf{p}_1,\ \mathbf{p}_2|\mathbf{p}_n,\ldots) \\
& +i\theta(8\pi^3/\Omega) \sum_{n} [V_{|\mathbf{p}_1-\mathbf{p}_n+\mathbf{k}_1/2|}+\theta V_{k_1}]\hat{\rho}_{\mathbf{k}_1,\mathbf{k}_2}(\mathbf{p}_n-\tfrac{1}{2}\mathbf{k}_1,\ \mathbf{p}_2|\mathbf{p}_1+\tfrac{1}{2}\mathbf{k}_1,\ldots) \\
& +i\theta(8\pi^3/\Omega) \sum_{n}\sum_{\mathbf{l}} [V_l+\theta V_{|\mathbf{p}_1-\mathbf{p}_n+\mathbf{l}-\mathbf{k}_1/2|}]\delta_{\mathbf{p}_n-\mathbf{p}_2+\mathbf{k}_2/2} \\
& \qquad \hat{\rho}_{\mathbf{k}_1-\mathbf{l},\mathbf{k}_2+\mathbf{l}}(\mathbf{p}_1+\tfrac{1}{2}\mathbf{l},\ \mathbf{p}_2-\tfrac{1}{2}\mathbf{l}|\mathbf{p}_n,\ldots) \\
& +i\theta(8\pi^3/\Omega) \sum_{n}\sum_{\mathbf{l}} [V_l+\theta V_{|\mathbf{p}_1-\mathbf{p}_n+\mathbf{l}-\mathbf{k}_1/2|}]\delta_{\mathbf{p}_2-\mathbf{p}_n+\mathbf{l}+\mathbf{k}_1/2} \qquad \text{(A.6.4)} \\
& \qquad \hat{\rho}_{\mathbf{k}_1-\mathbf{l},\mathbf{k}_2+\mathbf{l}}(\mathbf{p}_1+\tfrac{1}{2}\mathbf{l},\ \mathbf{p}_2+\tfrac{1}{2}\mathbf{l}|\mathbf{p}_n-\mathbf{l},\ldots) \\
& +2i\theta^2 \sum_{n} [V_{|\mathbf{p}_2-\mathbf{p}_1+(\mathbf{k}_1+\mathbf{k}_2)/2|}+\theta V_{k_1}]\delta_{\mathbf{p}_2-\mathbf{p}_n+\mathbf{k}_1+\mathbf{k}_2/2} \\
& \qquad \hat{\rho}_{\mathbf{k}_1+\mathbf{k}_2}(\mathbf{p}_2+\tfrac{1}{2}\mathbf{k}_1|\mathbf{p}_1+\tfrac{1}{2}\mathbf{k}_1,\ \mathbf{p}_n-\mathbf{k}_1,\ldots) \\
& +2i\theta^2 \sum_{n} [V_{|\mathbf{p}_1-\mathbf{p}_2+(\mathbf{k}_1+\mathbf{k}_2)/2|}+\theta V_{k_1}]\delta_{\mathbf{p}_2-\mathbf{p}_n-\mathbf{k}_2/2} \\
& \qquad \hat{\rho}_{\mathbf{k}_1+\mathbf{k}_2}(\mathbf{p}_2-\tfrac{1}{2}\mathbf{k}_1|\mathbf{p}_1+\tfrac{1}{2}\mathbf{k}_1,\ \mathbf{p}_n,\ldots) \\
& +2i\theta^2 \sum_{n} [V_{|\mathbf{p}_2-\mathbf{p}_1+(\mathbf{k}_1+\mathbf{k}_2)/2|}+\theta V_{k_2}]\delta_{\mathbf{p}_n-\mathbf{p}_2+\mathbf{k}_2/2} \\
& \qquad \hat{\rho}_{\mathbf{k}_1+\mathbf{k}_2}(\mathbf{p}_1-\tfrac{1}{2}\mathbf{k}_2|\mathbf{p}_2+\tfrac{1}{2}\mathbf{k}_2,\ \mathbf{p}_n,\ldots) \\
& + \text{terms containing } \delta_{\mathbf{p}_1-\mathbf{p}_2\pm(\mathbf{k}_1+\mathbf{k}_2)/2}
\end{aligned}
$$

Note that all terms obtained from the second contractions ($\sim \theta^2$) appear twice; they must be divided by 2 because each contraction must be counted only once.

The remaining terms of (A6.1) are treated in the same fashion. Finally all the terms are grouped according to the nature of the Fourier components appearing on the r.h.s.: one thus obtains all matrix elements of the type $\langle \mathbf{k}_1\mathbf{k}_2|\mathscr{L}'|\ldots\rangle$. For instance, the three terms of (A6.3) are respectively contributions to the vertices E, B, B (Table 63.1); the terms of (A6.4) contribute respectively to vertices D, H, G, F, E, E, B, B, B.

APPENDIX 7

Proof of Equation (81.1)

The density matrix in equilibrium is well known:

$$\boldsymbol{\rho}^{\mathrm{eq}} = \exp\left[-\beta\mathscr{H}\right]/\mathrm{Tr}\exp\left(-\beta\mathscr{H}\right) \tag{A7.1}$$

Applying eq. (61.20), the average energy per particle is:

$$\begin{aligned} E &= \mathrm{Tr}\{\mathscr{H}\exp\left(-\beta\mathscr{H}\right)\}/N\,\mathrm{Tr}\exp\left(-\beta\mathscr{H}\right) \\ &= -N^{-1}(\partial/\partial\beta)\ln\mathrm{Tr}\exp[-\beta\mathscr{H}_0-\beta e^2\mathscr{V}] \end{aligned} \tag{A7.2}$$

Differentiate this expression with respect to e^2:

$$\begin{aligned} \frac{\partial E}{\partial e^2} &= -\frac{\partial^2}{\partial\beta\,\partial e^2}N^{-1}\ln\mathrm{Tr}\exp\left(-\beta\mathscr{H}_0-\beta e^2\mathscr{V}\right) \\ &= \frac{\partial}{\partial\beta}\beta\frac{\mathrm{Tr}\{\mathscr{V}\exp\left(-\beta\mathscr{H}\right)\}}{N\,\mathrm{Tr}\exp\left(-\beta\mathscr{H}\right)} \end{aligned} \tag{A7.3}$$

Substituting $\mathscr{V}$ from eq. (61.12) we obtain:

$$\frac{\partial E}{\partial e^2} = \frac{1}{N}\frac{\partial}{\partial\beta}\beta\frac{1}{2}\int d\mathbf{x}\,d\mathbf{x}'\,V(\mathbf{x}-\mathbf{x}')\frac{\mathrm{Tr}[e^{-\beta\mathscr{H}}\boldsymbol{\psi}^+(\mathbf{x})\boldsymbol{\psi}^+(\mathbf{x}')\boldsymbol{\psi}(\mathbf{x}')\boldsymbol{\psi}(\mathbf{x})]}{\mathrm{Tr}\exp\left(-\beta\mathscr{H}\right)} \tag{A7.4}$$

But the two-body distribution function integrated over the momenta, $n_2(\mathbf{x}, \mathbf{x}')$, defined by eq. (1.17), is easily calculated from eq. (61.28) by using the theorems of Appendix 5, with the result:

$$n_2(\mathbf{x}, \mathbf{x}') = \mathrm{Tr}\,\{\boldsymbol{\rho}\boldsymbol{\psi}^+(\mathbf{x})\boldsymbol{\psi}^+(\mathbf{x}')\boldsymbol{\psi}(\mathbf{x}')\boldsymbol{\psi}(\mathbf{x})\} \tag{A7.5}$$

Hence eq. (A7.4) can be rewritten as

$$\begin{aligned} \frac{\partial E}{\partial e^2} &= \frac{1}{N}\frac{\partial}{\partial\beta}\beta\frac{1}{2}\int d\mathbf{x}\,d\mathbf{x}'\,V(|\mathbf{x}-\mathbf{x}'|)n_2^{\mathrm{eq}}(|\mathbf{x}-\mathbf{x}'|;\beta, e^2) \\ &= \frac{\Omega}{N}\frac{1}{2}\int d\mathbf{r}\,V(r)\frac{\partial}{\partial\beta}\beta n_2^{\mathrm{eq}}(r;\beta, e^2) \end{aligned} \tag{A7.6}$$

We now integrate eq. (A7.6) with respect to e^2 (which we rename η) and fix the integration constant in such a way that for $e^2 \to 0$

we recover the internal energy of a perfect Fermi gas of the same temperature, E_F:

$$E = E_F + \tfrac{1}{2}c^{-1} \int_0^{e^2} d\eta \int d\mathbf{r}\, V(r) \frac{\partial}{\partial \beta} \beta n_2^{\mathrm{eq}}(r; \beta, \eta) \qquad \text{(A7.7)}$$

This formula is identical with (81.1).

APPENDIX 8

Non-markoffian Kinetic Equation for a Stable Classical Plasma

In the main part of this book, the various types of kinetic equations describing the evolution of the velocity distribution have been derived by making use of a "long-time approximation". This approximation has been defined and discussed in detail in Chapter 6. There are, however, some problems for which this approximation is insufficient. One of these, which is of great importance, is the study of the behavior of a plasma in an oscillating electromagnetic field of very high frequency $\omega = O(\omega_p)$. In this case, the approximation $t \gg \omega_p^{-1}$ wipes out all the interesting details of the evolution process. Another more academic but also very important problem is the more rigorous justification of the occurrence of the effective short-time scale ω_p^{-1} in the statistical mechanics of charged particles.

For all these problems, we need a kinetic equation valid over the whole range of times. On the other hand, we want to retain our criteria of choice of diagrams based on the e^2- and c-dependence. We therefore retain to each order of approximation the same diagrams as in the long-time study, but evaluate the time dependence of the diagrams exactly. In particular, we shall now derive the kinetic equation for a homogeneous plasma within the ring approximation, without making the "leading universal term" approximation.

We start from the general equation (5.26). We are interested in the velocity distribution at time t, i.e. we take $k = 0$; moreover within the ring approximation, we retain only the contribution of $\rho_0(|v;0)$ hence eq. (5.26) becomes:

$$\rho_0(|v;t) = (2\pi)^{-1} \sum_{n=0}^{\infty} \int_C dz\, \mathrm{e}^{-izt}(-e^2)^n \langle 0|\, \mathscr{R}^0(z)[\mathscr{L}'\, \mathscr{R}^0(z)]^n |0\rangle \rho_0(|v;0) \tag{A8.1}$$

where the resolvent matrix elements contain only the successions of rings defined in Chapter 9. To be more explicit, let us introduce the notation:

$$\mathfrak{R}(z) = \sum_{\substack{n=1 \\ \text{(all rings)}}}^{\infty} (-e^2)^{n+1} \langle 0 | \mathscr{L}' [\mathscr{R}^0(z) \mathscr{L}']^n | 0 \rangle \tag{A8.2}$$

where the summation is carried out over all the rings. Then (A8.1) can be rewritten as [see also eq. (30.3)]:

$$\begin{aligned} \rho_0(|v; t) &= (2\pi)^{-1} \sum_{m=0}^{\infty} \int dz \, \mathrm{e}^{-izt} \frac{1}{-iz} \left\{ \mathfrak{R}(z) \frac{1}{-iz} \right\}^m \rho_0(|v; 0) \\ &\equiv (2\pi)^{-1} \int dz \, \mathrm{e}^{-izt} \tilde{\rho}_0(|v; z) \end{aligned} \tag{A8.3}$$

By the last equality, the Laplace transform $\tilde{\rho}_0(|v; z)$ of the velocity distribution is defined as an expansion in rings. We now take the time derivative of both sides of (A8.3) and integrate the result over all velocities but $\mathbf{v}_\alpha$ to obtain an equation for the reduced distribution function $\varphi(\mathbf{v}_\alpha; t) \equiv \varphi(\alpha; t)$

$$\partial_t \varphi(\mathbf{v}_\alpha; t) = (2\pi)^{-1} \int_{(\alpha)} (d\mathbf{v})^{N-1} \int_C dz \, \mathrm{e}^{-izt} \, \mathfrak{R}(z) \tilde{\rho}_0(|v; z) \tag{A8.4}$$

At this point, we substitute for $\tilde{\rho}_0(|v; z)$ its Laplace transform in order to come back to the time-dependent velocity distribution:

$$\begin{aligned} \partial_t \varphi(\alpha; t) &= (2\pi)^{-1} \int_C dz \, \mathrm{e}^{-izt} \int_{(\alpha)} (d\mathbf{v})^{N-1} \mathfrak{R}(z) \int_0^\infty d\tau \, \mathrm{e}^{iz\tau} \rho_0(|v; \tau) \\ &= (2\pi)^{-1} \int_0^t d\tau \int_C dz \int_{(\alpha)} (d\mathbf{v})^{N-1} \, \mathrm{e}^{-iz(t-\tau)} \mathfrak{R}(z) \rho_0(|v; \tau) \end{aligned} \tag{A8.5}$$

In going from the first to the second form, use was made of the property

$$\int_C dz \, \mathrm{e}^{-iz(t-\tau)} \, \mathfrak{R}(z) = 0 \qquad \text{for} \quad t - \tau < 0$$

which is an expression of the causality condition (4.9). The relative order of the z- and τ-integrations has been changed which can be shown to be permissible.

The transformation leading to (A8.5) is an important step, for the following reason. If $\tilde{\varphi}_s\,(1, \ldots, s; z)$ is the Laplace transform of the reduced s-particle velocity distribution, this function is *not factorized* into s one-particle velocity distributions. Indeed, the factorization theorem (2.29) implies that $\tilde{\varphi}_s(1, \ldots, s; z)$ is a *convolution* of s one-particle functions $\tilde{\varphi}(j; z)$. But the factorization theorem (2.29) is crucial to the summation procedure of Chapter 9. By making the partial Laplace transformation leading to (A8.5) we recover an equation in terms of $\rho_0(|v; \tau)$ in which the factorization theorem can again be applied. This equation can be rewritten as follows by making a change of variables from t to $t-\tau$:

$$\partial_t \varphi(\alpha; t) = \int_0^t d\tau\, (2\pi)^{-1} \int_C dz\, e^{-iz\tau} \int_{(\alpha)} (d\mathbf{v})^{N-1} \Re(z) \rho_0(|v; t-\tau) \quad \text{(A8.6)}$$

This equation is a typical *non-markoffian equation*. The evolution of the distribution function at time t depends on the whole past history of the system, from time zero up to the present time t.

This non-markoffian character (i.e. the memory of the past) is characteristic of any general kinetic equation valid for short times.* The markoffian limiting case is obtained by neglecting the change in ρ_0 during the effective duration of the memory. This means that one replaces $\rho_0(|v; t-\tau)$ by $\rho_0(|v; t)$ on the r.h.s. of (A8.6). The following result is then easily obtained

$$\begin{aligned}\partial_t \varphi(\alpha; t) &= \int_0^t d\tau (2\pi)^{-1} \int_C dz\, e^{-iz\tau} \int_{(\alpha)} (d\mathbf{v})^{N-1} \Re(z) \rho_0(|v; t) \\ &= \int (2\pi)^{-1} \int_C dz \frac{e^{-izt} - 1}{-iz} \int_{(\alpha)} (d\mathbf{v})^{N-1} \Re(z) \rho_0(|v; t) \qquad \text{(A8.7)}\end{aligned}$$

It follows from the definition (A8.2) and the discussion of Chapter 6 that $\Re(z)$ has singularities only in the lower half-plane, at an average distance ω_p. It then follows that the second term in the expression $(e^{-izt}-1)$ gives a vanishing contribution; indeed, as it contains no exponential, the contour C can be closed in the upper half-plane where the integrand is regular. Also, for times much longer than ω_p^{-1}, the contribution of these singularities can be neglected and the z-integral can be replaced by the residue of the integrand at the pole $z = 0$ (or, more precisely, $0+$)

* I. Prigogine and P. Résibois, *Physica,* **27**, 629 (1961).

$$\partial_t \varphi(\alpha; t) = \int_{(\alpha)} (d\mathbf{v})^{N-1} \Re(0+) \rho_0(|v; t) \tag{A8.8}$$

This equation is exactly the ring equation (39.7) in the leading universal term approximation.

We now go over to the explicit summation of the "*non-markoffian rings*". We note that the summation procedure of § 40 depends only on the topological structure of the rings and not on the precise form of the symbols d_j, d_{jn}, and δ_-^{jn}. Hence the summation method for (A8.8) can be taken over with only trivial changes for the more general expression:

$$J(z) = \int_{(\alpha)} (d\mathbf{v})^{N-1} \Re(z) \rho_0(|v; t-\tau)$$

One merely has to make the substitutions

$$\pi\delta_-[\mathbf{l} \cdot (\mathbf{v}_j - \mathbf{v}_n)] \to 1/i(\mathbf{l} \cdot \mathbf{v}_j - \mathbf{l} \cdot \mathbf{v}_n - z) \tag{A8.9'}$$

$$\varphi(j; t) \to \varphi(j; t-\tau) \tag{A8.10}$$

in all the expressions of § 40. We now note that in eq. (A8.6) we only need the values of $J(z)$ on the Laplace contour C. The latter can be drawn down to the real axis. Hence, it is sufficient to evaluate $J(z)$ for real values of z; the result can then be continued analytically into the whole complex plane. For convenience, we call $z \equiv lw$ and take in (A8.9′) the limit $z = l(w+i\varepsilon)$ for real w, $0 < \varepsilon \to 0$:

$$\pi\delta_-(\mathbf{l} \cdot \mathbf{v}_j - \mathbf{l} \cdot \mathbf{v}_n) \to \pi\delta_-(\mathbf{l} \cdot \mathbf{v}_j - \mathbf{l} \cdot \mathbf{v}_n - lw) \tag{A.8.9}$$

Once we have given a new meaning to the symbols d_{jn}, d_j, δ_-^{jn} in terms of the substitutions (A8.9) and (A8.10), we can perform the summation exactly as in § 40, with the result:

$$\partial_t \varphi(\mathbf{v}; t) = -\frac{4\pi e^2 c}{m} \int d\mathbf{l} \frac{1}{l^2} i\mathbf{l} \cdot \partial \mathscr{F}_{\mathbf{l}}(\mathbf{v}; t) \tag{A.8.11}$$

$$\mathscr{F}_{\mathbf{l}}(\mathbf{v}; t) = \frac{l}{2\pi} \int_0^t d\tau \int_C dw \, e^{-ilw\tau} F_{\mathbf{l}}(\mathbf{v}; w, \tau) \tag{A8.12}$$

[In writing (A8.11) use was made of the Fourier expansion of the Coulomb potential, (9.8).] The function $F_{\mathbf{l}}(\mathbf{v}; w, \tau)$ obeys the following integral equation (we shall usually drop the letter τ

which appears merely as a parameter):

$$\varepsilon_l^-(\mathbf{l}\cdot\mathbf{v}/l-w)F_l(\mathbf{v};w)-\pi i d_l(\mathbf{v})\int d\mathbf{v}_1\delta_-(\mathbf{l}\cdot\mathbf{v}-lw-\mathbf{l}\cdot\mathbf{v}_1)F_{-l}(\mathbf{v}_1;w)$$
$$= q_l(\mathbf{v};w) \tag{A8.13}$$

where

$$\varepsilon_l^-(u)\equiv\varepsilon^-(u) = 1+4\pi^2ie^2cm^{-1}l^{-2}\int d\mathbf{v}_1\delta_-(lu-\mathbf{l}\cdot\mathbf{v}_1)\mathbf{l}\cdot\partial_1\varphi(\mathbf{v}_1;t-\tau)$$
$$\equiv\varepsilon_1(u)+i\varepsilon_2(u) \tag{A8.14}$$

$$d_l(\mathbf{v}) = 4\pi e^2cm^{-1}l^{-2}\mathbf{l}\cdot\partial\varphi(\mathbf{v};t-\tau) \tag{A8.15}$$

$$q_l(\mathbf{v};w) = (ie^2/2\pi ml^2)\int d\mathbf{v}_1\delta_-(\mathbf{l}\cdot\mathbf{v}-lw-\mathbf{l}\cdot\mathbf{v}_1)$$
$$\mathbf{l}\cdot(\partial-\partial_1)\varphi(\mathbf{v};t-\tau)\varphi(\mathbf{v}_1;t-\tau) \tag{A8.16}$$
$$\equiv q_1(\mathbf{v};w)+iq_2(\mathbf{v};w)$$

Note that $\varepsilon_l^-(\nu)$ is identical with the minus-dielectric constant defined earlier using a slightly different notation (23.15b, 40.6a):

$$\varepsilon_l^-(\nu)\equiv\varepsilon_-(\mathbf{l};l\nu)$$

In order to solve eq. (A8.13) the procedure adopted is a simple generalization of the one used in §§ 41 and 45. As stated there, the function F_l is complex, and hence eq. (A8.13) is actually a system of two equations for the two unknowns F_l and F_l^*. Moreover, $\mathbf{l}$ and w are fixed parameters in (A8.13), hence F_l is regarded as a function of the vector $\mathbf{v}$. But the argument of § 41 showed that, because of the form of the kernel, the equation can be reduced to a simpler equation, involving only one scalar variable (instead of three), by performing the "*barring operation*" [see eqs. (14.6), (41.3)]

$$\bar{f}_l(\nu) = \int d\mathbf{v}\,\delta(\nu-\mathbf{l}\cdot\mathbf{v}/l)f_l(\mathbf{v}) \tag{A8.17}$$

Note also that

$$\bar{f}_{-l}(\nu) = \int d\mathbf{v}\,\delta(\nu+\mathbf{l}\cdot\mathbf{v}/l)f_{-l}(\mathbf{v}) \qquad \text{(A8.17a).}$$

Multiplying both sides of eq. (A8.13) by $\delta(\nu-\mathbf{l}\cdot\mathbf{v}/l)$, integrating over $\mathbf{v}$ and using (41.7), we obtain the result:

$$\varepsilon_{\boldsymbol{l}}^{-}(\nu-w)\bar{F}_{\boldsymbol{l}}(\nu;w)-i\varepsilon_2(\nu)\int d\nu_1\delta_-(\nu-w-\nu_1)\bar{F}_{-\boldsymbol{l}}(-\nu_1;w)$$
$$=\bar{q}_{\boldsymbol{l}}(\nu;w) \qquad \text{(A8.18)}$$

We now need a second equation to close the system. Instead of writing one equation for F_1 and one for F_2, as was done before, the symmetry of the equation dictates that an equation for $\bar{F}_{-\boldsymbol{l}}(-\nu+w;w)$ be adjoined to (A8.18). Thus, changing $\mathbf{l}$ into $-\mathbf{l}$ and ν into $-\nu+w$ in eq. (A8.18), we obtain:

$$\varepsilon_{\boldsymbol{l}}^{+}(\nu)\bar{F}_{-\boldsymbol{l}}(-\nu+w;w)+i\varepsilon_2(\nu-w)\int d\nu_1\delta_+(\nu-\nu_1)\bar{F}_{\boldsymbol{l}}(\nu_1;w)$$
$$=\bar{q}_{-\boldsymbol{l}}(-\nu+w;w) \qquad \text{(A8.19)}$$

where we have made use of the property

$$\varepsilon_{-\boldsymbol{l}}^{-}(-u)=[\varepsilon_{\boldsymbol{l}}^{-}(u)]^*=\varepsilon_{\boldsymbol{l}}^{+}(u); \qquad u \text{ real} \qquad \text{(A8.20)}$$

$\varepsilon_{\boldsymbol{l}}^{+}(u)\equiv\varepsilon^{+}(u)$ is the plus-dielectric constant associated with $\varepsilon^{-}(u)$, a function which has been thoroughly studied in Chapters 4 and 5. We finally subtract (A8.19) from (A8.18) and use the definition of the functions $\delta_+(x)$ (A2.2.2–3):

$$\varepsilon_1(\nu-w)\bar{F}_{\boldsymbol{l}}(\nu;w)-\varepsilon_1(\nu)\bar{F}_{-\boldsymbol{l}}(-\nu+w;w)$$
$$-\varepsilon_2(\nu)\frac{\mathscr{P}}{\pi}\int d\nu_1\frac{\bar{F}_{-\boldsymbol{l}}(\nu_1;w)}{\nu-w+\nu_1}-\varepsilon_2(\nu-w)\frac{\mathscr{P}}{\pi}\int d\nu_1\frac{\bar{F}_{\boldsymbol{l}}(\nu_1;w)}{\nu_1-\nu}$$
$$=\bar{q}_{\boldsymbol{l}}(\nu;w)-\bar{q}_{-\boldsymbol{l}}(-\nu+w;w) \qquad \text{(A8.21)}$$

Equations (A8.21) and (A8.19) are the starting point of our treatment.

Several methods can be used to solve this system of equations. We shall depart here from the methods which have been published so far in the literature and which are of the type illustrated in § 25 or in Appendix 2, and give a simple and illuminating method which generalizes our approach of § 45. Let us use the shorter notations:

$$\begin{aligned}
&\bar{F}_{\boldsymbol{l}}(\nu;w)\equiv\bar{F}(\nu)\\
&\bar{F}_{-\boldsymbol{l}}(-\nu+w;w)\equiv\bar{G}(\nu)\\
&\bar{q}_{\boldsymbol{l}}(\nu;w)\equiv\bar{q}(\nu)\\
&\bar{q}_{-\boldsymbol{l}}(-\nu+w;w)\equiv\bar{p}(\nu)\\
&\bar{p}(\nu)-\bar{q}(\nu)\equiv\bar{\theta}(\nu)
\end{aligned} \qquad \text{(A8.22)}$$

Equation (A8.21) can be written in a very simple form in terms of the van Kampen–Case eigenfunctions of the Vlassov operator. Using the definition (23.16), eq. (A8.21) is rewritten as:

$$\int d\nu_1 \bar{\chi}_{\nu_1-w}(\nu-w)\bar{F}(\nu_1) = \int d\nu_1 \bar{\chi}_{\nu_1}(\nu)\bar{G}(\nu_1) - \bar{\theta}(\nu) \quad \text{(A8.23)}$$

It can be shown in the same way that eq. (A8.19) can be written as

$$-\int d\nu_1 \bar{\chi}_{\nu_1-w}(\nu-w)\bar{F}(\nu_1) + \varepsilon^-(\nu-w)\bar{F}(\nu) + \varepsilon^+(\nu)\bar{G}(\nu)$$
$$= \bar{q}(\nu) + \bar{\theta}(\nu) \quad \text{(A8.24)}$$

We now intend to solve (A8.23), regarded as an equation for $\bar{F}(\nu)$. For the first time, we shall need the *stability condition.* Indeed, *if the system is stable,* the set of eigenfunctions $\bar{\chi}_\nu(\nu')$, as well as the *set* $\tilde{\chi}_\nu(\nu')$ *of the adjoint Vlassov eigenfunctions, is complete* (*see* § 25). Hence, we can expand any function of ν in terms either of $\bar{\chi}_\nu(\nu')$ or of $\tilde{\chi}_\nu(\nu')$. In particular, we can expand $\bar{F}(\nu_1)$ in the form

$$\bar{F}(\nu_1) = (C_{\nu_1-w})^{-1} \int d\nu_2 \tilde{\chi}_{\nu_1-w}(\nu_2-w) f(\nu_2) \quad \text{(A8.25)}$$

Substituting this expansion in the l.h.s. of (A8.23) and making use of the closure relation (25.11), the former equation reduces to:

$$f(\nu) = \int d\nu_1 \bar{\chi}_{\nu_1}(\nu)\bar{G}(\nu_1) - \bar{\theta}(\nu)$$

Hence, substituting this "van Kampen–Case transform" back into (A8.25), we obtain the solution

$$\bar{F}(\nu) = (C_{\nu-w})^{-1} \int d\nu_1 \tilde{\chi}_{\nu-w}(\nu_1-w) \left\{ \int d\nu_2 \bar{\chi}_{\nu_2}(\nu_1)\bar{G}(\nu_2) - \bar{\theta}(\nu_1) \right\} \quad \text{(A8.26)}$$

We now use this expression for $\bar{F}$ in terms of $\bar{G}$ to eliminate $\bar{F}$ in eq. (A8.24). In order to do this, we need the following property, valid for an arbitrary function $P(\nu)$

$$\int d\nu_1 \int d\nu_2 \, \tilde{\chi}_{\nu-w}(\nu_1-w)\bar{\chi}_{\nu_2}(\nu_1)P(\nu_2)$$

$$= \pi \frac{\varepsilon_1(\nu-w)\varepsilon_1(\nu)+\varepsilon_2(\nu-w)\varepsilon_2(\nu)}{\varepsilon_2(\nu-w)} P(\nu) \qquad \text{(A8.27)}$$

$$+ \frac{\varepsilon_1(\nu)\varepsilon_2(\nu-w)-\varepsilon_1(\nu-w)\varepsilon_2(\nu)}{\varepsilon_2(\nu-w)} \mathscr{P}\int d\nu_1 \frac{P(\nu_1)}{\nu_1-\nu}$$

The proof of this relation involves a calculation similar to the one of Appendix 3. Using (A8.23) and (A8.27), eq. (A8.24) reduces to:

$$\frac{\varepsilon_2(\nu-w)\varepsilon^+(\nu)}{\pi\varepsilon^+(\nu-w)} \int d\nu_1 \, \tilde{\chi}_{\nu-w}(\nu_1-w)\bar{G}(\nu_1) = \bar{Q}(\nu) \qquad \text{(A8.28)}$$

with

$$\bar{Q}(\nu) = \bar{q}(\nu) + \frac{\varepsilon_2(\nu-w)}{\pi\varepsilon^+(\nu-w)} \int d\nu_1 \, \tilde{\chi}_{\nu-w}(\nu_1-w)\bar{\theta}(\nu_1) \qquad \text{(A8.29)}$$

We now use again the *stability condition,* which states that an arbitrary function can be expanded in terms of the eigenfunctions $\bar{\chi}_\nu(\nu_1)$. In particular, let

$$\bar{G}(\nu) = \int d\nu_1 \, \bar{\chi}_{\nu_1-w}(\nu-w)\, g(\nu_1)$$

Then, substituting into (A8.28) and using the orthogonality relations (24.3) together with (24.8), the former equation reduces to

$$\varepsilon^+(\nu)\varepsilon^-(\nu-w)g(\nu) = \bar{Q}(\nu)$$

and, hence, the solution is

$$\bar{G}(\nu) = \int d\nu_1 \, \bar{\chi}_{\nu_1-w}(\nu-w) \frac{\bar{Q}(\nu_1)}{\varepsilon^+(\nu_1)\varepsilon^-(\nu_1-w)} \qquad \text{(A8.30)}$$

This is the final solution of the system of equations: $\bar{G}(\nu)$ is expressed entirely in terms of known functions.

We note that eq. (A8.30) can be transformed in the following way by making use of the representation (23.17b) of the Vlassov eigenfunctions:

$$\bar{G}(\nu) = \tfrac{1}{2}\left\{\varepsilon^+(\nu-w)\int d\nu_1\,\delta_+(\nu-\nu_1)\frac{\bar{Q}(\nu_1)}{\varepsilon^+(\nu_1)\varepsilon^-(\nu_1-w)}\right.$$

$$\left.+\varepsilon^-(\nu-w)\int d\nu_1\delta_-(\nu-\nu_1)\frac{\bar{Q}(\nu_1)}{\varepsilon^+(\nu_1)\varepsilon^-(\nu_1-w)}\right\} \quad \text{(A8.31)}$$

$$\equiv \bar{G}_+(\nu)-\bar{G}_-(\nu)$$

As can be easily seen, this formula automatically provides the decomposition of $\bar{G}(\nu)$ into plus and minus parts according to (A2.2.1). But from eq. (A8.13) we have an expression for the unbarred function $F_l(\mathbf{v}; w)$ in terms of the minus part of the barred function $\bar{F}_{-l}(-\nu+w; w) \equiv \bar{G}(\nu)$:

$$F_l(\mathbf{v}; w) = \frac{1}{\varepsilon^-(\nu-w)}\left\{\bar{q}_l(\mathbf{v}; w)-\frac{2\pi i}{l}d_l(\mathbf{v})\bar{G}_-(\nu; w)\right\} \quad \text{(A8.32)}$$

Combining this equation with (A8.31) we obtain

$$F_l(\mathbf{v}; w) = \frac{q_l(\mathbf{v}; w)}{\varepsilon^-(\nu-w)}$$

$$+\frac{\pi i}{l}d_l(\mathbf{v})\int d\nu_l\delta_-(\nu-\nu_1)\frac{\bar{Q}(\nu_1; w)}{\varepsilon^+(\nu_1)\varepsilon^-(\nu_1-w)} \quad \text{(A8.33)}$$

In the *stable case* which we are considering here, this expression can be further simplified. Indeed, let us write

$$\frac{\bar{Q}(\nu)}{\varepsilon^+(\nu)\varepsilon^-(\nu-w)} \equiv \frac{\bar{P}(\nu)}{\varepsilon^+(\nu)\varepsilon^+(\nu-w)\varepsilon^-(\nu-w)}$$

Using (A8.29) together with (A2.2.1), (A8.14) and (A8.20), we obtain

$$\bar{P}(\nu) = \varepsilon^+(\nu-w)\bar{Q}(\nu) = \varepsilon^+(\nu-w)[\bar{p}_+(\nu)-\bar{p}_-(\nu)]$$

$$+[\varepsilon^-(\nu-w)-\varepsilon^+(\nu-w)][\bar{p}_+(\nu)-\bar{q}_+(\nu)]$$

$$= \varepsilon^+(\nu-w)[\bar{q}_+(\nu)-\bar{p}_-(\nu)]+\varepsilon^-(\nu-w)[\bar{p}_+(\nu)-\bar{q}_+(\nu)]$$

Hence

$$\frac{\bar{Q}(\nu)}{\varepsilon^+(\nu)\varepsilon^-(\nu-w)} = \frac{\bar{q}_+(\nu)-\bar{p}_-(\nu)}{\varepsilon^+(\nu)\varepsilon^-(\nu-w)}+\frac{\bar{\theta}_+(\nu)}{\varepsilon^+(\nu)\varepsilon^+(\nu-w)} \quad \text{(A8.34)}$$

But, *in the stable case, the second term on the right-hand side is a plus-function* [$\varepsilon^+(\nu)$ has no zeros in S_+]. Hence, when substituted in (A8.33), it gives a vanishing contribution, because the integral involving the δ_--function in the latter formula is equivalent to taking the minus part of the integrand. Thus (A8.33) reduces to

$$F_l(\mathbf{v}; w) = \frac{q_l(\mathbf{v}; w)}{\varepsilon^-(\nu - w)} + \frac{\pi i}{l} d_l(\mathbf{v}) \int d\nu_1 \delta_-(\nu - \nu_1) \frac{\bar{Q}_1(\nu_1; w)}{\varepsilon^+(\nu_1)\varepsilon^-(\nu_1 - w)} \tag{A8.35}$$

Using the definitions (A8.16), (A8.22) and the Poincaré–Bertrand theorem (A2.2.12), it is easily shown that

$$\bar{Q}_1(\nu; w) \equiv \bar{q}_+(\nu) - \bar{p}_-(\nu) = \tfrac{1}{2}(ie^2/2\pi m l^2) \int d\nu_1 \int d\nu_2 \delta_+(\nu - \nu_1)\delta_-(\nu - \nu_2)[\bar{\varphi}'(\nu_1)\bar{\varphi}(\nu_2 - w) - \bar{\varphi}(\nu_1)\bar{\varphi}'(\nu_2 - w)] \tag{A8.36}$$

Equation (A8.35), substituted into (A8.11–12), yields the kinetic equation

$$\partial_t \varphi(\mathbf{v}; t) = -\frac{2e^2 c}{m} \int_0^t d\tau \int_C dw \int d\mathbf{l} \frac{e^{-ilw\tau}}{l} \mathbf{l} \cdot \partial \left\{ \frac{q_l(\mathbf{v}; w)}{\varepsilon^-(\nu - w)} + \frac{\pi i}{l} d_l(\mathbf{v}) \int d\nu_1 \delta_-(\nu - \nu_1) \frac{\bar{Q}_1(\nu_1; w)}{\varepsilon^+(\nu_1)\varepsilon^-(\nu_1 - w)} \right\} \tag{A8.37}$$

where ν is to be interpreted as $\mathbf{l} \cdot \mathbf{v}/l$. The symbols q_l, d_l, ε^-, ε^+, and $\bar{Q}_1$ appearing in the equation are functions of $\varphi(\mathbf{v}; t - \tau)$ defined by (A8.14–16, 22, 29).

Equation (A8.37) is a very complicated non-linear equation, whose non-markoffian character is clearly apparent from the presence of the integration over τ and the occurrence of the retarded functions $\varphi(\mathbf{v}; t - \tau)$. This equation contains the maximum information we can obtain on the evolution of the velocity distribution within the ring approximation. It was first derived by Résibois,* who used a different method. We shall present his derivation in some detail in Appendix 10, because his method deserves special importance by itself. We may stress that, apart from the choice of the diagrams, no approximation whatsoever

* P. Résibois, *Phys. of Fluids*, **6**, (1963) (to be published).

has been made in deriving (A8.37); in particular, no linearization around a stationary state has been necessary. The latter point deserves special attention. In fact, Guernsey * has derived a particular case of eq. (A8.37). He started from the BGBKY hierarchy in which three-particle correlations were neglected. His method, however, depends crucially on his assumption that the system is close to equilibrium. Guernsey** extended his theory to cover slightly inhomogeneous systems; of course, a linearization around the equilibrium is also necessary in that case. The possibility of relaxing the assumption that the state is near equilibrium is important for the following reason.

We shall be very interested in the behavior of unstable systems (see Appendix 9), but such systems are very far from equilibrium: a double-humped distribution function cannot be obtained by a small perturbation of the maxwellian.

Let us also mention that the present method can be generalized to cover slightly inhomogeneous systems of the type studied by Guernsey, but of course in the more general situation where the system departs slightly from an arbitrary homogeneous *non-equilibrium* state.

The first attempt to derive a non-markoffian equation of evolution was made by Tchen. † He derived essentially the integral equation (A8.13), but did not solve it. Although his equation contains many essential features of the exact equation, a direct comparison is difficult because of his approximations.

The kinetic equation (A8.37), being quite recent and moreover very complicated, has not yet been much exploited. In his Boeing report quoted above, Guernsey studied the nature of the relaxation times. He showed in particular that the characteristic short-time scale is ω_p^{-1}, as expected, but that the decay of the transient terms is not exponential but varies as $(\omega_p t)^{-1}$ for long times, and is therefore much slower than expected. This may cause difficulties in the higher order approximations.

* R. L. Guernsey, *Relaxation Time for the 2-Particle Correlation Function in a Plasma*, Boeing Scient. Res. Lab., D1-82-0083, 1960.

** R. L. Guernsey, *Phys. of Fluids*, **5**, 322 (1962).

† C. M. Tchen, *Phys. Rev.*, **114**, 394 (1959).

In a quite recent paper, Oberman, Ron and Dawson * have used Guernsey's equation for a rigorous calculation of the high-frequency conductivity of a plasma. We have no space to reproduce these interesting calculations here.

We shall only show that eq. (A8.37) reduces to eq. (41.19) in the limit of long times. In order to take this "markoffian limit", we assume that we are interested in times much longer than the duration of the memory, i.e. ω_p^{-1}. In that case, we may replace the argument $t-\tau$ by t in all velocity distribution factors on the r.h.s.: $\varphi(\mathbf{v}; t-\tau) \to \varphi(\mathbf{v}; t)$. Hence, $F_l(\mathbf{v}; w)$ no longer depends on τ, and the time integration in (A8.12) can be performed explicitly. As a result, eq. (A8.11) becomes:

$$\partial_t \varphi(\mathbf{v}; t) = -\frac{2e^2 c}{m} \int d\mathbf{l} \int_C dw \frac{e^{-ilwt}}{-iw} \frac{1}{l^2} i\mathbf{l} \cdot \partial F_l(\mathbf{v}; w) \qquad \text{(A.8.38)}$$

We now perform the w-integration. We therefore use the fact that the singularities of the resolvent are, roughly speaking, at a large distance ω_p from the real axis. Disregarding these singularities in the limit of long times, we take account only of the residue at the pole at $w = 0$, with the result

$$\partial_t \varphi(\mathbf{v}; t) = -\frac{4\pi e^2 c}{m} \int d\mathbf{l} \frac{1}{l^2} i\mathbf{l} \cdot \partial F_l(\mathbf{v}; 0) \qquad \text{(A8.39)}$$

Comparing this equation with eq. (40.9), we see that our purpose is reduced to showing that $F_l(\mathbf{v}; 0)$ is identical to the function $F(\alpha)$ whose complete form was found in eq. (45.6). But, as $w \to 0$, $q_l(\mathbf{v}; 0) = q(\alpha)$, defined in (40.7). Moreover, it is easily shown that $\bar{q}_l(\mathbf{v}; 0) = [\bar{q}_{-l}(-\nu); 0)]^*$. But $\bar{q}_l(\nu; 0)$ is real [see (41.10)], therefore $\bar{p}(\nu; 0) = \bar{q}(\nu; 0)$ and, by eq. (A8.29), $\bar{Q}_1(\nu; 0) \to \bar{q}(\nu; 0)$. Thus, from (A8.33):

$$F_l(\mathbf{v}; 0) = \frac{q_l(\mathbf{v}; 0)}{\varepsilon^-(\nu)} + \frac{\pi i}{l} d_l(\mathbf{v}) \int d\nu_1 \delta_-(\nu - \nu_1) \frac{\bar{q}(\nu_1; 0)}{\varepsilon^+(\nu_1)\varepsilon^-(\nu_1)} \qquad \text{(A8.40)}$$

which is identical with (45.6).

* C. Oberman, A. Ron and J. Dawson, *Phys. of Fluids*, **5**, 1514 (1962).

Kinetic Equation for Unstable Classical Plasmas

The phenomenon of "two-stream instability" has been introduced in connection with the study of the evolution of the density excess in an inhomogeneous plasma, within the linearized Vlassov approximation (Chapters 3 and 4). It has been shown there that under certain conditions, *which depend only on the velocity distribution* $\varphi(\mathbf{v})$, the plasma responds to an initial excitation by producing oscillations whose amplitude grows exponentially in time. The general condition of instability is related, as was shown in § 19, to the position of the zeros of the dielectric constant $\varepsilon_{\mathbf{k}}^{+}(w)$, eq. (14.9),* in the complex plane. More precisely, if $\varepsilon_{\mathbf{k}}^{+}(w)$ has a zero ζ_+ in the upper half-plane S_+:

$$\varepsilon_{\mathbf{k}}^{+}(w) = (w-\zeta_+)\sigma^+(w), \qquad \zeta_+ \in S_+ \tag{A9.1}$$

the plasma is unstable. For simplicity, we shall assume from this point that there is only *one* such zero and that it is simple, i.e. $\sigma^+(\zeta_+) \neq 0$. Letting

$$\zeta_+ = w_0 + i\gamma_0 \tag{A9.2}$$

we recognize that w_0 is the frequency of the unstable oscillation, and γ_0 is its rate of growth. We note immediately that from the definition of the "minus-dielectric constant" $\varepsilon_{\mathbf{k}}^{-}(w)$, it follows that this function has a zero in S_-, and that this zero is the complex conjugate of ζ_+:

$$\begin{aligned} \varepsilon_{\mathbf{k}}^{-}(w) &= (w-\zeta_-)\sigma^-(w), \qquad \zeta_- \in S_- \\ \zeta_- &= (\zeta_+)^* = w_0 - i\gamma_0 \end{aligned} \tag{A9.3}$$

[Indeed, $\varepsilon_{\mathbf{k}}^{-}(w)$ is defined by the same Cauchy integral as $\varepsilon_{\mathbf{k}}^{+}(w)$, eq. (14.9), in which however $w \in S_-$; hence ε^+ and ε^- take complex conjugate values for complex conjugate values of w.]

It is obvious on the other hand that the unlimited growth of plasma oscillations cannot go on *ad infinitum*. There must exist

* For later convenience we denote the complex frequency by $kw \equiv z$, and write shortly $\varepsilon_{\mathbf{k}}^{+}(w) \equiv \varepsilon_+(\mathbf{k}; kw)$. When no confusion is possible, the subscript $\mathbf{k}$ will be omitted.

one or several mechanisms which stop the amplification. From this mathematical point of view, one or more of the hypotheses made in the linear Vlassov approximation must break down for unstable systems. One of these is obviously the assumption of small density excesses, which underlies the linearizations. However, there exists no general technique at the present time for solving the non-linear Vlassov equation to the same order of precision as the linear one. A recent attempt in this direction is due to Drummond and Pines* and is inspired by the theory of anharmonic solids. It introduces a dissipation by the interaction of normal modes among themselves, through the anharmonicity due to the non-linear terms in the Vlassov equation.

This rather obvious mechanism is not the only one which prevents the catastrophe. Another assumption, which also underlies the complete as well as the linearized Vlassov equation, concerns the *time scales*. We assumed that there were three natural time scales in the plasma: the hydrodynamic time scale, t_h, which does not introduce any limitation of the equation, the relaxation time, t_r, and the period of a plasma oscillation, t_p. The fundamental assumption $t_r \gg t_p$ allowed us to infer that the velocity distribution is regarded as a constant in the Vlassov approximation. The reason for this is that this function can change only by collisions, the only dissipative mechanism encountered so far. The collisional (or "correlational") dissipation acts on the time scale t_r. But in an unstable system there appears a new time scale which is not related to the previous ones: it is the inverse of the rate of growth, $t_i = (\bar{k}\gamma_0)^{-1}$ (where $\bar{k}$ is an average value of the wave-vector). The possibility exists that the velocity distribution changes over this new time scale even if $t_i < t_r$.

The possibility of such an additional path of evolution for $\varphi(\mathbf{v}; t)$ in the presence of instabilities is strongly suggested by the part played by the stability condition in the derivation of the kinetic equation. It has been stressed repeatedly in §§ 41, 45, 50, and 80, that the formulae derived within the ring approximation are valid only if $\varepsilon_{\mathbf{k}}^{+}(w)$ has no zeros in the upper half-plane. Hence,

* W. E. Drummond and D. Pines, *Proc. Conf. Ionization Processes in Gases*, Salzburg, 1961.

it appears quite reasonable that some other type of evolution prevails in unstable plasmas. We shall now show that this is indeed the case. Following a recent paper by the author,* we shall study the evolution of the velocity distribution $\varphi(\mathbf{v}; t)$ in a homogeneous system which is assumed to be initially unstable. The restriction to homogeneous systems is not very limiting because of the independence of homogeneous and inhomogeneous Fourier components. Even in an inhomogeneous plasma, the velocity distribution, and hence the dielectric constant, will evolve in the same way as below. It is, however, probable that in addition to the effect to be studied below, the evolution of the inhomogeneity factor $\rho_{\mathbf{k}}(\mathbf{v}; t)$ will be affected by similar correlational effects, which we shall however not study here.

Our starting point is the general equation (A8.11–12), which was derived without any reference to the stability condition. Hence the problem comes back to the evaluation of $F_{\mathbf{l}}(\mathbf{v}; w)$.

However, in writing the integral equation (A8.13), a step involving the stability condition has been taken, as will now be shown.

The equation which is obtained directly from the summation of the diagrams is derived by a generalization of the method of § 40, using the substitutions (A8.9′) and (A8.10). This results in the following fundamental integral equation:

$$\varepsilon_{\mathbf{l}}^{-}(\nu - w)F_{\mathbf{l}}(\mathbf{v}; w) = d_{\mathbf{l}}(\mathbf{v})\int d\mathbf{v}_1 \frac{F_{-\mathbf{l}}(\mathbf{v}_1; w)}{\mathbf{l}\cdot\mathbf{v} - lw - \mathbf{l}\cdot\mathbf{v}_1} + q_{\mathbf{l}}(\mathbf{v}; w) \tag{A9.4}$$

where

$$\begin{aligned} \varepsilon_{\mathbf{l}}^{-}(\nu) &= 1 + \frac{\omega_p^2}{l^2}\int d\nu_1 \frac{\bar{\varphi}'(\nu_1; t-\tau)}{\nu - \nu_1} \\ d_{\mathbf{l}}(\mathbf{v}) &= (\omega_p^2/l^2)\mathbf{l}\cdot\partial\varphi(\mathbf{v}; t-\tau) \\ q_{\mathbf{l}}(\mathbf{v}; w) &= \frac{\omega_p^2}{8\pi^3 c l^2}\int d\mathbf{v}_1 \frac{\mathbf{l}\cdot(\partial - \partial_1)\varphi(\mathbf{v}; t-\tau)\varphi(\mathbf{v}_1; t-\tau)}{\mathbf{l}\cdot\mathbf{v} - lw - \mathbf{l}\cdot\mathbf{v}_1} \end{aligned} \tag{A9.5}$$

These coefficients are the analytical continuations for $w \in S_+$ of those coefficients defined in (A8.14–16). Equation (A9.4) is

* R. Balescu, *J. Math. Phys.*, (1963) (in press).

derived without any recourse to the stability condition. It can, however, in general, not be solved directly: for complex w it is not a singular integral equation. The idea of Appendix 8 is to adjoin to this equation an auxiliary equation, by letting w move down to the real axis. Equation (A8.13) obtained in this way is a singular integral equation to which we can apply the standard methods of Appendix 2 or the simpler method used in Appendix 8. Knowing the solution for real w, a simple analytical continuation into the upper half-plane gives its value for $w \in S_+$, and it is readily verified that in the stable case the function obtained in this way is a solution of the fundamental equation (A9.4).

It will now be shown that for unstable plasmas this simple procedure is no longer justified: *the analytical continuation of the solution of eq.* (A8.13) *is not a solution of eq.* (A9.4) *if the plasma is unstable.*

The auxiliary equation (A8.13) can be solved in the unstable case too. The solution will, however, no longer be (A8.33), because the expansions (A8.33) and (A8.28) are no longer valid. The equation can, however, be solved by essentially the same method, if the complete system of van Kampen–Case eigenfunctions for unstable plasmas, introduced in § 25, is used. It turns out that the solution consists of the stable solution (A8.33) plus some additional terms, but for the present argument a discussion of the first term on the r.h.s. of (A8.33) is sufficient. This term [and hence $F_l(\mathbf{v}; w)$] has a pole at $w = \nu - \zeta_-$. The corresponding term in $F_{-l}(\mathbf{v}; w)$ is

$$\frac{q_{-l}(\mathbf{v}; w)}{\varepsilon_{-l}^{-}(-\nu - w)} \equiv \frac{q_{-l}(\mathbf{v}; w)}{\varepsilon_{l}^{+}(\nu + w)}$$

and it has a pole at $w = \nu + \zeta_+$. The Cauchy integral appearing on the r.h.s. of (A9.4) is therefore of the form

$$\int d\nu_1 \frac{f(\nu_1)}{(\nu_1 - \nu + w)(\nu_1 + w - \zeta_+)}, \qquad \operatorname{Im} w > |\gamma_0| \tag{A9.6}$$

This function has *two cuts*: one on the real axis, $\operatorname{Im} w = 0$, and one on the line $\operatorname{Im} w = \gamma_0$. In eq. (A9.4), w is assumed to be far up in S_+: we now see that it must lie above both cuts.

In the *stable case,* the second cut lies in the lower half-plane. Hence the expression

$$\pi i \int dv_1 \delta_-(v-w-v_1) \frac{f(v_1)}{v_1+w-\zeta_+}, \qquad w \text{ real} \qquad \text{(A9.6a)}$$

is the correct analytical continuation of (A9.6). Thus, (A8.13) and (A9.4) are analytical continuations of each other, and so are their solutions. The procedure of Appendix 8 is therefore justified.

In the *unstable case,* however, the cut lies *above* the real axis. Hence (A9.6a) is *not* the analytical continuation of (A9.6), neither is (A8.13) the analytical continuation of the fundamental equation. Therefore the solution of the unstable equation (A8.13) does not provide the solution of (A9.4) by a simple analytical continuation in w.

However, a knowledge of the *stable* solution enables us to obtain the solution in the unstable case too, by appropriate deformations of the contours of integration. Thus consider again the stable case, and let w be on the original contour C of the inverse Laplace transformation, i.e. above all singularities of F_l: eq. (A8.33) provides the solution in that case. Let us introduce the following abbreviation

$$\bar{R}(v; w) \equiv -i \frac{\omega_p^2}{4\pi^2 cl^2} \frac{\bar{\varphi}_-(v-w)\bar{\varphi}'_+(v)-\bar{\varphi}'_-(v-w)\bar{\varphi}_+(v)}{\sigma^+(v)\sigma^-(v-w)} \qquad \text{(A9.7)}$$

Then the solution is (see Fig. A9.1b):

$$F_l(\mathbf{v}; w) = \frac{q_l(\mathbf{v}; w)}{\varepsilon^-(v-w)} + \frac{\pi i}{l} d_l(\mathbf{v}) \int_{-\infty}^{\infty} dv_1 \delta_-(v-v_1) \frac{\bar{R}(v_1; w)}{(v_1-\zeta_+)(v_1-w-\zeta_-)}$$

$$(\text{stable; Im } w > |\gamma_0|) \qquad \text{(A9.8)}$$

Now let ζ_+ move into the upper half-plane and ζ_- into the lower half-plane, keeping w constant. As long as w is sufficiently far up in S_+, eq. (A9.4) changes continuously in this process. Hence, in the unstable case, the solution of eq. (A9.4) will be the analytical continuation of (A9.8) for $\zeta_+ \to S_+$. The latter is obtained by deforming the contour of the ν_1-integration as shown in Fig. A9.1c:

$$F_l(\mathbf{v}; w) = \frac{q_l(\mathbf{v}; w)}{\varepsilon^-(\nu - w)} + \frac{\pi i}{l} d_l(\mathbf{v}) \int_{\Gamma_1} d\nu_1 \delta_-(\nu - \nu_1) \frac{\bar{R}(\nu_1; w)}{(\nu_1 - \zeta_+)(\nu_1 - w - \zeta_-)}$$

$$(\text{unstable; Im } w > \gamma_0) \qquad \text{(A9.9)}$$

This equation is the solution to our problem. For convenience in further calculations it is usually convenient to have expressions

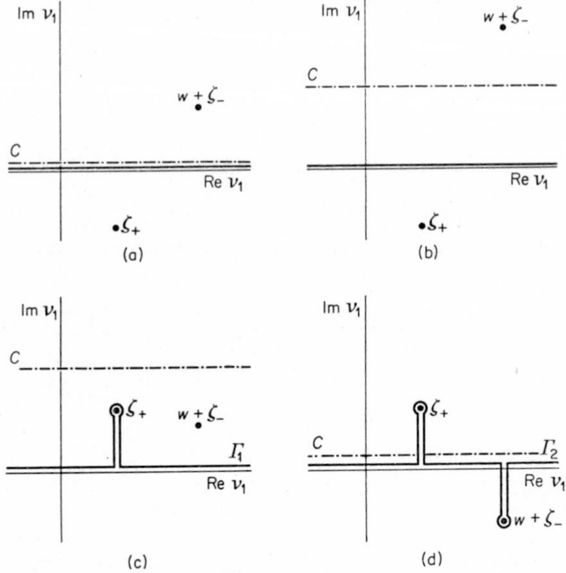

Fig. A9.1. Contours of integration over ν_1 in the expression for $F_l(\mathbf{v}; w)$. (a) Stable plasma; w real; (b) stable plasma, Im $w > |\gamma_0|$; (c) unstable plasma, Im $w > \gamma_0$; (d) unstable plasma, w real.

in terms of real w. We therefore move the contour C downwards and bring it onto the upper edge of the real axis, keeping ζ_+ and ζ_- constant. The pole $w + \zeta_-$ thus moves into the lower half-plane, and the analytical continuation of (A9.9) is obtained by deforming the contour of integration into the contour Γ_2 shown in Fig. A9.1d. The result, expressed in terms of an integral over the real axis, is therefore:

$$F_l(\mathbf{v}; w) = \frac{q_l(\mathbf{v}; w)}{\varepsilon^-(\nu - w)} + \frac{2\pi i}{l} d_l(\mathbf{v})$$

$$\left\{ \tfrac{1}{2} \int_{-\infty}^{\infty} d\nu_1 \delta_-(\nu - \nu_1) \frac{\bar{R}(\nu_1; w)}{(\nu_1 - \zeta_+)(\nu_1 - w - \zeta_-)} \right.$$

$$\left. + \frac{\bar{R}(\zeta_+; w)}{(w + \zeta_- - \zeta_+)(\nu - \zeta_+)} + \frac{\bar{R}(w + \zeta_-; w)}{(w + \zeta_- - \zeta_+)(\nu - \zeta_- - w)} \right\}$$

$$\text{(unstable; } w \text{ real)} \qquad \text{(A9.10)}$$

This is the final form of the solution. Our initial statement can now be verified *a posteriori*: (A9.10) is not a solution of eq. (A8.13).

Equation (A9.10), substituted into eqs. (A8.12) and (A8.11), provides the explicit form of the kinetic equation for unstable systems. It is an extremely complicated non-markoffian equation [remember that all functions $\varphi(\mathbf{v})$ occurring in (A9.10) are evaluated at time $t - \tau$]. The equation can, however, be appreciably simplified in the case of "*weakly unstable*" *plasmas*, a concept which will now be defined. Assume that the imaginary part of the "unstable" zero of the dielectric constant γ_0 is much smaller than the imaginary part of the stable zero closest to the real axis, which we denote by γ_m: the latter has been assumed to be of order $\omega_p/\bar{k}$, i.e. γ_m measures the short-time scale of the plasma:

$$\gamma_0 \ll |\gamma_m| \approx \omega_p/\bar{k} \qquad \text{(A9.11)}$$

We also assume (this must be verified on the result) that if (A9.11) is satisfied at the initial time it remains true at all later times, as long as $\gamma_0 > 0$.

We can now distinguish in (A8.12) a slow process (described by the residues at the unstable poles) and rapidly damped transient processes. Hence, if we are interested in times much longer than ω_p^{-1}, we can drop the latter contributions. Moreover, as was shown in Appendix 8, we can use the markoffian approximation (A8.7) by setting $\tau = 0$ in all factors $\varphi(\mathbf{v}; t - \tau)$ appearing in the function F_l:

$$\mathscr{F}_l(\mathbf{v}; t) = -\frac{1}{2\pi i} \int_C dw \frac{e^{-ilwt}}{w} F_l(\mathbf{v}; w) \qquad \text{(A9.12)}$$

However, we are now interested in times which can be of the order of $(\bar{l}\gamma_0)^{-1}$, which is much longer than the period of a plasma oscillation. We must therefore take into account not only the residue at 0+ but also the residues at the unstable poles.

In order to evaluate these residues, we first integrate the second term in eq. (A9.10) explicitly over ν_1. This is most easily done by using the decompositions of $\bar{R}(\nu_1; w)$ (regarded as a function of ν_1) into plus and minus parts according to the Plemelj formula (A2.2.1). Using also the expression for $\delta_-(x)$ given by (A2.2.8) and closing the contour of integration, we obtain

$$F_l(\mathbf{v}; w) = \frac{q_l(\mathbf{v}; w)}{\varepsilon^-(\nu - w)} + \frac{2\pi i}{l} d_l(\mathbf{v}) \left\{ - \frac{\bar{R}_-(\nu; w)}{(\nu - w - \zeta_-)(\nu - \zeta_+)} \right.$$
$$\left. - \frac{\bar{R}_-(\zeta_+; w)}{(w + \zeta_- - \zeta_+)(\nu - \zeta_+)} + \frac{\bar{R}_+(\zeta_- + w; w)}{(w + \zeta_- - \zeta_+)(\nu - \zeta_- - w)} \right\} \tag{A9.13}$$

Substituting this expression into (A9.12), the Laplace transform is easily calculated by the method of residues. Noting that $q_l(\mathbf{v}; w)$ and $\bar{R}_+$ have poles far down in the lower half-plane and thus give rapidly damped terms which can be neglected, we obtain

$$\mathscr{F}_l(\mathbf{v}; t) = \frac{q_l(\mathbf{v}; 0)}{\varepsilon^-(\nu)} + \frac{2\pi i}{l} d_l(\mathbf{v}) \left\{ - \frac{\bar{R}_-(\nu; 0)}{(\nu - \zeta_+)(\nu - \zeta_-)} \right.$$
$$\left. + \frac{1}{2i\gamma_0} \left[\frac{\bar{R}_-(\zeta_+; 0)}{\nu - \zeta_+} - \frac{\bar{R}_+(\zeta_-; 0)}{\nu - \zeta_-} \right] \right\}$$
$$+ \frac{e^{2l\gamma_0 t}}{2i\gamma_0} \frac{2\pi i}{l} d_l(\mathbf{v}) \frac{\bar{R}(\zeta_+; \zeta_+ - \zeta_-)}{\nu - \zeta_+} \tag{A9.14}$$
$$+ \frac{e^{-il(\nu - \zeta_-)t}}{\nu - \zeta_-} \left\{ - \frac{q_l(\mathbf{v}; \nu - \zeta_-)}{\sigma^-(\zeta_-)} - \frac{2\pi i}{l} d_l(\mathbf{v}) \frac{\bar{R}(\nu; \nu - \zeta_-)}{\nu - \zeta_+} \right\}$$

This expression, substituted into (A8.11), provides the general kinetic equation for weakly unstable systems. It consists of a term independent of time [except through $\varphi(\mathbf{v}; t)$], an exponentially growing term and an oscillating term which is exponentially ampified.

Before discussing this equation further, we note the following important fact. After stabilization of the plasma, the formerly unstable zero ζ_+ moves into the lower half-plane. However, it remains for a certain time much closer to the real axis than all the other zeros of $\varepsilon^+(w)$. Hence, if we want an asymptotic description of such a *"weakly stable" plasma* to the same degree of precision as (A9.14), we must retain in (A9.12) the residues at the poles related to ζ_+ in addition to the residue at $w = 0$ which is considered in "normal" stable plasmas. But in the stable case $F_{\mathbf{l}}(\mathbf{v}; w)$, regarded as a function of the parameter ζ_+, is the analytical continuation of this same function for the unstable case. Hence the expression for $\mathscr{F}_{\mathbf{l}}(\mathbf{v}; t)$ has the same analytical form (A9.14) in both cases, as can also be verified directly from (A8.33). Of course, in the stable case $\zeta_+ \in S_-$, $\zeta_- \in S_+$, $\gamma_0 < 0$.

We now proceed to simplify the kinetic equation. We first note that

$$-\frac{\bar{R}_-(\nu; 0)}{(\nu-\zeta_+)(\nu-\zeta_-)} + \frac{1}{2i\gamma_0}\left\{\frac{\bar{R}_-(\zeta_+; 0)}{\nu-\zeta_+} - \frac{\bar{R}_+(\zeta_-; 0)}{\nu-\zeta_-}\right\}$$

$$= \tfrac{1}{2}\int d\nu_1 \delta_-(\nu-\nu_1)\,\frac{\bar{R}(\nu_1; 0)}{(\nu_1-\zeta_+)(\nu_1-\zeta_-)} - \frac{1}{2i\gamma_0}\left[\frac{\bar{R}(\zeta_+; 0)}{\nu-\zeta_+} + \frac{\bar{R}(\zeta_-; 0)}{\nu-\zeta_-}\right]$$

It is easily shown that the bracketed term on the r.h.s. of this equation is an odd function of the wave-vector $\mathbf{l}$; hence it does not contribute to the kinetic equation (A8.11).

Consider now the last term in (A9.14), proportional to $\exp[-il(\nu-\zeta_-)t]$. There is a certain inconsistency in retaining rapid oscillations in an asymptotic equation valid for times $t \gg \omega_p^{-1}$. We must in some way smooth out these rapid oscillations, and retain only their slow systematic growth. This can be achieved by noting that $\varphi(\mathbf{v}; t)$ is actually a distribution in the sense of Schwartz: the physically relevant quantities are integrals of products of φ and of some function of $\mathbf{v}$. Let $U(\mathbf{v})$ be such a function, which we can assume to be an entire function (in all physical applications $U(\mathbf{v})$ is a polynomial). Let us call ν and $\mathbf{v}_\perp$ respectively the components of $\mathbf{v}$ parallel and perpendicular to $\mathbf{l}$. Whenever it is required for clarity, we shall write functions $f(\mathbf{v})$ of the vector $\mathbf{v}$ in the form $f(\mathbf{v}_\perp, \nu)$. Multiplying now both sides of the kinetic equation (A8.11) by the test function $U(\mathbf{v}_\perp, \nu)$ and integrating over $\mathbf{v}$, the last term in (A9.14) gives a contribution of the following form to $\partial_t\langle U\rangle$:

$$\int_{-\infty}^{\infty} d\nu \frac{e^{-il(\nu-\zeta_-)t}}{\nu-\zeta_-}\left\{f_1(\nu)+\frac{f_2(\nu)}{\nu-\zeta_+}\right\} \tag{A9.15}$$

For positive times the contour of integration is closed in the lower half-plane. In the *unstable case,* the pole $\nu = \zeta_-$ is thus within the contour, whereas $\nu = \zeta_+$ is outside. The functions $f_1(\nu)$ and $f_2(\nu)$ have poles far down in the lower half-plane, and thus give rapidly damped residues which are neglected in order to be consistent with our approximations. This asymptotic result of neglecting all the poles but the one in ζ_- is achieved by replacing the integrand in (A9.15) by

$$\frac{1}{\nu-\zeta_-}\left\{f_1(\zeta_-)+\frac{f_2(\zeta_-)}{\zeta_--\zeta_+}\right\}$$

Consider now the *stable case*. The pole $\nu = \zeta_-$ is now outside the contour whereas $\nu = \zeta_+$ has moved inside. An argument similar to the previous one shows that the asymptotic form of (A9.15) is obtained by replacing the integrand by

$$\frac{1}{\nu-\zeta_+}e^{-il(\zeta_+-\zeta_-)t}\frac{f_2(\zeta_+)}{\zeta_+-\zeta_-}$$

As a result of this discussion, the consistent asymptotic forms of the kinetic equation (A8.11) are:

$$\partial_t\varphi(\mathbf{v}; t) =$$

$$-\omega_p^2\int d\mathbf{l}\, l^{-2} i\mathbf{l}\cdot\partial\left\{\frac{q_l(\mathbf{v}; 0)}{\varepsilon^-(\nu)}+\frac{\pi i}{l}d_l(\mathbf{v})\int d\nu_1\delta_-(\nu-\nu_1)\frac{\bar{R}(\nu_1; 0)}{(\nu_1-\zeta_+)(\nu_1-\zeta_-)}\right.$$

$$+\frac{1}{\nu-\zeta_-}\left[-\frac{q_l(\mathbf{v}_\perp, \zeta_-; 0)}{\sigma^-(\zeta_-)}+\frac{2\pi i}{l}d_l(\mathbf{v}_\perp, \zeta_-)\frac{\bar{R}(\zeta_-; 0)}{2i\gamma_0}\right]$$

$$\left.+\frac{1}{\nu-\zeta_+}\frac{2\pi i}{l}d_l(\mathbf{v})\, e^{2l\gamma_0 t}\frac{\bar{R}(\zeta_+; \zeta_+-\zeta_-)}{2i\gamma_0}\right\} \quad \text{(unstable)} \tag{A9.16}$$

and

$$\partial_t \varphi(\mathbf{v}; t) =$$

$$-\omega_p^2 \int d\mathbf{l}\, l^{-2} i\mathbf{l} \cdot \partial \left\{ \frac{q_l(\mathbf{v}; 0)}{\varepsilon^-(\nu)} + \frac{\pi i}{l} d_l(\mathbf{v}) \int d\nu_1 \delta_-(\nu-\nu_1) \frac{\bar{R}(\nu_1; 0)}{(\nu_1-\zeta_+)(\nu_1-\zeta_-)} \right.$$

$$\left. + \frac{2\pi i}{l} \frac{1}{\nu-\zeta_+} \mathrm{e}^{2l\gamma_0 t} \frac{\bar{R}(\zeta_+; \zeta_+-\zeta_-)}{2i\gamma_0} [d_l(\mathbf{v})-d_l(\mathbf{v}_\perp, \zeta_+)] \right\}$$

(stable) (A9.17)

Let us now write these expressions more explicitly. We first note that the residue at $w = 0$ is exactly identical in form to the function $F_l(\mathbf{v}; 0)$ appearing in (A8.35): hence it gives precisely the "*normal collision term*" $\mathscr{C}\{\varphi\}$ defined in eq. (41.19). The other terms, rewritten explicitly by using (A9.7), give the following expressions:

$$\partial_t \varphi(\mathbf{v}; t) = \mathscr{C}\{\varphi\} + \mathscr{I}\{\varphi\} \qquad \text{(A9.18)}$$

with

$$\mathscr{I}\{\varphi\} = \frac{4e^4 c}{m^2} \int d\mathbf{l}\, l^{-4} \mathbf{l} \cdot \partial \left\{ -\frac{1}{2\pi i} \frac{l^2}{\omega_p^2} \frac{\varphi(\mathbf{v}_\perp, \zeta_-)}{(\nu-\zeta_-)\sigma^-(\zeta_-)} \right.$$

$$\left. + \frac{\bar{\varphi}(\zeta_-)}{2il\gamma_0(\nu-\zeta_-)\sigma^+(\zeta_+)\sigma^-(\zeta_-)} \mathbf{l} \cdot \partial\varphi(\mathbf{v}_\perp, \zeta_-) + \frac{D(t)}{\mathbf{l} \cdot \mathbf{v} - l\zeta_+} \mathbf{l} \cdot \partial\varphi(\mathbf{v}) \right\}$$

(unstable) (A9.19)

and

$$\mathscr{I}\{\varphi\} = \frac{4e^4 c}{m^2} \int d\mathbf{l}\, l^{-4} \mathbf{l} \cdot \partial \frac{D(t)}{\mathbf{l} \cdot \mathbf{v} - l\zeta_+} \mathbf{l} \cdot \partial \{\varphi(\mathbf{v}_\perp, \nu) - \varphi(\mathbf{v}_\perp, \zeta_+)\}$$

(stable) (A9.20)

The time-dependent coefficient $D(t)$ is defined in both cases by:

$$D(t) = \frac{\mathrm{e}^{2l\gamma_0 t}}{2i\gamma_0} \frac{\bar{\varphi}_+(\zeta_+) - \bar{\varphi}_-(\zeta_-)}{\sigma^+(\zeta_+)\sigma^-(\zeta_-)} \qquad \text{(A9.21)}$$

The discussion of the detailed mechanism of evolution of an initially unstable plasma has to await further investigation of the properties of eq. (A9.18), in particular the study of some simple examples. However, we may draw some qualitative conclusions from the form of this equation.

Equation (A9.18) has the general structure of a Fokker–Planck equation, i.e. it contains a friction term and a diffusion term. A characteristic feature of the equation is the exponential time-dependence of the diffusion and friction coefficients. The latter are moreover functionals of the distribution function; hence their form changes as the latter function evolves in time. The zero ζ_+ of the dielectric constant is itself a functional of $\varphi(\mathbf{v})$. Hence, the overall process is a very complex non-linear friction and diffusion phenomenon in velocity space.

We keep in mind (§ 19) the fact that an unstable plasma is characterized by a velocity distribution with two humps, which are sufficiently widely separated. Both friction and diffusion are stabilizing agents: the first brings the two maxima closer together, the second one broadens them. As a consequence, the zero ζ_+ will move down until the critical separation of the maxima is attained. At this point the zero ζ_+ attains the real value w_0 (neutral stability) and the plasma is stabilized. The exponential time-dependence of the friction and diffusion coefficients makes this process more effective the more unstable the plasma (i.e. the larger γ_0). The transition through the neutral point will be discussed below. The evolution of the weakly stable plasma resulting from this process then continues toward thermal equilibrium under the combined action of the normal collision term and of the extra term of eq. (A9.20). The latter becomes less and less effective as the plasma becomes more and more stable (i.e. as ζ_+ moves down into S_-). In the final stage, this term becomes negligible and the plasma evolves toward equilibrium under the action of the normal collision term alone.

We now study more closely the transition through the neutral stability and show that the two forms of the equation (A9.18) go continuously into one another. The discussion is most clearly performed on eqs. (A9.16) and (A9.17). We first note that although the first bracketed term on the r.h.s. has the same form in both equations, its behavior is different in stable and unstable cases. Indeed, introducing again a test function $U(\mathbf{v})$ and integrating over $\mathbf{v}$, as was done above, we obtain schematically the following contributions:

$$\frac{q_l(\mathbf{v}_\perp, \zeta_-; 0)}{\sigma^-(\zeta_-)} + \text{Sum of residues at poles of } q_l \text{ in } S_- \ldots$$

(unstable case)

Sum of residues at poles of q_l in $S_- \ldots$ (stable case)

We see that there is an extra term in the unstable case. But the contribution of this extra term is exactly canceled by the third term in (A9.16). Hence we may write the following equivalence relation in the sense of distributions:

$$\left\{\frac{q_l(\mathbf{v}_\perp, \nu; 0)}{\varepsilon^-(\nu)} - \frac{q_l(\mathbf{v}_\perp, \zeta_-; 0)}{(\nu-\zeta_-)\sigma^-(\zeta_-)}\right\}_{\text{unstable}} \sim \left\{\frac{q_l(\mathbf{v}_\perp, \nu; 0)}{\varepsilon^-(\nu)}\right\}_{\text{stable}}$$

It is easily seen that the integral terms in both equations tend toward the same limit as $\gamma_0 \to 0$. Consider now the last two terms in eq. (A9.16). They have the following limit as $\gamma_0 \to 0$

$$\frac{2\pi i}{l} \frac{\bar{R}(w_0; 0)}{2i\gamma_0} \{-\pi i \delta_+(\nu - w_0) d_l(\mathbf{v}_\perp, w_0)$$
$$+\pi i \delta_-(\nu - w_0) \cdot d_l(\mathbf{v}_\perp, \nu)(1 + 2l\gamma_0 t + \ldots)\}$$

It is easily verified that, for reasons of parity in $\mathbf{l}$, only the δ-part of the δ_+-functions contributes to the kinetic equation. Hence the divergent parts cancel each other and there remains a term proportional to $d_l(\mathbf{v}_\perp, w_0)$. This term, however, vanishes because $\bar{\varphi}'(w_0)$ (for real w_0) is the imaginary part of $\varepsilon^+(w_0)$, and this implies in turn that $\mathbf{l} \cdot \partial\varphi(\mathbf{v}_\perp, w_0) = 0$. The same argument shows that the exponential term in (A9.17) also vanishes at neutral stability. This achieves the proof that the two forms of the kinetic equation go over continuously into one another.

The type of "self-braking" mechanism of stabilization described here appears to be very general. We have no space to enter into the study of the similar processes occurring in inhomogeneous systems. Drummond and Pines * have studied systems in which there exist periodic inhomogeneities (i.e. waves). They have shown that the interaction among these waves induces an additional stabiliza-

* See reference p. 432.

tion mechanism whose overall effect (although not its origin) is rather similar to the one described here (Fokker–Planck equation with exponentially time-dependent coefficients). In more general unstable inhomogeneous situations, it is thought that not only the velocity distribution but also the inhomogeneity factor $\rho_{\mathbf{k}}(\mathbf{v}; t)$ undergoes a self-braking mechanism.

APPENDIX 10

An Alternative Summation Method for the Ring Diagrams

We shall present here another method of summation of the ring diagrams which was developed in a very recent paper by Résibois.* The characteristic feature of this method lies in the fact that it avoids the solution of an integral equation. This is an advantage, especially in the treatment of generalized problems in which the fundamental integral equation cannot be solved as simply as in the cases treated in this book. The price paid is a delicate handling of contours of integration in some cases (such as unstable plasmas). Moreover, it is our opinion that the fundamental integral equation reflects a basic structure of plasma physics (or, more generally, of the collective behavior), as can be seen from its universal occurrence in these problems.

Résibois' summation procedure is based on an important theorem which we shall now state. We must first define a certain type of diagram, called a *"general F-type diagonal fragment"*. This is defined as a diagonal fragment which contains, between two given vertices A and B, a set of vertices which can be split into two subsets such that no particle involved in the first subset appears in the second one. An example is given in Fig. A10.1.

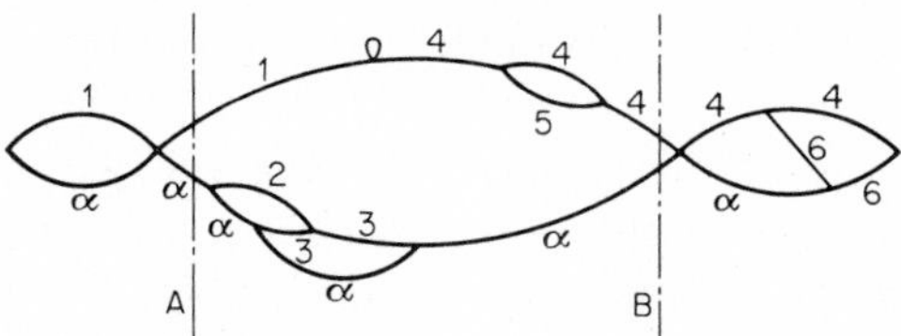

Fig. A10.1. A general F-type fragment.

Topologically such diagonal fragments can be characterized in a very simple way. If one cuts the lines entering to the right of A and those entering to the left of B, the portion between the two cuts is a disconnected diagram with two components. The ring

* P. Résibois, *Phys. of Fluids,* **6**, 817(1963).

diagrams are obviously diagrams of this type, the vertices A and B being the extreme right and left vertices. In order to simplify the notations we shall restrict ourselves in this appendix to the consideration of the rings, which are the only diagrams of interest to us.

Among all possible rings, consider those which have the following property: all the loops on the upper line are to the right of all the loops on the lower line (Fig. A10.2,P). Such diagrams will be called *primitive diagrams*.

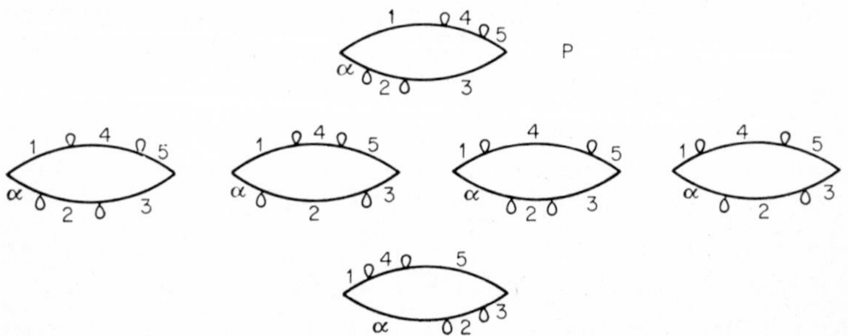

Fig. A10.2. A primitive ring (P) and the permutation class which it generates.

Each primitive ring generates a *permutation class* of rings if the following operations are performed: permute in all possible ways the positions of the upper loops with respect to the positions of the lower ones, by preserving the relative order of particle succession on each line. An example of a permutation class is given in Fig. A10.2.

The structure of the F-diagrams suggests two features:

a) The particles on the upper line are — in some way — independent of those on the lower line.

b) As a consequence, the positions of the vertices on the upper line relative to those of the vertices on the lower line should be — in some sense — irrelevant.

Résibois' theorem, which we are now going to state, gives a precise formulation to these properties.

We start from eq. (A8.6), which we rewrite here in the form:

$$\partial_t \varphi(\mathbf{v}_\alpha; t) = \int_0^t d\tau (2\pi)^{-1} \int dz \, \mathrm{e}^{-izt} Q(z) \tag{A10.1}$$

with

$$Q(z) = \sum_{\substack{n=1 \\ \text{rings}}} (-e^2)^{n+1} \int (d\mathbf{v})^{N-1} \langle 0|\mathscr{L}'[\mathscr{R}^0(z)\mathscr{L}']^n|0\rangle \rho_0(v; t-\tau) \quad \text{(A10.2)}$$

Let Q_P be the contribution of the primitive ring P having i inferior particles (numbered α, *2*, *3*, . . ., i) and $s-i+1$ superior particles (numbered *1*, $i+1$, $i+2$, . . ., s) (see Fig. A10.3). Ac-

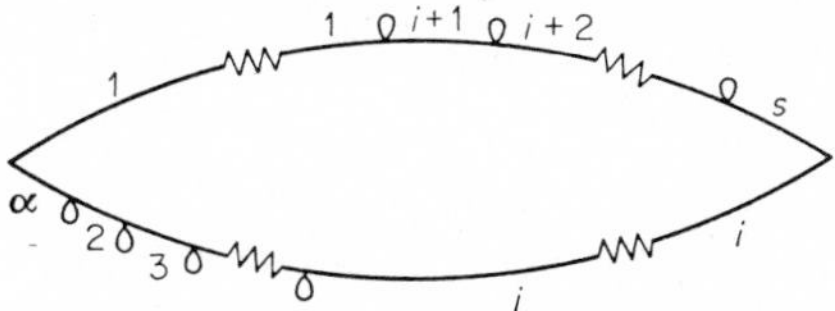

Fig. A10.3. Numbering of particles in a general primitive ring.

cording to the rules of Chapter 9 and Appendix 8, the contribution of this diagram to $Q(z)$ is:

$$Q_P(z) =$$

$$\int d\mathbf{l} \int (d\mathbf{v})^{N-1} d_\alpha \left\{ \frac{1}{i(\mathbf{l} \cdot \mathbf{g}_{\alpha 1} - z)} (-d_\alpha) \frac{1}{i(\mathbf{l} \cdot \mathbf{g}_{21} - z)} (-d_2) \frac{1}{i(\mathbf{l} \cdot \mathbf{g}_{31} - z)} \cdots \right.$$

$$\cdots (-d_{i-1}) \frac{1}{i(\mathbf{l} \cdot \mathbf{g}_{i,1} - z)} d_1 \frac{1}{i(\mathbf{l} \cdot \mathbf{g}_{i,i+1} - z)} d_{i+1} \frac{1}{i(\mathbf{l} \cdot \mathbf{g}_{i,i+2} - z)} \cdots$$

$$\left. \cdots d_{s-1} \frac{1}{i(\mathbf{l} \cdot \mathbf{g}_{i,s} - z)} \right\} d_{si} \rho_0(v; t-\tau) \quad \text{(A10.3)}$$

Résibois' theorem * states that the sum $\sum_{\{P\}} Q_P(z)$ of the contributions of all the rings belonging to the permutation class generated by P is:

$$\sum_{\{P\}} Q_P(z) =$$

$$\int d\mathbf{l} \int (d\mathbf{v})^{N-1} d_\alpha \left\{ (2\pi)^{-1} \int_{C'} dz' \, \mathscr{V}_{\mathbf{l}}^{\alpha i}(z') \mathscr{V}_{-\mathbf{l}}^{1s}(z-z') \right\} d_{si} \rho_0(v; t-\tau)$$

$$\text{(A10.4)}$$

* Actually, in his original paper, Résibois proved the corresponding theorem for the time-dependent Green's function. Of course, the convolution is then replaced (through a Laplace transformation) by the ordinary product. We prefer to state here the theorem within the resolvent formalism in order not to switch back and forth between the variables t and z.

where

$$\mathscr{V}_{l}^{\alpha i}(z) = \frac{1}{i(\mathbf{l}\cdot\mathbf{v}_\alpha - z)}(-d_\alpha)\frac{1}{i(\mathbf{l}\cdot\mathbf{v}_2 - z)}(-d_2)\ldots(-d_{i-1})\frac{1}{i(\mathbf{l}\cdot\mathbf{v}_i - z)}$$

$$\mathscr{V}_{-l}^{1s}(z) = \frac{1}{i(-\mathbf{l}\cdot\mathbf{v}_1 - z)}d_1\frac{1}{i(-\mathbf{l}\cdot\mathbf{v}_{i+1} - z)}d_{i+1}\ldots d_{s-1}\frac{1}{i(-\mathbf{l}\cdot\mathbf{v}_s - z)}$$

The contour C' passes above all the singularities of $\mathscr{V}_{l}^{\alpha i}(z')$ and below all the singularities of $\mathscr{V}_{-l}^{1s}(z-z')$.

The content of this theorem is as follows. The sum of contributions $\sum Q_{\mathrm{P}}(z)$ is obtained (roughly speaking) by replacing the propagator between the two terminal vertices (d_α, d_{si}) in eq. (A10.3) by the convolution of the propagators corresponding to the upper and lower lines, each being calculated independently of the other.

We shall not give a general proof of this theorem, but shall rather illustrate it in the especially simple case of the permutation class shown in Fig. A10.4.

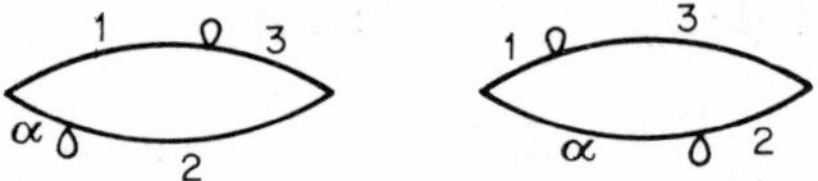

Fig. A10.4. A simple permutation class.

We note that the vertices do not play any role in the proof of the theorem; we therefore delete them in the mathematical expressions. Then the sum of the bracketed terms in (A10.3) corresponding to the two rings of the permutation class is:

$$J = \frac{1}{i(\mathbf{l}\cdot\mathbf{g}_{\alpha 1} - z)}\left\{\frac{1}{i(\mathbf{l}\cdot\mathbf{g}_{21} - z)} + \frac{1}{i(\mathbf{l}\cdot\mathbf{g}_{\alpha 3} - z)}\right\}\frac{1}{i(\mathbf{l}\cdot\mathbf{g}_{23} - z)}$$

By Résibois' theorem this expression should equal the following one:

$$J' = \frac{1}{2\pi}\int_{C'} dz' \frac{1}{i(\mathbf{l}\cdot\mathbf{v}_\alpha - z')}\,\frac{1}{i(\mathbf{l}\cdot\mathbf{v}_2 - z')}\,\frac{1}{i(-\mathbf{l}\cdot\mathbf{v}_1 - z + z')}\,\frac{1}{i(-\mathbf{l}\cdot\mathbf{v}_3 - z + z')}$$

The contour C' can be taken as a straight line parallel to and above the real axis, lying below the ordinate of z (which of course lies in S_+ by definition). It can be completed by a semi-circle at infinity in S_-. The value of the integral then equals $(-2\pi i)$ times the sum of the residues at the two poles $z' = \mathbf{l}\cdot\mathbf{v}_\alpha$, $z' = \mathbf{l}\cdot\mathbf{v}_2$:

$$J' = \frac{1}{i\mathbf{l}\cdot\mathbf{g}_{2\alpha}}\left\{\frac{1}{i(\mathbf{l}\cdot\mathbf{g}_{\alpha 3}-z)}\,\frac{1}{i(\mathbf{l}\cdot\mathbf{g}_{\alpha 1}-z)} - \frac{1}{i(\mathbf{l}\cdot\mathbf{g}_{23}-z)}\,\frac{1}{i(\mathbf{l}\cdot\mathbf{g}_{21}-z)}\right\}$$

It is readily verified that $J' = J$.

We now go over to the application of Résibois' theorem to the summation of the rings. Referring to eqs. (A8.11–12), we write the kinetic equation in the form

$$\begin{aligned}\partial_t\varphi(\mathbf{v}_\alpha; t) &= \int d\mathbf{l}\, d_\alpha\, \mathscr{F}_{\mathbf{l}}(\mathbf{v}_\alpha; t)\\ \mathscr{F}_{\mathbf{l}}(\mathbf{v}; t) &= \int_0^t d\tau\, \frac{l}{2\pi}\int_C dw\; e^{-ilw\tau}\, F_{\mathbf{l}}(\mathbf{v}; w)\end{aligned}\tag{A10.5}$$

Moreover, we write $F_{\mathbf{l}}(\mathbf{v}; w)$ in the form

$$F_{\mathbf{l}}(\mathbf{v}; w) = \frac{l}{2\pi}\int_{C'} dw'\, G_{\mathbf{l}}(\mathbf{v}; w, w')\tag{A10.6}$$

We shall now prove that Résibois' method gives for $F_{\mathbf{l}}(\mathbf{v}; w)$ the same expression as the solution of the singular integral equation (A8.13).

Applying our previous rules of construction of the diagrams, as well as the theorem (A10.4), we obtain $G_{\mathbf{l}}(\mathbf{v}; w, w')$ by a summation of the loops on each line separately (see also § 13)*

$$\begin{aligned}G_{\mathbf{l}}(\mathbf{v}_\alpha; w, w') = \int_s \frac{1}{i(\mathbf{l}\cdot\mathbf{v}_\alpha - lw')}&\left\{\frac{1}{i(-\mathbf{l}\cdot\mathbf{v}_s - lw + lw')}\right.\\ &+ \int_1 \frac{1}{i(-\mathbf{l}\cdot\mathbf{v}_1 - lw + lw')}\, d_1\varphi_1\, \frac{1}{i(-\mathbf{l}\cdot\mathbf{v}_s - lw + lw')}\\ &+ \int_1 \frac{1}{i(-\mathbf{l}\cdot\mathbf{v}_1 - lw + lw')}\, d_1\varphi_1 \int_2 \frac{1}{i(-\mathbf{l}\cdot\mathbf{v}_2 - lw + lw')}\\ &\left.\cdot\, d_2\varphi_2\, \frac{1}{i(-\mathbf{l}\cdot\mathbf{v}_s - lw + lw')} + \ldots\right\} d_{s\alpha}\varphi_s\varphi_\alpha +\end{aligned}$$

* The diagrams with no loop on the lower line have to be treated separately, because the fixed particle α is involved in the extreme right vertex $d_{s\alpha}$.

$$+\int_i\int_s\left\{\frac{1}{i(\mathbf{l}\cdot\mathbf{v}_\alpha-lw')}d_\alpha^*\varphi_\alpha\frac{1}{i(\mathbf{l}\cdot\mathbf{v}_i-lw')}\right.$$

$$+\frac{1}{i(\mathbf{l}\cdot\mathbf{v}_\alpha-lw')}d_\alpha^*\varphi_\alpha\int_2\frac{1}{i(\mathbf{l}\cdot\mathbf{v}_2-lw')}d_2^*\varphi_2\frac{1}{i(\mathbf{l}\cdot\mathbf{v}_i-lw')}$$

$$\left.+\ldots\right\}\left\{\frac{1}{i(-\mathbf{l}\cdot\mathbf{v}_s-lw+lw')}+\int_1\frac{1}{i(-\mathbf{l}\cdot\mathbf{v}_1-lw+lw')}\right.$$

$$\left.\cdot\, d_1\varphi_1\frac{1}{i(-\mathbf{l}\cdot\mathbf{v}_s-lw+lw')}+\ldots\right\}d_{si}\varphi_s\varphi_i \tag{A10.7}$$

The summation is now immediate, because we are left only with geometric progressions. Calling

$$J_l(w)=\int_1\frac{d_1\varphi_1}{il(\nu_1-w)}$$

the result is

$$G_l(\mathbf{v}_\alpha;w,w')=\int_s\frac{1}{il(\nu_\alpha-w')}\,\frac{1}{il(-\nu_s-w+w')}\,\frac{1}{1-J_l(w'-w)}\,d_{s\alpha}\varphi_s\varphi_\alpha$$

$$+\int_i\int_s\frac{d_\alpha^*\varphi_\alpha}{il(\nu_\alpha-w')}\,\frac{1}{il(\nu_i-w')}\,\frac{1}{il(-\nu_s-w+w')}$$

$$\cdot\,\frac{1}{1-J_l(w')}\,\frac{1}{1-J_l(w'-w)}\,d_{si}\varphi_s\varphi_i \tag{A10.8}$$

From this expression we now calculate $F_l(\mathbf{v};w)$ by substituting it into eq. (A10.6). The first term is calculated straightforwardly by taking the residue of the integrand at the pole $w'=\nu_\alpha$

$$F_l^{(1)}(\mathbf{v}_\alpha;w)=\int_s\frac{1}{il(\nu_\alpha-\nu_s-w)}\,\frac{1}{\varepsilon_l^-(\nu_\alpha-w)}\,d_{s\alpha}\varphi_\alpha\varphi_s$$

We have used the fact that for $w\in S_+$ (and hence $\nu-w\in S_-$), the expression $1-J_l(\nu-w)$ is nothing other than the minus-dielectric constant. Letting w approach the real axis from above, it is easily seen, by using the definitions (A8.14, 16), that

$$F_l^{(1)}(\mathbf{v};w)=\frac{q_l(\mathbf{v};w)}{\varepsilon_l^-(\nu-w)}$$

This expression is identical with the first term of eq. (A8.35).

In the second term we deform the contour C', bringing it onto the upper edge of the real axis. *In the stable case*, no singularities are encountered in the area swept during the deformation. After this operation we also bring w onto the real axis, letting it always be infinitesimally above w'. We now rename the real variable $w' \equiv \nu_r$. We thus obtain

$$F_l^{(2)}(\mathbf{v}; w) = \frac{\pi i}{l} d_l(\mathbf{v}) \int d\nu_r \delta_-(\nu - \nu_r) \frac{\bar{Q}_1(\nu_r; w)}{\varepsilon^+(\nu_r)\varepsilon^-(\nu_r - w)}$$

where

$$\bar{Q}_1(\nu_r; w) = \frac{1}{2i} \frac{e^2}{2\pi m l^2} \int d\nu_i \int d\nu_s \delta_+(\nu_r - \nu_i) \delta_-(\nu_r - \nu_s) [\bar{\varphi}'(\nu_s - w)\bar{\varphi}(\nu_i) - \bar{\varphi}(\nu_s - w)\bar{\varphi}'(\nu_i)]$$

which is identical to the second term of (A8.35). We have thus finished the proof of the equivalence of the two methods of summation.

Let us now consider the *unstable case*. Much care must be taken in order to obtain the correct branch of the multivalued functions which occur here. We therefore take the contour C in eq. (A10.5) *very far* up into S_+. We shall see presently what "very far" means. The contour C' will lie, according to its definition, somewhere in the upper half-plane, below the contour C. Hence,

$$w' \in S_+$$
$$w' - w \in S_-$$

These relations determine uniquely the branch of the Cauchy integrals which appears in (A10.8):

$$G_l(\mathbf{v}; w, w') = -\frac{e^2}{\pi m l^3} \frac{1}{\nu - w'} \left\{ \frac{\varphi'(\mathbf{v})\bar{\varphi}_-(w' - w) - \varphi(\mathbf{v})\bar{\varphi}'_-(w' - w)}{\varepsilon^-(w' - w)} \right.$$
$$\left. + \frac{2\pi i}{l} d_l(\mathbf{v}) \frac{\bar{\varphi}'_+(w')\bar{\varphi}_-(w' - w) - \bar{\varphi}_+(w')\bar{\varphi}'_-(w' - w)}{(w' - \zeta_+)(w' - w - \zeta_-)\sigma^+(w')\sigma^-(w' - w)} \right\} \qquad \text{(A10.11)}$$

We now calculate $F_l(\mathbf{v}; w)$ by substituting (A10.11) into (A10.6). The first term is trivial and gives the same result as in the stable case, namely (A10.9). In the contribution of the second term we must carefully analyze the positions of the poles of the integrand in order to determine the position of the contour C' of the convolution:

$$F_l^{(2)}(\mathbf{v}; w) =$$

$$-\frac{ie^2}{\pi ml^2}\frac{d_l(\mathbf{v})}{l}\int_{C'} dw' \frac{1}{\nu-w'} \frac{\bar{\varphi}'_+(w')\bar{\varphi}_-(w'-w)-\bar{\varphi}_+(w')\bar{\varphi}'_-(w'-w)}{(w'-\zeta_+)(w'-w-\zeta_-)\sigma^+(w')\sigma^-(w'-w)}$$

The contour C' must pass above the singularities of the functions of w' alone, and below those of functions of $w'-w$. The former of these are $w' = \nu$, on the real axis, $w' = \zeta_+ \in S_+$, the singularities of $\bar{\varphi}_+(w')$, $\bar{\varphi}'_+(w')$ and the zeros of $\sigma^+(w')$. Using the general properties of plus-functions and of the plus-dielectric constant, we know that the singularities of $\bar{\varphi}_+ (w')$, $\bar{\varphi}'_+ (w')$ and the zeros of $\sigma^+ (w')$ lie in S_-. In the second group, we have $w' = w+\zeta_-$, which lies below the contour C, and a set of singularities located above C. If the contour C is at a distance greater than $2\gamma_0$ above the real axis, i.e. Im $w > 2\gamma_0$, the two groups of singularities are separated by a strip (see Fig. A10.5a). The contour C', according to its definition, can be any line parallel to the real axis and located in the latter strip. But this contour can be deformed into the contour of Fig. A10.5b, which is exactly identical to the contour Γ_1 appearing in eq. (A9.9). The equivalence of both methods of summation is therefore proven in the unstable case too.

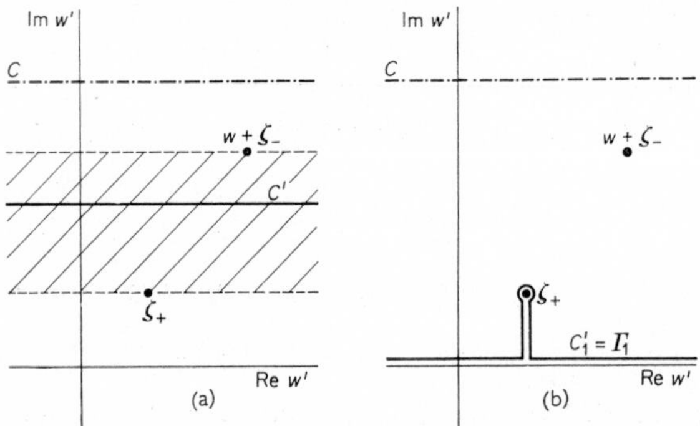

Fig. A10.5. Contour of integration C' in the unstable case.

APPENDIX 11

The Kinetic Equation for a Plasma in a Strong External Field

We consider, in this appendix, some very recent developments in this domain, following very closely a paper by Severne,* to which we refer the reader for additional details.

In Chapter 13, the two approaches to the study of the behavior of a plasma in an external field were discussed in some detail. The most general method is the one based on the direct calculation of the response of the system to an external field, because in this case we do not need any restrictive assumption. Formula (57.13), for instance, is a quite general expression, and (57.16) or (57.19) represent its complete linear approximation.

The alternative method, based on kinetic equations, suffered until recently from a number of drawbacks which have been listed under the headings B, C and D of § 56 (we exclude here the discussion of thermal constraints). The recent work, which we shall briefly consider below, has eliminated these difficulties and has brought the method of kinetic equations to the same degree of generality as the response-function method.

We consider the system defined by the Liouville equation (57.1–4). For simplicity, we shall assume here that the electric field $\mathbf{E}$ is constant in space and time; the more general case of time-dependent fields, as well as that of magnetic and gravitational fields, has been treated by Severne in the paper quoted above.

The equation of motion of particle j in the field, in the absence of interactions, is

$$m_j \dot{\mathbf{v}}_j = z_j e \mathbf{E} \tag{A11.1}$$

Let us write the solution of this equation in the form:

$$\mathbf{v}_j' = \mathbf{v}_j'' + (z_j e/m_j)(t'-t'')\,\mathbf{E}, \qquad t' > t'' \tag{A11.2}$$

where $\mathbf{v}_j'$ is the velocity at time t', and $\mathbf{v}_j''$ is the velocity at an earlier time t''. The position of the particle is then given by:

* G. Severne, *Physica*, **29** (1963) (to be published).

$$\mathbf{x}_j' = \mathbf{x}_j'' + (t'-t'')\mathbf{v}_j'' + \tfrac{1}{2}(z_j e/m_j)(t'-t'')^2\mathbf{E} \qquad \text{(A.11.3)}$$

Once we know the solution of the equations of motion, we can also solve the Liouville equation in the absence of interactions. In particular, let the Green's function $\mathscr{G}_E(xvt|x'v't')$ be defined as the solution of the following equation:

$$(\mathscr{L}^0 + e\mathscr{L}_E)\mathscr{G}_E(xvt|x'v't') = \delta(x-x')\delta(v-v')\delta(t-t') \qquad \text{(A11.4)}$$

with the causal condition

$$\mathscr{G}_E(xvt|x'v't') = 0, \qquad t < t' \qquad \text{(A11.5)}$$

Let us rewrite the definitions of $\mathscr{L}^0$ and $\mathscr{L}_E$:

$$\begin{aligned} \mathscr{L}^0 &= \sum_j \mathbf{v}_j \cdot \boldsymbol{\nabla}_j \\ \mathscr{L}_E &= \sum_j (z_j/m_j)\mathbf{E} \cdot \boldsymbol{\partial}_j \end{aligned} \qquad \text{(A11.6)}$$

The solution of (A11.4) is obtained in a way similar to that in § 5, with the result (written in matrix notation):

$$\begin{aligned} \langle xvt|\mathscr{G}_E|x'v't'\rangle = \theta(t-t') \prod_j \delta[\mathbf{x}_j - \mathbf{x}_j' - (t-t')\mathbf{v}_j' \\ -\tfrac{1}{2}(z_j e/m_j)(t-t')^2\mathbf{E}]\delta[\mathbf{v}_j - \mathbf{v}_j' - (z_j e/m_j)(t-t')\mathbf{E}] \end{aligned} \qquad \text{(A11.7)}$$

In the Fourier representation, this matrix goes over into

$$\begin{aligned} \langle kvt|\mathscr{G}|k'v't'\rangle = \theta(t-t') \prod_j e^{-i\mathbf{k}_j\mathbf{v}_j'(t-t') - \frac{1}{2}(z_j e/m_j)\mathbf{k}_j\mathbf{E}(t-t')^2} \\ \delta(\mathbf{k}_j - \mathbf{k}_j')\delta[\mathbf{v}_j - \mathbf{v}_j' - (z_j e/m_j)\,\mathbf{E}(t-t')] \end{aligned} \qquad \text{(A11.8)}$$

We note that the unperturbed Green's function is no longer diagonal in $\mathbf{v}$, $\mathbf{v}'$ as it is in the field-free case. It can, however, be written in the following form, in which the non-diagonal character is "hidden":

$$\begin{aligned} \langle kvt|\mathscr{G}_E|k'v't'\rangle = \theta(t-t') \prod_j \delta(\mathbf{k}_j - \mathbf{k}_j') \\ e^{-\frac{1}{2}e\mathscr{L}_E(t-t')}\, e^{-i\mathbf{k}_j\cdot\mathbf{v}_j(t-t')}\, e^{-\frac{1}{2}e\mathscr{L}_E(t-t')}\,\delta(\mathbf{v}-\mathbf{v}') \end{aligned} \qquad \text{(A11.9)}$$

It is indeed immediately verified that the action of the operator $\mathscr{G}_E$ on a function of k, v and t:

$$\mathscr{G}_E f = \int_0^\infty dt' \int dk' \int dv' \,\langle kvt|\mathscr{G}_E|k'v't'\rangle f(k'v't')$$

gives the same result with the two forms of $\mathscr{G}_E$. When written in the form (A11.9), the formulae bear a very close resemblance to the corresponding field-free expressions.

From here on, we can take over the perturbation theory developed in § 5. The essential point of this formalism is the fact that the field is included in the unperturbed hamiltonian and hence is treated exactly. The main difference from the field-free perturbation theory is the following:

In the general case of a time-dependent field, the Green's function is no longer a function of $(t-t')$ *only*; therefore the resolvent method, based on a Laplace transformation of the Green's function, can no longer be used: the perturbation theory must be developed in terms of the Greens' function itself. This is, however, not a serious difficulty. [In the case of a static field which we consider here, the difficulty does not appear, but we shall still use the Green's function technique, which, for a complicated time dependence of $\mathscr{G}_E$, as given by (A11.8), proves to be the only practicable method.] The diagram technique will be taken over without any essential changes. The only difference is that in the present case the lines represent matrix elements not of the resolvent but of the unperturbed Green's function, and that there will be additional integrations over the intermediate times.

As we consider only a spatially uniform field, it is sufficient to study homogeneous systems. We shall, moreover, restrict ourselves to the evaluation of the N-particle velocity distribution $\rho_0(|v; t)$.

It is very easy to obtain from eqs. (4.16), (5.4) and (2.8) a perturbation expansion quite similar to that of eq. (5.26):

$$\rho_{\{k\}}(v; t) = \sum_{n=0}^{\infty} (-e^2)^n \int dv' \sum_{\{k'\}} \left(\frac{8\pi^3}{\Omega}\right)^{\nu'-\nu} \langle kvt|\mathscr{G}_E[\mathscr{L}'\mathscr{G}_E]^n|k'v'0\rangle \, \rho_{\{k'\}}(v'; 0) \tag{A11.10}$$

In this formula, the matrix element of the perturbation is given by (5.24); we rewrite it, however, in order to adapt it to the present notation:

$$\langle kvt|\mathscr{L}'|k'v't'\rangle = \delta(t-t') \prod_{l=1}^{N} \delta(\mathbf{v}_l - \mathbf{v}'_l) \prod_{r(\neq j,n)} \delta_{\mathbf{k}'_r - \mathbf{k}_r}$$

$$\delta_{\mathbf{k}'_j+\mathbf{k}'_n-\mathbf{k}_j-\mathbf{k}_n} (8\pi^3/\Omega) V^{jn}_{|\mathbf{k}'_j-\mathbf{k}_j|} i(\mathbf{k}'_j - \mathbf{k}_j) \cdot (m_j^{-1} \partial_j - m_n^{-1} \partial_n) \quad \text{(A11.11)}$$

The complete expression for $\rho_0(v; t)$ given by (A11.10) can be split into three parts (see § 29)

a) The trivial part $n = 0$;

b) The sum of all contributions beginning at left with a diagonal fragment;

c) The sum of all destruction fragments (see § 29c).

Equation (A11.8) can then be rewritten as

$$\begin{aligned} \rho_0(|v; t) &= \mathrm{e}^{-e\mathscr{L}_E t} \rho_0(|v; 0) \\ &+ \int_0^t dt' \, \mathrm{e}^{-e\mathscr{L}_E(t-t')} \left\{ \int_0^{t'} dt'' K_E(t'|t'') \rho_0(|v; t'') \right\} \\ &+ \sum_{\{\mathbf{k}''\}} (8\pi^3/\Omega)^{\nu''} D_E(t'|0) \rho_{\{\mathbf{k}''\}}(v; 0) \end{aligned} \quad \text{(A11.12)}$$

where K_E is defined as the sum of all possible diagonal fragments:

$$K_E(t|t') = \sum_{n=2}^{\infty} (-e^2)^n \langle 0vt|[\mathscr{L}' \mathscr{G}_E]^{n-1} \mathscr{L}'|0v't'\rangle_{\text{diag. fr.}} \quad \text{(A11.13)}$$

and $D_E(t|t')$ is the sum of all destruction fragments

$$D_E(t|t') = \sum_{n=1}^{\infty} (-e^2)^n \langle 0vt|(\mathscr{L}' \mathscr{G}_E)^n |k'v't'\rangle_{\text{destr. fr.}} \quad \text{(A11.14)}$$

We now take the time derivative of (A11.12):

$$\begin{aligned} \partial_t \rho_0(|v; t) + e\mathscr{L}_E \rho_0(|v; t) &= \int_0^t dt' K_E(t|t') \rho_0(|v; t') \\ &+ \sum_{\{\mathbf{k}'\}} (8\pi^3/\Omega)^{\nu'} D_E(t|0) \rho_{\{\mathbf{k}'\}}(v; 0) \end{aligned} \quad \text{(A11.15)}$$

This is the most general "*master equation*" describing the evolution of a large system in a uniform field of arbitrary strength for all times. It is clearly a *non-markoffian* equation of the type discussed in Appendix 8. The kernel $K_E(t|t')$ defines a non-instantaneous collision process, and also takes correctly into account the change in velocities due to the action of the external field during the collision time $(t-t')$.

We now proceed to simplify our model. We shall restrict ourselves to the Landau approximation, i.e. retain in the sum defining $K_E(t|t')$ only the cycle diagram.

Let $\mathscr{C}_E \equiv \sum_{j<n} \mathscr{C}_{E;jn}$ denote the term in (A11.15) corresponding to the cycle:

$$\mathscr{C}_{E;jn}(t-t') = (8\pi^3 e^4/\Omega) \int d\mathbf{k}\, V_k^{jn} \mathbf{k} \cdot (m_j^{-1} \partial_j - m_n^{-1} \partial_n)$$

$$\mathrm{e}^{-\frac{1}{2}e\mathscr{L}_E(t-t')}\, \mathrm{e}^{-i\mathbf{k}\cdot\mathbf{g}_{jn}(t-t')}\, \mathrm{e}^{-\frac{1}{2}e\mathscr{L}_E(t-t')}$$

$$V_k^{jn} \mathbf{k} \cdot (m_j^{-1} \partial_j - m_n^{-1} \partial_n) \tag{A11.16}$$

The master equation of evolution of our simpler system thus becomes:

$$\partial_t \rho_0(v;t) + e\mathscr{L}_E \rho_0(v;t) = \sum_{j<n} \int_0^t dt'\, \mathscr{C}_{E;jn}(t|t') \rho_0(|v;t') + \mathscr{D}_E[t;\{\rho_{k'}(0)\}] \tag{A11.17}$$

We do not write down the explicit expression for the term related to the destruction fragments because we shall not study it in detail here. It will be sufficient to note that this term contains the memory of the initial correlations. One can prove that $\mathscr{D}_E[t;\{\rho_{k'}(0)\}] \to 0$ for $t \gg t_p$; this is actually the only property of this term which we shall use.

From this master equation, we can easily derive kinetic equations for the reduced velocity distributions of the ions and of the electrons. Performing explicitly the displacement operations, we finally obtain for the electrons [distribution: $\varphi(\mathbf{v};t)$, mass: m]

$$\partial_t \varphi(\mathbf{v}_\alpha;t) - (e/m)\mathbf{E}\cdot\partial_\alpha \varphi(\mathbf{v}_\alpha;t) = 4\pi^3 e^4 c \int_0^t d\tau \int d\mathbf{v}_n \int d\mathbf{l}$$

$$V_l^2 m^{-1} \mathbf{l}\cdot\partial_\alpha \{\mathrm{e}^{-i\mathbf{l}\cdot\mathbf{g}_{\alpha n}\tau} m^{-1} \mathbf{l}\cdot\partial_{\alpha n} \varphi(\mathbf{v}_\alpha + em^{-1}\mathbf{E}\tau; t-\tau) \tag{A11.18}$$

$$\varphi(\mathbf{v}_n + em^{-1}\mathbf{E}\tau; t-\tau) + \mathrm{e}^{-i\mathbf{l}\cdot\mathbf{g}_{\alpha n}\tau - ie\mathbf{l}\cdot\mathbf{E}\tau^2/2\mu}$$

$$\mathbf{l}\cdot(m^{-1}\partial_\alpha - M^{-1}\partial_n)\varphi(\mathbf{v}_\alpha + em^{-1}\mathbf{E}\tau; t-\tau)\Phi(\mathbf{v}_n - eM^{-1}\mathbf{E}\tau; t-\tau)\}$$

A similar equation holds for the ions [distribution: $\Phi(\mathbf{v};t)$, mass: M]; μ is the reduced mass of ions and electrons: $\mu^{-1} = m^{-1} + M^{-1}$.

This equation was first obtained by Silin,* who used a different method of derivation. The general equation (A11.15), as well as the present method of derivation, are due to Severne.

We shall now investigate the behavior of this equation for long times. This is actually the more important problem from the practical point of view. Indeed, in the case of a constant field, we are mainly interested in the existence and the properties of the stationary state.** A difficulty appears here as compared with the field-free case treated in Appendix 8. Indeed, in eq. (A11.18) the "retardation" appears *twice* in the argument of φ (or Φ) in the collision term: once associated with t in the form $t-\tau$, and once associated with $\mathbf{v}$ in the form $\mathbf{v}+em^{-1}\mathbf{E}\tau$. One would be tempted to set $\tau = 0$ in both arguments in order to get the asymptotic equation; this, however, is not a correct procedure, as will be shown now.

The long-time approximation is based on the following basic theorems, whose proof we omit here:

$$\left.\begin{array}{l}\mathscr{C}_{E;jn}(\tau) \to 0 \\ \mathscr{D}_E(\tau) \to 0\end{array}\right. \quad \text{as } \tau \to \infty \qquad (\tau \gg t_p)$$

For such long times, eq. (A11.17) reduces to

$$(\partial_t + e\mathscr{L}_E)\rho_0(|v; t) = \sum_{j<n} \int_0^\infty d\tau\, \mathscr{C}_{E;jn}(\tau)\rho_0(|v; t-\tau) \tag{A11.19}$$

Let us expand $\rho_0(|v; t-\tau)$ into a Taylor series around $\tau = 0$:

$$(\partial_t + e\mathscr{L}_E)\rho_0(|v; t) = \sum_{n=0}^{\infty} \Psi_E^{(n)}\, \partial_t^n\, \rho_0(|v; t) \tag{A11.20}$$

$$\Psi_E^{(n)} = \left[\frac{(-1)^n}{n!}\right] \int_0^\infty d\tau\, \tau^n\, \mathscr{C}_E\,(\tau) \tag{A11.21}$$

The term $\Psi_E^{(0)} \equiv \Psi_E$ in this expansion corresponds to the "leading universal approximation", whereas the other terms of

* V. P. Silin, *J. Exptl. Theoret. Phys. U.S.S.R.*, **38**, 1771 (1960); transl. *Sov. Phys. JETP*, **11**, 1277 (1960).

** The situation is different in the case of a time-dependent external field, in which case the non-markoffian property of eq. (A11.15) plays an important role.

the series correspond to the "non-leading universal terms" discussed in § 27. The "specific" terms have disappeared here through our asymptotic approximation which consisted in going from (A11.17) to (A11.19). Equation (A11.20) can be made more explicit by successive substitutions of $\partial_t^n \rho_0$. The result is

$$\partial_t \rho_0(|v; t) = \Omega_E (\Psi_E - e\mathscr{L}_E) \rho_0(|v; t) \tag{A11.22}$$

where

$$\Omega_E = 1 + \Psi_E^{(1)} + \Psi_E^{(1)} \Psi_E^{(1)} + \Psi_E^{(2)} (\Psi_E - e\mathscr{L}_E) + \dots \tag{A11.23}$$

Equation (A11.22) is equivalent to the asymptotic equation (A11.19). It is clear that the "leading universal approximation" or, equivalently, the *general Boltzmann-like equation*, according to the definition of § 56, is obtained by setting $\Omega_E = 1$. It is also clear that, except for the usual simple cases (dilute gases, Landau approximation, ring approximation) it would be inconsistent to set $\Omega_E = 1$: the evolution toward equilibrium of a system other than the simple systems mentioned above is *not* dictated by the leading universal equation. However, it is immediately seen that the stationary state characterized by $\partial_t \rho_0 = 0$, if it exists, is determined by *

$$(\Psi_E - e\mathscr{L}_E) \rho_s(v) = 0 \tag{A11.24}$$

This property generalizes the basic theorem of § 59 for the case of arbitrarily strong fields: in all cases, the Boltzmann-like equation determines the stationary state completely, and the operator Ω_E related to the non-instantaneous character of the collisions disappears in this state.

Equation (A11.24) obtains directly from (A11.19) by replacing $t-\tau$ by t in the argument of ρ_0 on the r.h.s. of (A11.19). It should be noted that, in spite of its appearance, eq. (A11.24) is still an *integral equation* (*with respect to* τ), relating the value of $\rho_s(v)$ to all the values it had for velocities at an earlier time.

We shall now study, in some detail, the stationary state in the *linear approximation*. This case is particularly interesting because

* This statement implies that the stationary state, if it exists, is unique. This property is closely related to the H-theorem; see the remark in the footnote on p. 276.

it relates to the concept of *electrical conductivity*. We expand the asymptotic collision operator Ψ_E and retain only terms which are linear in the field:

$$\Psi_E = \Psi_0 + \Xi_E + \ldots \qquad \text{(A11.25)}$$

From eq. (A11.16) is immediately follows that

$$\Xi_E = 4\pi^4 e^4 \Omega^{-1} \sum_{j<n} \int d\mathbf{l}\, V_l^2 \mathbf{l} \cdot (m_j^{-1} \partial_j - m_n^{-1} \partial_n)(2i)^{-1}$$
$$\{e\mathscr{L}_E \delta'_-(\mathbf{l} \cdot \mathbf{g}_{jn}) + \delta'_-(\mathbf{l} \cdot \mathbf{g}_{jn}) e\mathscr{L}_E\} \mathbf{l} \cdot (m_j^{-1} \partial_j - m_n^{-1} \partial_n) \qquad \text{(A11.26)}$$

It is easily shown that

$$\tfrac{1}{2}\{e\mathscr{L}_E \delta'_-(\mathbf{l} \cdot \mathbf{g}_{jn}) + \delta'_-(\mathbf{l} \cdot \mathbf{g}_{jn}) e\mathscr{L}_E\} = -\pi i \delta_-(\mathbf{l} \cdot \mathbf{g}_{jn}) e\mathscr{L}_E \delta_-(\mathbf{l} \cdot \mathbf{g}_{jn})$$

But $\pi\delta_-(\mathbf{l} \cdot \mathbf{g}_{jn})$ is nothing other than a matrix element of the field-free unperturbed resolvent for $z = 0$: $\mathscr{R}^0(0)$. Hence eq. (A11.26) can be written as

$$\Xi_E = -e^4 \langle 0|\mathscr{L}' \mathscr{R}^0(0) e\mathscr{L}_E \mathscr{R}^0(0) \mathscr{L}'|0\rangle \qquad \text{(A11.27)}$$

We now solve the linearized equation (A11.24) for the stationary state:

$$(\Psi_0 + \Xi_E - e\mathscr{L}_E)\rho_s = 0 \qquad \text{(A11.28)}$$

by assuming

$$\rho_s = \rho_0^0 + \bar{\rho}_s$$

The zeroth-order equation shows that ρ_0^0 is the equilibrium distribution in the absence of the field, and hence the first-order equation is:

$$e\mathscr{L}_E \rho_0^0 = \Psi_0 \bar{\rho}_s + \Xi_E \rho_0^0 \qquad \text{(A11.29)}$$

We now note that by the general theorem on creation fragments, §§ 43 and 52, the creation vertex acting on ρ_0^0 provides the equilibrium two-body correlation function in the Landau approximation:

$$\langle \mathbf{k}, -\mathbf{k}|\mathscr{R}^0(0)\mathscr{L}'_{jn}|0\rangle \rho_0^0 = \rho_{\mathbf{k},-\mathbf{k}}^0(j, n| \ldots)$$

Hence, the formal solution of eq. (A11.29) is

$$\bar{\rho}_s = \frac{1}{\Psi_0} e\mathscr{L}_E \rho_0^0 + \frac{1}{\Psi_0} \sum_{j<n} \int d\mathbf{k}\, \langle 0|\mathscr{L}'_{jn} \mathscr{R}^0(0)|\mathbf{k}, -\mathbf{k}\rangle$$
$$e\mathscr{L}_E \rho_{\mathbf{k},-\mathbf{k}}^0(j, n| \ldots) \qquad \text{(A11.30)}$$

We now calculate the stationary current from this formula:

$$\mathbf{J}^s = e\int d\mathbf{v} \sum_m z_m \mathbf{v}_m \bar{\rho}_s = e^2\beta \sum_{mn} z_m z_n \mathbf{E} \cdot \int dv\, \mathbf{v}_m \frac{1}{-\Psi_0} \left\{ \mathbf{v}_n \rho_0^0(v) \right.$$

$$\left. + \sum_{r<s} \int d\mathbf{k} \langle 0|\mathscr{L}'_{rs}\mathscr{R}^0(0)|\mathbf{k}, -\mathbf{k}\rangle \mathbf{v}_n \cdot \rho^0_{\mathbf{k},-\mathbf{k}}(r, s| \ldots) \right\} \tag{A11.31}$$

One recognizes in the first term the *kinetic part of the stationary current* as given in eq. (59.1). It is easily shown that the second term is the *correlational part* of this current. Indeed, starting from (57.19) and performing the calculations of § 58 on the term connected with the correlations, one obtains:

$$\mathbf{j}_{\text{corr}}(z) = e^2\beta \sum_m \sum_n z_m z_n \sum_k \int dv\, \mathbf{v}_m \{(-iz)^{-1}\langle 0|\mathscr{R}(z)|k\rangle\} \mathbf{v}_n \tilde{\rho}_k^0(v) \tag{A11.32}$$

The general matrix element of the resolvent connecting 0 to k consists of a succession of diagonal fragments followed at right by a destruction fragment. Let us call $\mathscr{D}_k$ the sum of all destruction fragments defined as follows:

$$\mathscr{D}_k(z) = \sum_{\substack{\text{dest.}\\ \text{fr.}}} (-e^2)^n \langle 0|[\mathscr{L}'\mathscr{R}^0(z)]^n|k\rangle$$

Then

$$\frac{1}{-iz}\langle 0|\mathscr{R}(z)|k\rangle = \frac{1}{-z}\,\frac{1}{z - i\Psi^{(+)}(z)}\,\mathscr{D}_k(z)$$

and hence the asymptotic value, obtained by taking the residue of the integrand at $z = 0$, is

$$\mathbf{j}^s_{\text{corr}} = e^2\beta \sum_m \sum_n z_m z_n \sum_k \mathbf{E} \cdot \int dv\, \mathbf{v}_m \frac{1}{-\Psi^{(+)}(0)} \mathscr{D}_k(0) \mathbf{v}_n \rho_k^0(v) \tag{A11.33}$$

It is now immediately seen that (A11.31) is just a particular case of eq. (A11.33) (cycle approximation). We have thus proven the following remarkable theorem:

The asymptotic (long-time) kinetic equation

$$\partial_t \varphi(\mathbf{v}_\alpha; t) + (ez_\alpha/m_\alpha)\mathbf{E} \cdot \partial\varphi(\mathbf{v}_\alpha; t) = \int_\alpha (d\mathbf{v})^{N-1} \Psi_E \rho_0(|\mathbf{v}; t) \tag{A11.34}$$

linearized in the field **E** *provides the exact value of the current in the stationary state, (including the correlational contribution).*

This theorem is the generalization of the theorem proven in § 59, which dealt with the kinetic part of the current. The significance of this theorem is two-fold. First, it shows that *one can write a kinetic Boltzmann-like equation which provides a complete description of the stationary state, including the contribution of the correlations.* Until very recently the possible existence of such an equation was not known. It was thought that the correlational part of the electrical conductivity could only be obtained by a response-function method, i.e. eq. (57.19). The present result shows that *the kinetic equation approach is exactly equivalent to the response-function method.* On the other hand, it shows that the stationary state linear in the field is described by a markoffian asymptotic equation, which can therefore be called a generalized Boltzmann-like equation, in the sense defined on p. 267.

The lack of space prevents us from developing further these results. We refer the interested reader to the paper by Severne quoted above for further discussion of this theorem.

Symbol Index

(see also Table 2.1, p. 13, and Table 2.2, p. 18)

Vectors are always printed in heavy type. Their absolute values are denoted by the corresponding italic letters.

Symbol	Description	Page where first defined
Latin symbols		
$a^{+}(\hbar\mathbf{k})$	Creation operator	288
$a(\hbar\mathbf{k})$	Destruction operator	288
B	Characteristic factor in Landau equation	174
c	Average number density	9
C_{rs}	Collision operator (Landau equation)	174
$d(\alpha) \equiv d_{\boldsymbol{l}}(\mathbf{v}_{\alpha})$	Coefficient in plasma integral equation	
	classical	200
	quantum-statistical	358
d_j, d_{jm}	Abbreviations in ring approximation	
	classical	196
	quantum-statistical	357
e	Absolute value of electron charge	1
$E(\mathbf{x}; t)$	Local energy density	7
$\mathbf{E}$	External electric field	262
$f_s(\mathbf{x}_1, \ldots, \mathbf{x}_s, \mathbf{v}_1, \ldots, \mathbf{v}_s; t)$	Reduced s-particle distribution function	6
	Reduced s-particle Wigner function	294
$f_N(x, v; t)$	N-particle distribution function	5
$f_N^W(x, v; t)$	N-particle Wigner function	294
$F(\alpha) \equiv F_{\boldsymbol{l}}(\mathbf{v}_{\alpha})$	Fundamental function in ring approximation	199
$F_{\boldsymbol{l}}(\mathbf{v}; w)$	Laplace transform of fundamental function in non-markoffian ring approximation	422
F_r	Dynamical friction vector (Fokker–Planck equation)	183
$\mathscr{F}_{\boldsymbol{l}}(\mathbf{v}; t)$	Fundamental function in non-markoffian ring approximation	422
$g(\mathbf{x}, \mathbf{x}'; t)$	Density correlation function	7
$\mathbf{g}_{jn} = \mathbf{v}_j - \mathbf{v}_n$	Relative velocity of particles j and n	131
$G_k(\tau)$	Equilibrium propagator (quantum ring approximation)	369
$\mathscr{G}(xvt\|x'v't')$	Retarded Green's function of Liouville operator	30

Symbol	Meaning	Page
$h(\mathbf{x}; t)$	Local density excess	15
$h_{\mathbf{k}}(t)$	Fourier transform of local density excess	15
H	Classical hamiltonian	2
$\mathscr{H}$	Hamiltonian operator (second quantization)	288
$\mathbf{j}(z)$	Laplace transform of electric current	272
$\mathbf{j}_K(z)$	Laplace transform of kinetic part of electric current	274
$\mathbf{j}_{\text{corr}}(z)$	Laplace transform of correlational part of electric current	461
$\mathbf{J}(t)$	Electric current	270
$\mathbf{k}$	Wave vector	11
$\mathbf{l}$	Wave vector associated with interaction energy	131
L_h	Hydrodynamic length scale	22
L_m	Molecular length scale	22
$\mathscr{L}$	Classical Liouville operator (in absence of external fields)	29
$\mathscr{L}'$	Perturbation of Liouville operator due to interactions	29
$\mathscr{L}^0$	Unperturbed Liouville operator	29
$\mathscr{L}^e$, $\mathscr{L}_E$	Perturbation of Liouville operator due to external field	269
m	Mass of particles	1
m_j	Mass of particle of species j	1
$n(\mathbf{x}; t)$	Local density	7
$n_s(\mathbf{x}_1, \ldots, \mathbf{x}_s; t)$	Reduced s-particle position distribution	8
N	Total number of particles	1
$\mathbf{p}_j$	Momentum of particle j	1
$p \equiv \{\mathbf{p}_j\} \equiv \{\mathbf{p}_1, \ldots, \mathbf{p}_N\}$	Set of all momenta	4
$\mathscr{P}$	Cauchy principal part	392
$q(\alpha)$	Coefficient in plasma integral equation	
	classical	200
	quantum-statistical	358
$q_{\mathbf{l}}(\mathbf{v}; w)$	Coefficient in plasma integral equation	423
$q_{\mathbf{k}}(\mathbf{v})$	Initial inhomogeneity factor (Vlassov equation)	73
$R(xv\|x'v'; z)$	Resolvent of Liouville operator	32
$\mathscr{R}(z)$	Resolvent operator	35
$\langle k\|\mathscr{R}(z)\|k'\rangle$	Matrix element of resolvent operator in Fourier representation	35
$\langle k\|\mathscr{R}^0(z)\|k'\rangle$	Matrix element of unperturbed resolvent operator in Fourier representation	38

$s(t)$	Entropy density	
	classical	171
	quantum-statistical	344
t	Time	3
t_c	Duration of a collision	56
t_h	Hydrodynamic time scale	56
t_p	Short-time scale in plasmas	60
t_r	Relaxation time	56
T_{rs}	Diffusion tensor (Fokker–Planck equation)	183
$\mathbf{u}(\mathbf{x}; t)$	Local (hydrodynamic) velocity	7
v_F	Fermi velocity	328
$\mathbf{v}_j$	Velocity of particle j	4
$v \equiv \{\mathbf{v}_j\} \equiv \{\mathbf{v}_1, \ldots, \mathbf{v}_N\}$	Set of all velocities	4
$\mathbf{v}_\perp$	Component of velocity perpendicular to wave-vector	77
$V_{jn}(\lvert\mathbf{x}_j - \mathbf{x}_n\rvert)$	Interaction energy	2
V_l	Fourier transform of interaction energy	58
$\mathscr{V}$	Interaction energy operator (second quantization)	288
w_0	Real part of ζ_+	431
$\mathbf{x}_j$	Position (cartesian coordinate) of particle j	1
$x \equiv \{\mathbf{x}_j\} \equiv \{\mathbf{x}_1, \ldots, \mathbf{x}_N\}$	Set of all positions	4
z	Laplace variable	385
z_j	Valence of particle j	1

Greek symbols

$\beta = (kT)^{-1}$	Inverse of temperature (in energy units)	57
γ	Dimensionless plasma parameter (quantum statistics)	317
γ_0	Imaginary part of ζ_+	431
Γ	Dimensionless plasma parameter (classical statistics)	59
$\delta(x)$	Dirac delta-function	
$\delta(\mathbf{k}) = \delta(k_x)\delta(k_y)\delta(k_z)$	Vector delta-function	
δ_n	Kronecker delta symbol $\delta_n \begin{cases} = 0, n \neq 0 \\ = 1, n = 0 \end{cases}$	
$\delta_\mathbf{k} = \delta_{k_x}\delta_{k_y}\delta_{k_z}$	Vector delta symbol	
$\delta_+(x), \delta_-(x)$	Plus- and minus-parts of Dirac delta-function	397
δ_-^{jm}	Abbreviation in ring approximation	196

∂_t	$\partial/\partial t$	5
$\boldsymbol{\partial}_j$	$\partial/\partial \mathbf{v}_j$	5
$\boldsymbol{\partial}_{jn}$	$\partial/\partial \mathbf{v}_j - \partial/\partial \mathbf{v}_n$	5
$\varepsilon_+(\mathbf{k}; z) \equiv \overset{+}{\varepsilon}_{\mathbf{k}}(z/k) \equiv \varepsilon_1 - i\varepsilon_2$	Plus-dielectric constant of wave-vector $\mathbf{k}$ and complex frequency z	
	classical	75
	quantum-statistical	323
$\bar{\varepsilon}_{\mathbf{k}}(\nu_\alpha) \equiv \varepsilon(\alpha) \equiv \varepsilon_1 + i\varepsilon_2$	Minus-dielectric constant of wave-vector $\mathbf{k}$ and real frequency $k\nu_\alpha$; coefficient in plasma integral equation	200
$\zeta_\pm$	Unstable zero of plus(minus)-dielectric constant	431
θ	Quantum-statistical factor	288
$\theta(x)$	Heaviside function	32
$\varkappa$	Inverse Debye length	85
$\varkappa_q$	Inverse quantum Debye length	347
Λ	Liouville operator in presence of an external field	269
$\nu = \mathbf{k} \cdot \mathbf{v}/k$	Component of velocity parallel to wave-vector	77
$\boldsymbol{\rho}$	von Neumann density matrix	291
$\rho_{\mathbf{k}_1,\ldots,\mathbf{k}_s}(\mathbf{v}_1, \ldots, \mathbf{v}_s \| \ldots; t)$	Explicit Fourier component	
	N-particle classical distribution function	13
	N-particle Wigner function	297
$\rho_{\mathbf{k}_1,\ldots,\mathbf{k}_s}(\mathbf{v}_1, \ldots, \mathbf{v}_s; t)$	Fourier component	
	reduced s-particle distribution function	17
	reduced s-particle Wigner function	297
$\tilde{\rho}_{\{\mathbf{k}\}}(v; t)$	Compact Fourier component	
	N-particle classical distribution function	10
	N-particle Wigner function	296
$\hat{\rho}_{\mathbf{k}_1,\ldots,\mathbf{k}_s}(\mathbf{v}_1, \ldots, \mathbf{v}_s \| \ldots; t)$	Non-singular Fourier component of Wigner function	299
$\rho_{\mathbf{k}}(\mathbf{v}; t)$	Fourier transform of inhomogeneity factor	15
$\sigma_{\mathbf{k}}(t)$	Fourier transform of charge density	89
$\sigma^\pm(w)$	$(\pm)$ dielectric constant divided by $(w - \zeta_\pm)$	431
Σ	Degree of quantum-statistical degeneration	317
v_p	Reduced plasma frequency	118
$\varphi(\mathbf{v}; t)$	Reduced velocity distribution function	8
$\bar{\varphi}(\nu; t)$	Barred velocity distribution function	97
$\Phi_{\mathbf{k}}(z)$	Fourier–Laplace transform of average potential (Vlassov approximation)	89

$\Phi(w)$	Error function	184
$\bar{\chi}_v(\nu)$	van Kampen–Case eigenfunction of Vlassov equation	118
$\tilde{\chi}_v(\nu)$	van Kampen–Case eigenfunction of adjoint Vlassov equation	122
$\Psi(y)$	Error function of a complex argument	110
ω_p	Plasma frequency	75
Ω	Volume of system	9
∇_j	$\partial/\partial \mathbf{x}_j$	5

Operations

$[h, g]$	Poisson bracket	4
$[A, B]_\theta$	Commutator $(\theta = -1)$ Anticommutator $(\theta = +1)$	288
$\bar{f}(\nu)$	Barred function associated with $f(\mathbf{v})$	77
$\bar{f}'(\nu)$	Derivative of $\bar{f}(\nu)$ with respect to ν	97
$\Phi^{(\pm)}(z)$	Plus- (minus-) branch of Cauchy integral	393
$F_\pm(x)$	Plus- (minus-) function	398

Author Index

Abe, R., 269
Allis, W. P., 188, 267
Andrews, F. C., 28, 36
Auer, P. L., 96

Backus, G., 96
Balescu, R., 10, 27, 166, 206, 210, 220, 258, 267, 272, 359, 433
Bateman, H., 110, 138, 139
Berz, F., 96
Bird, R. B., 163, 176, 259
Blochintzev, D. J., 287
Bogoliubov, N. N., 27, 58, 71, 176, 206, 228
Bohm, D., 103, 330, 332, 365
Boltzmann, L., 167
Brout, R., 27, 166, 167, 330, 378
Brueckner, K. A., 330, 378
Buneman, O., 96

Case, K. M., 117, 125, 128
Chandrasekhar, S., 169, 186, 187, 190, 191
Chapman, S., 163, 166, 259
Chen-Chun-Sian, 330
Choh-Shih-Hsun, 330
Chuck, W., 188
Cohen, R. S., 188, 259
Conte, S. D., 110
Courant, R., 371, 404
Cowling, T., 163, 166, 259
Curtiss, C. F., 163, 176, 259

Dawson, J., 430
de Boer, J., 22
Debye, P., 250
Defay, R., 248
de Gottal, Ph., 210, 281
de Groot, S. R., 173, 258
Delcroix, J. L., 180, 259, 281
de Witt, H., 370
Dirac, P. A. M., 287
Doetsch, G., 385
Dreicer, H., 266
Drummond, W. E., 432, 443

Edwards, S. F., 269
Ehrenfest, P. and T., 167
Emde, F., 110

Fano, U., 365
Feshbach, H., 371, 404
Fock, V. A., 287
Fowler, R., 58
Fried, B. D., 110
Fujita, S., 269, 301, 309, 372, 375
Fukuda, N., 330, 378

Gakhov, F. D., 391, 405
Gell-Mann, M., 354, 378
Ghertsenstein, M. E., 96
Ginzburg, V. L., 92
Goldstein, H., 4
Grad, H., 167, 259
Green, M. S., 167, 268, 273
Guernsey, R. L., 71, 206, 229, 240, 359, 429
Guggenheim, E. A., 58

Härm, R., 183, 259
Heisenberg, W., 311
Heitler, W., 37
Henin, F., 166, 267
Hilbert, D., 371, 404
Hill, T. L., 251
Hirota, R., 301, 372, 375
Hirschfelder, J. O., 163, 176, 259
Hubbard, J., 330, 379
Hückel, E., 250

Ichimaru, S., 244
Irving, J. H., 7
Ivanenko, D., 36

Jackson, J. D., 96, 102, 106, 112
Jahnke, E., 110
Judd, D. L., 188

Kac, M., 167
Kaufman, A. N., 259
Khinchine, A. I., 5
Kihara, T., 258
Kirkwood, J. G., 7, 167
Klimontovitch, Iu., 359
Kohn, W., 269
Konstantinov, O. V., 359
Kubo, R., 269, 273
Kudriavtsev, V. S., 188

Landau, L. D., 75, 92, 103, 106, 128, 165, 169, 174, 248, 372
Landsberg, P. T., 248
Landshoff R. K. M., 259
Lenard, A., 206, 223, 229, 230, 238
Lifshitz, E. M., 92, 248, 372
Lighthill, M. J., 119
Lindhard, J., 330
Linhart, J. G., 180
London, F., 301
Luttinger, J. M., 269

Macke, W., 354
Massignon, D., 7, 293, 311
Mayer, J., 250, 354
Mazur, P., 173, 258
McDonald, W. M., 188
McLennan, Jr., J. A., 269, 273
McRoutly, P., 188, 259
Meeron, E., 251
Messiah, A., 287
Mikhlin, S. G., 391
Milner, R., 58
Montroll, E. W., 58, 269, 309, 316, 370, 378

Mori, H., 269, 273, 293
Morse, P. M., 371, 404
Muskhelishvili, N. I., 391

Nozières, P., 330
Nyquist, H., 96, 99

Oberman, C., 430
Ono, S., 310
Oppenheim, I., 293

Penrose, O., 96, 101
Perel', V. I., 359
Pines, D., 103, 309, 330, 332, 334, 365, 432, 443
Placzek, G., 301
Prigogine, I., 5, 10, 19, 27, 140, 166, 173, 212, 216, 256, 258, 268, 276, 309, 310, 420

Résibois, P., 5, 27, 166, 267, 309, 420, 429, 445, 447
Ritchie, R. H., 330, 365
Ron, A., 430
Rosenbluth, M. N., 71, 188, 206, 233
Ross, J., 293
Rostoker, N., 71, 206, 233

Salpeter, E. E., 251
Sanderson, J. J., 269
Sawada, K., 330
Schiff, L. I., 287
Schlögl, F., 30
Schwartz, L., 119
Severne, G., 266, 453
Silin, V. P., 266, 458
Sokolov, A., 36
Sommerfeld, A., 178, 180, 345
Spitzer, Jr, L., 177, 188, 259, 281

Taylor, H. S., 220
Tchen, C. M., 206, 429
Tchersky, Iu. I., 398

Temko, S. V., 359
Titchmarsh, E. C., 16
Tolman, R. C., 163, 166, 291

Uehling, E. A., 343
Uhlenbeck, G. E., 343

van Hove, L., 269, 309
van Kampen, N. G., 87, 117, 128
von Neumann, J., 291
Vlassov, A. A., 68, 103

Waldmann, L., 259
Ward, J. C., 58, 269, 309, 316, 370, 378
Weyl, H., 290
Wigner, E., 292

Subject Index

Analytical continuation
  Cauchy integral, 78, 131, 393, 395, 433, 435
  and integral equations, 433, 435
  Laplace transform, 387
Analytical function, 393
Assumptions in general theory of irreversibility
  finiteness of local quantities, 10, 12, 297
  infinite distance correlations, 16, 297
Asymptotic behavior
  $N \to \infty$, 14, 295
  $t \to \infty$, 94, 133, 140, 155, 192, 276, 335
Average potential (Vlassov), 89, 229

Barring operation, 77, 97, 118, 203, 220, 226, 325, 423
Boltzmann equation, 163, 259
Boltzmann-like equation, 165, 259, 267, 276, 278, 459, 462
Boltzmann statistics, 288, 310
Bose–Einstein statistics, 287, 288
Boundary conditions, 29
Branch points, 136, 138, 338
Brownian motion, 181, 233

Cauchy integral
  analytical continuation, 78, 131, 393, 395, 433, 435
  definition, 390
  discontinuity, 392
  and long-time behavior, 131, 338
  and short-time behavior, 78, 91, 324
  and singular integral equations, 402
  singularities, 395
Causality condition, 30, 270, 454
Chapman–Enskog expansion, 162, 168, 263
Classical limit, 293, 311
Cluster expansion, 251
Collective behavior, 83, 135, 332
Collision operator, 163, 260, 267, 441
Collisional invariants, 178, 180, 224
Collisions
  collective, 225, 359
  duration of, 56, 132, 213
  instantaneous, 164, 260
Commutator, 291, 412
Conductivity
  electrical, 262, 460
  thermal, 263, 273
Conservation
  of collisional invariants, 178, 180 224
  of wave-vector, 39
Contraction, 299, 415
Convolution
  complex, 448
  time, 389
Correlation energy, 378
Correlation function, density
  definition, 7, 210
  in equilibrium, 244, 247, 417
  ring approximation, 226
Correlation function, two-particle
  definition, 15, 23, 210
  in equilibrium, 242, 367
  ring approximation, 221, 226, 367
Correspondence principle, 316
Creation fragment, 141, 215, 217, 245
Cut-off, 135, 176, 195
Cycle, 129, 133, 336, 457 (*see also*: Landau equation)

Damping
free flow, 82
Landau, 103, 105, 114, 128, 324, 332
de Broglie wave-length, 339
Debye–Hückel theory, 71
Debye length, 85, 136, 243, 247
Debye potential, 94, 135, 230, 244, 347, 361
Degeneration, degree of, 317
Density excess, 15, 76, 86, 323
Density matrix, 291, 303, 311, 417
Destruction fragment, 141, 456
Diagonal fragment, 130, 140, 212, 267, 336, 408, 445
Diagrams
classification, 141
connected, 43, 50, 52
dependence on N, Ω, e^2, c, 46, 54
labeling, 53
long-time, 130, 140, 144, 146, 336
Mayer, 251
quantum-statistical, 308, 312, 319
reduced distribution function, 49
ring approximation, 192, 217
semiconnected, 50, 147
short-time, 60
time dependence, 133, 138, 141
Dielectric constant
Fermi–Dirac, 328, 348, 362
Maxwell, 109, 233
minus-d. c., 120
plus-d. c., 75, 323
ring approximation, 200, 205, 225, 238, 358, 360, 423
zeros, 80, 96, 125, 326, 428, 431
Dispersion equation, 79, 83, 101, 118, 324
Distribution (Schwartz), 119, 218, 243, 397, 399, 439
Distribution function
N-particle, 3
normalization, 5, 14
reduced, 6, 8, 17, 29, 49, 52
Divergence
long-distance, 135, 229, 255, 353
short-distance, 134
Dynamical friction, 188, 237, 348, 360, 365

Eigenfunctions (van Kampen–Case)
adjoint Vlassov operator, 120
completeness, 125
orthogonality, 122, 406
and ring approximation, 220, 425, 434
for unstable plasmas, 125
Vlassov operator, 117
Electric current, 270, 461
Energy
correlation, 378
exchange, 377
free, 248
ground state, 375
internal, 248, 419
Ensemble, 3
Entropy, 171, 223, 249, 344
Exchange interaction matrix element, 289, 320, 355
External lines, 43

Fermi–Dirac distribution
finite temperature, 344
zero temperature, 328, 368, 375
Fermi–Dirac statistics, 287, 288, 352
Field
external, 262
self-consistent, 70
strong external, 264
Fokker–Planck equation, 186, 190, 233 (*see also*: Landau equation, Cycle)
Forward scattering, 308, 320
Fourier components
distribution function, 11, 12, 17, 20, 22
Wigner function, 297, 311

Fundamental solution (Hilbert problem), 402

Green's function
external field, 266, 270, 454
Fourier representation 37
integral equation, 34
of Liouville equation, 28, 30
in quantum statistics, 312
unperturbed, 36
Green's theorem, 30
Ground-state energy, 375

H theorem, 164, 171, 223, 259, 344
Hamiltonian, 1, 2, 288
Heisenberg's principle, 287
Hilbert problem
homogeneous, 400
inhomogeneous, 402
Hodograph, 97, 326
Homogeneous systems, 8, 20, 210
Hydrodynamic approximation, 22, 151, 168, 224
Hydrodynamic length scale, 22
Hydrodynamic time scale, 56

Index of singular integral equation, 204, 400, 403
Individual particle behavior, 79, 80
Inhomogeneity factor
definition, 15
long-time, 156, 208
shape, 22
short-time, 62, 65, 319, 321
Inhomogeneous systems, 8, 22, 24, 26
Initial condition for Liouville equation, 10, 12, 20
Initial-value problem
linearized Vlassov equation, 73, 126, 321
Liouville equation, 31
Integral equation, Fredholm, 371, 404
Integral equation, plasmas
long-time, 201, 203
non-markoffian, 422, 433, 437
quantum statistics, 358
Integral equation, singular, 123, 204, 220, 403
Interaction, effective, 90, 324, 361
Interaction energy, Fourier expansion, 39, 58, 136

Kinetic equation
classical Landau approximation, 150, 159, 162, 260
classical ring approximation, 206, 260
external field, 262, 264
general leading universal term approximation, 408
non-markoffian, 419, 421, 428
quantum cycle approximation, 342
quantum ring approximation, 358
strong external field, 453, 457
and transport coefficients, 259, 260, 279, 460
unstable plasma, 431, 440
weakly stable plasma, 441
Kramers–Kronig relation, 98, 244

Landau damping: *see* Damping
Landau equation, 145, 150, 174, 230, 258, 260, 457
Laplace transformation
analytical continuation, 387
convolution theorem, 389
definition, 385
half-plane of convergence, 386
inversion formula, 387
singularities, 387, 389
Liouville equation
classical, 4
external field, 269, 454
quantum-statistical, 308, 415
Liouville operator, 30

Liouville operator (*cont.*)
  adjoint, 30
  external field, 269
Lipschitz condition, 390
Local equilibrium, 172, 263
Lorentz gas, 280

Markoffian limit, 430, 437, 458
Maxwell distribution, 107, 109, 170, 223, 247
Maxwell–Boltzmann distribution, 21
Minus-function
  definition, 396
  and dielectric constant, 120
  and solution of singular integral equation, 123, 427

Neutral stability, 442
Non-markoffian equation, 240, 268, 278, 456 (*see also*: Kinetic equation)
Non-uniform systems: *see* Inhomogeneous systems
Normal product, 291, 295, 411

Parameters
  classical plasma, 58
  electric field, 266
  ordinary gas, 57
  quantum degeneration, 317
  quantum plasma, 317
Permutation class, 446
Perturbation theory, 28, 33, 35
Phase space, 3, 293
Plasma frequency, 75, 84, 333
Plasmons, 333
Plemelj formulae, 150, 393, 397, 438
Plus-function
  definition, 396
  and dielectric constant, 91
  Kramers–Kronig relation, 98
  and solution of singular integral equation, 123, 427

Poincaré–Bertrand theorem, 399, 406, 428
Poisson equation, 71, 89
Polarization, 90
Pressure
  equilibrium, 249, 251
  tensor, 179
Principal part, 392
Prototype, 154
Pseudo-diagonal fragment, 151, 154, 156

Relaxation time, 56
Resolvent
  external field, 272
  integral equation, 34
  Liouville equation, 28, 32, 419
  operator, 35
  quantum statistics, 312
  unperturbed, 38
Riemann surface of Cauchy integral, 395
Ring approximation
  correlations, 216, 367
  generalities, 223
  kinetic equations, 192, 259, 260, 358, 428, 441
Ring diagrams
  classical, 192
  Mayer, 254
  non-markoffian, 420
  quantum statistics, 353
  rules of construction, 192, 356, 422
  summation, 198, 219, 449
Runaway effect, 258, 266, 277, 351

Screening, 93, 229, 230
Second quantization, 287
Sectionally regular function, 393, 400, 402, 403
Selection rules: *see* Conservation rules

Singularities
  Cauchy integral, 395
  Fourier components of Wigner function, 299
  inhomogeneity factor, 79
  Laplace transform, 389
  resolvent, 132, 275, 421
Specific heat, 249
Stability
  condition, 124
  criterion, 95
  quantum statistics, 325, 374
  ring approximation, 195, 205, 221, 238, 425, 432
  weak, 439
Stabilization, 442
Stationary state, non-equilibrium, 257, 274, 278, 461

Test particle, 181, 233, 346, 360
Time scales, 56, 335, 432
Topological index, 47
Transport coefficients
  correlational part, 264, 461
  dynamical, 260
  kinetic equation method, 259, 283, 460
  kinetic part, 274
  response function method, 269, 283
  thermal, 260, 264, 273

Uehling–Uhlenbeck equation, 343
Uncertainty principle, 287
Universal contributions, 133, 216, 239, 240, 335
  leading, 133, 140, 239, 267, 278, 340, 408, 418, 458
Unstable plasma, 101, 112, 431, 437 449

Velocity, hydrodynamic, 7
Velocity distribution function, 15, 17, 62, 65, 146, 192, 340, 353, 419
Vertices, basic
  classical statistics, 44
  quantum Boltzmann statistics, 310
  quantum statistics, 308, 316, 355
Vlassov equation
  derivation, 65
  linearized, 69
  quantum statistics, 320
  solution, 73, 323
von Neumann equation, 291

Weyl's rule, 290
Wigner function
  Boltzmann statistics, 311
  equation of evolution, 303
  Fourier components, 297, 311
  N-particle, 292
  reduced, 294
YBGBK hierarchy, 27, 227